The Cytokine Network and Immune Functions

The Cytokine Network and Immune Functions

Jacques Thèze, M.D., Ph.D.

Professor and Head of the Division of
Cellular Immunology and Immunogenetics,
Pasteur Institute, Paris

OXFORD
UNIVERSITY PRESS

UNIVERSITY PRESS

Great Clarendon Street, Oxford OX2 6DP

Oxford University Press is a department of the University of Oxford
and furthers the University's aim of excellence in research, scholarship.
and education by publishing worldwide in

Oxford New York
Athens Auckland Bangkok Bogotá Buenos Aires Calcutta
Cape Town Chennai Dar es Salaam Delhi Florence Hong Kong Istanbul
Karachi Kuala Lumpur Madrid Melbourne Mexico City Mumbai
Nairobi Paris São Paulo Singapore Taipei Tokyo Toronto Warsaw

and associated companies in Berlin Ibadan

Oxford is a registered trade mark of Oxford University Press

Published in the United States
by Oxford University Press, Inc., New York

A catalogue record for this book is available from the British Library

Library of Congress Cataloging in Publication Data

The cytokine network and immune function / [edited by] Jacques Theze.
1. Cytokines. 2. Immunity. 3. Immunopathology. I. Theze, Jacques.
  [DNLM: 1. Cytokines–immunology. 2. Immunity–physiology.
3. Immunologic Diseases–physiopathology. QW 568 C9944 1999]
QR185.8.C95C94 1999 616.07'9–dc21 98-50202

ISBN 0 19 850136 6

Typeset by Best-set Typesetter Ltd., Hong Kong
Printed in Great Britain by
The Bath Press, Avon, UK

To Nicole, Veronique and Lionel

Foreword

The great progress that has been made in our understanding of the mechanisms of immune regulation in the last 15 years has been driven by a virtual explosion of research on a set of small proteins and their receptor molecules. These molecules, which have been known under many guises, but are now confidently referred to as *cytokines*, link the specificity arm of the cellular immune system to its effector mechanisms, much as complement and Fc receptors link the specificity of antibodies to the effector mechanisms of humoral immunity.

It was the application first of careful biochemistry and then of molecular biology that moved the study of cytokines and cytokine biology from what had been an almost magical system, in which scientists studied the effects of complex mixtures of proteins on equally complex biological functions, to one in which precisely defined molecules cause specific (although often not simple) outcomes in manners that can be precisely measured both in terms of changes in existing cellular proteins and in the induction of new proteins that endow cells with new properties.

Of course, not all is perfect in the cytokine world. To begin with, the single rubric 'cytokine' covers a wide range of molecular types that act with different mechanisms and often with a different logic. At the very least, we now recognize five major cytokine families: (i) the type I cytokines, including many of the interleukins; (ii) the interferons and related cytokines; (iii) the tumour necrosis factor (TNF) family of cytokines; (iv) the immunoglobulin supergene family of cytokines; and (v) the chemokines. The members of these families and their related receptors display a set of self-consistent, logical rules through which they mediate their functions and through which they participate in the regulation of immune and inflammatory responses.

One of the greatest triumphs of modern cytokine biology is the demonstration that these molecules may be grouped into families and that the individual members of a given family show substantial commonality in the principles underlying their activity. The study of these molecules as a system provides knowledge that not only advances our understanding of the cytokine under study but often illuminates the means through which the action of other members of the family is controlled. Cytokine biology is now a prime research area, not a simple mantle to justify the joint meetings of individuals studying what often seemed to be quite independent phenomena, as was all too frequent in the early days of the subject.

The great scientific progress in our understanding of the genetics, chemistry, biology and pathophysiology of the cytokines has made clear the need for a comprehensive discussion of these molecules as part of a unified system. *The Cytokine Network and Immune Functions* provides just such a treatment. It represents the efforts of many of the most prominent scientists studying these molecules. Not only does it present a general discussion of each of the major cytokines or sets of cytokines, it deals in detail with how these molecules effect all aspects of immune function and how they contribute to a wide range of pathological conditions. As the title of this work clearly indicates, this work does not treat individual cytokines in isolation; it recognizes that they act in a complex web of synergic and counter-regulatory effects so as to provide opportunities for very fine control of immune responses.

Cytokine biology is not only a fascinating scientific discipline, giving insight into how the immune and inflammatory systems are linked and regulated; it is also a subject of profound importance in modern medicine. As the chapters in the section of this book entitled 'Cytokines in Pathology' indicate, virtually all insults to the organism, be they acute or chronic, call upon cytokine responses and virtually every disease entity has a component involving the function of cytokines. In some diseases, such as allergy and asthma and many of the autoimmune disorders, 'abnormal' cytokine responses are at the very heart of the disease process; in others, their impact may be more peripheral. There can be no doubt, however, that the study of cytokine biology has contributed greatly to the growing power of molecular medicine to transform our approach to disease and to build a new armamentarium of drugs and other treatments that promise to revolutionize our capacity to control a wide range of disease states.

I am confident that readers of *The Cytokine Network and Immune Functions* will benefit greatly from the insights provided and from the detailed consideration of this remarkable field. Moreover, all who have contributed to this book hope that the readers will be among those who will advance this field in the future and that their research will have been facilitated through the organized information provided in this volume. Indeed, the true test of the value of this and other such books will be that it contributes to the progress of this field and to the increasingly wider application of the understanding of cytokine biology to scientific advancement and to the prevention and treatment of disease. In this regard, *The Cytokine Network and Immune Functions* should be of particular value to young investigators now entering the field of cytokine biology and should be of great benefit to medical school and university professors who have the responsibility to teach this important but sometimes daunting subject.

NIH, Bethesda, Maryland, USA   William E. Paul

Acknowledgements

The preparation of this book required the efforts of many contributors who have devoted some of their precious time to this collaborative work. I wish to express my gratitude to all of them. I particularly wish to emphasize the roles of Abul Abbas, Jacques Banchereau, Didier Fradelizi, Alberto Mantovani and staff at Oxford University Press who provided help and stimulating advice at different steps along the way. I also wish to acknowledge gratefully the friendly support of Mrs Corinne Baran, Mrs Christiane Corel, and all of my colleagues at the Division of Cellular Immunology and Immunogenetics at the Pasteur Institute.

Paris, July 1998

J. T.

Contents

Jean-Luc Taupin, Stéphane Minvielle, Jacques Thèze, Yannick Jacques and Jean-François Moreau

Marina Pretolani, Patrick Stordeur and Michel Goldman

B. CYTOKINES AND IMMUNE FUNCTIONS

C. CYTOKINES IN PATHOLOGY

Contributors

Ken-ichi ARAI
Department of Molecular and Developmental Biology, The Institute of Medical Science, The University of Tokyo, and CREST, Japan Science and Technology Corporation (JST), 4-6-1 Shirokanedai, Minato-ku, Tokyo 108-8639, Japan.

Andris AVOTS
Department of Molecular Pathology, Institute of Pathology, University of Würzburg, Josef Schneider Strasse 2, D-97080 Würzburg, Germany.

A. BADOU
INSERM U28 and Université Paul Sabatier, Hôpital Purpan, Place du Dr Baylac, 31059 Toulouse Cedex, France.

Jacques BANCHEREAU
Baylor Institute for Immunology Research, 3434 Live Oak, Dallas, TX 75204, USA.

Tamas BARTFAI
Department of Neurochemistry and Neurotoxicology, Stockholm University, S-106 91 Stockholm, Sweden.

I. BERNARD
INSERM U28 and Université Paul Sabatier, Hôpital Purpan, Place du Dr Baylac, 31059 Toulouse Cedex, France.

Jacques BERTOGLIO
INSERM U461, Faculté de Pharmacie Paris XI, 5, rue Jean-Baptiste-Clément, 92296 Chatenay-Malabry, France.

Jacek BIGDA
Department of Histology and Intercollegiate Faculty of Biotechnology, Medical University of Gdansk, Debinki 1, 80-211 Gdansk, Poland.

J.-Y. BLAY
Cytokines and Cancer Unit, Centre Leon Berard, 69008 Lyon, France.

Fionula M. BRENNAN
Kennedy Institute of Rheumatology, 1 Aspenlea Road, Hammersmith, London W6 8LH, UK.

Francesca BRUGNOLO
Clinical Immunology Department, Istituto di Medicina Interna e Immuno-allergologia, Policlinico Careggi, 50134 Firenze, Italy.

Jean-Laurent CASANOVA
INSERM U429, Hôpital Necker Enfants Malades, Pavillon Kirmisson, 149 rue de Sèvres, 75743 Paris Cedex 15, France.

B. CAUTAIN
INSERM U28 and Université Paul Sabatier, Hôpital Purpan, Place du Dr Baylac, 31059 Toulouse Cedex, France.

Jean-Marc CAVAILLON
Institut Pasteur, 28 rue Dr Roux, 75015 Paris, France.

J. CHEHIMI
Division of Immunology and Infectious Diseases of the Children's Hospital of Philadelphia, Philadelphia, PA 19104, USA.

S. CHOUAIB
INSERM U487, Laboratory of Cytokines and Human Antitumor Immunity Institute, Gustave Roussy, 94805 Villejuif, France.

Robert L. COFFMAN
Department of Immunobiology, DNAX Research Institute, 901 California Avenue, Palo Alto, CA 94304, USA.

Mario P. COLOMBO
Gene Therapy Program, Istituto Nazionale Tumori, Via Venezian, 1-20133 Milan, Italy.

Margaret J. DALLMAN
Sir Alexander Fleming Building, Department of Biology, Imperial College of Science Technology and Medicine, Imperial College London SW7 2AZ, UK.

Jean-Michel DAYER
Division of Immunology and Allergy, University Hospital, 1211 Geneva 14, Switzerland.

Geneviève De Saint BASILE
INSERM U429, Hôpital Necker Enfants Malades, Pavillon Kirmisson, 149 rue de Sèvres, 75743 Paris Cedex 15, France.

Jean-Pierre De VILLARTAY
INSERM U429, Hôpital Necker Enfants Malades, Pavillon Kirmisson, 149 rue de Sèvres, 75743 Paris Cedex 15, France.

James DISANTO
INSERM U429, Hôpital Necker Enfants Malades, Pavillon Kirmisson, 149 rue de Sèvres, 75743 Paris Cedex 15, France.

P. DRUET
INSERM U28 and Université Paul Sabatier, Hôpital Purpan, Place du Dr Baylac, 31059 Toulouse Cedex, France.

Gordon DUFF
Section of Molecular Medicine, University of Sheffield, Royal Hallamshire Hospital, Sheffield, UK.

Scott K. DURUM
Laboratory of Immunoregulation, National Cancer Institute, Frederick, MD, USA.

Michel DY
Université René Descartes–Paris V, CNRS URA 1461, Hôpital Necker, Paris, France.

Dominique EMILIE
INSERM U131, Institut Paris-Sud sur les Cytokines, 32 rue des Carnets, 92140 Clamart, France.

Hartmut ENGELMANN
Institute of Immunology, University of Munich, 8000 Munich 2, Germany.

Marc FELDMANN
Kennedy Institute of Rheumatology, 1 Aspenlea Road, Hammersmith, London W6 8LH, UK.

Fred D. FINKELMAN
Department of Medicine, College of Medicine, University of Cincinnati, Cincinnati, OH 45267, USA.

Alain FISCHER
INSERM U429, Hôpital Necker Enfants Malades, Pavillon Kirmisson, 149 rue de Sèvres, 75743 Paris Cedex 15, France.

Didier FRADELIZI
INSERM U477, Pavillon Hardy A, Hôpital Cochin, 27 rue du Faubourg Saint Jacques, 75674 Paris Cedex 14, France.

Pierre GALANAUD
INSERM U131, Institut Paris-Sud sur les Cytokines, 32 rue des Carnets, 92140 Clamart, France.

Michel GOLDMAN
Service d'Immunologie–Transfusion, Hôpital Erasme, Route de Lennik 808, B-1070 Brussels, Belgium.

Hervé GROUX
INSERM U343, Hôpital de l'Archet, Route de Saint Antoine de Ginestiere, 06200 Nice, France.

Thomas HÜNIG
Institute for Virology and Immunobiology, University of Würzburg, Versbacherstrasse 7, D 97078 Würzburg, Germany.

Tohru ITOH
Department of Molecular and Developmental Biology, The Institute of Medical Science, The University of Tokyo, 4-6-1 Shirokanedai, Minato-ku, Tokyo 108-8639, Japan.

Yannick JACQUES
Groupe Cytokine-Récepteurs-Transduction, Unité INSERM 463, Institut de Biologie, 9 Quai Moncousu, 44035 Nantes Cedex 01, France.

Daniel H. KAPLAN
Center for Immunology, Department of Pathology, Washington University School of Medicine, 660 S. Euclid Avenue, Box 8118, St Louis, MO 63110, USA.

Stefan KLEIN-HESSLING
Department of Molecular Pathology, Institute of Pathology, University of Würzburg, Josef Schneider Strasse 2, D-97080 Würzburg, Germany.

Sophie LAYÉ
INSERM U394, rue C. Saint-Saëns, 33077 Bordeaux, France.

Robert J. LECHLEIDER
Laboratory of Cell Regulation and Carcinogenesis, National Cancer Institute, Bethesda, MD 20892-5055, USA.

Françoise LE DEIST
INSERM U429, Hôpital Necker Enfants Malades, Pavillon Kirmisson, 149 rue de Sèvres, 75743 Paris Cedex 15, France.

Chong-kil LEE
Laboratory of Immunoregulation, National Cancer Institute, Frederick, MD, USA.

Johan LUNDKVIST
Department of Neurochemistry and Neurotoxicology, Stockholm University, S-106 91 Stockholm, Sweden.

Enrico MAGGI
Clinical Immunology Department, Istituto di Medicina Interna e Immunoallergologia, Policlinico Careggi, 50134 Firenze, Italy.

Ravinder MAINI
Kennedy Institute of Rheumatology, 1 Aspenlea Road, Hammersmith, London W6 8LH, UK.

Alberto MANTOVANI
Istituto di Ricerche Farmacologiche 'Mario Negri', via Eritrea 62, Milano, Italy.

J. D. MARSHALL
The Wistar Institute of Anatomy and Biology, Philadelphia, PA 19104, USA.

Cecilia MELANI
Gene Therapy Program, Istituto Nazionale Tumori, Via Venezian, 1-20133 Milan, Italy.

Adrian MINTY
SANOFI Recherche, Labege Innopole, voie no. 1, BP 137, 31676 Labège Cedex, France.

Stéphane MINVIELLE
INSERM U463, Plateau technique, Centre Hospitalier Universitaire, 9 quai Moncousu, 44035 Nantes Cédex, France.

Jean-François MOREAU
Laboratoire Cytokines et Transplantation, Université de Bordeaux 2, CNRS-UMR 5540, 146 rue Léo Saignat, 33076 Bordeaux Cédex, France.

Kathrin MUEGGE
SAIC, Frederick, MD, USA.

Laurent P. NICOD
Pulmonary Division, University Hospital, 1211 Geneva 14, Switzerland.

Giorgio PARMIANI
Gene Therapy Program, Istituto Nazionale Tumori, Via Venezian, 1-20133 Milan, Italy.

Paola PARRONCHI
Clinical Immunology Department, Istituto di Medicina Interna e Immunoallergologia, Policlinico Careggi, 50134 Firenze, Italy.

L. PELLETIER
INSERM U28 and Université Paul Sabatier, Hôpital Purpan, Place du Dr Baylac, 31059 Toulouse Cedex, France.

Marina PRETOLANI
Unité de Pharmacologie cellulaire/Unité associée INSERM U485, Institut Pasteur, Paris, France.

Paul PROOST
Rega Institute for Medical Research, Laboratory of Molecular Immunology, University of Leuven, B-3000 Leuven, Belgium.

Anita B. ROBERTS
Laboratory of Cell Regulation and Carcinogenesis, National Cancer Institute, Building 41, Room C629, 41 Library Drive, MSC 5055, Bethesda, MD 20892-5055, USA.

Donald A. ROWLEY
Department of Pathology, The University of Chicago, 5841 S. Maryland Avenue, Chicago, IL 60637, USA.

Salvatore SAMPOGNARO
Clinical Immunology Department, Istituto di Medicina Interna e Immunoallergologia, Policlinico Careggi, 50134 Firenze, Italy.

A. SAOUDI
INSERM U28, Place du Dr Baylac, 31059 Toulouse Cedex, France.

M. SAVIGNAC
INSERM U28 and Université Paul Sabatier, Hôpital Purpan, Place du Dr Baylac, 31059 Toulouse Cedex, France.

Anneliese SCHIMPL
Institute for Virology and Immunobiology, University of Würzburg, Versbacherstrasse 7, D 97078 Würzburg, Germany.

Elke SCHNEIDER
Université René Descartes–Paris V, CNRS URA 1461, Hôpital Necker, Paris, France.

Hans SCHREIBER
Department of Pathology, The University of Chicago, 5841 S. Maryland Avenue, Chicago, IL 60637, USA.

Robert D. SCHREIBER
Center for Immunology, Department of Pathology, Washington University School of Medicine, 660 S. Euclid Avenue, Box 8118, St Louis, MO 63110, USA.

Edgar SERFLING
Department of Molecular Pathology, Institute of Pathology, University of Würzburg, Josef Schneider Strasse 2, D-97080 Würzburg, Germany.

Lisa P. SEUNG
Department of Obstetrics, Gynecology, and Reproductive Sciences, University of California San Francisco, 505 Parnassus Avenue, San Francisco, CA 94143, USA.

Clifford M. SNAPPER
Department of Pathology, Uniformed Services University of the Health Sciences, 4301 Jones Bridge Road, Bethesda, MD 20814, USA.

Patrick STORDEUR
Service d'Immunologie–Transfusion, Hôpital Erasme, Brussels, Belgium.

Sofie STRUYF
Rega Institute for Medical Research, Laboratory of Molecular Immunology, University of Leuven, B-3000 Leuven, Belgium.

Jean-Luc TAUPIN
Laboratoire Cytokines et Transplantation, Université de Bordeaux 2, CNRS-UMR 5540, 146 rue Léo Saignat, 33076 Bordeaux Cédex, France.

Jacques THÈZE
Immunogénétique Cellulaire, Institut Pasteur, 28 rue du Docteur Roux, 75724 Paris Cedex 15, France.

Giorgio TRINCHIERI
The Wistar Institute, 3601 Spruce Street, Philadelphia, PA 19104, USA.

Jo Van DAMME
Rega Institute for Medical Research, Laboratory of Molecular Immunology, University of Leuven, B-3000 Leuven, Belgium.

Aimé VAZQUEZ
INSERM U131, 32 rue des Carnets, 92140 Clamart, France.

David WALLACH
Department of Biological Chemistry, The Weizmann Institute of Science, 76100 Rehovot, Israel.

Anja WUYTS
Rega Institute for Medical Research, Laboratory of Molecular Immunology, University of Leuven, B-3000 Leuven, Belgium.

Hans YSSEL
INSERM U454, Hôpital Arnaud de Villeneuve, 375 Avenue du Doyen Gaston Giraud, 34295 Montpellier, France.

Overview

General aspects of cytokine properties and functions

Michel Dy, Aimé Vazquez, Jacques Bertoglio and Jacques Thèze

1. Introduction

Most biological systems require cellular interactions for their development and regulation. The cross-talk between cells generally involves a variety of cytokines which are induced by a large panel of stimuli and secreted by many cell types. These mediators are particularly important in promoting the proliferation and differentiation of hematopoietic cells and in regulating and determining the nature of the immune response. In addition they participate in various inflammatory processes. Each cytokine exerts a number of effects on several different target cells, and its action can be modulated by the presence of other cytokines (cytokine interactions can be either synergic or inhibitory). Moreover, each cytokine can either induce or inhibit the production of others. These observations have given rise to the concept of a cytokine network, controlling all physiological processes involving cellular interactions. In addition, they have revealed two fundamental hallmarks of these molecules, namely their pleiotropy and their redundancy. The biological effects of cytokines are generally quite diverse (for example, cytokines can promote survival, proliferation, differentiation, activation or cell death) and apply to more than one responder cell population; however, each cytokine possesses a certain degree of specificity in terms of activity in function of the target cells. Furthermore, different cytokines can act on the same cell type to induce similar biological activities. The recent understanding of the multimeric nature of cytokine receptors as well as the determination of the various signals transduced after binding of the ligand have provided some explanations for these two properties.

Many cytokines have now been cloned, sequenced and characterized. No primary sequence homology has been consistently found among all these molecules. However, some similarities in terms of spatial conformation have been found. For example, cytokines of the hematopoietin family comprise four long helices arranged in antiparallel fashion (up–up–down–down) whereas cytokines of the tumor necrosis factor (TNF) family occur as trimers and for each ligand protomer the polypeptide chain folds in the form of three packed groups of eight antiparallel sheets arranged in a β-jellyroll topology.

The study of cytokine receptors has also contributed to the understanding of the regulation of cytokine functions, another field of intensive research. The discovery of natural receptor antagonists and natural soluble receptors has led investigators to address the capacity of these molecules to modulate the biological activities of cytokines.

This chapter will describe some general aspects of cytokines and their receptors, and present recent research on cytokine receptors and their signal transduction which has given new insight into the cytokine network.

2. Properties of cytokines

2.1 Cytokines are classified based on the structure of their receptors

With the cloning of cytokine receptors, it has become clear that these molecules can be classified into various families characterized by primary sequence homologies in their extracellular domains, as shown in Figure 1.

One of the largest families is the hematopoietin receptor family, also called the type I cytokine receptor family. It includes the receptors for

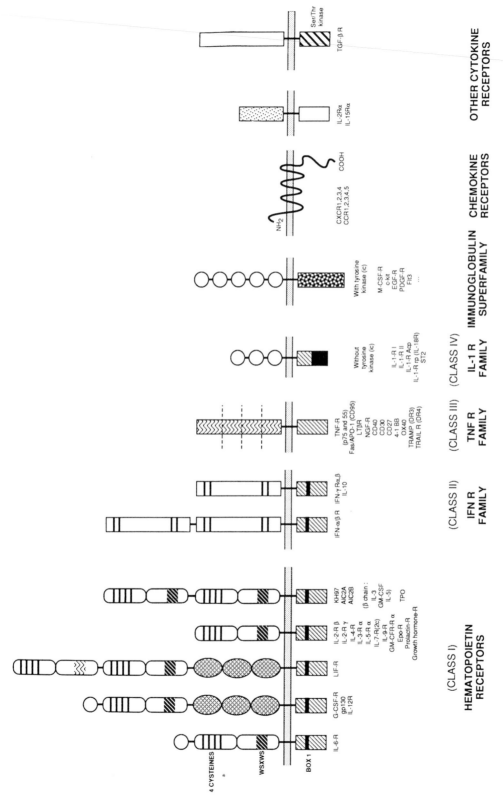

■ **Figure 1.** Schematic representation of the various classes of cytokine receptors.

interleukin 2 (IL-2), IL-3, IL-4, IL-5, IL-6, IL-7, IL-9, IL-11, IL-12, IL-15, granulocyte–macrophage colony-stimulating factor (GM-CSF), granulocyte colony-stimulating factor (G-CSF), leukemia inhibitory factor (LIF), oncostatin M (OSM), ciliary neurotrophic factor (CNTF), cardiotrophin 1 (CT-1), thrombopoietin, erythropoietin, growth hormone and prolactin. The general features shared by these cytokine receptors are concentrated in an approximately 200 amino acid stretch located in the extracellular part of the receptor. This region can be subdivided into two domains each of about 100 amino acids: the N-terminal domain, containing four positionally conserved cysteine residues, and the C-terminal domain, which includes a Trp-Ser-X-Trp-Ser (WSXWS) motif near the membrane-spanning domain. This homologous region is composed of two globular structures consisting of seven antiparallel β-sheets which are arranged so as to form a barrel-like structure. A hinge region connecting the two barrel-like modules has been predicted to function as a ligand-binding pocket. This basic unit can be duplicated in some cytokine receptors or complemented by the addition of type III fibronectin domains and/or immunoglobulin-like domains. All these receptors have a single transmembrane domain composed of 22–28 amino acids (except for the CNTF receptor, which is anchored to the membrane by a glycosyl-phosphatidylinositol linkage). The cytoplasmic domains of these cytokine receptors contain no consensus catalytic region but share some limited similarity in membrane-proximal sequences called box 1 and box 2. Box 1 is characterized by a proline-rich motif (Al-Ar-Pro-X-Pro-X-Pro-X-Pro or Ar-X-X-X-Al-Pro-X-Pro, where 'Al' indicates an aliphatic amino acid residue, 'Ar' an aromatic amino acid residue and 'X' any amino acid residue). Box 2 is only conserved in about half of these cytokine receptors; it is composed of several hydrophobic amino acid residues followed by negatively charged amino \acid residues and ending in one or two positively charged residues.

The class II cytokine receptor family includes receptors for interferon (IFN) α/β, IFNγ and IL-10. Members of this family have similar organization in their 210 amino acid extracellular domain, containing cysteine pairs at both the N- and C-termini as well as several proline and tyrosine residues in a conserved position. This domain is duplicated in the IFNα/β receptor. Class I and class II receptors show some structural similarities, suggesting that these two types evolved from a common ancestor with evolutionary links to the fibronectin type III domains.

Members of the class III cytokine receptor family (also called the TNF receptor family) share several cysteine-rich motifs of approximatively 40 residues in their extracellular domains. They can be divided into two groups. The first group comprises the death receptors (DRs), namely Fas/Apo-1 (CD95), TNFR1 or p55, DR3 (also called Apo3, TRAMP, Wsl or LARD), TRAIL receptors (DR4 and DR5) as well as the TRAIL decoy receptor (TRID). The second group includes all other members, such as the low-affinity nerve growth factor (NGF) receptor, TNFR2 (or p75), the lymphotoxin (LT) β-receptor, CD40, CD30, CD27, 4-1 BB, OX40. They show no significant homology in their intracytoplasmic domain, except in a motif called the 'death domain', which is found in Fas, TNFR1, DR3, DR4 and DR5 and is responsible for the transduction of apoptotic signals.

The class IV cytokine receptor family (also called the IL-1 receptor family) comprises the two receptors for IL-1 (the type that mediates IL-1 effects and type II which was found to be a decoy receptor), the IL-1 receptor accessory protein (IL-1-RAcP), and the IL-1 receptor-related protein (IL-1-Rrp) which could be part of the IL-1-8R. The extracellular part of these receptors contains three immunoglobulin-like motifs. In addition, all these members share conserved motifs in their intracellular part, which is devoid of any kinase activity.

Another class of receptors is appointed to the immunoglobulin-like superfamily because their intracellular domain comprises five successive immunoglobulin-like motifs. Their intracytoplasmic regions have intrinsic tyrosine kinase activity. It comprises the receptor for the macrophage colony-stimulating factor (M-CSF), platelet-derived growth factor (PDGF), epidermal growth factor (EGF) as well as the proto-oncogene products c-Kit (stem-cell factor (SCF) receptor) and Flt-3 (receptor for Flt3-ligand).

Chemokine receptors can be divided into two large groups according to the nature of their ligand: the Cys–X–Cys (CXC) receptors, such as CXC-R1, -R2, -R3 and -R4 bind α-chemokines, while the Cys–Cys (CC) receptors CC-R1, -R2A and B, -R3, -R4 and -R5 bind β-chemokines. Both types belong to the large family of G-protein-coupled, seven-transmembrane-domain or serpentine receptors. The Duffy blood group antigen is an erythrocyte chemokine receptor that binds selected CXC (α) as well as CC (β) chemokines.

There are also several cytokine receptors with distinct features. For example, the α-chain of the IL-2 receptor contains two 'Sushi domains' and the α-chain of the IL-15 receptor contains one such domain (the Sushi domain is a motif also found in molecules such as some complement components). Lastly, the transforming growth factor beta (TGFβ) receptor is characterized by the presence of a serine/threonine kinase in its cytoplasmic domain.

2.2 Functional redundancy and the multimeric composition of cytokine receptors

Elucidation of the make-up of hematopoietin receptors has led to a better understanding of the functional redundancy observed in the biological activities of these cytokines. High-affinity receptors are usually composed of two chains: α and β. The α-chain is the cytokine-specific subunit, while the β-chain cannot on its own bind any cytokine, but forms a high-affinity receptor when associated with the α-chain, and then transduces the signal. The fact that the same β-chain is often shared by various receptors provides a basis for common intracellular signals induced by these cytokines, which could explain some of their shared biological activities (redundancy). For example, IL-3, GM-CSF and IL-5 exert a number of similar effects on a variety of cells. In addition, they often compete with each other in binding to their receptors. Each receptor is constituted of a cytokine-specific subunit (IL-3Rα, IL-5Rα and GM-CSFRα) with low affinity (K_d = 2–100 nM) and a common β-subunit (βc) which associates with one of the three α-chains to form the high-affinity receptor (K_d = 100 pM). The cross-competition in receptor binding and the functional redundancy are thus explained by competition of the different α-subunits for βc. This conclusion applies to other complex receptor systems with a common signal-transducer chain, such as gp130 (receptors for IL-6, LIF, OSM, CNTF, IL-11 and CT-1) or γc (receptors for IL-2, IL-4, IL-7, IL-9 and IL-15) (Figure 2).

The multimeric composition of class I receptors is thus very useful for further cytokine classification based on the presence of a common signal-transducer chain:

*   the gp140 or βc family: IL-3, GM-CSF, IL-5;
*   gp130 family: IL-6, LIF, OSM, CNTF, IL-11 and CT-1; and
*   the γc family: IL-2, IL-4, IL-7, IL-9 and IL-15.

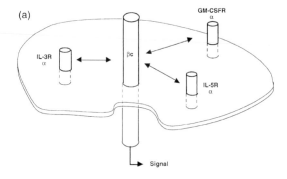

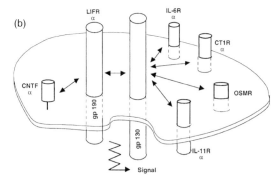

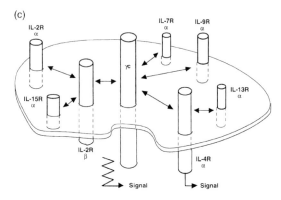

▮ **Figure 2.** Representation of the multimeric composition of cytokine receptors. (a) IL-3 family sharing βc (KH 97 in human, AIC2B in mouse); (b) IL-6 family sharing gp130; (c) IL-2 family sharing γc.

(See Figure 2 for details.) The other class I cytokine receptors (G-CSF, erythropoietin, thrombopoietin, prolactin and growth hormone) are homodimeric and the high-affinity IL-13 receptor is quite particular since a specific IL-13R α-chain is associated with the IL-4R α-chain. The composition of this receptor again explains the high functional redundancy between these two cytokines.

The overlapping biological activities of some cytokines cannot be explained entirely by the

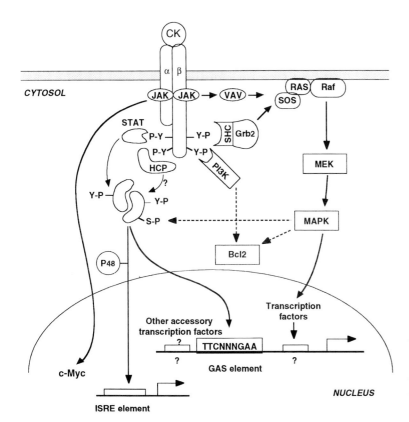

∎ **Figure 3**. Schematic representation of the various intracellular pathways induced by cytokine receptors. Distinct combinations of these signals could explain the functional cell-specific pattern of cytokine response.

sharing of a common signalling chain. Indeed, it seems that some redundancies rather reflect the use of a similar transducer pathway by two receptors lacking a common element. G-CSF, for example, shares some biological activities but no common receptor subunit with IL-6. Nevertheless, both cytokines activate STAT3, which might account for their functional redundancies.

2.3 Specificity and pleiotropy of cytokines and combination of various transduction signals

Most cytokine receptors do not have intrinsic protein tyrosine kinase activity, yet receptor stimulation usually involves rapid tyrosine phosphorylation of intracellular proteins. It is now clear that these receptors are capable of recruiting and/or activating a variety of non-receptor protein tyrosine kinases to induce downstream signalling pathways. Thus, the intracytoplasmic structure of cytokine receptors has evolved so as to allow the combined action of different protein tyrosine kinase family members expressed in different cell

types, which may ultimately determine the activity of a cytokine.

Cytokines bind to their specific receptor chains, inducing their dimerization or oligomerization. Jak kinases, which are constitutively associated with the 'box 1' region in the majority of cytokine receptors can then phosphorylate each other, as well as tyrosine residues present in the intracytoplasmic domain of the receptor, creating numerous docking sites for proteins that contain the specialized, phosphotyrosine-interacting Src homology 2 (SH2) domains. Adaptor or effector molecules are then recruited and their phosphorylation leads to the activation of numerous signal transduction pathways (e.g. Ras, Raf, MAP kinase, PI 3-kinase, Src kinases, STAT) which explain the diversity of cytokine effects. A variation upon this basic mechanism is the recruitment and phosphorylation of docking proteins, e.g. insulin receptor substrate 2 (IRS-2) by cytokines such as IL-4, which could increase the level of diversification by increasing the number of SH2-binding sites and then diversifying the signal output from this receptor.

Individual Jaks (Jak1, Jak2, Jak3 and Tyk2) can

be activated by multiple cytokines and each cytokine uses different combinations of Jaks; for example, IL-2 uses Jak1 and Jak3; IL-6 uses Jak1, Jak2 and Tyk2; and interferon gamma (IFNγ) uses Jak1 and Jak2. This suggests that Jaks alone do not determine the specificity of cytokine action; instead this depends upon recognition of phosphotyrosine docking sites on the receptor complex by effector molecules such as STATs. STATs are a family of latent cytoplasmic transcription factors which, after tyrosine phosphorylation, homodimerize or heterodimerize and translocate to the nucleus where they interact with a specific DNA-binding site within the promoter region of target genes the GAS (gamma activation sequence). Recent data demonstrate that some STAT proteins are very selective, mediating the specific action of particular cytokines as shown by studies of STAT4- and STAT6-deficient mice which show profound deficits in IL-12 and IL-4 functions, respectively. Similarly, STAT1 knockout mice have a defective response to interferons. These data suggest that, for some of the biological activities of the cytokine, a specifc STAT is required, while on others there is redundancy. Indeed, results obtained with STAT4 and STAT6 knockout mice are easily explained by the exquisite specific activation of STAT4 by IL-12 and of STAT6 by IL-4, while STAT1 is only vital for interferon-induced signals, in spite of its implication in the transduction of various cytokine receptors.

There are far fewer STATs than cytokine receptors and the specificity of action of each cytokine probably results from more complex mechanisms, such as the following:

(1) different affinities for the GAS sequence, according to the dimer composition of the STATS;

(2) different expression of STATs in different cell types, as exemplified by myeloid cells, where STAT expression varies with the state of differentiation;

(3) quantitative variation in signal transduction, meaning that two ligands activating the same STATs might do so at different levels and not for the same period of time, thus changing the signal emanating from similarly activated STATs;

(4) interactions with additional intracellular signals or proteins, such as serine phosphorylation (possibly mediated by MAP kinases), modifying the functions and affinities of STATs for the GAS sequence, or association with other proteins (such as STAT1 or STAT2 with p48), altering their DNA-binding specificity;

(5) selective interactions between different STATs and other simultaneously induced transcription factors, which bind either upstream or downstream of the GAS sequence.

Cytokines induce both overlapping and unique biological effects and their action can be strikingly different according to the nature of the target cells. If overlapping is often explained by common subunit shared by several cytokine receptor complexes, the unique biological effect must reflect the implication of either a specific pathway or a specific combination of pathways in signal transduction, leading to the expression of genes required for specific biological activities. Cytokine specificity and/or pleitropy results probably from the involvement of different signal transduction pathways which interact with each other in order to give rise to an integrated signal from which the target cell deduces a cell-specific pattern of response.

3. Basic cellular functions of cytokines

Cytokines produced in response to immunological stimuli are mainly involved in regulating cell viability, proliferation, differentiation and death (Figure 4). This task is performed through their specific cell surface receptors, which are multimeric complexes activating various signal transduction pathways in response to ligand binding. One of the most exciting and promising fields in cytokine research concerns the identification of the exact intracellular events leading to each of these main functions.

3.1 Cytokines and cell survival

It is well established that the viability of hematopoietic cells depends on the stromal environment or cytokines such as hematopoietic growth factors (e.g. IL-3, GM-CSF or G-CSF). In the absence of such growth factors, hematopoietic precursor cells rapidly die by apoptosis, indicating that cytokines promote cell survival by suppressing the apoptotic process. This protective effect is observed in other conditions. For example:

• IL-3 inhibits apoptosis induced by ionizing irradiation or DNA-damaging agents;

• IL-2 protects T lymphocytes from glucocorticoid-induced apoptosis; and

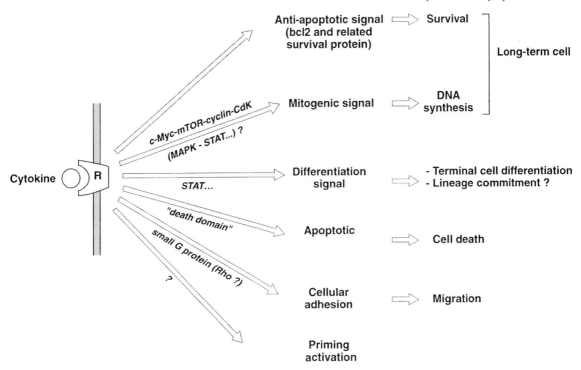

∎ **Figure 4.** Biological activities of cytokines.

- IL-6 prevents p53-induced apoptosis in myeloid leukemic cells.

Similarly, cytokines like IL-1β, TNFα, IFNγ, GM-CSF and G-CSF prolong the survival of mature circulating polymorphonuclear cells by interfering with the physiological process of apoptosis which normally induces rapid death of these cells *in vitro*. This finding has been confirmed by the increased susceptibility to apoptosis of the small number of neutrophils isolated from G-CSF receptor-deficient mice.

Thus, prevention of apoptosis is an important function of several cytokines, particularly those acting via type I cytokine receptors (or hematopoietin receptors). The mechanism by which these cytokines interfere with the apoptotic pathway is not completely understood. However, several data argue in favor of a role for the anti-apoptotic oncoprotein Bcl-2 or related products in this activity. It has been shown that IL-3/GM-CSF receptors suppress apoptotic cell death in hematopoietic cells by activating a signalling pathway distinct from the one which induces DNA synthesis. Indeed, Ras activation does not transduce mitogenic signals in IL-3-dependent cell lines, even though it plays a pivotal role in inhibiting apoptotic death by up-regulating the expression of Bcl-2 and Bcl-XL (a related survival protein), without affecting Bax (an antagonist of Bcl-2). Similarly, chimeric receptors composed of the external domain of G-CSF and the transmembrane and cytoplasmic domains of gp130 induce an anti-apoptotic signal mediated by increased Bcl-2 expression, in the absence of proliferation. At present all data point to a pivotal role of Bcl-2 in the anti-apoptotic signal induced by cytokines. It should, however, be noted that other events have also been implicated in cytokine-induced suppression of apoptosis, such as the protein kinase C signalling pathway, suggesting that the nature of cytokine-induced anti-apoptotic signals may depend on the cell type.

3.2 Cytokines and cell proliferation

The effects of many cytokine/cytokine receptor systems on the regulation of cell proliferation have been examined, but the exact nature of their target genes and the mechanisms whereby they promote cell cycle progression are not yet clearly established. Most investigations have focused on the $G_1{\rightarrow}S$ transition, since this boundary is known to be an important control point at which cells 'decide' whether to proliferate or to remain quiescent. Studies on cytokine-induced long-term proliferation of cells have demonstrated that stimulation of DNA synthesis is insufficient for cells to proliferate continually. Additional activation of anti-apoptotic pathways is required and can be provided by an increased expression of Bcl-2 or related survival proteins (see preceding paragraph). This kind of cooperation between a mitogenic signal and an anti-apoptotic signal has been clearly established both for the IL-3/GM-CSF and the gp 130 receptor families.

The notion that multiple signals are involved in the growth signalling mediated through the type I cytokine receptors is supported by many recent data. There may be slight variation in the combination of pathways leading to cytokine-induced cell proliferation depending on the target cells or the cytokine receptor. Cell proliferation induced by IL-2 and IL-2R involves at least three distinct signal pathways, namely c-Myc, c-Fos/c-Jun, and Bcl-2, all of which are activated by interaction of effector molecules with distinct regions of the intracytoplasmic IL-2Rβ. Induction of c-Myc is mediated by the S region (serine-rich region) which binds the two protein tyrosine kinases Jak1 and Syk. Induction of Bcl-2 requires the same S region of IL-2Rβ but depends on a different pathway which is blocked by rapamycin, unlike c-Myc expression. Lastly, the A region (acidic region) of IL-2Rβ is needed for the c-Fos/c-Jun induction which results from p56^{lck} activation. The three pathways do not interact with one another and, importantly, expression of any two of them is sufficient to promote cell growth. The determination of the exact implication of the *bcl-2* proto-oncogene in the regulation of cytokine-induced cell growth needs further investigation. Nevertheless, different groups using *bcl-2* transgenic mice reported that overexpression of *bcl-2* increased the length of G_1 phase and thus delayed S phase entry.

3.3 Cytokines and cell differentiation

3.3.1 Th1/Th2 differentiation

Naïve T cells (Thp: T helper precursor) which produce only IL-2 after stimulation, differentiate either into Th1 or Th2 cells. Because they produce IL-2 and IFNγ, Th1 cells are more suited for the enhancement of cellular immunity, while Th2 cells,

which secrete IL-4, IL-5, IL-6, IL-10 and IL-13, induce immune responses that are better qualified in helping B cells to develop into antibody-producing cells. Cytokines present in the microenvironment of naïve T cells at the onset of their stimulation are the major factors determining the differentiation of Thp cells into the Th1 or Th2 phenotype. Thus, the presence of IL-4 at the time of T-cell priming drives differentiation towards the Th2 type, while IL-12 favors Th1 differentiation. Specific activation of transcription factors by IL-12 and IL-4 seems to represent the molecular basis of their effect on Th1 or Th2 differentiation. Indeed, STAT4 is one member of this family that is only activated in response to IL-12, and STAT4-deficient mice obtained by gene targeting do not respond to IL-12 and consequently cannot develop Th1 responses. Similarly, STAT6 is mainly activated in response to IL-4, and in mice with a disrupted STA 6 gene IL-4-induced Th2 differentiation and IgE class switching are no longer observed. These data clearly demonstrate that cytokines play an instructive role (see below) in the differentiation of the T-helper cell subset.

3.3.2 B-cell differentiation

B-cell activation, proliferation and differentiation to immunoglobulin synthesis are greatly influenced by cytokines. IL-2 enhances immunoglobulin production in T-cell-independent and T-cell-dependent immune responses. A number of *in vitro* studies using β-cell polyclonal activates or activated fixed T cells suggest that IL-2 is essential for the induction of immunoglobulin synthesis. Nevertheless, B cells activated through CD40 in the presence of IL-4 or IL-10 can secrete immunoglobulin in the absence of IL-2, sug-gesting redundant pathways for B-cell maturation. Some investigations suggest that IL-2 may be more important in promoting B-cell proliferation, whereas IL-10 is a more potent differentiation factor.

Immunoglobulin class switching is not a random process but can be directed towards expression of particular immunoglobulin classes and subclasses by distinct modes of B-cell activation and cytokines. For instance, IL-4 induces IgG4 and IgE production by B cells derived from blood, tonsils and spleen. Since IL-5 stimulates eosinophils, IL-4 plus IL-5 (Th2 cytokines) are involved in allergy and in responses to some parasites. TGFβ is the main cytokine inducing IgA production and therefore participates to the regulation of mucosal immunity.

3.3.3 Hematopoietic progenitor cell differentiation

Hematopoietic progenitors depend on various cytokines for their survival and proliferation. The extent to which these cells require cytokines for their commitment into a specific cell lineage is still a matter of debate. On the one hand, the 'instructive model' assumes that cytokines play a deterministic role in inducing cell differentiation into a specific lineage by unique signals originating from their receptors (as shown for Th1/Th2 cell differentiation). On the other hand, it has been postulated that cytokines play only a permissive role, inducing the development of cells already committed to a specific cell lineage by a stochastic process.

Insights into the role of cytokines in hematopoietic cell differentiation have been recently provided by the analysis of mice carrying targeted null mutations of cytokine receptors. Mice lacking erythropoietin receptor die *in utero* from a failure of definitive erythropoiesis. However, committed BFU-E (burst-forming unit — erythroblast) and CFU-E (colony-forming unit — erythroblast) progenitors are present, suggesting that erythropoietin is not required for erythroid lineage commitment but is critical for proliferation, survival and terminal differentiation of CFU-E progenitors. Likewise, thrombopoietin receptor-deficient mice have few megakaryocytes and are severely thrombocytopenic but show also a marked deficiency in committed megakaryocytic progenitors. Similarly, disruption of the G-CSF receptor gene results in strikingly decreased numbers of phenotypically normal circulating neutrophils associated with a decreased frequency of hematopoietic progenitors in the bone marrow of these mice. Terminal differentiation into neutrophils is impaired, confirming that G-CSF is a major regulator of granulopoiesis. However, the presence of some functionally normal neutrophils in G-CSF receptor-deficient mice can be taken either as an argument for a stochastic process regulating neutrophil lineage commitment or for the redundancy of the G-CSF-induced process.

3.4 Cytokines and cell death

Some specialized cytokines, such as Fas-L, TNFα, TRAIL or the as yet unidentified ligand of TRAMP (or DR3), induce apoptosis in various cell types. Their receptors, including Fas/Apo-1 (or CD95), the TNF receptor type I (TNFRI or p55), DR3, DR4 and DR5 (the newly identified TRAIL receptors) belong to the class III cytokine receptor family. All these receptors possess an intracytoplasmic 'death domain' enabling interaction with other death-domain-containing intracytoplasmic molecules like TRADD and FADD/MORT-1, which are associated with TNFRI and Fas respectively. TRADD is an adaptor molecule mediating the interaction between TNF-RI and FADD. FADD subsequently interacts, through the death effector domain (DED) in its N-terminal region, with the homologous domain contained in the effector protease FLICE (or caspase 8). This is the first element of a protease cascade leading to apoptosis (more details are provided in Chapter IV).

3.5 Other cellular functions of cytokines

In addition to their well-known effect on cellular survival, proliferation, differentiation or apoptosis, cytokines also have roles in the control of migration, activation and priming. Although various cytokines act on cell migration, including IL-3 (which increases basophil mobility), this property is exemplified mainly by chemokines (α and β), a superfamily of cytokines which are chemoattractants for monocytes, T and B cells, basophils, eosinophils, dendritic cells and NK cells (see Chapter IX).

Cytokines can also modify the response of cells to other stimuli, without having any apparent effect by themselves. This is the case for IL-3, which primes basophils, increasing the amount of mediator they release in response to degranulating agents. IL-3 can also induce the conversion of basophils that do not release histamine in response to IgE/anti-IgE into cells that do release histamine. A possible explanation for such a phenomenon might be that, in non-releasing basophils, the cytosolic concentration of Ca^{2+} is too low for granulation to occur (Ca^{2+} is required for the signal transduction after cross-linking of FcεRI), and that IL-3 increases the concentration of this ion. However, further investigations are needed to confirm this hypothesis.

4. Cytokine regulation

In contrast to hormones, cytokines act mainly locally in an autocrine or paracrine manner, but some of them may also exert systemic effects. They are operative at very low concentrations, requiring occupancy of only a fraction of their membrane receptors to induce maximal biological effects. This high potentiality combined with the wide range of biological activities carried out by cytokines implies that tight control is needed *in vivo* in order to maintain homeostasis. Indeed, there are multiple ways of regulating cytokine functions, at the level of both the producer cell (by means of transcriptional, translational and post-translational regulation) and the target cell level (where the accessibility of the specific receptor or its signal transduction can be modulated). The binding capacity of a given cytokine can be modified in several ways, namely by decreasing the number or affinity of its receptors, by competing with decoy receptors (such as IL-1R type II), receptor antagonists (such as IL-1ra), non-receptor binding proteins (such as uromodulin), anti-cytokine autoantibodies and soluble cytokine receptors.

4.1 Mode of cytokine action

Each cytokine and its multimeric receptor form complexes able to interact in the cytoplasm with numerous protein kinases.

Cytokine production is controlled not only at the genetic level but also at the level of cellular release. Most cytokines are processed in the usual way for secreted proteins, whereas IL-1 needs to be cleaved by caspases before transport through the membrane. The activation of cytokine genes may depend on specific signals or non-specific signals, or both. IL-2 secretion is observed after T-cell receptor (TCR) binding but optimal production is obtained after additional binding of the cell surface molecules B7.1 and B7.2 to CD28.

The sensitivity of cells to a given cytokine is dependent on the controlled expression of the different chains of its receptor. As the cytokine binds to the receptor the complex is internalized and the cells are desensitized to further stimulation by the same cytokine. In the case of cytokines using a common chain, there may be some competition as this chain may be limiting both at the level of the activation mechanism and at the level of the desensitization process. Under some conditions cells may express receptors but may be insensitive to a cytokine. The molecular parameters defining the competence of a cell to respond to a cytokine are not yet understood.

4.2 Soluble cytokine receptors

The majority of cytokine receptors are also generated *in vivo* in a soluble form which results from the translation of an alternatively spliced pre-mRNA molecule lacking the transmembrane and the cytoplasmic domain or from limited proteolytic cleavage of the membrane-bound receptor. These soluble cytokine receptors can bind specific ligands and are made at low levels under resting conditions, but increase dramatically under conditions associated with active immunological or inflammatory processes. Usually, stimulation of T cells leads to enhanced soluble receptor production for cytokines involved in T-cell activation, while monocytic cell activation results in a preferential up-regulation of their pro-inflammatory counterpart. Production of soluble cytokine receptors in response to cellular activation might be envisioned as a feedback mechanism by which a cytokine induces production of a soluble form of its receptor in order to limit its own biological activity both temporally and spatially. However, conditions where soluble cytokine receptors are increased in response either to unrelated cytokines or to other factors, such as co-stimulatory molecules, have also been described.

The immunoregulatory function of soluble cytokine receptors is still a matter of debate because of the paradox that even though most soluble cytokine receptors act as competitive inhibitors of their respective cytokine binding on membrane-bound receptors *in vitro*, they can exert various physiological roles *in vivo*, including antagonist, carrier or agonist functions. Most of the soluble cytokine receptors are potent specific cytokine inhibitors if they are present in sufficient amounts. Indeed, relatively high concentrations of the soluble form are required because of the competitive nature of the inhibition (with increasing amounts of cytokine, more and more soluble cytokine receptors are needed to block its binding to membrane-bound receptors) and the difference in affinity between the soluble form and the membrane-bound receptor. The latter is formed by the two or three chains constituting the high-affinity receptor complex, while the soluble form represents only the cytokine-specific subunit of this receptor (the α-chain), which binds the ligand with a lower affinity.

This antagonistic activity has been demonstrated *in vivo* for various cytokines and could therefore represent a physiological means of down-regulating the immune response and a way to prevent excessive cytokine activity. In favor of such a possibility, it should be mentioned that several viruses express genes coding for cytokine-binding proteins which are soluble cytokine receptor homologues. In all cases, infection with viruses in which such genes have been disrupted results in a dramatic decrease in virulence *in vivo*, suggesting that these proteins have antagonistic activity and indeed modulate the host response during infection. Despite this competitive inhibitory activity, several soluble cytokine receptors can act as carrier proteins for their specific cytokine, potentiating their biological activity. This effect is explained by increased cytokine stability and protection from proteolysis, increasing the half-life and decreasing the clearance of the cytokine. The overall effects of these soluble forms of cytokine receptors result from a balance between their antagonistic and agonistic actions, which is influenced by the molecular ratio of the soluble form and the cytokine itself. This is exemplified by IL-4, whose soluble IL-4 receptor potentiates the *in vivo* effect of the cytokine on murine IgE production at a relatively low molar ratio, while at a high ratio the same effect is inhibited.

A completely different situation has been reported for soluble IL-6 and CNTF receptors, whose effect is exclusively agonistic. This special situation results from the nature of the cytokine-specific binding subunits of these two receptors. They do not play any role in signal transduction resulting from either gp130 homodimerization (in the case of the IL-6 receptor) or formation of gp130–gp190 heterodimers (in the case of the CNTF receptor). Consequently, the soluble receptors of these cytokines bind to their specific ligand, form complexes which can interact with gp130 (or gp130 + gp190), leading to the generation of the biological signal even on cells which do not usually express IL-6 or CNTF receptors.

4.3 Regulatory molecules for cytokine signal transduction

One of the characteristics of cytokines is the transient expression of their biological effect, suggesting that negative feedback regulation must operate in order to switch off cytokine signals. The main candidates for this function are the SH2-containing phosphatase SHP-1 and the cytokine-inducible SH2 containing protein CIS, which is a 257 amino acid polypeptide containing one SH2 domain which is induced in hematopoietic cells by a subset of cytokines, including IL-2, IL-3, GM-CSF and erythropoietin. CIS production requires STAT5 activation CIS then binds to the membrane-distal region of the cytokine receptor intracellular domain that contains phosphotyrosine residues constituting the docking sites for STAT5. Over-expression of CIS inhibits cell proliferation in

response to IL-3 and thus is considered as a possible mechanism for negative feedback of STAT5 activation. Ever since the discovery of CIS, investigators have been looking for similar proteins with specificities for other STATs. Recently, two such proteins were shown to be involved in the regulation of STAT3 after activation by IL-6. These two new mediators were named SOCS-1 (suppressor of cytokine signal 1) and SSI-1 (STAT-induced STAT inhibitor 1); it is possible that these two proteins are in fact identical. They contain an SH2 domain and inhibit both IL-6-induced receptor phosphorylation and STAT activation. The SOCS molecules (SOCS-1, -2 and -3), together with SSI-1, JAB and CIS, represent a new family of proteins whose principal function seems to consist in turning off cytokine signalling.

5. Cytokines in pathology

The knowledge accumulated over the years on cytokines, their receptors and signalling mechanisms has provided the molecular basis to understand a number of pathological disorders of the immune and hematopoietic systems. Three examples are described in this chapter.

5.1 Cytokines and genetic immunodeficiencies

One of the best examples of recently discovered genetic defect of the immune system is illustrated by the X-linked Severe Combined Immunodeficiency Syndrom (SCID-XI). This pathology is due to mutations of the common γ-chain which, as described above, is shared by receptors for several important cytokines (IL-2, IL-4, IL-7, IL-9 and IL-15). SCID-XI is a rare form of immunodeficiency characterized by faulty T and NK lymphocyte development whereas mature B cells are usually present in the periphery. A number of mutations of the γc gene have been described in SCI-XI patients. They mostly affect the extra-cellular domain of the molecule either, preventing membrane γc expression or appropriate cytokine binding. A few γc mutations were also described that results in a inability to recruit JAK-3. In SCID-XI patients the defect of T cell development is mainly explain by the impairment of a prothymocyte proliferation wave mediated by IL-7 whereas the NK cell differentiation block mainly results from faulty IL-15/IL-15 receptor interaction.

5.2 Th1 and Th2 responses in infectious diseases

It has also become obvious that by influencing immunocytes differentiation cytokines participate to the outcome of some infectious diseases. The consequences of producing anti-pathogen responses dominated by either Th1 or Th2 subsets are clearly illustrated in the response to infection with protozoan parasites like *Leishmania*. These organisms live and replicate intracellularly and do so primarily in cells of the monocyte-macrophage lineage. The most common disease form in man is cutaneous leishmaniasis, a skin lesion with low numbers of parasites that are confined largely to the local lesion. A second form of disease is visceral leishmaniasis with very high number of parasites distributed throughout the body, especially visceral organs, such as spleen and liver. This is a progressive disease that is usually fatal if left untreated. The dominant T cell response in cutaneous leishmaniasis is a typical Th1 response whereas the response in visceral leishmaniasis has many of the features of a Th2 response, although many patients do not have elevations in IL-4 or IgE responses.

The consequences of the different T cell responses to *Leishmania* infection dramatically illustrate the differences in the effector functions mediated by Th1 and Th2 cells. The principal effector mechanism for control of this parasite is IFN-γ mediated macrophage activation, leading to the production of microbiocidal molecules, especially nitric oxide (NO). Th2 responses are not simply ineffective because of the absence of IFN-γ production, but also strongly inhibit Th1-mediated parasite killing. This is due both to the inhibition of Th1 cells, and to the inhibition of macrophage activation and NO production by IL-4 and IL-10. The protective effects of Th1 cytokines, principally IFN-γ, and the counterprotective effects of the Th2 cytokines IL-4, IL-10 and IL-13 are general features of the host responses to a wide range of non-viral intracellular infections, including bacterial, protozoan and fungal infections.

5.3 Chemokines and HIV infection

Recently, an important discovery has been made concerning the role of chemokine receptors in cellular infection by HIV. Chemokines are cytokines with chemoattractant activities. A number of cells express a receptor for one or the other chemokines, and particularly cells of the immune system. Chemokines and their receptors thus play an important role in the development of any immune and inflammatory response, through the recruitment of circulating immune cells.

It was demonstrated several years ago that CD4 expression by a cell was not sufficient to allow its infection by HIV. Recent results show that two groups of chemokine receptors are also involved in such a process. One is the CXCR4 receptor, initailly named fusin, which is the receptor for the chemokine SDF-1 (stromal-derived factor-1). Murine cells expressing CD4 and CXCR4 both of human origin can be infected with HIV, thus identifying CXCR4 as one of the co-receptors for the virus. CXCR4 is also involved in the fusion between infected and uninfected cells, which accounts for its initial denomination as « fusin ». Importantly, this coreceptor is selective for viral strains with a tropism for T cell lines (T-tropic strains), able to form syncytia and predominant at advanced stages of patient infection. Addition of high concentrations of SDF-1 prevents infection of cells by T-tropic strains, but not by M-tropic (infecting cells of the monocyte/macrophage lineage). This is explained because CXCR4 does not allow entry of M-tropic strains, which predominates early in the infection. Such M-tropic strains, unabel to form syncytia, are also responsible for most contaminations.

This observation indicated that there may be other co-receptors for HIV in addition to CXCR4. It was first shown that some chemokines (RANTES, MIP-1α and MIP-1β), when present at high concentrations, inhibited cell infection by M-tropic strains of HIV. This led to demonstrate that the receptor common to these 3 chemokines, CCR-5, is the main co-receptor for M-tropic strains of HIV. In addition to CCR-5, other chemokine receptors, CCR3 and CCR2b, also allow infection by HIV, although their involvement is restricted to few strains by a broader cellular tropism. T-tropic strains using the CXCR4 receptor can also use CCR5.

These observations were strenghtened by the demonstration that, *in vivo* in humans, chemokine receptors played a critical role in HIV infectivity. It was known for many years that a number of individuals, although repeatidly exposed to HIV-infected people, remained free of infection, and even developed an anti-HIV cellular immunity. This resistance to HIV infection largely relies on an inability of HIV to infect cells from such individuals. It has been recently shown that a fraction of such exposed uninfected individuals display a 32 base pair deletion of the CCR5 chemokine receptor. Approximately 1% of the caucasian population is homozygous for this deletion. No homozygotic individuals has been identified among HIV-infected patients, showing that expression of a functional CCR5 is absolutely required for HIV infection.These results also emphasize the role for M-tropic strains in transmission of the virus. This apparently applies for all modes of contamination, as an homozygotic CCR5 deletion is never find in HIV-infected people regardless of the way of infection. It should be added that such an homozygous deletion only accounts for a fraction (in the range of one out of 10) of individuals that remain uninfected after HIV exposure. This deletion is apparently missing in african and possibly in asiatic populations. Therefore, other phenomenons may explain the natural resistance against infection, some of which may rely on yet unidentified mutations in chemokines receptors.

5.4 Involvement of cytokines in the evolution of different pathology

Cytokine defects or dysregulation of cytokine functions may be the direct cause of diseases but in many other pathological conditions cytokines participate to the evolution of the disease especially for all syndromes were strong inflammatory processes are involved such as allergic responses, autoimmune disorders, graft rejection and cytotoxic responses against tumor cells.

Altogether the critical findings concerning the involvement of cytokines in differents pathologies open many new therapeutical hopes.

References

Abbas, A.K., Murphy, K.M., and Sher, A. (1996). Functional diversity of helper T lymphocytes. *Nature*, **383**, 787–93.

Colotta, F., Re, F., Polentarutti, N., Sozzani, S., and Mantovani, A. (1992). Modulation of granulocyte survival and programmed cell death by cytokines and bacterial products. *Blood*, **80**, 2012–2020.

Demaison, C., Chastagner, P., Moreau, J.-L., and Thèze J. (1996). Ligand-induced autoregulation of IL-2 receptor α chain expression in T cell lines. *Int. Immunol.*, **8**, 1521–8.

Doree, M., and Galas, S. (1994). The cyclin-dependent protein kinases and the control of cell division. *Faseb Journal*, **8**, 1114–20.

Fernandez-Botran, R., Chilton, P.M., and Ma, Y. (1996). Soluble cytokine receptors: their roles in immunoregulation, disease and therapy. *Advances in Immunology*, **63**, 269–336.

Friedman, M.C., Migone, T.S., Russell, S.M., and Leonard, W.J. (1996). Different interleukin 2 receptor β-chain tyrosines couple to at least two signaling pathways and synergistically mediate interleukin 2-induced proliferation. *Proceedings of the National Academy of Sciences of the United States of America*, **93**, 2077–82.

Fukata, T., Hibi, M., Yamanaka, Y., Takahashi-Tezuka, M., Fujitani, Y., Yamaguchi, T., Nakajima, K., and Hirano, T. (1996). Two signals are necesssary for cell proliferation induced by a cytokine receptor gp130: involvement of STAT3 in anti-apoptosis. *Immunity*, **5**, 449–60.

Heim, M.H. (1996). The Jak-STAT pathway: specific signal transduction from the cell membrane to the nucleus. *European Journal of Clinical Investigation*, **26**, 1–12.

Ihle, J.N., Witthuhn, B.A., Quelle, F.W., Yamamoto, K., and Silvennoinen, O. (1995). Signaling through the hematopoietic cytokine receptors. *Annual Review of Immunology*, **13**, 369–98.

Leonard, W.J. (1996). STATs and cytokine specificity. *Nature Medicine*, **2**, 968–9.

Lotz, M., Setareh, M., Von Kempis, J., and Schwarz, H. (1996). The nerve growth factor/tumor necrosis factor receptor family. *Journal of Leukocyte Biology*, **60**, 1–7.

Miyazaki, T., Liu, Z.J., Kawahara, A., Minami, Y., Yamada, K., Tsujimoto, Y., Barsoumian, E.L., Perlmutter, R.M., and Taniguchi, T. (1995). Three distinct IL-2 signaling pathways mediated by bcl-2, c-myc, and lck cooperate in hematopoietic cell proliferation. *Cell*, **81**, 223–31.

Moreau, J.-L. , Chastagner P., Tanaka, T., Miyasaka, M., Kondo, M., Sugamura, K. and Thèze, J. (1995). Control of the IL-2 responsiveness of B lymphocytes by IL-2 and IL-4. *J. Immunol.*, **155**, 3401–8.

Naka T;, Narazaki M., Hirata M., Matsumoto T., Minamoto S., Aono A., Nishimoto N., Kajita S., Taga T., Yoshizaki K., Akira S., Kishimoto T. (1997). Structure and function of a new STST-induced STAT inhibitor. *Nature*, **387**, 924–8.

Nourse, J., Firpo, E., Flanagan, W. M., Coats, S., Polyak, K., Lee, M. H., Massague, J., Crabtree, G. R., and Roberts, J. M. (1994). Interleukin-2-mediated elimination of the p27Kip1 cyclin-dependent kinase inhibitor prevented by rapamycin. *Nature*, **372**, 570–6.

Orkin, S.H. (1996). Development of the hematopoietic system. *Current Opinion in Genetics & Development*, **6**, 597–602.

Schindler, C., and Darnell, J.E. (1995). Transcriptional responses to polypeptide ligands: the JAK-STAT pathway. *Annual Review of Biochemistry*, **64**, 621–51.

Sherr, C. J. (1996). Cancer cell cycles. *Science*, **274**, 1672–4.

Starr, R., Willson, T.A., Viney, E.M., Murray, L.J.L., Rayner, J.R., Jenkins, B.J., Gonda, T.J., Alexander, W.S., Metcalf, D., Nicola, N.A., Hilton, D.J. (1997).A family of cytokine-inducible inhibitors of signalling. *Nature*, **387**, 917–21.

Taniguchi, T. (1995). Cytokine signaling through nonreceptor protein tyrosine kinases. *Science*, **268**, 251–5.

Thèze, J., Alzari, P. and Bertoglio, J. (1996). IL-2 and IL-2 Receptors: recent advances and new immunological functions. *Immunology Today*, **17**, 481–6.

Williams, G.T., Smith, C.A., Spooncer, E., Dexter, T.M., and Taylor, D.R. (1990). Haemopoietic colony stimulating factors promote cell survival by suppressing apoptosis. *Nature*, **343**, 76–9.

Wu, H., Liu, X., Jaenisch, R., and Lodish, H.F. (1995). Generation of committed erythroid BFU-E and CFU-E progenitors does not require erythropoietin or the erythropoietin receptor. *Cell*, **83**, 59–67.

Yoshimura, A., Ohkubo, T., Kiguchi, T., Jenkins, N.A., Gilbert, D.J., Copeland, N.G., Hara, T., Miyajima, A. (1995). A novel cytokine-inducible gene CIS encodes an SH2-containing protein that binds to tyrosine-phosphorylated IL-3 and EPO receptors. *EMBO Journal*, **12**, 2816–26.

Part A

Cytokines and their receptors: molecular aspects

Interleukins 2, 4, 7, 9, 13 and 15

Yannick Jacques, Adrian Minty, Didier Fradelizi and Jacques Thèze

1. Introduction

Interleukin-2 (IL-2) was one of the first cytokines to be discovered. It was initially identified as a soluble factor secreted by mitogen-activated peripheral blood lymphocytes and able to promote the cellular expansion of T lymphocytes from human bone marrow.

The discovery of this factor, along with its subsequent purification, large-scale production as recombinant protein and elucidation of the structure of its specific receptors, has led to major advances in immunotechnology. Firstly, it has enabled T cell lines and normal human T cell clones to be cultured *in vitro* for considerable periods of time, thereby allowing the development of new areas of research, such as the analysis of the T cell repertoire at the clonal level. Secondly, it has led to the development of bioreagents (agonists, antagonists) useful in tumour immunology, graft rejection and autoimmune diseases.

It was subsequently found that IL-2 is a key factor in the regulation of the immune responses, acting on both lymphoid (T and B) and myeloid components of the immune system. IL-2 is involved both in the cellular expansion of antigen-activated cells and in the feedback negative regulatory control of this expansion. Gene knockout experiments have shown that IL-2 is not essential for some of these activities, indicating that other factors can substitute for IL-2 *in vivo*. In accordance with this, a number of other cytokines have been discovered which elicit biological activities similar to those of IL-2 within the lymphoid compartment of the immune system, namely IL-4, IL-7, IL-9, IL-13 and IL-15. These cytokines, and IL-2, show structural similarities, both at the cytokine and receptor levels, and form a sub-group of cytokines within the haematopoietic cytokine family. In this chapter the structure and function of this group of cytokines are discussed.

2. Molecular aspects of cytokines

2.1 IL-2

Natural IL-2 has been purified and sequenced, and its cDNA cloned. Crystallization of IL-2 from *Escherichia coli* has led to the determination of its three-dimensional structure at a resolution of 3 Å. Human IL-2 cDNA encodes a precursor protein of 153 amino acids, the first 20 of which constitute the signal peptide (Table 1). Natural human IL-2 is a 15.5 kDa glycoprotein. Variable *O*-glycosylation (sialic residues) at threonine 3 is responsible for its heterogeneous isoelectric point (between 6.8 and 8.0). This *O*-glycosylation has no effect on the biological activities of the cytokine, but improves its half-life *in vivo*. Gibbon and mouse IL-2 are 100% and 60% homologous to human IL-2 respectively. Mouse IL-2 differs from its human homologue in that it has a much lower isoelectric point, it contains of a repeat of 12 glutamine residues in the N-terminal region, and it behaves as a homodimer when subjected to electrophoresis. Both human and mouse IL-2 contain three conserved cysteine residues, two of which form a disulfide bond which is crucial for the biologically active conformation of the cytokine. Human and mouse IL-2 both have similar activity on mouse cells, whereas mouse IL-2 is 10- to 100-fold less active than its human homologue on human cells.

■ **Table 1.** Cytokine biochemical features

	Amino acids		Molecular mass (kDa)		N-glycosylation sites	Disulfide bonds	pI (poly peptide)
	Precursor	Mature	Polypeptide	Glycosylated			
human IL-2	153	133	15.4	15.5	0	1	7.1
mouse IL-2	169	133	17.2	30 (dimer)	0	1	4.9
human IL-15	162	114	12.8	14–15	3	2	4.5
mouse IL-15	162	114	13.3		4	2	4.6
human IL-4	153	129	14.9	15–19	3	3	9.3
mouse IL-4	140	120	13.6	15–19	3	3	8.3
human IL-13	132	112	12.3	14	4	2	8.8
mouse IL-13	131	111	12.1		3	2	8.6
human IL-7	177	152	17.4		3	3	8.7
mouse IL-7	154	129	14.9	25	2	3	8.7
human IL-9	144	126	13.7	32–39	4	5	8.7
mouse IL-9	144	126	14.2	30–40	4	5	9.0

2.2 IL-15

IL-15 was recently (in 1994) identified in the super-natant of the CV-1/EBNA monkey kidney epithe-lial cell line on the basis of its ability to promote the growth of the IL-2-dependent mouse cell line CTL-L2 and the inability of neutralizing anti-IL-2 anti-bodies to block this response. Simian IL-15 purified from CV-1/EBNA cells has a molecular mass of 14–15 kDa whereas human IL-15 subsequently purified from a T-cell leukaemia cell line (HUT-102) has a molecular mass of 12–16 kDa with an isoelectric point of 4.5. Human and mouse IL-15 cDNAs both encode 162 amino acid precursor pro-teins which, after cleavage of a signal peptide of unusual length (48 amino acids), lead to mature peptides of 114 amino acids. Human and simian IL-15 differ only by five residues whereas mouse and human IL-15 show 73% identity at the amino acid level. IL-15 contains four cysteine residues which are postulated to form two disulfide bridges. By a mechanism involving the use of an alternative enhancer/promoter in intron 2, another human IL-15 transcript can be generated that encodes a human IL-15 precursor protein with an alternative and shorter (21 amino acid) leader peptide. Whereas IL-15 associated with the long signal peptide is observed in the endoplasmic reticulum and is largely secreted, the short signal sequence drives IL-15 to the nucleus and cytoplasmic compartments, where it is stored rather than secreted.

2.3 IL-7

IL-7 was originally identified (in 1988) as a B cell growth and differentiation factor (lymphopoietin-1; LP-1) produced by bone marrow stromal cells. Subsequently, it was established that IL-7 also acts on thymocytes, monocytes/macrophages and NK cells and is a potent T cell growth factor.

Human IL-7 cDNA encodes a 177 amino acid peptide beginning with a signal peptide of 25 amino acids and containing two potential sites of N-glycosylation. The mature protein has a molecular mass of 25 kDa. Another cDNA clone has been described which lacks a 44 amino acid sequence as a result of exon splicing but whose physiological relevance has not been established. Human IL-7 contains six cysteine residues which are believed to make three intramolecular disulfide bridges. There is 60% sequence identity at the amino acid level between human and murine IL-7. The two N-glycosylation sites and six cysteine residues are conserved in mouse IL-7. Functional inter-species cross-reactivity is observed between human and mouse IL-7.

2.4 IL-4

IL-4 was first identified in 1982 as a B lymphocyte proliferation factor, and called B cell growth factor 1 (BCGF 1) then B cell stimulatory factor 1 (BSF-1). The term IL-4 was adopted when the murine and human cDNAs were cloned in 1986. Mouse IL-4 purified from the EL-4 thymoma cell line is a 20 kDa glycoprotein. Its cDNA encodes a 140 amino acid polypeptide including a 20 amino acid signal sequence. Human IL-4 cDNA codes for a 153 amino acid precursor containing a 24 amino acid signal peptide. Human IL-4 has a molecular weight of 15–19 kDa depending on its glycosylation state. Human and mouse IL-4 have two and three *N*-glycosylation sites respectively. Glycosylation does not affect the bioactivity or the receptor binding affinity of the molecules. Glycosylated human IL-4 has an isoelectric point of 10.4 and is moderately hydrophobic. Although they show 50% homology, human and mouse IL-4 are species-specific in their activity.

2.5 IL-13

IL-13, a recently described cytokine, is a potent modulator of monocyte and B cell functions. The molecular cloning of IL-13 and the first publications describing its biological activity were in 1993. Human IL-13 cDNA was found to display 58% homology with a mouse cDNA clone (S600) described in 1989 and corresponding to mouse IL-13. Mouse and human IL-13 cDNAs encode proteins of 131 and 132 amino acids respectively, each including a 20 amino acid leader peptide. The five cysteine residues are conserved at the same position in each protein, as well as three of the four *N*-glycosylation sites of human IL-13. Human and mouse IL-13 have four and three *N*-glycosylation sites respectively. Human IL-13 is mainly produced in a non-glycosylated form. Human and mouse IL-13 have similar specific activities on human cells, whereas human IL-13 has a 10-fold lower specific activity than mouse IL-13 on mouse cells.

2.6 IL-9

IL-9 was first identified (in 1988) in the supernatant of activated murine T cells by its growth-promoting activity on a murine T-helper cell clone in the absence of antigenic and accessory signals. Purified native murine IL-9 is a highly glycosylated cytokine, the molecular mass of which (32–39 kDa) is reduced to 15–16 kDa after removal of N-linked carbohydrates. Accordingly, its cDNA encodes a precursor protein consisting of a 18 amino acid signal sequence followed by a 126 amino acid mature protein with a predicted molecular mass of 14 150 Da and containing four potential *N*-glycosylation sites. Native human IL-9 has not been purified yet. Its cDNA has been cloned by cross-hybridization with an IL-9 mouse probe. Human

and mouse IL-9 have the same number of amino acid residues and display 55% homology. Mouse IL-9 is active on human cells, whereas human IL-9 is not active on mouse cells.

2.7 Common features

There are few direct amino acid sequence similarities within this family of cytokines. IL-4 and IL-13, which are functionally related, display 20% overall homology, whereas IL-2 and IL-15, which are also functionally related to one another, show no significant similarity. Despite this, structure-based alignments and molecular modelling prediction studies show that all members of the family share a common topology in which the core of the protein is made up of a bundle of four amphipathic α-helices (designated A, B, C and D) packed in an antiparallel up/up/down/down configuration. This topology was first described from the crystal structure of porcine growth hormone. It has been confirmed by X-ray crystallography and nuclear magnetic resonance spectroscopy for IL-2 (Figure 1) and IL-4.

Comparison of available helical cytokines folds has revealed that they can be divided into two main subgroups, namely the short-chain and long-chain cytokines. IL-2, -15, -7, -4, -13 and -9 all belong to the short-chain subgroup, characterized by molecules of 100–140 amino acids, helices of 15 amino acids , large (35°) packing angles between the A–D and B–C helix layers and long AB and CD loops each containing a short β-strand.

The number and positions of cysteine residues and disulfide bonds among these cytokines vary, ranging from three cysteine residues and one disulfide bond in IL-2 to 10 cysteine residues and five potential disulfide bonds in IL-9. However, one disulfide bond seems to be conserved; it tethers helix B to the C–D loop, and corresponds to residues Cys58 and Cys105 in hIL-2, Cys42 and Cys88 in hIL-15, Cys46 and Cys99 in hIL-4, and Cys44 and Cys70 in hIL-13. This bond appears most important in preserving the biologically active conformations of these cytokines.

2.8 Genes

The mouse IL-2 gene is located at the centromere-proximal end of chromosome 3 (band B-C, 19.2 cM from the centromere) (Table 2). Mouse chromosome 3 also contains the genes encoding IL-7 (6.6 cM from the centromere) and IL-12 (37 cM from the centromere). The mouse IL-9 gene is located on chromosome 13 whereas the mouse

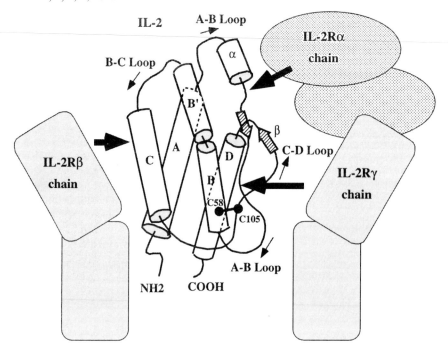

∎ **Figure 1.** High-affinity IL-2/IL-2R complex (extracellular part): The two 'sushi' domains of the IL-2Rα chain are shown as ovals. The IL-2Rβ and IL-2Rγ chains each contain two fibronectin type III modules. The IL-2 structure is made up of four α-helices separated by loops. The AB loop also contain an extra short α-helix (α). Two short β-sheet structures are also present in the AB and CD loops. The position of the disulfide bond (Cys58–Cys105) is shown. Arrows indicate the regions of IL-2 involved in the interaction with the different receptor chains.

∎ **Table 2.** Cytokine chromosomal clustering

Mouse		Human	
chromosome	cytokine	chromosome	cytokine
3	IL-7/IL-2/ IL-12		
8	IL-15	4q	IL-2/IL-15
11	IL-4/IL-13/ IL-15/IL-3 GM-CSF	5	IL-13/IL-4/ IL-5/IL-3/ GM-CSF/ IL-9
13	IL-9	8q	IL-7

IL-15 gene maps in the central region of chromosome 8, 37 cM from the centromere. This region has some homology with human chromosome 4q. Accordingly, human IL-15 gene has been shown to map on chromosome 4 (band q13), a region which contains a cluster of genes encoding growth factors and chemokines. Within this cluster, the human IL-15 gene is relatively near the human IL-2 gene located at 4q26-q27. The human IL-4 and IL-13 genes are also close together, separated by approximately 12 kb, on the long arm of human chromosome 5 (q23–q31), in a region containing a cluster of genes for cytokines and cytokine receptors and which has been shown to be involved in a number of pathologies. The human IL-13 gene is at the centromere-proximal end of this cluster which includes IL-4, IL-5, IL-3 and granulocyte–macrophage colony-stimulating factor (GM-CSF). The human IL-9 gene is also located on chromosome 5, about 2 cM from the GM-CSF gene. Similarly, the mouse IL-4 and IL-13 genes have been mapped to the central region of chromosome 11, in a cluster region also containing the IL-5, IL-3 and GM-CSF genes.

Many cytokine genes have a similar structure, made up of four introns separated by three introns, with each exon encoding an α-helix-forming

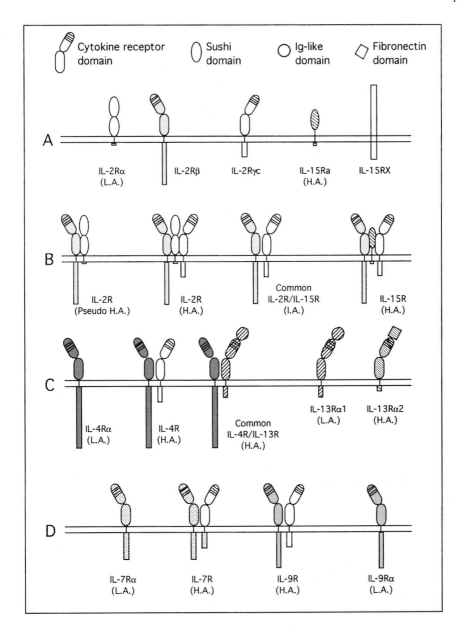

∎ **Figure 2.** Diagram illustrating the different forms of cytokine receptors (L.A., low affinity; I.A., intermediate affinity; H.A., high affinity) and sharing of common subunits. The IL-15RX has not yet been identified. (A and B) IL-2 and IL-15 receptors. (C) IL-4 and IL-13 receptors. (D) IL-7 and IL-9 receptors.

domain. This four exon/three intron structure is found in the genes of human and mouse IL-2, IL-4 and IL-13. In the case of IL-4 and IL-13, intron positions in the two genes are conserved but introns are considerably smaller in the IL-13 gene. The eight exon/seven intron structure of the human and mouse IL-15 genes apparently contrasts with this structure; however, the translated region of IL-15 mRNA corresponds mainly to four exons (4 to 7) with an organization similar to that of the four exons of IL-2 in that each of these exons codes for an α-helical structural domain. The human IL-7 gene contains six exons whereas murine IL-7 lacks exon 5, which explains the difference in size of the proteins. The human and mouse IL-9 genes each contain five exons and four introns.

3. Receptors

Cytokines act through specific cell-surface receptors. Molecular components involved in the ligand binding and functioning of these receptors were identified either by biochemical means or by expression cDNA cloning. A general feature seemed to be that these receptors were formed of multiple subunits assembling in a cooperative manner to provide specificity, high affinity and signal transduction.

Another feature commonly found within cytokines is that different cytokines acting through specific receptors can evoke identical biological responses. This redundancy can be partially explained by the discovery that different cytokine receptors can share common subunits which are involved in binding and/or signal transduction.

3.1 Common structural features

Most of the subunits involved in the structure of the IL-2, -4, -7, -9, -13 and -15 receptors belong to the type I haematopoietic growth factor receptor family (Figure 2). This family is characterized by the presence in the extracellular domain of a 'haematopoietic domain' of about 210 amino acids. This domain can be divided into two homologous fibronectin-type III subdomains of about 100 amino acids separated by a proline-rich sequence. The first subdomain is characterized by the presence of a pair of disulfide bridges in its N-terminal part, while the second contain a consensus Trp-Ser-X-Trp-Ser (WSXWS) sequence at its C-terminal end. The three-dimensional structure of the growth hormone receptor (GHR) has served to build a model for the haematopoietic domain. Both subdomains are composed of seven β-sheets closely packed to form a barrel-like structure.

Another family of receptor subunits is emerging; as yet it contains only two members, the α-chains of the IL-2 and IL-15 receptors (IL-2Rα and IL-15Rα). The extracellular parts of these chains do not contain a haematopoietic domain but instead characteristic domains of about 70 amino acids that are structurally homologous to the short consensus repeats (SCRs) or 'sushi' domains found in some adhesion or complement proteins.

3.2 IL-2 and IL-15 receptors

IL-2 receptors are mainly expressed by immune cells, particularly those from the lymphoid and myeloid lineages: activated but not resting T and B cells, NK cells, monocyte/macrophages (including Kupffer cells in liver and Langerhans cells in skin), granulocytes, and a variety of neoplastic cells (adult T cell leukaemia, cutaneous T cell lymphoma, chronic B cell leukaemia, hairy cell leukaemia, acute lymphoblastic leukaemia, B lymphoma, myeloid leukaemia, granulocytic leukaemia, megakaryocytic leukaemia, histiocytic leukaemia, Hodgkin disease Reed–Stenberg cells).

To date, three membrane-anchored glycoproteins have been shown to participate in the structure of the IL-2 receptors (Table 3). The first chain identified, IL-2Rα, is a 55 kDa glycoprotein the extracellular part of which contains two sushi domains. The intracellular domain of IL-2Rα is very short (11 amino acids). Two additional chains (β and γ) were subsequently identified which participate in ligand binding and are required for signal transduction. Il-2Rβ and IL-2Rγ are 75 and 64 kDa glycoproteins of the haematopoietic cytokine receptor family and contain large intracellular domains of 286 and 86 amino acids respectively.

The IL-2R α-chain is a low affinity IL-2 receptor ($K_d = 10$ nM) and the β- and γ-chains on their own bind IL-2 with even lower affinity. The three chains can assemble non-covalently in the membrane to form complexes of higher affinity. Three main complexes have been described: the β/γ complex is an intermediate-affinity receptor ($K_d = 1$ nM), whereas the ternary $\alpha/\beta/\gamma$ complex is a high-affinity receptor ($K_d = 10$ pM). A third α/β complex has been described which binds IL-2 with 'pseudo-high' affinity ($K_d = 200$ pM). Only the β/γ and $\alpha/\beta/\gamma$ complexes have been reported to transduce a biological signal.

A soluble form of the IL-2R α-chain has been described which results from cleavage of the membrane-anchored form following an as yet unknown proteolytic mechanism. This soluble receptor, which consists of the first 192 amino acids or the extracellular domain, retains the IL-2 binding characteristics of the membrane form (low affinity). In agreement with the high expression level of the membrane form, soluble IL-2Rα is secreted in large amounts following lymphocyte activation. Its function is not yet known.

Monoclonal antibody mapping studies, site-directed mutagenesis and molecular modelling based on the structure of the growth hormone/

▮ **Table 3.** Cytokine receptor chains biochemical features

	No. of amino acids				Molecular mass (kDa)		*N*-glyc sites	Disulfide bonds	pI
	SP signal peptide	extracellular	trans memb.	intracellular	peptide	glycosylated			
h-IL2Rα	21	219	19	13	28.5	55	2	4	6.4
m-IL2Rα	21	215	21	11	28.4	55	4	4	8.5
h-IL2Rβ	26	214	25	286	58.4	70–75	4	2	4.9
m-IL2Rβ	26	214	28	271	57.7	70–75	6	2	5.4
h-IL2Rγ	22	240	21	86	39.9	64	6	2	5.8
m-IL2Rγ	22	241	21	85	39.8	64	6	2	6.3
h-IL15Rα	30	175	21	41	24.8		1	2	6.5
m-IL15Rα	32	173	21	37	26.4	58–60	1	2	9.0
h-IL4Rα	25	207	24	569	87.1	140	7	2	5.0
m-IL4Rα	25	208	24	553	85.1	140	4	2	4.9
h-IL13Rα1	21	325	24	60	46.7	65	11	3	5.6
m-IL13Rα1	26	314	24	60	45.7	55–65	6	3	6.1
h-IL13Rα2	26	317	20	17	41.5	58	4	2	4.8
m-IL13Rα2	21	313	22	27	42.1	56	4	1	4.7
h-IL7Rα	20	219	25	195	49.5		6	1	5.3
m-IL7Rα	20	219	25	195	49.6		3	1	8.0
h-IL9Rα	40	230	21	231	52.9		2		5.4
m-IL9Rα	37	233	21	177	48.1		2		5.4

GHR complex have made it possible to delineate the topography of assembly of the IL-2/IL-2R complex. Distinct regions of the cytokine interact with corresponding areas on the different receptor chains, resulting in a cooperative increase in affinity. Residues (Lys35, Arg38, Phe42 and Lys43) located in the loop between helices A and B/B' are involved in the interaction with the first sushi domain of the α-chain, whereas helix A (residues Asp20 and Leu17) interacts with the β-chain and helix D (Glu126) with the γ-chain. The dynamics of assembly of the high-affinity receptor are still being investigated. One model proposes that α/β heterodimers are preassociated in the membrane in the absence of IL-2 and that IL-2 binding to these heterodimers promotes the recruitment of the third, γ-chain. In this context, high-level expression of the α-chain by activated lymphocytes has a double advantage, to favour the formation of α/β preassociated complexes and to concentrate IL-2 at the cell surface.

The high functional redundancy between IL-2 and IL-15 suggested that the two cytokine recep-

tors might use common structural elements. This was confirmed by the demonstration that some anti-IL-2Rβ monoclonal antibodies inhibited the biological effects of IL-15, whereas anti-IL-2Rα antibodies did not. Transfection experiments have further shown that the IL-2Rγ chain is required for IL-15 signal transduction.

The distribution of IL-15 receptors has not yet been analysed in detail. Preliminary data indicate that it largely overlaps that of IL-2 receptors on lymphocytic cells and extends to numerous other cell types, such as stromal cells derived from the thymic, liver or bone marrow compartments and endothelial cells. This more widespread cellular distribution suggested the existence of a specific IL-15 binding component. This notion was also supported by the observation that B cells from X-linked severe combined immunodeficiency (X-SCID) patients having a defect in IL-2R γ-chain expression bind IL-15 with high affinity. Accordingly, an IL-15-specific receptor component was cloned as the IL-15R α-chain. It resembles the IL-2R α-chain, with a sushi domain in its extracellular part and a short cytoplasmic tail (37 amino acids). Furthermore, IL-15Rα and IL-2Rα have similar intron/exon organization and are closely linked in the human (chromosome 10) and mouse (chromosome 2) genomes. Three isoforms of the IL-15R α-chain resulting from alternative exon splicing have been described; each of these is capable of high-affinity IL-15 binding.

Like IL-2Rα, IL-15Rα has no demonstrated signal-transducing capacity in its own right. However, in contrast to IL-2Rα, it is a high-affinity receptor ($K_d = 10\,$pM). It may thus serve as a scavenging molecule, removing excess IL-15 at inflammatory sites. The IL-2R β/γ complex is also an intermediate affinity receptor for IL-15 ($K_d = 4\,$nM); it is not known what are the respective contributions of the three chains in the formation of the functional high affinity $\alpha/\beta/\gamma$ IL-15R.

The fact that IL-2 and IL-15 both use the β/γ complex is in line with the observation that some residues on IL-15 are homologous to residues on IL-2 that are known to interact with the β- and γ-chains of IL-2R. Asp8 on human IL-15 is homologous to Asp20 on human IL-2 and are both involved in the binding of the respective cytokines to the IL-2R β-chain. Similarly, Gln108 on human IL-15 is homologous to Gln126 on human IL-2 (and Gln141 in murine IL-2) with respect to binding to the IL-2R γ-chain. However, monoclonal antibody blocking studies indicate that the two cytokines may use different epitopes for binding. In addition, recent data have indicated that IL-15 creates a less stable bridge between the β- and γ-chains than does IL-2.

The observation that some functions of IL-15, namely proliferation of murine mast cell lines and bone marrow derived mast cells, are not shared with IL-2 suggests the existence of a distinct IL-15-specific receptor. In these cells, neither IL-2Rβ, nor IL-2Rγ nor IL-15Rα participates in ligand binding or signal transduction. Cross-linking experiments have shown the existence of a novel receptor, provisionally designated IL-15RX, with a molecular weight of 60–65 kDa.

3.3 IL-4 and IL-13 receptors

IL-4 and IL-13 bind with high affinity (with an estimated K_d of 40–120 pM for IL-4 and 300 pM for IL-13) to receptors which are present only in small numbers (200–800 sites/cell) on most cell types, an exception being a number of solid tumour-derived cell lines (renal carcinoma, melanoma, glioma) which express higher receptor densities (2000–4000

sites/cell). This finding has been exploited to develop IL-13-based immunotoxins with relative specificity for these tumour cells. Otherwise, it seems that cell types expressing IL-13 receptors are a subset of those cells expressing IL-4 receptors. In particular, IL-13 receptors appear to be absent from activated T cells, T cell lines and certain B cell lines. Cross-competition experiments have shown that IL-4 can usually completely compete for IL-13 binding (except in those tumour cell lines which express large numbers of IL-13 receptors).

The IL-4 receptor (IL-4Rα or gp140), which was cloned in 1990, does not bind IL-13. The cross-competition between IL-4 and IL-13 was thus initially taken to indicate the presence of an additional shared component between IL-13 and IL-4 receptors. This conclusion was also drawn from studies showing that an IL-4 antagonist (Y124D) would antagonize the effects of IL-13. Upon the discovery that the γ-chain of the IL-2R could be part of the IL-4 receptor, it was postulated that the shared component of the IL-4 and IL-13 receptors would be the γ-chain. However, a large amount of evidence now suggests that IL-13 does not use the γ-chain-associated Jak-3 signalling pathway:

(1) X-SCID patients lacking the γ-chain show normal B cell responses to IL-13 and IL-4. Indeed, in B cells from some patients STAT-6, (Signal transclucer and activator of transcription) activation by IL-13 seems to be *enhanced*, while STAT-6 activation by IL-4 is impaired.

(2) IL-4/IL-13 cross-competition occurs in non-haematopoietic cells lacking the γ-chain.

(3) Antibodies against the γ-chain antagonize proliferation of the MC/9 cell line induced by IL-4 but not that induced by IL-13, and transfection of the γ-chain into the B9 plasmacytoma cell line increases proliferation only with IL-4.

(4) Insulin receptor substrates IRS-1/IRS-2 phosphorylation in L-cells transfected with the γ-chain is increased in response to IL-4 but not in response to IL-13. Low-level IRS-1/IRS-2 phosphorylation by IL-4 and IL-13 can be seen in these cells in the absence of the γ-chain.

(5) (Janus kinase) Jak-3 is activated by IL-4 but not by IL-13, although this is controversial.

Thus IL-13 does not use the γ-chain for signalling, although it has been shown to increase the presence of IL-4Rα in anti-γ-chain immunoprecipitates from B cells, suggesting that the γ-chain may be physically associated with IL-13R. The nature of the shared component between IL-4 and IL-13 receptors was shown to be the IL-4Rα/gp-140 receptor subunit, using antibodies to this subunit which inhibit both IL-4 and IL-13 activities.

Cross-linking studies with IL-13 show a broad band corresponding to a 55–70 kDa protein. A 424 amino acid IL-13-binding protein (IL-13Rα1) has been cloned from mouse genomic DNA using degenerate oligonucleotides encoding the Trp-Ser-Asp-Trp-Ser motif found in many members of the haematopoietic receptor family. Simultaneously, starting from the human CAKI-1 adenocarcinoma, which expresses large quantities of IL-13R, a 380 amino acid IL-13-binding site (IL-13Rα2) has been cloned using a COS cell expression system.

Both IL-13 receptors are part of the haematopoietic cytokine receptor superfamily, but they are only moderately homologous (approximately 30%). Subsequent isolation of the human α1 chain has shown that both α1 and α2 mRNAs are expressed in many human cell types. The affinity of IL-13 for the α1 chain alone is low ($K_d = 4$ nM) and is increased by association with the IL-4 gp140 ($K_d = 30$ pM). The affinity of IL-13 for the α2-chain alone is much higher ($K_d = 250$ pM). IL-13Rα1 is clearly involved in signal transduction, whereas this has yet to be shown for IL-13Rα2, which contains only a very short (17 amino acid) cytoplasmic region. Complexes of IL-4Rα and IL-13Rα1 show the expected cross-competition between IL-4 and IL-13 for receptor binding and constitute a functional receptor.

The current model for IL-4 and IL-13 receptors is thus that IL-13 transduces signals through a receptor comprising the IL-4Rα subunit and an IL-13 α1-chain, whereas IL-4 transduces signals through receptors comprising either the γ-chain or the IL-13 α1-chain associated with the IL-4Rα subunit. The role of the IL-13 α2-chain remains to be elucidated.

3.4 IL-7 receptors

The cellular distribution of IL-7R is in agreement with the known biological spectrum of IL-7. In mouse, IL-7R is expressed by pre-B lymphocytic cell lines but not by mature B cell subsets. It is also found on thymocytes, some T cell lines and bone marrow macrophages. On human cells, IL-7R is expressed by peripheral T lymphocytes (including resting T cells) and by lymphoblastoid T and B cell precursors.

Low-, intermediate- and high-affinity IL-7 binding sites have been described. Human and mouse IL-7-specific α-chains were identified first; their cDNAs encode homologous (64% identity) mature transmembrane proteins each containing 439 amino acids with a 219 amino acid extracellular domain and a 195 amino acid intracellular domain. This IL-7R α-chain belongs to the haematopoietic receptor family, although the N-terminal part of its haematopoietic domain contains only two cysteine residues rather than four. The molecular mass of the glycosylated receptor, as detected in different cell lines, is about 75 kDa, 25 kDa more than the mass predicted from the cDNA (50 kDa). A soluble form of IL-7Rα corresponding to the extracellular domain has also been described which results from differential splicing of the mRNA. Transmembrane IL-7Rα expressed alone and soluble IL-7Rα both bind IL-7 with a similar low affinity ($K_d = 10$ nM). Dimerization of the α-chain has been proposed to account for the formation of receptors with increased affinity. Subsequently, co-precipitation studies have demonstrated that the functional high-affinity species involves heterodimerization of the IL-7R α-chain with the common IL-2R γ-chain. The existence of an additional IL-7R β-chain of molecular mass close to that of IL-7Rα has been suggested which could combine with the IL-7R α-chain to form a receptor with intermediate affinity for IL-7.

3.5 IL-9 receptors

High-affinity ($K_d = 100$ pM) IL-9 receptors were first identified using radiolabelled recombinant mouse IL-9. Such receptors are expressed by T lymphocyte clones and up-regulated upon mitogen stimulation. They are also detected on mast cells, macrophages, thymocytes and certain T lymphoma cell lines. The cDNA encoding mouse IL-9R was isolated in 1992 and enabled the human IL-9R cDNA to be identified by cross-hybridization; there was 55% homology between the two species. The IL-9Rα chain is a transmembrane glycoprotein, which has an extracellular part (234 amino acids) containing a haematopoietic cytokine receptor domain, and a large cytoplasmic domain of 177 amino acids (mouse) or 231 amino acids (human). This cytoplasmic domain shows some homology with the intracellular part of the IL-2R β-chain, particularly within a proline-rich region called 'box 1' which is involved in the recruitment of Jak tyrosine kinases implicated in signal transduction.

The human IL-9Rα gene is located in a pseudoautosomal region on the long arm of chromosomes X/Y and contains 11 exons. It consists of 10 exons spread over approximately 13.7 kb of DNA. At least five different transcripts, resulting from alternative gene splicing, have been described both in human and mouse which encode membrane receptor isoforms truncated in their cytoplasmic domain and a soluble receptor isoform. The reason for this diversity is not yet understood but might indicate complex post-transcriptional regulation of receptor expression and function.

The IL-9Rα cDNA, when expressed in IL-9R-negative COS cells, binds IL-9 with high affinity ($K_d = 100$ pM). However, it is not sufficient to transduce a biological signal. It was subsequently shown that IL-2Rγ association with IL-9Rα is required for IL-9-driven signal transduction, without inducing significant increase in IL-9 binding affinity.

4. Cellular sources and regulation of expression

4.1 IL-2/IL-2R

Activated T-helper (Th) lymphocytes are the major source of IL-2. Whereas immature CD4 Th precursor and Th0 cells all produce IL-2, only a fraction of differentiated Th cells produce this cytokine, namely the Th1 phenotype (defined by their ability to produce IL-2 and IFNγ); Th2 cells produce IL-4, IL-6 and IL-10, but not IL-2. Some CD8 T cell clones have also been reported to produce IL-2. This IL-2 production is tightly regulated and requires a two-signal induction process involving CD3/T cell receptor (TCR) cross-linking by the antigen/MHC complex and triggering of CD28/CTLA-4 (cytotoxic T lymphocyte-associated antigen) accessory components by B7/2 and B7/1 molecules. Other adhesion molecules and cytokines (IL-1 and IL-6) display co-stimulatory functions. In contrast, the anti-inflammatory cytokines IL-4 and IL-10 produced by Th2 cells or macrophages can inhibit IL-2 production by Th1 cells. B cells have also been reported to produce IL-2. This is the case for B cells transformed by Epstein–Barr virus and for normal peripheral B cells pre-stimulated through CD40 and activated by phorbol myristate acetate (PMA) and calcium ionophore. This production is, however, 10-fold lower than that of T cells and is a late event.

IL-2 gene transcription is controlled by a 300 bp promoter/enhancer region 5' flanking the transcription initiation site. This region contains multiple docking sequences for several transcription factors. Among them are two sites for the binding of nuclear factor of activated T cells (NF-AT) and activator protein-1 (AP-1) (Fos–Jun heterodimer) and one site for the binding of AP-1 and octamers Oct-1 and -2. The promoter activity of these sites is under the control of the TCR/CD3 pathway (via a calcium/calmodulin/calcineurin pathway). The immunosuppressive drugs cyclosporin A and FK506 act mainly (through calcineurin) by inhibiting this pathway. Two motifs within the IL-2 promoter have been shown to bind Rel/NF-κB (nuclear factor κB) transcription factors (the κB site and the CD28 response element). The promoter activity of these sites is controlled by the co-stimulatory CD28 signalling pathway.

The expression of the IL-2R α-chain is also tightly regulated, being absent on resting T cells and present at high density on activated T cells (about 100 000 surface molecules per cell). A variety of factors have been shown to induce the expression of the IL-2R α-chain, including antigen/MHC complex, B7/CD28 interaction, mitogens, phorbol esters, and some cytokines like IL-1, IL-2 itself, tumour necrosis factor alpha (TNFα) or interferon gamma (IFNγ) in T cells, IL-5 and IL-10 in B cells. IL-2Rα expression is also induced by transactivator proteins of T lymphotropic viruses (Human T lymphotropic viruses HTLV-I, HTLV-II and human immunodeficiency virus HIV). Conversely, cAMP inducers and other cytokines like IL-4 down-regulate IL-2Rα.

The promoter region of IL-2Rα contains a TATA box characteristic of inducible genes. It also contains three main regulatory regions (PRRI, PRRII and PRRIII) which bind a number of factors participating in a coordinated fashion in the induction of transcription. PRRI and PRRII are together required for mitogenic stimulation of the IL-2Rα gene. PRRI binds NF-κB/c-Rel and serum response factor (SRF), while PRRII binds the lymphoid/myeloid specific Ets family protein Elf-1 and the high mobility group proteins HMG-1(Y). Intermolecular interactions between these transcription factors also contribute to optimal induction of the promoter. IL-2 induction of IL-2Rα involves not PRRI and PRRII but a third region designated PRRIII (or IL-2RE). This region, located about 3 kb upstream of PRRI and PRRII, is a complex element containing binding sites for multiple transcription factors including Elf-1, HMG-I(Y) and STAT-5.

The β-chain is also inducible by antigen, TCR/CD28 activation, and high concentrations of IL-2.

Its promoter region does not contain a TATA motif but contains several transcription factor binding sites. An Ets binding site (EBS) and GGAA binding proteins (GABP) are critical for promoter inducibility by phorbol esters.

The γ-chain is constitutively expressed by many cell types within the lymphoid system and is only modestly induced upon stimulation by antigen (T cells) or IL-2 (NK cells). Its promoter region contain a GC-rich region characteristic of house-keeping genes. It also contains several potential motifs for fixation of transcription factors.

4.2 IL-15/IL-15R

In sharp contrast to IL-2 transcripts, expression of which is restricted to activated lymphoid cells (mainly T cells), IL-15 transcripts are not detected in lymphoid cells even after activation but are found instead in activated monocytes and a wide variety of cell types and tissues, including epithelial cells, fibroblasts, placenta, skeletal muscle, heart, kidney and lung. Another major difference is that, whereas IL-2 production and secretion usually correlate with mRNA levels, IL-15 protein is not found in propor-tional quantities in many cells that contain IL-15 mRNA. Low levels of IL-15 have been detected in the culture supernatants of activated monocytes and human blood-derived dendritic cells. A possible explanation for this unusual behaviour has been provided by the discovery that the 5' untranslated region of IL-15 mRNA contains multiple initiation (AUG) codons, a structure which is known to be responsible for a profound reduction in the efficiency of translation. The unusual 48 amino acid signal peptide appears also to contribute to the inefficiency of IL-15 synthesis and secretion. IL-15 secretion seems therefore to be naturally impeded and it is speculated that it might occur following various activation signals. In HTLV-I-associated adult T cell leukaemia (ATL) cell lines, a 5' untrans-lated region (5'-UTR) of the virus is spliced to the 5'-UTR of IL-15, leading to the removal of eight of the 10 AUG codons in the IL-15 5'-UTR. This leads to a marked increase in both IL-15 transcription and translation. Translationally inactive IL-15 mRNAs may be maintained in cells, so that they can be readily activated to yield secreted IL-15. This could represent a mechanism by which cells can respond efficiently to infectious agents. In accordance with this, high concentrations of IL-15 were found in the synovial fluids of patients with rheumatoid arthritis. High IL-15 expression has also been reported in chronic diseases such as pulmonary sarcoidosis,

leprosy and ulcerative colitis. *In vitro*, optimal IL-15 production by macrophages is obtained after priming with IFNγ and triggering with various microbial stimuli (such as lipopolysaccharide, myco-bacteria, *Toxoplasma gondii*, or ultraviolet irradia-tion). Similarly, human herpesvirus-6 enhances NK cell cytotoxicity by stimulating IL-15 production by peripheral mononuclear cells, and IL-15 release by dendritic cells is triggered upon phagocytosis.

4.3 IL-4 and IL-13

IL-4 and IL-13 are produced by activated T lym-phocytes, basophils and mast cells. IL-4 and IL-13 are both produced by the Th2 subset, but not the Th1 subset, of mouse T lymphocyte clones. IL-13 can, however, be produced by human T lymphocyte clones with a Th1-like phenotype. IL-13 production by human peripheral blood lymphocytes can be increased by IL-4 and inhibited by IL-12, indicating that IL-13 is primarily produced by Th2 type lym-phocytes, and *in vivo* studies have also detected IL-13 production in Th2 type pathologies such as atopic asthma. IL-13 is produced by both 'naïve' $CD4^+$ $CD45RO^-$ and 'memory' $CD4^+$ $CD45RO^+$ T lymphocytes, whereas IL-4 is produced only by the latter subset.

A number of other differences have been noted between the expression of IL-4 and IL-13. IL-13 expression is more prolonged after T cell activation, representing both a prolonged expression of mRNA and the depletion in the culture medium by the activated T cells of IL-4, but not of IL-13. IL-13 expression is more pronounced in the $CD8^+$ cell population than is IL-4 expression, and IL-13 can be maximally induced in T lymphocytes by combi-nations of inducers such as PMA and anti-CD28 which lead only to minimal IL-4 production. IL-13 mRNA expression has been reported by B lym-phocyte cell lines and in normal B lymphocytes after CD40 cross-linking and subsequent exposure to PMA and ionomycin. The latter B lymphocytes do not express detectable IL-4 mRNA.

In conclusion, IL-13 and IL-4 are generally produced by similar cell types (T lymphocytes, basophils and mastocytes). In most cases, IL-13 is produced in greater quantities than IL-4. In addi-tion, IL-13 is produced preferentially in response to certain stimuli, and also in certain cellular subsets.

4.4 IL-7

IL-7 mRNA is expressed by a variety of stromal cells originating from bone marrow, spleen and thymus, and also by epidermal keratinocytes. This expression is constitutive and it seems that it cannot be modulated by activating agents, except IFNγ. In accordance with this, no classical TATA or CAAT boxes are found in the IL-7 gene promoter, whereas it does contain an interferon-stimulated response element (ISRE). T lymphocytes fail to express IL-7 transcripts, even after strong stimulation.

As is the case for IL-15, IL-7 mRNA expression is usually not associated with detectable secretion of IL-7 protein in cell culture supernatants. Interestingly, the 5′-UTR of the IL-7 mRNA contains, like that of the IL-15 transcript, multiple initiation (AUG) codons which may be responsible for blocking translation. The only known cellular sources of IL-7 protein are some human and murine medullary and thymic stromal cell lines as well as keratinocyte cell lines.

4.5 IL-9

The IL-9 gene promoter region contains multiple consensus binding sequences for transcription factors (AP-1, AP-2, SP-1 and NF-κB), suggesting that its expression might be under the control of inducible factors.

IL-9 expression, like that of IL-2, seems to be restricted to activated T lymphocytes. Unlike IL-2, IL-9 is produced by Th2 cells. It is also produced much later than IL-2 during the course of lymphocyte activation. Therefore a cascade of cytokine induction steps and cooperation between Th1 and Th2 subsets has been proposed to be necessary for IL-9 secretion: IL-2 induces IL-4 secretion, IL-4 and IL-2 synergize to induce IL-10 production (also by Th2 cells), and finally, the combination of IL-4 and IL-10 allows IL-9 production.

5. Target cells and biological activities

5.1 IL-2

The main cellular targets of IL-2 are activated lymphocytes. Naïve, resting T cells (CD4 or CD8) only express the IL-2R γ-chain, albeit at low levels, and do not respond to the cytokine. TCR/CD3 activation, by turning on both IL-2, IL-2Rα and IL-2Rβ gene expression, leads to autocrine IL-2/high affinity receptor signal transduction, resulting in cell proliferation. Naïve B cells also only express the γ-chain and require B-cell receptor (BCR) activation for acquisition of IL-2 sensitivity. In contrast

to T and B cells, resting NK cells (either CD56[bright] or CD56[dim]) do express IL-2Rβ and γ. By interacting with these intermediate affinity receptors, IL-2 induces their cytotoxic activity. IL-2 also turns on the synthesis of the α-chain, leading to enhanced sensitivity to the cytokine and induction of NK cell proliferation. CD56[bright], which represent 2% of NK cells, also express the α-chain constitutively and can readily proliferate in response to IL-2.

The γ-chain is expressed by all thymocytes. Immature CD3⁻ thymocytes also express IL-2Rα and some double-negative CD3⁺ thymocytes express IL-2Rβ but the functions of these chains are unknown.

Monocytes also respond to IL-2 through intermediate β/γ affinity receptors, with induction of cytotoxic, antibacterial and antitumour activities. Other haematopoietic cells which respond to IL-2 include neutrophils with increased antifungal activity, and various neoplastic cells such as adult T cell leukaemia, hairy cell leukaemia, acute lymphoblastoid leukaemia and chronic B lymphocytic leukaemia. Malignant Reed–Sternberg cells in Hodgkin disease also display high-affinity IL-2 receptors without demonstrated biological activity.

Non-haematopoietic cell lines have also been described which express IL-2 receptors. IL-2 stimulates the growth of human glioblastoma cells and rat oligodendrocytes, while inducing growth hormone secretion by pituitary adenocarcinoma cells. Some melanoma cells also respond to IL-2 by increased expression of adhesion molecules and proliferation. Recent studies have indicated that exogenous IL-2 could enhance epithelial restitution *in vitro*, suggesting that it may preserve epithelium integrity following injury.

5.2 IL-15

As a result of common usage of transducing receptor chains, IL-15 and IL-2 share numerous biological activities. These include induction of proliferation of activated T cells (including CD4⁺, CD8⁺ and γ/δ subsets), induction of proliferation and differentiation into immunoglobulin-producing cells of B cells activated with CD40 ligand or anti-IgM, promotion of T lymphocyte chemokinesis and chemotaxis, induction of the cytotoxic activity of effector cells, including cytotoxic T cells and lymphokine-activated killer cells. IL-15 also promotes the proliferation and cytotoxic activity of NK cells. It acts in synergy with IL-12 to increase the production of IFNγ and TNFα by these cells. Several reports have recently demonstrated a pivotal role for IL-15 in thymic development of T cells, particularly the NK lineage. It is speculated that IL-15, rather than IL-2, is the main physiological growth and differentiation factor for NK cells.

The fact that IL-15 and IL-15Rα mRNAs are found in many tissues and cell types suggest that this cytokine system may operate at multiple levels within and beyond the immune system. Three examples of such non-immune activities have thus far been demonstrated: (i) the description of IL-15 as an anabolic agent on skeletal muscle, by augmenting parameters associated with muscle fibre hypertrophy; (ii) stimulation of intestinal cell proliferation; (iii) the induction of proliferation of murine mast cell lines, an activity also shared with IL-7. Concerning the immune response, IL-15 is proposed to be a factor used by host tissues to contribute to the early phase of this response, prior to IL-2 production and action. IL-15 itself can promote IL-2 action by up-regulating the IL-2R α-chain.

5.3 IL-4 and IL-13

IL-4 and IL-13 act on monocytes, dendritic cells, endothelial cells and a variety of other non-haematopoietic cell types (including keratinocytes and fibroblasts). IL-13 acts only on human B lymphocytes. Their joint activities on these cells can be broadly described as inhibiting inflammatory responses and augmenting humoral immunity and IgE responses, although the anti-inflammatory activity is more subtle than that of IL-10. The principal difference between the two cytokines is in their activity on T lymphocytes . As noted above, binding studies on T lymphocytes have failed to reveal IL-13 receptors, and IL-13 is not a T lymphocyte proliferation factor, unlike IL-4. Signalling of IL-13 on T lymphocytes is controversial with conflicting reports on whether or not IL-13 causes STAT-6 activation in these cells. A potentially important difference between IL-4 and IL-13 is in the switch to a Th2 phenotype and a humoral immune response, where IL-4 has for some time been known to be a key cytokine, but here the effects of IL-13 are still largely unknown.

5.4 IL-7

IL-7 is an important factor involved in B lymphocyte ontogenesis. It is a growth and differentiation factor for precursor B cells of the pro-B/pre-B phenotype. It also acts in concert with IL-3 to promote the proliferation of activated mature human B lymphocytes.

In line with the expression of IL-7 mRNA by epithelial/thymic stromal cells, IL-7 was subsequently shown to be a growth factor for mouse and human thymocytes. On adult murine cells its effects are direct and do not require additional cytokine or activating agent. IL-7 acts on both immature (CD3⁻) and mature (CD3⁺) populations. On mature cells, its effect is more pronounced on the double negative (CD4⁻,CD8⁻) subset. On immature cells, its effect is restricted to the CD4⁻,CD8⁻ subset. IL-7 is also a growth factor for fetal thymocytes. Human IL-7 has also been reported to be a growth promoting factor for immature and mature thymocytes. In combination with IL-2, IL-7 favours the growth of more differentiated CD4⁺ and CD8⁺ subsets as well as TCRγ/δ T cells.

Like IL-2, IL-7 alone has no demonstrated action on resting peripheral T lymphocytes but promotes the proliferation of T cells (helper or cytotoxic, CD4⁺ or CD8⁺) activated with mitogens or by the CD3/TCR pathway. IL-7 is also a potent T cell growth factor when associated with CD2 or CD28 triggering. IL-7 is a better stimulus for memory T cells than for unprimed cells. Whether IL-7-driven proliferation is dependent on IL-2 production depends on the nature of the stimulation; when the stimulus is concanavalin A (ConA), it is IL-2-dependent, but when the stimulus is phytohaemagglutinin (PHA), CD3 or CD2, it is not. Like IL-2, IL-7 also stimulates LAK cell activity, and this action is independent of IL-2. IL-7 is able to stimulate the cytotoxicity of CD8 cytolytic effector cells as efficiently as IL-2. In contrast to IL-2, IL-7 is unable to induce the proliferation and cytotoxicity of NK cells.

IL-7 acts on the monocytic lineage. It augments IL-1 and TNFα production by monocytes. IL-7 produced by epidermal keratinocytes has been proposed to participate in the activation of skin-resident dendritic cells.

It has recently been shown that IL-7 is trophic for embryonic neurones and is expressed in developing brain. It is also produced by follicular dendritic cells and vascular cells, suggesting a potential role for IL-7 in the germinal centre reaction.

In contrast with IL-2 or IL-4 gene knock-out experiments which do not result in a profound alteration of the size and proportions of the thymic and peripheral lymphocytic cell compartments, inactivation by homologous recombination of the mouse IL-7 gene results in major T and B lymphopenia, highlighting the key role of IL-7 in lymphopoiesis, and demonstrating the existence of major IL-7 activities *in vivo* that cannot be replaced by redundant cytokines.

5.5 IL-9

The major role of IL-9 resides in its ability to promote the growth and differentiation of masts cells. *In vitro*, IL-9 alone is a growth factor for murine mast cell lines and synergizes with IL-3 or Steel factor to maintain mouse bone marrow derived mast cells. IL-9 induces mast cells to produce a number of specific granule-associated proteases, to produce and secrete IL-6 and to express high-affinity IgE receptors. IL-9 gene transgenesis has confirmed this major role: Transgenic mice exhibit massive systemic mastocytosis. Recent studies have identified the IL-9 gene as a candidate gene in allergic asthma and suggest a role for IL-9 in the pathogenesis of bronchial hyperresponsiveness as a risk factor for asthma.

The potential role of IL-9 in T cell growth remains unclear. Unlike IL-2, IL-4 or IL-7, IL-9 has no activity on freshly isolated peripheral blood T cells. T cell growth response to IL-9 requires pre-activation of T cells with lectins or antigen and IL-2. Mouse and human IL-9-dependent T cell lines have been obtained after prolonged culture in IL-9-containing medium. The only direct activity of IL-9 on freshly isolated T cells concerns the proliferation of fetal (but not adult) murine thymocytes (in synergy with IL-2).

Constitutive expression of IL-9 has been found in a number of human lymphomas including large cell anasplastic lymphomas and Hodgkin lymphoma. Proliferation of one Hodgkin lymphoma cell line has been demonstrated to involve an IL-9 autocrine loop. Thymic lymphomas were found to develop in about 7% of IL-9 transgenic mice. Together, these results suggest that IL-9 dysregulation might be involved in some T cell malignancies.

IL-9 has recently been found to stimulate the proliferation of human myeloid leukaemia cells. IL-9, by stimulating leukaemia cells to enter S phase, may play a role in concert with stem cell factor in the development of acute myeloid leukaemia (AML).

On B cells, IL-9 potentiates the effect of IL-4 on IgG, IgE and IgM production *in vitro*. Accordingly, in IL-9 transgenic mice, abnormal high levels of immunoglobulins are found.

Within the haematopoietic compartment, IL-9 acts on precursor cells from the erythroid lineage and promote the effect of erythropoietin to induce the development of red cells. Other potential targets of IL-9 are immature neuronal cell lines.

References

Banchereau, J. (1994). Interleukin-4. In: *The Cytokine Handbook*, 2nd edn (Thompson, A.W., ed.), pp. 99–126. London, Academic Press.

Bazan, J.F. (1990). Structural design and molecular evolution of a cytokine receptor superfamily. *Proc. Natl Acad. Sci. USA*, **87**, 6934–8.

Brandhuber, B.J., Boone, T., Kenney, W.C. and McKay, D.B. (1987). Three-dimensional structure of interleukin-2. *Science*, **238**, 1707–9.

Callard, R.E, Matthews, D.J. and Hibbert, L. (1996). IL-4 and IL-13 receptors: are they one and the same? *Immunol. Today*, **17**, 108–10.

Costello, R., Imbert, J. and Olive, D. (1993). Interleukin-7, a major T-lymphocyte cytokine. *Eur. Cyt. Netw.*, **4**, 253–62.

Gaulton, G.N. and Williamson, P. (1994). Interleukin-2 and the interleukin-2 receptor complex. *Chem. Immunol.*, **59**, 91–114.

Grabstein, K.H., Eisenman, J., Shanebeck, K., Rauch, C., Srinivasan, S., Fung, V. *et al.* (1994). Cloning of a T cell growth factor that interacts with the beta chain of the interleukin-2 receptor. *Science*, **264**, 965–8.

Jacques, Y., Chérel, M., Sorel, M., Godard, A. and Minvielle, S. (1996). Interleukine-2 et interleukine-15. In *Les Cytokines*, (Cavccillon, J.M., ed.), pp. 119–136. Éditions Masson, Paris.

Leonard, W.J., Noguchi, M., Russel, S.M. and McBridge, O.W. (1994). The molecular basis of X-linked severe combined immunodeficiency: the role of the interleukin-2 receptor γ chain as a common γ chain, γc. *Immunol. Rev.*, **138**, 61–86.

Lemoli, R.M., Fortuna, A., Tafuri, A., Grande, A., Amabile, M., Martinelli, G., Ferrari, S. and Tura, S. (1997). Interleukin-9 in human myeloid leukemia cells. *Leuk. Lymphoma*, **26**, 563–73.

McInnes, I.B. and Liew, F.Y. (1998). Interleukin-15: a proinflammatory role in rheumatoid arthritis synovitis. *Immunol. Today*, **19**, 75–9.

Minty, A., Chalon, P., Derocq, J.M., Dumont, X., Guillemot, J.C., Kaghad, M. *et al.* (1993). Interleukin-13 is a new human lymphokine regulating inflammatory and immune responses. *Nature*, **362**, 248–50.

Murray, R. (1996). Physiologic roles of interleukin-2, interleukin-4, and interleukin-7. *Curr. Opin. Hematol.*, **3**, 230–34.

Paul, W.E. (1991). Interleukin-4: a prototypic immunoregulatory lymphokine. *Blood*, **77**, 1859–70.

Plum, J., De Smedt, M., Leclercq, G., Verhasselt, B., and Vandekerckhove, B. (1996). Interleukin-7 is a critical growth factor in early human T-cell development. *Blood*, **88**, 4239–45.

Renauld, J.C. (1995). Interleukin-9: structural characteristics and biologic properties. *Cancer Treat. Res.*, **80**, 287–303.

Tagaya, Y., Bamford, R.N., DeFilippis, A.P. and Waldmann, T.A. (1996). IL-15: a pleiotropic cytokine with diverse receptor/signaling pathways whose expression is controled at multiple levels. *Immunity*, **4**, 329–36.

Thèze, J., Alzari, P.M. and Bertoglio, J. (1996). Interleukin-2 and its receptors: recent advances and new immunological functions. *Immunol. Today*, **17**, 481–6.

Welch, P.A., Namen, A.E., Goodwin, R.G., Armitage, R. and Cooper, M.D. (1989). Human IL-7: a novel T cell growth factor. *J. Immunol.*, **143**, 3562–7.

Zurawski, G. and de Vries, J.E. (1994). Interleukin 13, an interleukin 4-like cytokine that acts on monocytes and B cells, but not on T cells. *Immunol. Today*, **15**, 19–26.

The interleukin-6 family of cytokines and their receptors

Jean-Luc Taupin, Stéphane Minvielle, Jacques Thèze, Yannick Jacques and Jean-François Moreau*

1. Introduction

The family of cytokines discussed in this chapter includes interleukins 6 and 11 (IL-6 and -11), leukaemia inhibitory factor (LIF), oncostatin M (OSM), ciliary neurotrophic factor (CNTF) and cardiotrophin 1 (CT-1). This family belongs to the four-helix, long-chain cytokine group which also includes granulocyte colony-stimulating factor (G-CSF), erythropoietin, interleukin-12 (IL-12), growth hormone, prolactin and leptin. Their receptors belong to the superfamily of receptors characterized by a haematopoietin-binding homologous domain harbouring four cysteine residues and the canonical stretch of amino acids Trp-Ser-X-Trp-Ser (WSXWS).

This cytokine family is better defined by the type of receptors to which they bind (Figure 1). They all trigger a signal in target cells through a multichain receptor which always includes glycoprotein 130 (gp130). For IL-6 and, perhaps, IL-11, the signal originates from the homodimerization of gp130 by a heterodimeric complex of the cytokine bound to a specific chain (gp80 or IL-6Rα for IL-6, and IL-11R for IL-11). These specific chains exist either anchored on the target cell membrane or in a soluble form, and therefore do not require an intracellular segment to be operational; they are called 'alpha-type' or α-type chains.

Glycoprotein 130, in contrast, absolutely requires an intact intracellular domain in order to transduce the signal properly, and so by analogy is called 'beta-type' or β-type. For the four remaining cytokine members so far discovered, the cellular functional receptor results from the heterodimerization of gp130 with gp190 (which is also of the β-type) upon ligation with either the cytokines alone, as is also the case with LIF and OSM or with a het-erodimeric complex of a specific α-chain and the cytokine (CNTF and, putatively, CT-1). Thus there are a variety of ways in which a given ligand can interact productively with a multichain receptor, making this cytokine family an attractive area of research. These systems are highly redundant and their regulation finely tuned, as we will see later. The specificity of the biological effect on one cell type appears to rely not only on the proper expression of the receptor chains, but also on achieving the right balance on the inner face of the membrane between all the components of the signalling pathway; this subject is discussed elsewhere in this book.

Several pieces of the puzzle are still missing, so it is not yet possible to give a complete and perfect account of these systems. Thus it seems that there are as yet undiscovered ligands, since the existence of 'new' ligands seem to be necessary in order to explain the observed phenotype of transgenic 'knockout' mice, as well as unknown receptor chains (PCR-based strategies using consensus sequences have detected so-called 'orphan' receptors).

2. Interleukin 6

Interleukin 6 (IL-6), the prototype cytokine for this subfamily, was initially identified by a number of independent groups, and hence has had a variety of names. It was first described as a fibroblast-derived interferon β-like protein, IFN-β2, but the recombinant protein does not show this antiviral activity. It was also subsequently isolated as B-cell stimulating factor-2 (BSF-2), a factor that caused terminal maturation of activated B cells to form immunoglobulin-producing cells, as hybridoma plasmocytoma growth factor (HPGF), a factor required for the growth of B cell hybridoma lines, and as hepatocyte

* Corresponding author.

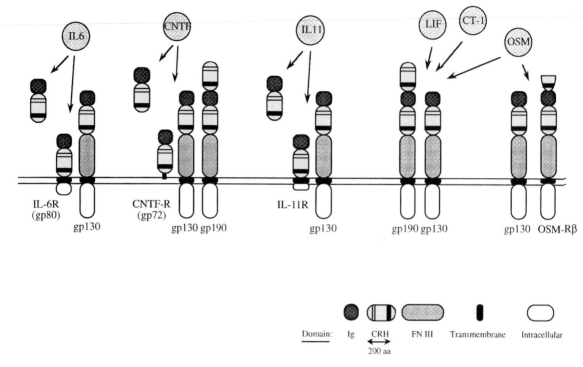

■ **Figure 1.** Model of the interactions of interleukins 6 and 11 (IL-6 and -11), ciliary neurotrophic factor (CNTF), leukaemia inhibitory factor (LIF) and cardiotrophin-1 (CT-1) receptors with gp130 and/or gp190. The CNTF receptor (gp72) is GPI-linked. gp80, gp72 and IL-R11α are also found in soluble forms, which retain their ability to bind ligands. The CT-1 receptor may include an as yet undiscovered α-chain. Abbreviations: Ig, immunoglobulin domain; CRH, cytokine receptor homologous domain; FN III, fibronectin type III domain.

stimulating factor (HSF), a factor inducing acute phase protein secretion by hepatocytes. Other biological activities of this pleiotropic cytokine include induction of thymocyte cell growth and neuronal differentiation. As expected from these multiple sites of action, IL-6 has been found to be produced by various types of lymphoid and non-lymphoid cells, such as B cells, T cells, monocytes, endothelial cells and fibroblasts.

The human IL-6 gene has been assigned to chromosome 7p21 and comprises five exons and four introns. The human IL-6 cDNA is 1.1 kb long with an open reading frame of 636 bp. It encodes a glycoprotein of 212 amino acids including a peptide leader sequence of 22 amino acids, two potential N-glycosylation sites and four cysteine residues. The calculated molecular mass of the protein core is 21 kDa. Human and murine IL-6 show 42% similarity at the protein level, much lower than the similarities between the human and murine forms of LIF, CT-1, CNTF and IL-11 (78%, 80%, 81% and 88% respectively).

IL-6 gene expression is tightly controlled and the promoter region contains many transcription regulation sites including a glucocorticoid responsive element (GRE), activator protein 1 binding site (AP-1), c-Fos serum responsive element (SRE), cyclic AMP responsive element (CRE) and a 14 bp palindromic sequence within an IL-1 responsive element. The transcription factor (NF-IL6) that binds to this latter sequence contains a leucine zipper motif and is a member of a liver-specific transcriptional factor (C/EBP) family. NF-IL6 also mediates expression of IL-6-inducible genes.

The crystal structure for IL-6 has been reported recently. IL-6 belongs to the long-chain group of helical cytokines with a four-helix bundle consisting of two long helices (A and D; 25 and 27 residues long, respectively) and two shorter helices (B and C; 23 and 21 residues, respectively). Helix A is connected to helix B by a long loop stabilized by two disulfide bonds located at both extremities of the loop. The four helices are arranged with a characteristic 'up-up-down-down' topology, the

∎ **Table 1.** Molecular characteristics of the human cytokines sharing the gp130 chain as receptor

	LIF	OSM	CNTF	CT-1	IL-6	IL-11
chromosome location	22q12	22q12	11q12	16p11	7p21	19q13.3
gene size (kb)	7.6	9	3.2	7	5	7.5
number of exons	3	3	2	3	5	5
mRNA (kb)	4 and 1.8	2 and 1	1.8	1.7	1.3	2.5 and 1.5
amino-acids length	180	252	200	201	184	199
signal peptide	yes	none	none	none	yes	yes
molecular mass (kDa)	38 to 67	20 to 28	23	22	25 to 30	23
number of *N*-glycosylation sites	7	2	0	0	2	0
number of cystein residues	6	5	1	2	4	0

internal core of the bundle being maintained by a network of hydrophobic interactions.

IL-6 exerts its many actions through a heteromeric receptor involving two glycoproteins; an 80 kDa IL-6-binding α-chain, IL-6Rα and gp130, which by itself does not bind IL-6 but is responsible for signal transduction and stabilization of the IL-6/IL-6Rα complex. IL-6 binds to IL-6Rα with a dissociation constant (K_d) of 10^{-9} M; this step triggers the subsequent association of gp130 with IL-6/IL-6Rα to form a multimeric high affinity complex with a K_d of 10^{-11} M.

The human IL-6Rα gene has been assigned to chromosome 1q21. In agreement with the pleiotropic nature of the cytokine, the IL-6Rα mRNA is expressed by numerous tissues and cell lines, including a myeloma cell line (U266), a histiocytic cell line (U937), an NK-like cell line (YT) and an Epstein–Barr virus-transformed B cell line (BMNH). However, IL-6Rα mRNA is not detected in the Jurkat T cell line or in the Daudi Burkitt's lymphoma line.

The human IL-6Rα cDNA encodes a protein precursor consisting of 468 amino acids with a predicted molecular mass of 80 kDa. Its extracellular region contains 340 amino acids with five potential *N*-glycosylation sites and has two domains: (i) a domain of approximately 100 amino acids that, based on its structure, has been classified as belonging to the immunoglobulin superfamily; (ii) a domain composed of two fibronectin type III modules, each arranged in a barrel-like structure made up of seven β-sheet strands and characterized by four conserved cysteine residues in the N-terminus of the first module and a strictly conserved Trp-Ser-X-Ser-Trp (WSXWS) sequence near the C-terminus of the second module. As has been demonstrated with the erythropoietin receptor, the WSXWS box is important for protein folding and for the intracellular stability of the receptor.

The IL-6Rα cytoplasmic domain is about 82 amino acids in length and lacks a tyrosine kinase or serine/threonine kinase domain; it has been shown to be dispensable for signal transduction. A natural soluble form of IL-6Rα generated by proteolytic cleavage from the membrane-anchored form has been identified which displays agonist properties towards IL-6. However, because it retains the ability to bind IL-6, thus forming a soluble binary complex, it can also associate with and trigger gp130 signalling on cells that do not bear the IL-6Rα–chain on their surface.

Molecular cloning of the cDNA encoding gp130 revealed an open reading frame of 918 amino acids with a single transmembrane domain. The extracellular region contains 597 amino acids and is composed of a typical 200 amino acid cytokine receptor homologous domain and three fibronectin type III modules. Its intracytoplasmic region encompasses 277 amino acids and does not contain any kinase domains, but does contain proline-rich conserved motifs (boxes 1, 2 and 3) known to be involved in the triggering of the intracellular signalling cascade.

In contrast with IL-6Rα expression, gp130 mRNA is present in all organs and cell lines so far examined. The human gp130 gene has been assigned to chromosome 5q11 and there is a

∎ **Table 2.** Molecular characteristics of human receptor chains shared by the cytokines of the IL-6 subfamily

	gp130	gp190	OSM-Rβ	IL-6Rα	IL-11Rα	CNTFRα
chromosome location	5q11	5p12/p13	—	1q21	9p13	9p13
gene size (kb)	—	70	—	—	10	35
number of exons	—	>20	—	—	13	10
mRNA size (kb)	7	6 and 4.5	—	5	1.9 and 2.8	2
amino-acids length*	918	1097	979	468	422	372
molecular mass (kDa)#	101	121	108	51	46	41
number of N-glycosylation sites	10	19	15	5	2	4

* = the number of amino-acids comprising the mature molecule is given for the membrane-bound form only.
\## = the molecular mass is deduced from the amino-acid composition and does not take into account the glycosylations. It is given for the proteic core and for the transmembrane protein only.

homologous pseudogene on chromosome 17p11. The intron–exon organization of IL-6Rα and gp130 genes is not yet known.

Mice lacking gp130 display a lethal and deficient phenotype showing multiple defects affecting predominantly ventricular myocardial development and haematopoiesis. In addition, recent *in vivo* studies of transgenic mice overexpressing a dominant-negative form of gp130 showed that this chain is essential for antigen-specific antibody production; this result is consistent with data previously obtained from studying the biological effects of IL-6 on the B cell compartment *in vitro*. Impairment of antigen-specific antibody production has also been observed in mice lacking IL-6 with a milder phenotype due to compensatory mechanisms by other cytokines sharing the gp130 signal transducer.

Recent X-ray structural analysis has shown that the X-ray structures of IL-6 and growth hormone are very similar. Together with IL-6 site-directed mutagenesis studies, this led to the proposal that an IL-6/IL-6Rα/gp130 signalling complex is assembled in an ordered and sequential process, by analogy with the model based on the crystal structure of growth hormone bound to two molecules of growth hormone receptor. The three proposed steps in the assembly are:

(1) the binding of IL-6 to IL-6Rα through its site 1, composed of the C-terminal part of helix D and the loop connecting helices A and B;

(2) the binding of the IL-6/IL-6Rα binary complex to gp130, mediated through IL-6 site 2, which comprises some of the amino acid residues in helices A and C;

(3) dimerization of two heterotrimeric IL-6/IL-6Rα/gp130 complexes.

This dimerization, leading to a final hexameric complex, (IL-6/IL-6Rα/gp130)$_2$, is made possible through the participation of two additional IL-6 binding sites. One of these, site 3, interacts with a second molecule of gp130 and comprises residues in the loop following helix A and in the loop linking helices C and D. The second, site 4, contributes to IL-6 dimerization and to the stabilization of the hexameric complex. Site 4 comprises residues Glu106–Arg 113. Disulfide bond formation between the two gp130 molecules is also suggested to stabilize the hexameric complex.

This oligomerization initiates cytoplasmic signalling cascades by activating associated intracytoplasmic members of the Jak family of tyrosine kinases and signal transducers and activators of transcription (STATs). Elucidation of the stoichiometry and the cytokine sites mediating ligand/receptor assembly may be the first step on the way to derivatizing agonist or antagonist molecules for use as new tools to interfere with situations where this system is known to be dysregulated. The list of physiological and pathological processes in which IL-6 may be involved is long (including, for example, autoimmune disorders and post-menopausal osteoporosis) and so is the list of the potential applications of these agents. As an example, IL-6 plays a central role in the pathogenesis of mutiple myeloma, acting as both a growth factor and a survival factor for the tumour cells. There is no effective treatment for such B-cell malignancies, but some IL-6 receptor antagonists, obtained by point mutation of IL-6, have been shown to inhibit growth and promote apoptosis of IL-6-dependent myelomas. Their capacity to induce cell death was related to their affinity for IL-6Rα, the degree of impairment of gp130 binding and the efficiency with which they inhibited intracellular events.

3. Leukaemia inhibitory factor

The cloning of the cytokine leukaemia inhibitory factor (LIF) was reported by two independent laboratories following the characterization of a soluble factor acting in opposite ways on two different murine myeloid cell lines. A factor produced by murine Krebs II ascites cells induced myeloid leukaemic M1 cells to differentiate into macrophage-like cells, while stopping their proliferation. The name leukaemia inhibitory factor originated from this feature. Simultaneously, a cytokine produced by activated human T cell clones was described, which was able to support the growth of the IL-3-dependent myeloid cell line DA-1, and was named HILDA (human interleukin for DA cells).

The cloning of the murine cDNA encoding LIF in 1987 allowed the corresponding human genomic sequence to be identified by cross-hybridization the following year. Independently, the human cDNA encoding HILDA was also isolated, and this cytokine was found to be identical to LIF.

LIF is encoded by a single gene, which is located on human chromosome 22q12 and on murine chromosome 11, in a region which is syntenic to human 22q12. The gene organization of the human and murine LIF genes is very similar. The human gene spans over 7.6 kb, and contains three exons, two introns and a 3.2 kb long 3'-untranslated region. The first exon is very short, encoding the first six amino acids of the signal peptide of LIF; the other two exons encode 60 and 136 amino acids. The protein is 202 amino acids long, including a 22 amino acid signal peptide. Although the molecular mass of its protein core is only 23 kDa, the apparent molecular mass of the naturally occurring form of LIF varies between 37 and 65 kDa, as a result of *N*-glycosylations on four of the seven putative *N*-glycosylation sites. Seventy-nine per cent of the amino acids in murine and human LIF are identity; there are insertions or deletions, and half of the

amino acid differences involve conservative changes. The LIF gene has also been cloned in sheep, rat, pig and cattle, and the proteins encoded by these genes are 79–88% identical to human LIF.

Although human LIF only exists as a secreted protein, a matrix-associated isoform has been identified in mouse, which is thought to be targeted to the extracellular matrix by an unknown mechanism. It differs from the diffusible LIF only by the usage of an alternative first exon located within the first intron, therefore leaving the mature protein unchanged. This exon is not conserved in the other species where the gene has been cloned. This matrix-associated LIF is generated from the triggering of a different and unknown promoter.

The three-dimensional structure of murine LIF has been solved by X-ray crystallography. It belongs to the growing family of haematopoietic cytokines whose prototype is growth hormone, and which are characterized by four antiparallel α-helices (A to D), interconnected by long loops between helices A and B, and C and D, and a short loop between helices B and C. A short A' mini-helix is found at the beginning of the AB loop. The interaction with gp190 involves two sites on the cytokine, one located at the N-terminus of helix D and one encompassing the second half of helix D. A third region, located on helices A and C, is predicted to bind gp130 in the high affinity receptor.

LIF is produced by various cell types, including fibroblasts, endothelial cells, T lymphocytes, monocytes, synoviocytes, astrocytes and the thymic epithelium. Its synthesis and secretion are enhanced by the inflammatory cytokines TNF and IL-1, by lipopolysaccharide and by TGF-β, but decreased by IL-4, glucocorticoids and cyclosporin A.

The biological effects of LIF on its target cells are mediated through interaction with a transmembrane high affinity receptor comprising two different chains, gp190 and gp130. The cDNA encoding the gp190 chain has been cloned in mouse and human. The corresponding gene is located on mouse chromosome 15 and human chromosome 5p12-13. The two receptors are homologous, and belong to the expanding family of the receptors for the haematopoietic cytokines. The members of this family are characterized by the 'cytokine receptor homologous' (CRH) domain, which comprises two modules of around 100 amino acids, each consisting of seven antiparallel β-sheets. The N-terminal module shelters four conserved cysteine residues involved in two intrachain disulfide bridges, whereas the C-terminal module contains a consensus Trp-Ser-X-Trp-Ser motif. They interact with cytoplasmic tyrosine kinases of the Jak family which recruit the STAT transcription factors, and activate the Ras/MAP kinase pathway.

Although the gp190 protein is 1097 amino acids long, its apparent molecular weight is 190 kDa, owing to the presence of 19 potential *N*-glycosylation sites. Its extracellular region contains 789 amino acids and displays 25% identity with gp130. Unusually for a member of this family, gp190 contains two CRH domains. They are separated by an immunoglobulin-like loop of around 100 amino acids, and the second CRH domain is followed by three modules homologous to the type III fibronectin encompassing around 100 amino acids each. These features are not found in all members of this family. The gp130 protein extracellular region is shorter because it has only one CRH domain. The intracellular region of gp190 encompasses 238 amino acids, and is phosphorylated on tyrosine residues upon activation of the LIF high affinity receptor by the cytokine. However, the transduction of the activation signal requires the intracellular regions from both gp130 and gp190. As a consequence, a soluble form of gp190 would act as an antagonist of the membrane receptor upon ligand binding. Such a soluble gp190 has been found in mouse and human. It is generated by alternative splicing and three kinds of cDNAs with different intronic sequences introducing a termination codon before the transmembrane domain have been isolated in human. They correspond to gp190 truncated forms in the second and the third of the type III fibronectin domains. They all bind LIF with a comparable affinity and are expressed by the human liver, placenta and a choriocarcinoma cell line.

Binding experiments performed with iodinated LIF on cells expressing LIF receptor components have demonstrated that the gp190 behaves as a LIF low affinity binding receptor, with a dissociation constant (K_d) of 2–5 nM for this ligand. LIF does not seem to exhibit any capacity to interact significantly with gp130, in the absence of gp190, whereas the LIF/gp190 complex does bind to gp130. The K_d of this trimeric complex is much lower, 30–100 pM. Therefore, the gp130 receptor acts as a high affinity converter, as for IL-6 and IL-11. However, the situation may be more complex, since cross-linking experiments on cells have shown that a still unknown third component of 150 kDa is part of the high affinity receptor.

LIF is a pleiotropic cytokine with multiple bio-logical effects on a wide array of cell types and tissues. It can maintain the totipotency of embryonic stem (ES) cells, and was found to be identical to the DIA/DRF (differentiation inhibiting activity/differentiation retarding factor) produced by feeder cells which were used to delay differentiation of ES cells for transgenesis experiments. Inactivation of the LIF gene in mouse has no drastic effect on males, but females are sterile because their fertilized eggs do not implant in the uterus. However, LIF–/– embryos are viable, showing that embryo-derived LIF is not necessary for their development. Therefore, a major action of LIF takes place at the implantation step, and is maternally derived.

In early *in vivo* experiments, injection of LIF-producing cells into mice induced cachexia, hypermotility, thymus atrophy, biological signs of inflammation, hypercalcaemia with calcifications in muscles, extramedullary haematopoiesis, anarchic bone remodelling and, finally, death. This fatal syndrome reflects the variety of LIF functions. LIF acts additively to other growth factors such as as IL-3, stem cell factor (SCF), or granulocyte–monocyte colony-stimulating factor (GM-CSF), in the renewal of the haematopoietic progenitors. LIF–/– mice display a significant decrease in spleen and granulocyte/monocyte colony-forming units and in erythroid burst forming units. Administration of low doses of LIF to animals induces an increase in platelet count, showing that LIF stimulates thrombopoiesis. LIF is also involved in the regulation of the inflammatory reaction. It was known, as hepatocyte-stimulating factor III (HSF III), to stimulate the synthesis by hepatocytes of several acute phase proteins, including fibrinogen and haptoglobin. It also strongly increases the secretion of adrenocortinotrophic hormone (ACTH) by pituicytes, alone and in synergy with corticotrophin-releasing factor (CRF), thereby controlling the inflammatory reaction at the neuroendocrine interface.

LIF has biological effects on neurones. A soluble factor called cholinergic neuronal differentiation factor (CNDF) was identified with LIF, which could induce a switch from the adrenergic to the cholinergic phenotype in embryonic sympathetic neurones. LIF is also a survival and regeneration factor for sensitive and motor neurones. LIF–/– mice do not show functional motor defects around the time of birth, but neurone regeneration is altered. In contrast, mice homozygous for the inactivation of the gp190 gene die on the day of birth from motor neurone deficiency.

LIF also stimulates bone metabolism, by activating both osteoblasts and osteoclasts, but only osteoblasts have receptors for this cytokine. The actions of LIF on the immune system are not clearly understood, but overexpression of this cytokine selectively in the thymocytes of transgenic mice induced profound disorganization of the thymic epithelium, which became unable to sustain normal thymocyte differentiation. These cells accumulated in lymph nodes at the $CD4^+CD8^+$ stage, whereas B lymphocytes migrated to the thymus.

4. Oncostatin M

Oncostatin M (OSM) was originally identified by its ability to inhibit proliferation of the human melanoma cell line A375. This biological activity was produced by U937 human histiocytic leukaemia cells which had been induced to differentiate into macrophage-like cells by treatment with phorbol 12-myristate 13-acetate. It was also found in the supernatants of activated human T lymphocytes and subsequently purified to homogeneity from U937 cells, microsequenced, and cloned using a cDNA library prepared from U937 activated cells.

The gene encoding OSM is located on human chromosome 22q12 and murine chromosome 11, less than 20 kp from the gene encoding LIF. The organization of the two genes is very similar, with three exons and two introns. As for LIF, the first exon is very short, encoding only the first 11 amino acids of the protein; the other two exons code for 48 and 193 amino acids. These characteristics led to the conclusion that these two genes originated from the duplication of a common ancestor.

The mRNA encoding human OSM is 1.8 kb long and contains a 1 kb 3′-untranslated region. The coding sequence of 756 nucleotides is translated in a 252 amino acid protein, which has a hydrophobic leader sequence of 25 residues in its N-terminus. The protein contains two potential *N*-glycosylation sites and five cysteine residues, four of which are at the same positions as in LIF and form two disulfide bridges identical to those in LIF. Despite the proposed common origin of LIF and OSM, sequence alignment shows only limited similarity between these two cytokines, with only 22% amino acid identity in human. Another divergent feature lies in the C-terminus of the OSM protein, which is very hydrophilic and contains several paired dibasic residues that may be proteolytic cleavage sites. Indeed, when expressed in COS cells, the full-length cDNA gave rise to two secreted proteins of 32 and 36 kDa. The shorter form lacks a stretch of

around 30 amino acids in the C-terminus of the protein, which can also be cleaved by trypsin in the 36 kDa precursor. Site-directed mutagenesis of both the two arginine residues at positions 220 and 221 of the immature protein impaired the production of the 32 kDa form. The natural OSM produced by U937 cells corresponds to the shorter form, is 196 amino acids long and has a molecular mass of 28 kDa. In growth inhibition assays, the truncated isoform was five to ten times more active than the precursor. Therefore, OSM is produced as a pro-protein, which is subsequently secreted following removal of a C-terminal domain, the function of which seems to be the regulation of OSM activity, although its mechanism is still unknown. Murine and human OSM are less similar than are murine and human LIF, since they only share 48% amino acid identity, while the two LIF proteins are 80% identical. Murine OSM is 263 amino acids long; compared with its human homologue, it lacks a stretch of 10 amino acids at around the beginning of putative helix C and has six more amino acid deletions spread over the protein and one insertion. Its hydrophilic C-terminal domain is 26 amino acids longer, with a cleavage site located at the same residue as in human OSM, and also needs to be removed to allow the secretion of a biologically active cytokine.

Two kinds of high affinity receptors have been characterized for OSM. Preliminary experiments had shown that OSM could bind to cells displaying high affinity receptors for LIF, but not to cells only expressing the LIF low affinity receptor gp190. The high affinity converter for LIF, which is gp130, was shown to be the OSM low affinity receptor. It is noteworthy that, of the cytokines that use gp130, OSM is the only one that can bind directly to gp130. However, gp130 alone is insufficient for OSM to induce a response in permissive cells: gp190 must be present too. The LIF low affinity receptor thus behaves as a high affinity converter for OSM. Besides this type I OSM receptor, it is thought that there may be a type II receptor in several cell types, such as the A375 melanoma cells and other solid tumour cell lines. Despite high affinity binding sites for OSM, these cells could not bind LIF, suggesting that gp190 was missing and could be replaced with another component that was incapable of interacting with LIF. Therefore, this alternative receptor was specific for OSM. The cloning of this receptor chain, called OSM-Rβ, has recently been reported in human. It is a 979 amino acid transmembrane protein, with a long intracellular region of 218 residues, and it belongs to the family of the receptors for the haematopoietins. It is most closely related to gp190 (32% amino acid identity) and to gp130 (23% identity). The organization of its extra-cellular domain is similar to gp190, harbouring two CRH domains, but the membrane-distal domain (unlike all other members of this family) is trun-cated, lacking the N-terminal module comprising the two consensus intrachain disulfide bridges.

OSM is produced mainly by activated T lymphocytes, but also by activated monocytes. The murine OSM gene promoter includes about 100 bases upstream of the transcription initiation site and a putative STAT5 binding site, which is respon-sible for the induction of OSM synthesis by cytokines using this transcription factor, such as IL-2, IL-3 and erythropoietin. Therefore, numerous cell lines of the myeloid and lymphoid lineages are expected to produce OSM. Recently, freshly explanted myeloma cells were also shown to secrete OSM constitutively, but only became responsive to it and to LIF following stimulation by IL-10. In fact, IL-10 induced expression of gp190 in these cells, which already express gp130 since they depend on IL-6 for their proliferation. Therefore OSM can also be an autocrine growth factor for myeloma cells.

OSM is a growth-inhibitory factor for and induces morphological changes in many cell lines derived from solid tumours of various origins such as skin, lung, breast, ovary and stomach, whereas it seems to *promote* the growth of Kaposi's sarcoma (KS) spindle-cells, which are found mainly in severely immunosuppressed patients such as those with acquired immune deficiency syndrome (AIDS) following infection with human immunod-eficiency virus type 1. Examination of fresh tumour samples has shown that OSM is also produced by these cells, thereby acting in an autocrine manner. OSM also increases the production by these cells of IL-6, which is another growth factor for this type of tumour. These OSM activities are mediated through interaction with the type II receptor, since these cells seem to lack the gp190. Interestingly, whereas the soluble form of the IL-6 receptor (gp80) along with IL-6 induces the proliferation of these cells, the soluble counterpart of the gp130 chain receptor inhibits AIDS–KS cell growth, which may well fit with what is known on the role of gp130 in OSM binding. OSM also stimulates the growth of fibroblasts and of endothelial cells, except endothelial cells from large vessels. This latter prop-erty is shared with LIF, and these two cytokines could favour atherosclerosis by inhibiting the replacement of injured endothelial cells.

Many other functions of OSM are shared with LIF, through the type I receptor. For example, OSM can induce the differentiation of murine myeloid leukaemic M1 cells and block the differentiation of ES cells, triggering the synthesis of acute-phase proteins by hepatocytes and of neuropep-tides in neurones, stimulating the hypothalamus–pituitary–adrenal axis leading to an increase in cir-culating cortisol, or regulating bone metabolism through activation of osteoblasts. Injection of OSM into mice induces extramedullary haematopoiesis in the spleen and elevation of platelet counts in blood, but no systemic effects were observed at the doses used. This thrombocytopoietic activity was confirmed *in vitro*, but required the presence of IL-3. Overexpression of OSM under the control of a non-tissue-specific promoter during mouse devel-opment in transgenic mice is deleterious. The sole survivor after birth displayed chronic tremor, no spermatogenesis, and bone disorganization with excess bone formation. When OSM coupled to a skin-specific promoter was introduced, no tran-s-genic animals were obtained, suggesting that expression of OSM in skin is lethal. Expression of OSM in neurones was also lethal, but some animals survived for several weeks, with tremor, ataxia and progressive weakness. The localized expression of OSM in the T lymphocyte lineage under the control of the p56^{lck} promoter, caused dramatic changes in the distribution of thymocytes and B lymphocytes, and obvious signs of autoimmunity. As shown with LIF, such mice displayed interconversion of thymic and lymph node morphologies, and accumulation of CD4$^+$CD8$^+$ double-positive T cells at the peri-phery. Therefore, many of the effects of LIF *in vivo* are shared by OSM, as expected from the usage of a common high affinity receptor.

5. Ciliary neurotrophic factor

Ciliary neurotrophic factor (CNTF) was discovered and partially characterized by several groups in the late 1970s. It was originally described as a protein that supported the survival of cultured neurones isolated from chicken embryonic parasympathetic ciliary, sensory and sympathetic ganglia. Unusually, its biological activity is not species restricted: the chicken protein is active on human cells, for example.

The development and functions of the nervous system rely on a set of proteins originally defined by their ability to support the survival of neuronal cells. Based on the type of chain receptor they use to transduce their intracellular signal, they can be divided into two subgroups. The first one includes, among others, nerve growth factor (NGF), brain-derived neurotrophic growth factor and the neutrotrophins. These proteins bind a heterodimeric receptor comprising the product of the proto-oncogene *trk* which belongs to the tyrosine-kinase receptor family as do the receptors for macrophage colony-stimulating factor (M-CSF), epidermal growth factor and platelet derived growth factor, and a second component earlier designated NGF receptor belonging to the tumour necrosis factor receptor family the second group includes CNTF which has no structural homology with the above neurotrophic factors, and binds to an unrelated multimeric receptor as we will see later.

CNTF from chick ocular tissues, bovine heart and sciatic nerves of rats and chicken has been described as an acidic protein with a molecular mass of around 20 kDa. It promotes the survival and differentiation not only of sympathetic neurones but also of primary sensory neurones, motor neurones, hippocampal neurones, basal forebrain neurones and type 2 astrocytes. CNTF has many other, non-neurological effects: it is an endogenous pyrogen, and induces acute-phase protein expression by hepatocytes and an increase in β-amyloid precursor protein mRNA in a rat glioma cell line. It exerts myotrophic effects by attenuating the morphological and functional changes associated with denervation of rat skeletal muscle. It is widely distributed in mammalian tissues but relatively concentrated in peripheral nerves where it is located in Schwann cells.

The full-length cDNA encoding human CNTF is 1855 bp long, with an open reading frame of 600 bp coding for a protein of 200 amino acids and a deduced molecular mass of 23 kDa. There is no evidence of *N*-glycosylation sites in the sequence and only one cysteine is found. There is one transcription start site, which in the human CNTF gene is located 81 bp upstream of the initiation methionine codon. The 3′-untranslated region is 1207 bp long and is terminated by only one polyadenylation sequence. Four ATTTA motifs are scattered in this untranslated sequence, of which two are located within an AT-rich region, as is the case for several other cytokines' cDNA. These short nucleotide stretches are thought to reduce mRNA half-life, leading to the conclusion that the CNTF mRNA is short-lived.

Like basic fibroblast growth factor, acidic fibroblast growth factor, interleukin-1 and cardiotrophin-1, CNTF lacks a signal peptide. It is not known how it is secreted in the absence of a signal peptide. This feature, along with the cytosolic localization of the protein and the observations that primary astrocytes and transfected COS or HeLa cells do not release substantial quantities of CNTF into the culture medium, has led to the conclusion that CNTF is indeed a typical non-secreted cytosolic molecule. Perhaps, though, an unconventional mechanism of release is operating and, because of its high potency, only a small proportion of the factor is needed *in vivo* for the maintenance of motor neurones.

The open reading frame is encoded in two exons separated by one intron which is 1165 bp long. Fluorescent *in situ* hybridization analysis on metaphase cells has localized CNTF on chromosome 11q12, confirming previous studies done on somatic cell hybrids and cosmid clones. No inherited neurological diseases are known to be linked to 11q12, although a null mutation in the human CNTF gene has been found. Deficiency of human CNTF is not causally related to amyotrophic lateral sclerosis (ALS), although beneficial effects were noted upon administration of recombinant CNTF in *pmn/pmn* (progressive motor neuronopathy) mice which bear an autosomal recessive mutation leading to progressive caudocranial motor neurone degeneration. Unfortunately it was not effective in ALS patients where it was administered by different routes. Attempts have been made to introduce it by the intrathecal route, but without success.

Subsequent studies demonstrated that CNTF is an inducer of reactive gliosis, a condition associated with a number of neurological diseases. In studies conducted on several samples of normal individuals from variable ethnic extractions, around 25–30% are heterozygous for the null mutation and around 3% are homozygous, with no visible consequences for their bearers. Knowing the biological roles of CNTF *in vitro*, a debate arose on the possible involvment of null mutations in psychiatric diseases. It seems now that where associations have been found between null mutation and psychiatric disease, this is probably fortuitous or may result from a reduced level of CNTF which may still represent a predisposing factor leading to disturbed development and function of the central nervous system. In normal subjects, homozygous for the null mutation, CNTF was completely suppressed in peripheral nerves, brain and muscle tissues which are the main sites for its expression. Abrogation of CNTF gene expression by homologous recombination and its effects on motor neurones had been reported earlier in mice. These transgenic CNTF null mice had a normal number of motor neurones during embryonic development and in the first postnatal weeks. In contrast, with increasing age, spinal motor neurones progressively atrophied and degenerated, reflected by a significant reduction of muscle strength. These results raise two points:

(1) The cytokine network is redundant, so the physiological effects of CNTF deficiency may be compensated for by other members of the family sharing both gp190 and gp130 as receptor chains. Cross-breeding of transgenic mice null for LIF and CNTF may bring new insights into this, since the effects of LIF on cultured sympathetic neurones are similar to those of CNTF.

(2) Other possible functions for CNTF may become apparent later in postnatal development, so detailed analyses are needed to identify alterations that are subtle and appear gradually in the postnatal phase.

The CNTF cDNA has been cloned and expressed in bacteria; a fully active mutant of it lacking 13 C-terminal residues allowed the crystal structure of human CNTF to be determined. The X-ray structure reveals that CNTF is dimeric; the association constant is weak (dimers form in solution at concentrations above $40\,\mu$M) even though the CNTF dimer interface encompasses a large area of the molecule. Contacts within each dimer are made mainly within the C helix and several residues located on the B helix through hydrophobic, polar and charged residues. Interestingly, nearly all interface side chains are conserved amongst CNTF sequences from different species, suggesting a conserved propensity to dimerize. Such sequences have not been found in other related cytokines, such as LIF, IL-6 and OSM, which are always identified as monomers. Other cytokines known to act as dimers use different helices or intermolecular disulfide bridges between connecting loops to stabilize their respective dimers (M-CSF or interleukin-5, for example).

Like LIF, the CNTF monomer is a four-helix-bundle cytokine comprising four antiparallel helices (A, B, C and D) connected by two long loops and one short one. When compared with granulocyte colony-stimulating factor (G-CSF), LIF and growth hormone, which all have helices of comparable length, similar packing arrangements and interhelical angles, the best overall superimposition of CNTF is with G-CSF. The short mini-helix A' found at the beginning of the AB loop in LIF seems to be preserved in CNTF. Both cytokines have a

shorter AB loop than growth hormone or G-CSF. Helix D extends 20 residues beyond helix D at the C-terminus of which 13 can be removed without loss of activity. Helix D, which is in contact with the gp190 chain of the receptor for LIF, is comparable in length to G-CSF and growth hormone but is approximately two turns longer than LIF. CNTF is a highly flexible molecule, probably because there are no disulfide bridges tethering pieces of the molecule together, unlike LIF which contains disulfide bridges and is much more compact.

A specific human CNTF low affinity receptor (with a K_d in the nanomolar range) has been cloned using a tagged-ligand panning procedure. This receptor (CNTFRα) is expressed exclusively within the nervous system and skeletal muscle. It is anchored to the cell membranes by a glycosyl-phosphatidylinositol (GPI) linkage, and has no intracellular region to transduce the signal. It is also found as a soluble entity which can bind its ligand with the same efficiency as the membrane form. The open reading frame encodes a 372 amino acid protein with a molecular mass of 41 kDa. Four potential N-glycosylation sites explain why cross-linking experiments have detected a mature protein with a molecular mass of 72 kDa. The protein can be divided in two domains: an immunoglobulin-like domain whose function is unclear, and another domain with the consensus residues defining the classical family of the haematopoietin-binding receptor. In the central nervous system, the cerebellum exhibits the largest amount of the 2 kb transcript; elsewhere, low amounts of CNTFRα are detected in the adrenal gland and the sciatic nerve. The CNTFRα gene has been assigned to chromosome 9p13 close to the specific IL-11 receptor; since there are structural similarities between these two genes, it is thought they may have had a common ancestor. The CNTF/CNTFRα complex recruits and interacts with two signal-transducing β-receptor subunits, gp130 and gp190, the heterodimerization of which in the target cell membrane initiates the signal cascade by activating cytoplasmic tyrosine kinases (Jak/Tyk family of proteins) pre-associated with the cytoplasmic domain of the β components.

Transgenic mice with a null mutation for the CNTFRα have been generated. Unlike mice lacking CNTF, they die perinatally and display severe motor neurone deficit, clearly suggesting that a second important CNTF-like factor also binding the CNTFRα remains to be discovered. The newborn pups could not feed because they could not move their jaws; they had dramatically decreased numbers of motor neurones in their brainstem motor nuclei and in the spinal cord.

The complex of the ligand bound to the soluble CNTFRα itself undergoes dimerization in the presence of gp130. In addition, gp130 and gp190, which can bind independently to the CNTF/CNTFRα complex, never form homodimers but always heterodimers. It therefore appears that CNTF assembles a hexameric receptor complex composed of two CNTF, two CNTFRα, one gp130 and one gp190. This conclusion is based on similarities with the IL-6 hexameric complex which is made possible because IL-6 is trivalent and could bind two distinct sites on two separate gp130 and one site on IL6Rα. CNTF initially would bind to CNTFRα with a 1:1 stoichiometry, then with either gp130 or gp190 giving asymmetric, non-functional complexes. One gp130 would bind two subcomplexes of CNTF/CNTFRα, forming a pentamer, whereas one gp190 would bind only one subcomplex, giving a trimer. The combination of both assemblies would result in the formation of a functional hexamer, in which heterodimerization of the β-subunit is stabilized by multiple protein–protein interactions. The apparent affinity of the CNTF for this multimolecular complex (K_d in the nanomolar range) classes it as a high affinity receptor.

6. Interleukin-11

Interleukin-11 was first identified from a permanent stromal cell line derived from the medullary cavity of non-human primates and was characterized by its ability to stimulate the proliferation of an IL-6-dependent plasmocytoma cell line. This assay was used for expression cloning of primate IL-11 cDNA and human IL-11 was subsequently cloned from a human fetal lung fibroblast cell line.

Whereas IL-11 does not stimulate early progenitors by itself, it exhibits synergic activity in combination with other cytokines such as IL-3, IL-4 and steel factor. IL-11 acts at multiple stages of megakaryopoiesis, stimulates erythropoiesis, enhances immunoglobulin secretion, may play a role in the maturation/activation of macrophages and promotes osteoclastogenesis. It also acts on various cell types outside the lymphohaematopoietic system: it inhibits lipoprotein lipase activity in preadipocyte cells, stimulates acute phase protein synthesis by hepatocytes and regulates neuronal differentiation. Many of the biological activities of IL-11 are also elicited by IL-6 (which has effects on haematopoietic progenitors, megakaryocytes, B cells, hepatocytes and adipocytes) and, to a lesser extent, by LIF, OSM and CNTF.

IL-11 elicits its activities by binding to a multi-subunit receptor complex expressed on the surface of target cells. Like IL-6, IL-11 binds with low affinity (K_d = 10 nM) to its specific α-chain (IL-11Rα) and with a high affinity (K_d = 275 pM) to a complex formed by IL-11Rα and the common signal transducer gp130, explaining why many of the biological effects of IL-11 are shared with other cytokines that also interact with gp130.

The human IL-11 cDNA contains an open reading frame of 597 nucleotides encoding a 199 amino acid sequence with a predicted molecular mass of 23 kDa and no *N*-glycosylation sites. Expression of IL-11 mRNA, as well as that of many other cytokine mRNAs, is regulated at the post-transcriptional level through complex events of mRNA stabilization. Recent studies have shown that AU-rich motifs, such as the AUUUA repeats which are present in the 3′-noncoding region of IL-11 mRNA, are not universal destabilizers shortening mRNA half-life and that multiple different regions within the IL-11 mRNA are required for this stabilization process; correct RNA folding may be a key element in the regulation of mRNA stability.

RNA analysis by Northern blotting revealed the expression of two transcripts of 1.5 and 2.5 kb generated by the use of alternative polyadenylation sites. The IL-11 gene contains five exons and four introns and has been assigned to chromosome 19q13.3, its intron–exon organization is similar to that of IL-6. In the absence of X-ray structure and multidimensional nuclear magnetic resonance data, a model of four-helix-bundle topology, based on secondary structure similarities and genomic organization between IL-6 and IL-11, was tested. Limited proteolysis and mutagenesis experiments support the model of a compact core and revealed, by analogy to growth hormone, two functionally important regions: region 1, involving Met58 in the loop connecting helices A and B and the C-terminal residues, and region 2, involving residues in the loops AB and BC in addition to an arginine residue in helix C.

Molecular cloning of the specific α-chain receptor for IL-11 revealed the presence of two isoforms referred to as IL-11Rα1 and IL-11Rα2. The first isoform is a transmembrane protein of 422 amino acids, with a predicted molecular weight of 43 kDa and two potential *N*-glycosylation sites. The extracellular region can divided into three domains, namely (from membrane-distal to -proximal) an immunoglobulin-like domain and cytokine receptor-like modules 1 and 2. Within the subfamily of human haematopoietic receptors sharing gp130, the IL-11Rα extracellular region shows the greatest homology scores with the IL-6- and CNTF-specific receptors (33.5% and 31.5% respectively). The transmembrane domain is followed by a short cytoplasmic domain of 32 amino acids devoid of tyrosine kinase or serine/threonine kinase domains.

The second isoform has extracellular and transmembrane domains identical to those of the first, but lacks the entire cytoplasmic domain. As is generally the case for α-type receptors, this intracellular domain is dispensable for IL-11 binding and signalling. Although not formally proven, the functions of IL-11Rα2 may be similar to those of CNTFRα and could be attached to the plasma membrane through a GPI linkage. A soluble form of murine IL-11Rα has been shown to exist; as for IL-6 and CNTF, it can bind IL-11, thereby potentiating the activity of exogenous IL-11 on cells responsive to IL-11.

The human IL-11Rα gene has been assigned to chromosome 9p13 and is unique. It spans 10 kb, and is composed of 13 exons; its intron–exon organization is that of the haematopoietin receptor superfamily. Both isoforms are generated by an alternative splicing mechanism. The short IL-11Rα transcript has not been identified in mice, but a second splice variant containing an alternative exon 1 with restricted expression in embryonic tissues has been characterized. It has recently been reported that there is a second murine gene locus; the mRNA transcript from the second IL-11Rα gene is 99% identical to the first and differs only in the 5'-noncoding region. Its expression is restricted to testis, lymph nodes and thymus. The two loci are less than 200 kb apart.

As already discussed, recent biochemical studies have proposed that functional IL-6 and CNTF ligand–receptor complexes are hexameric. The stoichiometry of the high affinity IL-11 receptor complex is not known but a fully functional high affinity receptor was reconstituted by expressing IL-11Rα and gp130 in heterologous cells, leading to the conclusion that the IL-11 complex could be similar to the IL-6 one. However, recent *in vitro* immunoprecipitation experiments were unable to reconstitute such a hexameric complex in solution (this can be done for IL-6 and CNTF); instead a pentameric complex was formed, consisting of two IL-11, two IL-11Rα and one gp130. If one assumes that mobilization of β-chains through ligand-induced dimerization is the key event leading to the activation cascade below the membrane, these data might suggest that an unidentified transducing receptor chain may be part of the high affinity receptor for IL-11 and that it was endogenous to the recipient cells in which it was functionally reconstituted.

7. Cardiotrophin-1

Cardiotrophin-1 (CT-1) is the most recently described member of the family of cytokines that using gp130, its cloning having been reported in 1995. It was isolated as a result of the search for soluble factors responsible for myocardial cell hypertrophy, using neonatal rat ventricular cardiac myocytes in an *in vitro* assay system. When totipotent mouse ES cells are cultured in the absence of LIF or fibroblast feeder layer, they differentiate into multicellular, cystic embryoid bodies which spontaneously beat and display cardiac-specific markers. Supernatants of such embryoid bodies contain a factor capable of inducing in these ventricular cardiac myocytes organization of myosin light chain in sarcomeric units, atrial natriuretic factor secretion, and increase in cell size in the absence of cell proliferation. Since these criteria are good evidence of a hypertrophic response, the cloning of the causative agent(s) was undertaken using a cDNA library prepared from 7 day differentiated embryoid bodies, by expression in human embryonic kidney 293 cells.

The cDNA encoding murine CT-1 is 1.4 kb long, with a 609 bp open reading frame encoding the protein, and a 750 bp long 3'-untranslated region. The human cDNA is 1.7 kb long, and the corresponding gene is located on chromosome 16p11.1–16p11.2. In both species, the gene encompasses 6–7 kb, and comprises three exons separated by two introns located after the triplet encoding amino acids 9 and 48. In both species, the mRNA encoding CT-1 is abundantly expressed in heart, skeletal muscle, lung, kidney and liver in mouse, and at lower level in prostate, ovary, thymus, brain and human liver. Peripheral blood leucocytes do not express it.

The murine and human CT-1 proteins contain 203 and 201 amino acids, respectively, and there is 80% amino acid identity between these two species. Although there is no hydrophobic N-terminal secretion sequence, the cytokine is efficiently secreted in the culture medium, by an unknown mechanism. The human protein has two cysteine residues and no potential *N*-glycosylation sites, while the mouse CT-1 has one cysteine and one potential *N*-glycosylation site. The mature cytokine migrates on polyacrylamide gels as a 21.5 kDa protein, suggesting that it is very lightly glycosylated. Its predicted tertiary structure is consistent with the classical four-helix-bundle model reported above.

The receptor for CT-1 has been partially characterized. Preliminary data showed that CT-1 was active in seven biological assays where LIF was active, and vice versa. Although CT-1 was also active in assays for CNTF, the converse was not always the case, and CT-1 was inactive in IL-6-specific assays. These results suggested that CT-1 bound and transduced its biological effects via the LIF receptor. CT-1 is able to bind directly to the LIF low affinity receptor gp190, with a K_d of 2 nM, which is identical to the affinity of LIF for gp190. Moreover, CT-1 and LIF cross-compete for the binding to LIF receptor on murine M1 cells. In contrast to what was observed for LIF, where murine LIF does not bind the human receptor, CT-1 is not species-restricted. Like LIF, CT-1 does not interact with gp130 by itself, but in the presence of gp130 the binding of CT-1 to gp190 is significantly increased, and antagonistic anti-gp130 antibodies also inhibit CT-1 signalling. Therefore, gp130 acts as an affinity converter for CT-1, presumably by the formation of a heterotrimeric complex, similarly to what happens for LIF. However, recent data complicate this analysis, and strongly suggest that a third receptor component should exist, in the form of a membrane-bound GPI-linked protein. Indeed, treatment of embryonic motor neurones with phosphatidylinositol phospholipase C (PI-PLC), which specifically cleaves GPI anchors, decreases the survival effect mediated by CT-1 on these cells. This third component would behave as a *α*-type chain, since in the absence of intracellular region, it will not be able to transduce a signal by itself. CNTF also possesses a GPI-linked receptor *α*-chain, gp72, which is distinct from its CT-1 counterpart, since CT-1 does not bind to gp72. However, this additional component may not be required for all CT-1-responsive cells since, for example, PI-PLC does not impair CT-1 function on M1 cells. Perhaps two kinds of receptors exist, or this uncharacterized chain may also be found in a transmembrane isoform, therefore insensitive to this enzyme, as a result of an alternative splicing from the same gene.

The biological properties of CT-1 on cardiac muscle cells include the hypertrophy response, which is morphologically distinct from the effects mediated by other agents such as the *α*-adrenergic agonists, endothelin-1 and angiotensin II, which all act via G-protein-coupled receptors. CT-1 is also able to promote cardiac myocyte survival and DNA synthesis in these cells. The gp130–/– mice die *in utero* from cardiac failure associated with an abnormally thin ventricle wall. By contrast, LIF–/– and CNTF–/– mice do not display any cardiac defects. In mouse embryo, and by contrast to other gp130 users, CT-1 mRNA is expressed in a relatively cardiac-restricted manner in the primitive mouse heart from day 8.5 *post coitum* to day 13.5. CT-1 expression continues in this organ at high levels after birth, suggesting a role for this cytokine in heart maintenance. Therefore, CT-1 is a good candidate to explain the particular phenotype of gp130–/– mice. Like LIF and CNTF, CT-1 also enhances skeletal muscle cell survival following muscle injury.

The biological effects of CT-1 are not limited to myocytes. CT-1 can inhibit the differentiation of mouse embryonic stem cells, induce a phenotypic switch in rat sympathetic neurones, promote the survival of rat dopaminergic and chick ciliary neurones, increase corticosterone circulating level through the activation of the hypothalamus–pituitary–adrenal axis, and trigger the synthesis of the acute-phase proteins fibrinogen, serum amyloid A and haptoglobin by hepatocytes. Administration of CT-1 to mice induces an increase in heart weight to body ratio, hypertrophy of the liver, the kidney and the spleen, while the thymus is atrophied. Platelet production was also strongly stimulated, which suggests that CT-1 may have direct effects on haematopoiesis. Since many of these properties have already been previously described for other cytokines that use gp130 and gp190, CT-1 appears highly redundant with these cytokines, but seems to exhibit more specific effects on muscle cell development, especially in the myocardium.

References

Akita, S., Webster, J., Ren, S.G., Takino, H., Said, J., Zand, O., and Melmed, S. (1995). Human and murine pituitary expression of leukemia inhibitory factor. Novel intrapituitary regulation of adrenocorticotropin hormone synthesis and secretion. *J. Clin. Invest.* **95**, 1288–98.

Bazan, J.F. (1990). Structural design and molecular evolution of a cytokine receptor superfamily. *Proc. Natl Acad. Sci. USA* **87**, 6934–8.

Boulton, T.G., Stahl, N. and Yancopoulos, G.D. (1994). Ciliary neurotrophic factor/leukemia inhibitory factor/interleukin-6/oncostatin-M family of cytokines induces tyrosine phosphorylation of a common set of proteins overlapping those induced by other cytokines and growth factors. *J. Biol. Chem.* **269**, 11648–55.

Conquet, F., Peyrieras, N., Tiret, L. and Brûlet, P. (1992). Inhibited gastrulation in mouse embryos overexpressing the leukemia inhibitory factor. *Proc. Natl Acad. Sci. USA* **89**, 8195–9.

Chrousos, G.P. (1995). The hypothalamic–pituitary–adrenal axis and immune-mediated inflammation. *N. Engl. J. Med.* **332**, 1351–62.

Clegg, C.H., Rulffes, J.T., Wallace, P.M. and Haugen, H.S. (1996). Regulation of an extrathymic T-cell development pathway by oncostatin M. *Nature* **384**, 261–3.

Davis, S., Aldrich, T.H., Valenzuela, D.M., Wong, V.V., Furth, M.E., Squinto, S.P. and Yancopoulos, G.D. (1991). The receptor for ciliary neurotrophic factor. *Science* **253**, 59–63.

De Chiara, T.M., Vejsada, R., Poueymirou, W.T., Acheson, A., Suri, C., Conover, J.C. *et al.* (1995). Mice lacking the CNTF receptor, unlike mice lacking CNTF, exhibit profound motor neuron deficits at birth. *Cell* **83**, 313–22.

Fry, R.C. (1992). The effect of leukemia inhibitory factor (LIF) on embryogenesis. *Reprod. Fertil. Dev.* **4**, 449–58.

Gearing, D.P., Thut, C.J., Van den Bos, T., Gimpel, S.D., Delaney, P.B., King, J. *et al.*, (1991). Leukemia inhibitory factor receptor is structurally related to the IL-6 signal transducer, gp130. *EMBO J.* **10**, 2839–48.

Kishimoto, T., Akira, S., Narazaki, M. and Taga, T. (1995). Interleukin-6 family of cytokines and gp130. *Blood* **86**, 1243–54.

Metcalf, D. (1997). Another way to generate T-cells? *Nature Med.* **3**, 18–19.

Pennica, D., Wood, W.I. and Chien, K.R. (1996). Cardiotrophin-1: a multifunctional cytokine that signals via LIF receptor-gp130 dependent pathways. *Cytokine Growth Factor Rev.* **7**, 81–91.

Robinson, R.C., Grey, L.M., Staunton, D., Vankelecom, H., Vernallis, A.B., Moreau, J.F. *et al.* (1994). The crystal structure and biological function of leukemia inhibitory factor: implications for receptor binding. *Cell* **77**, 1101–16.

Somers, W., Stahl, M. and Seehra, J.S. (1997). 1.9 Å crystal structure of interleukin 6: implications for a novel mode of receptor dimerization and signaling. *EMBO J.* **16**, 989–97.

Stewart, C.L., Kaspar, P., Brunet, L.J., Bhatt, H., Gadi, I., Kontgen, F. and Abbondanzo, J. (1992). Blastocyst implantation depends on maternal expression of leukemia inhibitory factor. *Nature* **359**, 76–9.

Taupin, J.L., Pitard, V., Dechanet, J., Miossec, V., Gualde, N. and Moreau, J.F. (1997). Leukemia inhibitory factor: part of a large ingathering family. *Int. Rev. Immunol.* **16**, 239–426.

Wells, J.A and de Vos, A.M. (1996). Hematopoietic receptor complexes. *Annu. Rev. Biochem.* **65**, 609–34.

III Interleukin 10

Marina Pretolani, Patrick Stordeur and Michel Goldman*

1. Interleukin 10 and its receptor: molecular aspects

1.1 Human and mouse IL-10 proteins

Human interleukin 10 (hIL-10) is a protein of 160 amino acids (molecular mass 18.5 kDa) containing two intramolecular disulfide bonds. It is acid-labile and appears in soluble form as a homodimer. Its structure is predominantly α-helical, suggesting that it belongs to the α-helical cytokine class of proteins such as interleukins 2 and 4 (IL-2 and -4), and interferons beta and gamma (IFNβ and γ) (IL-10 shows particular topological similarity to this latter cytokine). The mature murine IL-10 (mIL-10) contains 157 amino acids and shows 73% identity with its human homologue. In spite of this degree of homology, and the consequent structural similarity of these two proteins, mIL-10 acts only on mouse target cells, while hIL-10 acts on both human and mouse cells.

IL-10 shares common sequences with BCRF1, an open-reading frame product of the Epstein–Barr virus (EBV), now considered to be a viral form of IL-10. Sequence homologies have also been found in the genome of the poxvirus orf and equine herpesvirus 2.

1.2 Human and mouse IL-10 genes

The mIL-10 gene consists of five exons spanning ~5.1 kb of genomic DNA; like the human gene, it is located on chromosome 1. Both genes contain several non-coding sequences which are likely to be involved in the regulation of the transcription and of the stability of the corresponding mRNA. These include the TATA and CAT boxes, the glucocorticosteroid-responsive element, AP-1-binding sites, the cAMP-responsive element, NF-κB and STAT1 binding sites, CD28-, IFNγ-, IL-6- and GM-CSF-responsive elements and (in the 3′-untranslated region) several ATTA motifs. Recent studies demonstrated polymorphism in the IL-10 promoter. In one of them, a direct correlation between a simple base substitution and the levels of IL-10 produced by stimulated lymphocytes has been established. Additionally, two polymorphic dinucleotide repeats located around 1200 and 4000 bp upstream of the start codon seem to be involved in susceptibility to systemic lupus erythematosus. Moreover, the existence of these polymorphic motifs provides a molecular basis for the important inter-individual variations in IL-10 synthesis.

1.3 The IL-10 receptor

The human IL-10 receptor (hIL-10R) is composed of two different chains, hIL-10R1 (or α-chain), which was discovered first, and hIL-10R2 (or β-chain) which was described recently.

The gene encoding hIL-10R1 maps to chromosome 11q23.3; it is expressed mainly in haematopoietic cells. This expression can be modulated by different compounds: anti-CD3 monoclonal antibodies and phorbol ester decrease the IL-10R1 mRNA levels in human T cell clones, while lipopolysaccharide (LPS) and 1,25-$(OH)_2$-vitamin D_3 increase the amount of IL-10R1 mRNA in murine fibroblasts and human epidermal cells, respectively. Isolation and expression of cDNAs encoding the human and mouse R1 receptor disclosed a structural homology with the IFNγ receptor, supporting the hypothesis of an interaction between IL-10 and IFNγ signalling pathways. Accordingly, both IL-10 and IFNγ induced the expression of the high-affinity receptor for IgG (FcγRI) in human peripheral blood mononuclear cells or basophils, by activating DNA-binding

* Corresponding author.

proteins complexes recognizing the IFNγ response region located in the FcγRI promoter. These DNA-binding complexes contained the STAT1α transcription factor.

Besides the STAT1α transcription factor, other Jak-STAT proteins such as Jak1 and Tyk2 are involved in IL-10 signal transduction. Furthermore, IL-10 promotes tyrosine phosphorylation of STAT1α, STAT3 and STAT5 in different cell types, in such a way that distinct homodimeric and heterodimeric transcriptionally active complexes are formed in a cell- and/or promoter-specific manner.

The inability of viral IL-10 to compete effectively for IL-10 binding to IL-10R1 led to the discovery of the second receptor chain, IL-10R2. This latter has been identified as the transmembrane protein CRF2-4, an orphan receptor encoded on chromosome 21 and belonging to the class II cytokine receptor family. The common architecture of IL-10 and IFN receptors, the observation that IL-10 induces the phosphorylation of the Jak1 and Tyk2 tyrosine kinases and the fact that the intracellular domain of CRF2-4 associates with Tyk2 all suggest that IL-10 and IFNγ signalling pathways are closely related. It has been proposed that the binding of IL-10 homodimers to the IL-10R1 chains, which could be associated with Jak1 kinases, results in a non-functional receptor complex. The additional binding of IL-10 to the IL-10R2 chains, which could be associated with Tyk2 kinases, seems to be required to assemble an active receptor complex initiating signal transduction events.

2. Cellular origins of IL-10 and control of IL-10 synthesis

IL-10 was originally characterized as a factor, generated by mouse T-helper 2 (Th2) cells, that inhibited cytokine synthesis by Th1 cells. However, several other cell types have been further identified as a source of this cytokine, including CD4$^+$ and CD8$^+$ T lymphocytes, monocytes/macrophages, B cells, mast cells, eosinophils, keratinocytes, hepatocytes, epithelial cells, astrocytes, pituitary and hypothalamic cells, cytotrophoblasts, mesangial cells and various tumour cells. Thus, IL-10 is not strictly a Th2-specific cytokine and its pattern of expression resembles that of IL-6 more than that of IL-4 or IL-5. In most inflammatory disorders, cells of the monocytic lineage represent the major source of IL-10.

IL-10 is induced by a number of microbial pathogens causing activation of monocytes/macrophages (e.g. bacterial wall components , intracellular parasites, fungi , human immunodeficiency virus (HIV), HIV-1 Nef protein), T cells (e.g. human T cell lymphotropic virus type 1 (HTLV-1)) or B cells (EBV). Moreover, cytokines, hormones and arachidonic acid derivatives that are released in the course of severe infections as well as in other stress conditions contribute to amplifying IL-10 synthesis. Indeed, tumour necrosis factor alpha (TNF-α), IL-6, IL-12, type I interferons, glucocorticoids, adrenaline, prostaglandin E_2 (PGE$_2$) and other endogenous mediators have been shown to up-regulate IL-10 synthesis in macrophages and T cells.

Hypoxia is a good example of a cellular stress condition associated with IL-10 synthesis. Ischaemia–reperfusion is associated with overproduction of the endogenous purine nucleoside adenosine and the reactive oxygen species hydrogen peroxide, which are both responsible for IL-10 overexpression. Reactive oxygen species could also mediate the induction of IL-10 in other conditions such as exposure to UV light, which results in the accumulation of IL-10 in keratinocytes and macrophages.

A number of drugs modulate IL-10 production. These include glucocorticosteroids and cyclosporin A, which have been shown to up-regulate IL-10 levels both *in vitro* and *in vivo*. Up-regulation of IL-10 synthesis is also induced by the anti-psychotic agent chlorpromazine and the antidepressant Rolipram, a phosphodiesterase type IV inhibitor. Indeed, compounds that increase intracellular cAMP levels (pentoxylline, epinephrine, PGE$_2$) enhance IL-10 gene transcription in monocytes and macrophages. Furthermore, CD23 ligation induces IL-10 synthesis in human keratinocytes through a cAMP-dependent mechanism. Besides cAMP-responsive elements, NF-κB binding sites in the 5'-flanking region of the human IL-10 gene also seem to control its activation. This is consistent with the induction of IL-10 gene expression by several NF-κB activators including HIV-1, EBV, reactive oxygen species, UV light and TNF-α. NF-κB transcription factors could, in fact, play a pivotal role in IL-10 synthesis consecutive to cellular stress.

On the other hand, the antitumour tellurium-based compound AS101 down-regulates IL-10 synthesis. This may contribute to the beneficial effect of AS101 in murine systemic lupus erythematosus, a disease associated with IL-10 overproduction.

3. Target cells for the biological activity of IL-10

3.1 Effects on monocytes, macrophages and dendritic cells

The concept that IL-10 acts as an anti-inflammatory molecule emerged primarily from studies showing inhibition of the synthesis of a broad spectrum of pro-inflammatory cytokines by different cells, particularly of the monocytic lineage. Thus, IL-10 prevents the production of pro-inflammatory cytokines such as IL-1α, IL-1β, IL-6, IL-8, TNF-α, granulocyte colony-stimulating factor (G-CSF) and granulocyte–macrophage colony-stimulating factor (GM-CSF), and chemokines, including IL-8 and MIP-1α, from human LPS-activated monocytes, at both the mRNA and protein levels. These cytokines and chemokines play a crucial role in the activation of granulocytes, monocytes/macrophages, NK cells, T cells and B cells and in their recruitment to sites of inflammation.

Human monocytes produce large amounts of IL-10 upon activation and this endogenous IL-10 down-regulates the generation of cytokines. Production of IL-10 by monocytes develops later than that of pro-inflammatory cytokines and chemokines. This explains how IL-10 promotes the resolution of inflammatory responses, a phenomenon which was best documented in mice with a targeted disruption of the IL-10 gene. Interestingly, IL-10 prevents its own generation in LPS-stimulated monocytes, indicating an autoregulatory feedback mechanism controlling IL-10 expression.

The anti-inflammatory properties of IL-10 are not restricted to the inhibition of cytokine production, since addition of IL-10 to purified monocytes or neutrophils promotes the release of the IL-1 receptor antagonist, which displays anti-inflammatory activities on its own. IL-10 also prevents release of oxygen free radicals and nitric oxide-dependent microbicidal activity of macrophages and decreases their ability to generate prostaglandins. Additionally, IL-10 exerts similar inhibitory activities on neutrophils.

Due to its ability to prevent antigen-specific T cell responses, IL-10 was first considered as an immunosuppressive molecule. Indeed, IL-10 inhibits the antigen-presenting cell (APC) function of monocytes/macrophages by down-regulating their expression of MHC class II and co-stimulatory molecules including B7 and ICAM-1. The allostimulatory potential of human peripheral blood dendritic cells is also markedly influenced by IL-10, which inhibits the expression of MHC class II and B7-2 molecules on their surface. Furthermore, IL-10 inhibits the synthesis by APC of IL-12, a critical factor for the induction of Th1-type and cytolytic T cell responses.

3.2 Direct effects on lymphocytes

IL-10 also exerts direct deactivating properties on T cells, since it prevents IL-2, IL-5 and IL-6 generation from CD4$^+$ T lymphocytes. Moreover, chronic activation of CD4$^+$ T cell clones in the presence of IL-10 gives rise to antigen-specific clones with low proliferative capacity and producing high levels of IL-10 and TGF-β. This particular T cell subset, designated T regulatory cells 1 (Tr1), could be responsible for peripheral immune tolerance induced by IL-10. The multiple levels at which IL-10 interferes with T cell responses are indicated in Figure 1.

IL-10 also displays immunostimulatory activities, since it enhances the proliferation and cytolytic activity of CD8$^+$ T lymphocytes and promotes their chemoattraction. In parallel, IL-10 was shown to stimulate NK cell activation and to increase IL-2-induced NK cell proliferation, cytotoxicity and cytokine release. Finally, IL-10 is a potent proliferation and differentiation factor for B lymphocytes, promoting the synthesis of IgM, IgG and IgA. This property, associated with the up-regulation of Fcγ receptor expression on monocytes, explains how IL-10 enhances antibody-mediated cellular cytotoxicity.

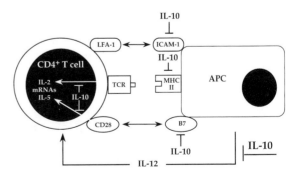

∎ **Figure 1.** Immunosuppressive properties of interleukin 10 (IL-10).

4. IL-10 in the context of allergic inflammation

Beside its well known inhibitory actions on inflammatory events triggered by microbial products and Th1-mediated experimental diseases, IL-10 might also control inflammatory processes of allergic origin. This was first suggested by the observation that IL-10 down-regulates the production by human T cells of IL-5, a critical cytokine for the differentiation and activation of eosinophils. A direct effect of IL-10 on eosinophil functions has also been demonstrated. Thus, IL-10 suppresses LPS-induced GM-CSF generation by eosinophils and thereby prevents their survival, suggesting that IL-10 may play an important role in the control of eosinophil accumulation in inflamed tissues. Recent findings indicate that human eosinophils express functional CD40 at their surface and that its ligation with a specific antibody or with its natural ligand, CD40L, prolongs their survival. Low concentrations of IL-10 were practically as active as the glucocorticosteroid budesonide in decreasing CD40 expression and in accelerating eosinophil death, again supporting a role for IL-10 in the resolution of eosinophilic inflammation.

Like the eosinophil, the mast cell is also considered as an important effector cell implicated in the allergic response. This results mostly from its ability to generate a number of cytokines acting directly and indirectly on eosinophil recruitment and activation in target tissues, particularly IL-3, IL-4, IL-5, GM-CSF and TNF-α. Although IL-10 has been shown to extend the viability of mouse mast cell lines *in vitro* and to influence mast cell development and differentiation, recent studies have proposed important inhibitory activities of this cytokine on mast cell functions. Accordingly, IL-10 inhibits TNF-α and IL-6 generation from mouse bone marrow-derived mast cells and rat peritoneal mast cells, respectively, in response to IgE cross-linking with a specific antigen. Since IL-10 also prevented the release of GM-CSF, a cytokine directly involved in the homing and activation of eosinophils and neutrophils in inflamed tissues, it was suggested that IL-10 may influence the antigen-dependent mast cell activation and thus interfere with the initiation of leucocytic inflammation.

The effects of IL-10 compatible with its anti-allergic activities are summarized in Figure 2.

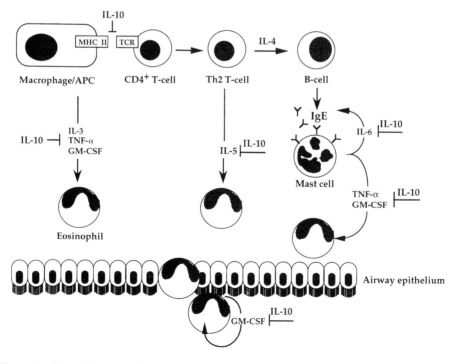

▮ **Figure 2.** Potential mechanisms involved in the anti-allergic activities of interleukin 10 (IL-10).

5. Molecular targets of IL-10

There is compelling that the inhibition of the synthesis of different cytokines by IL-10 results from a blockade of the transcription of the corresponding genes. Recent findings have shown that NF-κB proteins are molecular targets for IL-10 in the regulation of cytokine gene transcription. Accordingly, anti-CD3 monoclonal antibody-induced p50 and RelA NF-κB proteins in T lymphocytes are down-regulated by IL-10. Furthermore, IL-10 selectively prevents the nuclear localization of NF-κB in LPS- or TNF-α-stimulated human monocytes by inhibiting the formation of DNA-binding NF-κB dimers. In an IgG immune complex model of lung injury in rats, it has been shown that IL-10 (and IL-13) suppresses NF-κB activation by preservation of the cytoplasmic NF-κB inhibitor, IκBα. Another transcription factor that might be a target for IL-10 is STAT5, since its deactivation has been proposed as the main mechanism involved in the protection by IL-10 against LPS-induced cyclooxygenase-2 gene transcription.

IL-10 may also prevent cytokine synthesis by post-transcriptional mechanisms, as demonstrated in LPS-stimulated human macrophages where the inhibition of IL-1α, IL-1β and TNF-α release was related to degradation of the corresponding mRNAs. Since the protein synthesis inhibitor cycloheximide antagonized these inhibitory effects of IL-10, it was concluded that IL-10 may be responsible for *de novo* synthesis of a ribonuclease acting on cytokine transcripts. The ability of IL-10 to decrease mRNA stability has also been shown for other cytokines, including IL-6, IL-8, G-CSF, GM-CSF and IL-10 itself.

Although its effect on apoptosis is still controversial, IL-10 augments cell survival by increasing the level of the anti-apoptotic protein Bcl-2. Finally, the transporters associated with antigen processing (TAPs) also represent cellular components affected by IL-10, leading to a decreased expression of MHC class I molecules which accumulate in the endoplasmic reticulum. As far as expression of MHC class II molecules is concerned, IL-10 was recently shown to act by affecting arrival and recycling of MHC class II complexes at the cell membrane. Since this effect was not associated with a modification in the transcription rate of human leucocyte-associated antigen (HLA) class II genes, proteins that participate in peptide–class II assembly such as HLA-DM could represent other candidates for IL-10 action.

6. Conclusions

IL-10 plays a critical role in the regulation of inflammatory reactions and T cell-mediated immunity by deactivating macrophages and by inhibiting both the induction and the effector phases of T cell responses. Further characterization of the mechanisms controlling IL-10R expression and the molecular targets of IL-10 could lead to new concepts of the natural control of T cell-mediated pathological processes such as organ-specific autoimmunity, transplant rejection and allergic inflammation.

References

Bogdan, C., Paik, J., Vodovotz, Y. and Nathan, C. (1992). Contrasting mechanisms for suppression of macrophage cytokine release by transforming growth factor-beta and interleukin-10. *J. Biol. Chem.* **267**, 23301–8.

Buelens, C., Willems, F., Delvaux, A., Pierard, G., Delville, J.P., Velu, T. and Goldman, M. (1995). Interleukin-10 differentially regulates B7-1 (CD80) and B7-2 (CD86) expression on human peripheral blood dendritic cells. *Eur. J. Immunol.* **25**, 2668–72.

de Waal Malefyt, R., Haanen, J., Spits, H., Roncarolo, M.G., Tevelde, A., Figdor, C. *et al.* (1991). Interleukin-10 (IL-10) and viral-IL-10 strongly reduce antigen-specific human T-cell proliferation by diminishing the antigen-presenting capacity of monocytes via downregulation of class-II major histocompability complex expression. *J. Exp. Med.* **174**, 915–24.

Eskdale, J., Kube, D., Tesch, H. and Gallagher, G. (1997). Mapping of the human IL1O gene and further characterization of the 5′ flanking sequence. *Immunogenetics* **46**, 120–28.

Finbloom, D.S. and Winestock, K.D. (1995). IL-10 induces the tyrosine phosphorylation of Tyk2 and Jak1 and the differential assembly of STAT1 alpha and STAT3 complexes in human T cells and monocytes. *J. Immunol.* **155**, 1079–1090, 1995.

Fiorentino, D.F., Bond, M.W., Mosmann, T.R. (1989). Two types of mouse T helper cell. IV. Th2 clones secrete a factor that inhibits cytokine production by Th1 clones. *J. Exp. Med.* **170**, 2081–95.

Groux, H., O'Garra, A., Bigler, M., Rouleau, M., Antonenko, S., de Vries, J.E. and Roncarolo, M.G. (1997). A CD4+ T-cell subset inhibits antigen-specific T-cell responses and prevents colitis. *Nature* **389**, 737–42.

Ho, A.S.Y., Liu, Y., Khan, T.A., Hsu, D.H., Bazan, J.F. and Moore, K.W. (1993). A receptor for interleukin-10 is related to interferon receptors. *Proc. Natl Acad. Sci. USA* **90**, 11267–71.

Kim, J.M., Brannan, C.I., Copeland, N.G., Jenkins, N.A., Khan, T.A. and Moore, K.W. (1992). Structure of the mouse IL-10 gene and chromosomal localization of the mouse and human genes. *J. Immunol.* **148**, 3618–23.

Koppelman, B., Neefjes, J.J., de Vries, J.E. and de Waal Malefyt, R. (1997). Interleukin-10 down-regulates MHC Class II $\alpha\beta$ peptide complexes at the plasma membrane of monocytes by affecting arrival and recycling. *Immunity* **7**, 861–71.

Kotenko, S.V., Krause, C.D., Izotova, L.S., Pollack, B.P., Wu, W. and Pestka, S. (1997). Identification and functional characterization of a second chain of the interleukin-10 receptor complex. *EMBO J.* **16**, 5894–903.

Kuhn, R., Lohler, J., Rennick, D., Rajewsky, K. and Muller, W. (1993). Interleukin-10-deficient mice develop chronic entero-colitis. *Cell* **75**, 263–74.

Lentsch, A.B., Shanley, T.P., Sarma, V. and Ward, P.A. (1997). *In vivo* suppression of NF-κB and preservation of IκBα by inter-leukin-10 and interleukin-13. *J. Clin. Invest.* **100**, 2443–8.

Levy, Y. and Brouet, J.C. (1994). Interleukin-10 prevents sponta-neous death of germinal center B cells by induction of the Bcl-2 protein. *J. Clin. Invest.* **93**, 424–8.

Moore, K.W., Ogarra, A., Malefyt, R.D., Vieira, P. and Mosmann, T.R. (1993). Interleukin-10. *Annu. Rev. Immunol.* **11**, 165–90.

Pretolani, M. and Goldman, M. (1997). IL-10: a potential therapy for allergic inflammation? *Immunol. Today.* **18**, 277–80.

Salazaronfray, F., Charo, J., Petersson, M., Freland, S., Noffz, G., Qin, Z.H., Blankenstein, T. *et al.* (1997). Down-regulation of the expression and function of the transporter associated with antigen processing in murine tumor cell lines expressing IL-10. *J. Immunol.* **159**, 3195–202.

Vieira, P., de Waal Malefyt, R., Dang, M.N., Johnson, K.E., Kastelein, R., Fiorentino, D.F. *et al.* (1991). Isolation and expres-sion of human cytokine synthesis inhibitory factor cDNA clones: homology to Epstein–Barr virus open reading frame BCRFI. *Proc. Natl Acad. Sci. USA* **88**, 1172–6.

Walter, M.R. and Nagabhushan, T.L. (1995). Crystal structure of interleukin 10 reveals an interferon gamma-like fold. *Biochemistry* **34**, 12118–25.

Wang, P., Wu, P., Siegel, M.I., Egan, R.W. and Billah, M.M. (1995). Interleukin (IL)-10 inhibits nuclear factor kappa B (NFκB) activation in human monocytes — IL-10 and IL-4 suppress cytokine synthesis by different mechanisms. *J. Biol. Chem.* **270**, 9558–63.

IV

The tumor necrosis factor (TNF) family and related molecules

David Wallach*, Jacek Bigda and Hartmut Engelmann

1. Common features of the family members

1.1 General

The tumor necrosis factor (TNF) ligand and receptor families are defined by the molecular structures that enable the ligands and receptors to recognize one another. All members of the TNF ligand family whose receptors have been identified so far, namely TNFα, lymphotoxins α and β (LTα and LTβ), LIGHT, CD95-ligand (CD95-L), TNF-related apoptosis-inducing ligand (TRAIL, or Apo-2-L), CD40-L, CD27-L, CD30-L, 4-1BB-L, OX-40-L, TWEAK and TRANCE (RANK-L) bind to members of the TNF/nerve growth factor (NGF) receptor family. Conversely, almost all the members of the TNF/NGF receptor family whose ligands have been identified, including CD120a (the p55 TNF receptor), CD120b (the p75 TNF receptor), herpes virus entry mediator (HVEM; also called ATAR and TR2), CD95 (Fas/Apo-1; TRAIL-R1 (DR4), TRAIL-R2 (DRS), TRAIL-R3 (DR6/TRID/DcR1), TRAIL-R4, Apo-3/WSL-1/DR3/TRAMP/LARD, CD40, CD27, CD30, 4-1BB (ILA), OX-40, OPG (OCIF), and RANK, bind to known members of the TNF ligand family, four receptors, CAR1, GITR, TACI, and Crinkly-4, and one ligand, APRIL, are still 'orphans', i.e. they have no known ligand or receptor. The low-affinity NGF receptor is the only known receptor which, even though its ligand-binding moiety is homologous to those of the TNF receptors, binds to ligands (NGF and other neurotrophins) that are structurally unrelated to TNF (see Figure 1).

Two novel receptors and one novel ligand of the TNF families on whose function there is no knowledge yet have just recently been described (see web site www.gene.ucl.ac.uk/users/hester/tnftop.himl for further details)

1.2 Occurrence of the ligands and receptors

The patterns of cellular expression of the various ligand and receptor family members, share some common characteristics. Firstly, almost all ligands of the TNF family (except TRAIL/Apo-2-L and TWEAK) are produced mainly by cells of hematopoietic origin, though they can also be produced at low levels by other cells. Some of the receptors of the TNF/NGF family are likewise expressed mainly in hematopoietic cells. Most members of both ligand and receptor families are expressed in T lymphocytes, though not necessarily at the same stages of activation and differentiation of these cells. Several of them are also expressed in B lymphocytes. In addition, several receptors of the TNF/NGF family are also expressed in non-hematopoietic cells. The most prevalent of these is the p55 TNF receptor (CD120a), which is expressed in practically all cell types.

Secondly, except in the case of TRAIL/Apo-2-L and TWEAK, which are expressed widely in a constitutive manner, expression of ligands of the TNF family usually depends on specific stimulation of their producing cells and occurs mostly in a transient fashion. Their inducing agents are mostly immune-related compounds, such as pathogen components, other antigens and cytokines. Constitutive formation of the ligands occurs in only a few cases. TNFα, for example, is formed constitutively in some cells of the thymus. As to the receptors, their production in lymphocytes depends, like

* Corresponding author.

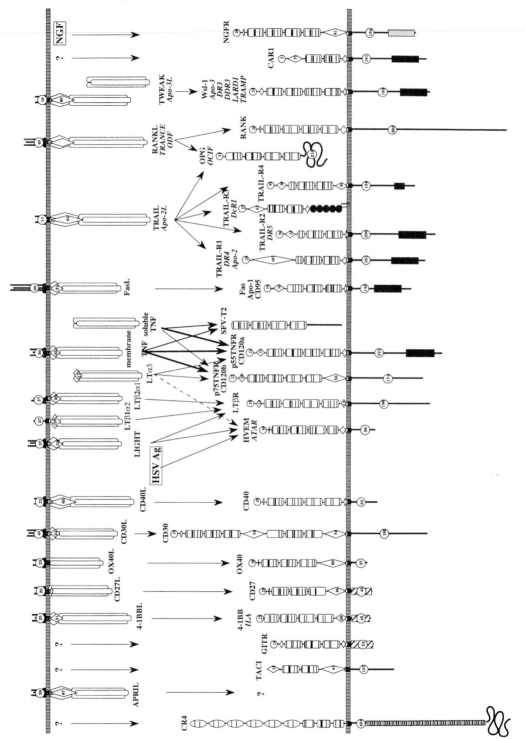

■ **Figure 1.** Diagrammatic representation of the known members of the TNF ligand and receptor families. Horizontal lines in the cysteine-rich domains symbolize conserved cysteine residues. Open circles indicate leader sequences in receptor molecules or lengths in amino acids. 'Spacer'-regions are indicated by diamonds. Death Domains (DD) are shown as black (or gray in the case of the NGFR) boxes. The GITR homology domain is symbolized by a hatched box. The hatched domain in CR4 (Crinkly-4) depicts a kinase-related structure.

that of their ligands, on cell activation. Resting lymphocytes do not express the ligand and express only few of the receptors (e.g. HVEM). They can be induced to express both by activation. Yet, in contrast to ligand expression, which is mostly transient, expression of the receptors in activated lymphocytes is prolonged. Receptors that occur in non-lymphoid cells are often found to be constitutively produced by these cells.

1.3 Common structural features of the TNF ligand family

Members of the TNF ligand family show sequence similarity in their receptor-binding regions. Overall, this similarity is moderate (as low as 12%), but in certain parts of the sequence, it is rather high, to the extent that it was possible for new family members to be cloned by applying a sequence consensus to search data banks.

All family members are formed as type II transmembrane proteins in which the receptor-binding moiety is located in the C-terminal part of the extracellular domain. With the exception of LTα, which is processed shortly after synthesis, they can therefore act as cell-surface proteins that bind and activate receptors on adjacent cells (juxtacrine signaling) or even on the same cell (autocrine signaling). At least some of these ligands can undergo proteolytic processing, resulting in the production of a soluble form of the receptor-binding portion of the ligand which can act remotely from its producing cells (paracrine and endocrine signaling). The proteolytic processing of the ligands occurs at a region that links the receptor-binding and transmembrane parts of these molecules (the 'spacer' region). In the cases of TNFα and CD95-L, the protease involved seems to be a metalloproteinase. An interesting feature of the cell-associated forms of the ligand molecules is the rather high interspecies sequence conservation within the intracellular domains of most of these molecules (except for TRAIL/APO-2 and 4-1BBL), which suggests that this region has functional significance. There are indications that these domains may indeed play some functional roles, perhaps retrograde signaling (upon binding of the ligands to the receptors) or control of the placing of the cell-associated ligand molecules at the cell surface.

Crystallographic fine structural analysis has been performed for only three of the ligands, namely TNFα, LTα and CD40-L. The results indicated that the soluble forms of these molecules occur as trimers. It further showed that in each ligand protomer the polypeptide chain folds in the form of two packed groups of eight antiparallel sheets arranged in β-jellyroll topology, as in some viral capsid proteins. Comparison of these structural analyses with the results of sequence alignment of the different members of the ligand family reveals that sequence conservation among members is highest at the sites that correspond to the protomer interface-regions of TNFα, LTα and CD40-L and to the cores of these molecules. This conservation, as well as molecular modeling of the ligands, indicates that other family members also have the trimeric structure found in TNF.

1.4 Common structural features of the TNF/NGF receptor family

1.4.1 The ligand-binding moiety in the extracellular domain and its interaction with the ligand

All members of the TNF/NGF family bind their ligands via a shared sequence motif in the extracellular domain. This motif (the cysteine-rich domain) is comprised of a repetitive unit of about 40 amino acid residues (the Cys motif), which contains several cysteines (usually six per motif unit) and certain other residues at conserved sites. The number of repeats varies between different receptors, from two to six. Some of the receptors (e.g. CD27, CD30, OX40 and 4-1BB) also contain, in addition to full-length copies of the motif, a copy that is half the length.

Only one of the TNF/NGF receptor family members, CD120a, has so far been subjected to crystallographic fine structural analysis. As described in detail in Section 2.4.4, the results indicated that each trimeric TNF molecule can bind three receptor molecules. This mode of binding provides a mechanism by which the ligand can induce association of two or three receptor molecules. Such induction is consistent with several indications that the principal, if not the only, mechanism by which TNF triggers signaling is by inducing juxtaposition of several receptor molecules. This seems to be the case with all receptors of the TNF/NGF family, as indicated by the ability of antibodies against their extracellular domains to trigger their signaling activity, in correlation with their ability to cross-link the receptor molecules.

However, in the absence of direct information about the structure of complexes formed by the binding of other members of the receptor family to their ligands, we cannot exclude the possibility that some of these complexes differ from those formed by the TNF receptors. The ability of the low-affinity NGF receptor to bind and be activated by the neurotrophins, and of HVEM/ATAR/TR2 and CAR1 to bind viral proteins (and, at least in the case of CAR1, be activated by them), clearly illustrates that the cysteine-rich domain can participate in the recognition of more than one kind of molecular structure. Also worth noting in this respect are the clear structural differences among the ligand-binding domains of the different members of this family. Variations are found not only in the number of Cys-motifs comprising these domains but also in their structure, most clearly manifested in differences in the number and placing of the cysteine residues within them. Moreover, there seems to be variation in the state of aggregation of the receptors. CD27 has been found to occur in two aggregation states: as monomers in activated T lymphocytes, and as disulfide-linked dimers in resting cells. Osteoprotegerin occurs primarily in the form of such dimers, and there is evidence that CD40 and CD120b also occur, in part, in cysteine-linked dimeric forms, while the other receptors apparently occur only as individual molecules. How these structural variations affect the interaction of the various members of the TNF family with their receptors remains to be clarified.

1.4.2 The 'spacer' region in the extracellular domain

The cysteine-rich domains are linked to the transmembrane domains of the receptors by spacer regions of varying length. As discussed below, in several receptors this region may be cleaved, resulting in the formation of soluble forms of the receptors. Comparison of the amino acid sequences in the vicinity of the cleavage sites in the different receptors did not disclose any common sequence feature that could account for their vulnerability to the protease(s) that mediate the shedding. There is, however, one common feature shared by amino acid sequences in the spacers of several of the family members, namely, an abundance of proline, serine and threonine residues, which most probably serve as target sites for O-linked glycosylation.

1.4.3 The receptors' transmembrane domains

It is currently believed that the most critical event in the initiation of signaling by the receptors of the TNF/NGF family is an induced association between the intracellular domains of two or more receptors. This association is thought to occur in a passive manner, as a consequence of juxtaposition of receptor molecules upon their binding to the trimeric ligand molecule. Such passive association could

imply that the transmembrane domain has a passive role in the initiation of signaling. It seems, however, that some features of the signaling process do depend on subtle structural features of this region, since its replacement with the transmembrane domain of an unrelated receptor abolishes signaling. Moreover, in at least some members of this receptor family the sequence of amino acids in the transmembrane domain is highly conserved through evolution, to an even higher degree than the sequence in other parts of the receptor.

1.4.4 Soluble forms of the receptors of the TNF/NGF family and the corresponding 'viroceptors'

Many, perhaps all, of the receptors of the TNF/NGF family occur in soluble forms, and at least one of them, osteoprotegerin, is known to be produced only as a soluble receptor. The structure of these soluble forms corresponds to that of the cytokine-binding, cysteine-rich, ligand-binding module in the corresponding cell-surface receptors. They also contain a portion of the spacer region. Some of these soluble receptors, e.g. those of TNF, are produced by proteolytic cleavage of the cell-surface forms. As in the case of the shedding of the TNF ligands, proteolytic cleavage of the receptors appears to be mediated by some cell-bound metalloproteinase. Other soluble receptors, e.g. those of CD95, have specific transcripts, which are derived by alternative splicing from the genes encoding the cell-surface receptors. In the soluble receptors that are derived proteolytically, the amino acid sequence in the part of the spacer region found in the soluble form is identical to that in the corresponding region within the cell-surface receptor. In the soluble receptors produced by alternative splicing mechanisms, however, the C-terminal sequence can be unique, reflecting a frameshift in reading that has resulted from the alternative splicing.

Most of the known soluble forms of the receptors of the TNF/NGF family can block the function of their ligands by competing for them with the cell-surface receptors. At least some, however, can also enhance the function of their ligands by stabilizing them, preventing the dissociation of their trimeric forms to inactive monomers.

Some viruses of the poxvirus family have exploited the inhibitory potential of the soluble receptors of the TNF family as a means of defense against antiviral activities of the ligands. They produce soluble forms of CD120b and CD30 ('viroceptors'), employing DNA sequences which they have probably usurped from host organisms.

More detailed descriptions of the structure and function of the cellularly and virally encoded soluble forms of the different receptors are provided under their specific headings.

1.4.5 The intracellular domains

In contrast to the extensive structural similarity among the extracellular domains of all the receptors of the TNF/NGF family, the structural similarity of their intracellular domains is limited. So far, only two shared structural domains have been identified in them: (i) the death domain (DD) motif, shared by CD120a, CD95, the TRAIL receptors, CAR-1, WSL-1/DR3/TRAMP/LARD and the low-affinity NGF receptor, and (ii) a sequence motif, as yet unnamed, shared by the intracellular domains of CD27, 4-1BB/ILA and GITR.

It appears, however, that the structural similarity between the intracellular domains of the receptors is greater than might be expected from a cursory inspection of their amino acid sequences. Some members with no apparent sequence similarity between their intracellular domains (CD120b, CD30, CD40 and the LTβ-R) were found to bind the same adaptor protein, TNF receptor-associated factor 2 (TRAF2), and some are capable of binding adaptor proteins related to TRAF2 (TRAF3, TRAF5, TRAF6). Mutational studies suggest that the motif PXQXT/D, found in the intracellular domains of CD30 and CD40, is partially responsible for their ability to bind TRAF2 and TRAF3.

1.5 Structure and chromosomal localization of the genes encoding the TNF receptor and ligand family members

The genes of several of the receptors of the TNF/NGF family are located in clusters. The human CD120a, LTβ-R and CD27 genes are located at 12p13, those of CD120b, CD30, 4-1BB, OX40, Apo-3/WSL-1/DR3/TRAMP/LARD and HVEM/ATAR/TR2 at 1p36 and the genes of all four known TRAIL/Apo-2-L receptors at 8p22-21 (about 49 cM from the telomere). The genes of several members of the TNF-ligand family are also located in clusters: the CD95-L and OX40-L genes are at 1q25, the TNFα, LTα and LTβ genes are within the major histocompatibility complex (MHC) locus in human 6, and the CD27-L and 4-1BB-L genes are at 19p13.3. Similar gene clustering can be observed in syntenic sites within the mouse genome (Table 1). The occurrence of TNF-ligand genes within the MHC locus may contribute to

coordination of the pro-inflammatory and immune regulatory functions of these cytokines with the immune defense functions controlled by other genes in this locus. Likewise, the clustering of other genes of the TNF receptor and ligand families may contribute to coordination of the expression of these functionally related receptors and ligands with each other, as well as with other nearby genes.

Irrespective of its functional significance, the mechanisms accounting for this clustering are probably the initial duplication events by which the ligand and receptor families evolved from their ancestral genes. Traces of these evolutionary processes can be detected in the fine structure of the genes. Similarities observed, for example, between the patterns of exon–intron boundaries

∎ **Table 1.** Chromosomal localization of the genes for the members of the TFN-ligand and receptor families

	Human	Mouse
RECEPTORS		
CD120a	12p13	6
LTβ-R	12p13	
CD27	12p13	
CD30	1p36	
CD120b	1p36.2-3	4 distal
4-1BB	1p36	4 distal
OX40	1p36	4 distal
DR3	1p36	
HVEM	1p36.22-36.3	
CD40	20q11-13	2 distal
CD95	10q24.1	19 distal
TRAIL-R1	8p22-21	
TRAIL-R2	8p22-21	
TRAIL-R3	8p22-21	
TRAIL-R4	8p22-21	
OPG	8q24	
RANK	18q22.1	
NGF-R	17q21-22	
LIGANDS		
CD95-L	1q25	1 distal
OX40-L	1q25	1 distal
TNFα	6 (MHC)	17 (MHC)
LTα	6 (MHC)	17 (MHC)
LTβ	6 (MHC)	17 (MHC)
TWEAK	17p38	
CD27-L	19p13.3	
4-1BB-L	19p13.3	17 central
CD30-L	9q33	4 central
CD40-L	Xq26-27	X proximal
TRAIL	3q26.1-2	
RANK-L	13q14	14
LIGHT	16	

within the cysteine-rich domains in the extracellular domains of CD120a, CD95, CD40, CD27, OX40 and 4-1BB, and their lack of similarities to the pattern in the low-affinity NGF receptor, are consistent with evolution of the former receptors from a common ancestral gene distinct from that of the latter receptor. The cysteine-rich domains themselves most probably evolved by duplication of an ancestral single motif. In some cases, analysis of the fine structure of the receptor genes indicates that the reverse process, namely deletion of part of the Cys-motifs comprising the cysteine-rich domains, has also occurred during evolution.

Certain functional elements within proteins are encoded by distinct exons, probably corresponding to distinct evolutionary elements that have fused in the creation of these proteins. This is the case with the DD. Both in CD120a and in CD95, as well as in the three known DD-containing adaptor proteins: mediator of receptor-induced toxicity/Fas-associated death domain protein(MORT1/FADD), TNF receptor-associated death domain protein (TRADD) and receptor interacting protein (RIP)) which bind to these receptors, the DDs are encoded by distinct exons . This is not so, however, for the ligand-binding cysteine-rich domain that defines the TNF/NGF family. Although crystallographic analysis of the region indicates that the individual Cys-motifs act as distinct structural units, the intron–exon borders in this region are invariably not between the motifs but within them. This apparent discrepancy might indicate that the actual evolutionary and functional unit from which the ligand-binding region evolved was not the Cys-repeat but a smaller unit containing only three of the six cysteines found in each. Alternatively, the discrepancy may reflect integration of introns into the Cys-motifs at a later time in evolution than the duplication events that resulted in formation of the repetitive cysteine-rich domains.

1.6 Common and distinct effects of the TNF ligand and receptor families

Most of the known effects of the TNF and NGF receptor family fall into four groups:

(1) *Co-mitogenic and activation effects, primarily on cells of the hematopoietic system.* These kinds of activities are shared by all TNF/NGF receptor family members. Notably, almost all of these receptors provide co-mitogenic stimuli in T lymphocytes. Several of them also provide co-mitogenic signals in other cells, including B lymphocytes, mononuclear phagocytes, fibroblasts and others.

(2) *Induction of cell death.* Initially, death induction appeared to be restricted to only two of the receptors, CD120a and CD95. Recently, however, additional receptors capable of triggering cytocidal effects, namely TRAIL-R1/DR4, TRAIL-R2/DR5, Apo-3/WSL-1/DR3/TRAMP/LARD and CAR1, were discovered. The LTβ-R, CD40, CD30, CD27 and the low-affinity NGF receptor were also found to trigger such effects, although in fewer kinds of target cells and with lower apparent effectiveness.

(3) *Induction of innate defense mechanisms, primarily those that participate in the inflammatory process.* The ability to induce such activities is most pronounced in two of the receptors, CD120a and (to a lesser degree) CD120b. There is some evidence, however, that such activities can also be induced to a limited extent by other receptors of the family, e.g. CD40.

(4) *Immune response and its organogenesis.* The ability of almost all receptors of the TNF/NGF family to provide co-mitogenic stimuli in lymphocytes suggests that they all contribute to the development of the immune response. Prolonged abolition of function, by the use of antibodies or soluble receptors that bind to the ligands or by targeted disruption of the ligand or receptor genes, was indeed found to result in severe impairment of immune functions in several cases. This is most clearly demonstrated in the case of CD40/CD40-L interaction, which is critical to several aspects of the development of the humoral immune response. Both in mice and in humans, mutations causing loss of function in the genes of CD40 or its ligand result in severe immune deficiency. Disruption of the genes of other ligands and receptors of the TNF families also affects the immune response, though in a more restricted manner. CD30-deficient mice display a gross defect in negative selection of thymocytes. Targeted disruption of the TNFα, LTα, CD120a, LTβ, CD40L or the CD40 gene results in defective or absent germinal center function. Consequently, isotype switching is severly impaired; for example, both LTα and LTβ deficient mice exhibit greatly reduced ability to produce IgA.

In addition to their functional implications, these activities of the TNF ligand and receptor families are also reflected in certain morphological aspects of the immune system. Deficiency in any one of several of these ligands and receptors, including TNFα, LTα, LTβ, CD40-L, CD40, OX-40, CD120a and the LTβ-R, results in disturbance of the micro-architecture within the lymphoid organs, manifesting abnormalities of some cellular interactions that are required for the normal immune response in these organs. Conversely, ectopic expression of LTα in mice was found to result in 'neo-organogenesis' of lymph nodes. Moreover, TNFα expression has an inducing role in the development of granuloma, an immune organ that develops on an *ad hoc* basis to entrap and destroy pathogens.

Analysis of mice with targeted disruption of LTα gave the first indication that the TNF ligand and receptor families also play a role in immune organ morphogenesis throughout ontogeny. Lack of peripheral lymphoid organs in these mice was interpreted as suggesting a critical role of the LTβ-R in the ontogenic development of these organs, a notion that was later confirmed by the more direct measure of blocking the function of this receptor and by knockout of its gene. It was further confirmed by a study of mice with targeted disruption of the LTβ gene. The latter study also pointed to a role for LTα, distinct from that of LTβ, in lymphoid organogenesis. Whether and to what extent other receptors of the TNF/NGF receptor family participate in the development of the lymphoid system *in utero* remains to be examined.

1.7 Common molecular activities of the TNF/NGF receptor family

The similarities in effects of the receptors of the TNF/NGF family are not necessarily correlated with the extent of similarity between their signaling activities. Thus, although almost all of these receptors can provide co-mitogenic stimuli in T cells, it may well turn out that the various receptors induce this common activity through different mechanisms. Conversely, although the ability to trigger pro-inflammatory functions appears to be largely restricted to the TNF receptors, at least one of the principal molecular events involved, activation of the transcription factor NF-κB, is shared by most if not all receptors of the family.

Emerging knowledge of the signaling mechanisms employed by these receptors suggests that their modes of action are more closely related than might appear from their known effects. Extensive interactions of the various receptors through their binding of common signaling proteins, as well

through association of different signaling molecules that bind to the various receptors, seem to allow rather extensive cross-talk between their signaling activities. It may therefore turn out that, as with some other receptor families (e.g. that of the epidermal growth factor (EGF)-related factors), many of the activities affected by the receptors of the TNF/NGF family are shared by all of these receptors. The observed differences in their functions may thus largely reflect differences not in the kinds of molecular activities that they affect, but rather in the relative effectiveness of their activation of the different functions, combined with differences in the patterns of cellular expression of the receptors.

2. The TNFs and their receptors

2.1 General description

TNFα and LTα were the first members of the TNF ligand family to be discovered. Like some other pleiotropic cytokines they were in fact 'discovered' more than once, and each time were given a different name, which was related to the kind of activity by which they were detected. The first discovery was documented in the 1940s, when Valley Menkin described the production of a mediator ('necrosin') that perpetuates tissue damage at sites of tissue injury. They were rediscovered in 1967, and described as factors with *in vitro* cytotoxic activity ('lymphotoxins') produced upon stimulation of lymphocytes. The name 'tumor necrosis factor' (TNF) was coined in 1975 in a study describing this cytokine as a serum factor, formed when bacillus Calmette-Guérin (BCG)-infected mice are challenged with endotoxin, and having the ability to cause selective destruction of the vasculature within some transplanted mouse tumor models. In other noteworthy studies TNF was identified as a macrophage-derived factor that affects adipocyte differentiation, raising the possibility that it contributes to the catabolic state and excessive loss of weight (cachexia) observed in certain parasitic diseases. Accordingly, in these studies it was dubbed 'cachectin'. TNF was also identified as a cytotoxic factor ('cytotoxin') that can induce cellular resistance to its own cytocidal effect in a protein synthesis-dependent manner and as a differentiation factor for human myelogenous leukemia cells ('differention-inducing factor', or DIF), formed by mitogen-stimulated human peripheral blood leukocytes.

As mentioned above, although the TNFs and their receptors are often considered as prototypic of their families, they are rather exceptional members in a number of respects:

(1) TNF and TNF receptors relate respectively to a number of ligand and receptor molecules that can interact in different ways (see Figure 1). In that respect they differ from all other members of the ligand and receptor families, except for TRAIL/Apo-2-L, which interacts with a multiplicity of receptors. Four structurally related ligands, TNFα, LTα (or TNFβ), LTβ and LIGHT, and four receptors are currently known to operate in this response system. Two of the receptors have received CD designations (though other names are also still in use): one of them is designated CD120a (or the type I receptor, p55 receptor or p60 receptor) and the other CD120b (or the type II receptor, p75 receptor or p80 receptor). A third is known as the LTβ receptor (LTβ-R), the type III TNF receptor or TNFR-RP. All of the above three receptors recognize interfaces between neighboring monomers in the trimeric TNF/LT molecule. CD120a and CD120b are able to bind the TNFα/TNFα and LTα/LTα interfaces and can thus be activated by both TNFα and LTα. The cell-bound form of TNFα is more effective than the soluble form in binding and activating CD120b. The LTβ-R recognizes the interface formed between LTα and LTβ and is therefore activated by heterotrimers (LTα1/β2 and LTα2/β1, see below) of these two molecules. LTα also binds to HVEM/ATAR/TR2, and so does LIGHT. LIGHT also binds to the LTβ-R. These recently identified interactions are described in Section 3 below.

(2) The range of cell types expressing CD120a (the receptor that appears to play the major role in signaling for most of the TNF effects) is apparently much wider than that expressing any other known TNF receptor family member.

(3) The range of cellular activities known to be regulated by the TNFs is much wider than that of any other member of the TNF ligand family. Like almost all other members of the family they provide co-mitogenic signals and, like most of them, they can trigger cell death. They are also able to activate a variety of innate defense mechanisms, particularly those involved in inflammation. Moreover, the LTβ-R and CD120a provide signals that are necessary for the embryonal development of the lymphoid organs (see below).

TNF ligands and receptors have been isolated and cloned by employing a variety of different methods, reflecting the progression of the approaches to the cloning of new genes throughout the last decade. TNFα and LTα were isolated (and then cloned) by chromatographic procedures by tracking their cytotoxic, cell death-regulating and adipocyte function-regulating activities. The receptors to which they bind, CD120a and CD120b, were only isolated 5 years later, either chromatographically or by affinity purification on TNF columns in studies tracking the inhibition of TNF function by the soluble forms of these receptors and the TNF-binding ability of the cell-surface forms. LTβ was isolated and cloned on the basis of its ability to associate with LTα. LTβ-R, although identified by its binding properties, was initially cloned by a 'reverse genetics' approach, in which new coding regions were sought at the vicinity of the CD120a gene (in human). Confirmation that this novel gene encodes the LTβ-R was, however, obtained only in later studies, in which the receptor was cloned by other procedures.

2.2 Cellular origins of the TNFs and their receptors

LTα is known to be produced only by lymphocytes: T lymphocytes, primarily those manifesting the T-helper cell 1 (Th1) pattern of cytokine formation and B lymphocytes at certain stages of their differentiation, and by natural killer (NK) cells. LTβ is expressed by the same cells, though the regulation of its expression seems to differ. In T-cell clones, expression of LTα (as well as of TNFα) is stimulus-dependent while that of the LTβ mRNA was found to be constitutive.

The main producers of TNFα are activated mononuclear phagocytes, T lymphocytes (Th1 and Th2 exhibiting similar effectiveness) and B lymphocytes. However, as opposed to the restricted formation sites of the LTs, TNFα can also be produced by a wide range of other cells, though in lower amounts than those produced by mononuclear phagocytes. Its producer cells include fibroblasts, keratinocytes, smooth and cardiac muscle cells, astrocytes, microglial cells and mast cells. The latter cells are the only ones known to accumulate the soluble form of TNFα within secretory granules and to secrete it exocytotically in response to specific inducers. TNF production is in most cases highly stimulus-dependent and can be observed only upon infection or injury. A variety of pathogen components (e.g. LPS, double-stranded RNA) and cytokines (e.g. TNF itself, interleukin 1 (IL-1) and granulocyte–macrophage colony-stimulating factor (GM-CSF)) can induce its synthesis. In adipocytes and in some thymus cells, however, TNF is produced continuously.

Our knowledge of the sites of TNF receptor expression is mainly based on their occurrence in cultured cells; little has been done so far to assess their expression *in vivo*. As opposed to the rather restricted pattern of the cells that produce the TNFs, expression of their receptors is observed in a wide range of different cell types, though in varying proportions on the different cells. As mentioned above, CD120a is constitutively expressed in all cell types; indeed, its promoter displays housekeeping features. In contrast, the expression of CD120b is quite restricted. It is constitutively expressed in certain hematopoietic cells, such as lymphocytes and mononuclear phagocytes. Expression of this receptor in other cells is low, but can be induced by certain stimuli (e.g. by TNF itself and by IL-1). LTβ-R mRNA is expressed by cell lines of monocytic and epithelial origin and is constitutively expressed *in vivo* in certain visceral and lymphoid tissues.

2.3 Functions of the TNFs

2.3.1 *In vivo* data

TNF is one of the most pleiotropic of the known cytokines. It affects all cells, altering their growth, differentiation and survival patterns and inducing a wide range of functional changes, some of them practically opposite in nature to others. Gaining a valid notion of the physiological and pathophysiological implications of such complex function may not be possible without direct assessment of their *in vivo* consequences. Studies in which the *in vivo* function of the TNFs or their receptors has been abolished by gene-targeted disruption, by injection of anti-TNF antibodies or by application of soluble receptor preparations indicate the following three principal roles for these cytokines:

(1) *a role of the TNFs in defense against pathogens, most notably against some intracellular pathogens*, manifested for example in the markedly increased vulnerability of mice deficient in CD120a or TNF to infection by the facultative intracellular pathogen *Listeria monocytogenes*.

(2) *a major role of TNFα in immunopathology, particularly in the pathology associated with inflammation*. This is manifested, for example, in the response of CD120a- or CD120b-deficient mice to bacterial lipopolysaccharide (LPS) injection. Mice deficient in either of these receptors exhibit decreased vulnerability to the toxic effects of LPS, though under different experimental conditions of induction of these effects, indicating involvement of the two receptors in different aspects of TNF function.

(3) *role of the LTβ-R and CD120a in the organogenesis of lymphoid organs*. Peripheral lymphoid organs are absent in mice with targeted disruption of LTα. Most of these organs are also missing in mice with targeted distruption of LTβ, as well as in mice exposed *in utero* to decoy LTβ-R. Conversely, ectopic expression of LTα results in *de novo* formation of organized lymphoid tissue in adult mice. Mice with targeted disruption of the LT-βR gene as well as mice made deficient in both LT-β and CD120a show greater deficiency in lymphoid organs than that found in mice deficient only in LT-β. These findings indicate that signaling by the LTβ-R and CD120a, and perhaps also by an unidentified additional ligand to which the LTβR binds (LIGHT?), play crucial and somewhat distinct roles in the ontogenic development of these organs. Signaling by these receptors is also required for proper cellular interactions within the lymphoid organs. Their deficiency results in lack of primary B-cell follicles, as well as the lack of organized follicular dendritic cell networks and germinal centers in this organ.

2.3.2 *In vitro* functions

(a) Functions of TNFα

(i) *Effects that appear to contribute to defense at the level of the individual cell.* Many of the effects of TNF seem to be directed towards providing defense against intracellular pathogens. Several of these effects provide protection at the level of the afflicted cell. The most extensively studied *in vitro* effect of TNF, namely cytotoxicity, probably serves such a function. This cytocidal effect appears to be exerted selectively against virus-infected cells and may thus help to arrest spreading of the virus. In addition, TNF has a number of other effects that can restrain the growth of various pathogens including viruses and bacteria; for example, it enhances the generation of reactive oxygen and nitrogen intermediates in macrophages. Many of the effects of this kind are synergic with those of interferon gamma (IFNγ).

(ii) Effects that appear to contribute to defense at the level of the afflicted tissue. TNF plays a major role in the induction of inflammation, a coordinated multicellular change involved in defense against injury and various pathogens. Its pro-inflammatory and other effects often closely overlap and synergize with the effects of another pro-inflammatory cytokine, IL-1. Most prominent are the effects of TNF that endow endothelial cells with a pro-thrombic phenotype. The TNF-induced changes in expression of cell-surface and secretory proteins by the endothelial cells promote leukocyte extravasion and coagulation. Some of the proteins induced by TNF are themselves cytokines, whose multiple actions further extend the range of pro-inflammatory effects. The effects of TNF on leukocyte migration, for example, are largely mediated through induction of chemokines like IL-8.

TNF makes a particular contribution to the various tissue-damaging manifestations of the inflammatory process. Besides its direct cytocidal effect, TNF can harm tissues by inducing damage to the vasculature and dissolution of extracellular matrices, cartilage and bone. Some of these TNF-induced destructive effects occur preferentially in certain tumor models as well in certain solid tumors of humans.

The TNF-induced coagulation and damage of the vasculature at the site of inflammation presumably act to restrict the spreading of pathogens that have elicited the inflammation. Pathogen spread can also be restricted by the formation of granuloma walls around the afflicted site by cells of the mononuclear phagocyte lineage (epitheloid cells). TNF plays a central role in the induction of granuloma.

As well as damaging tissues, TNF induces effects that promote recovery from the damage. While inducing the death of diseased cells it also stimulates the growth of normal fibroblasts, which play an important role in scar formation and in tissue fibrosis. TNF effectively destroys the vasculature within tumors and granulation tissues, yet it also acts as a potent angiogenic factor. It thus seems that, in general, the damage caused by TNF action is induced for the sake of tissue remodeling.

(iii) Effects that appear to contribute to defense at the level of the whole organism. Wide-ranging activation in the body of those TNF effects that are observed locally in inflammation can be extremely harmful. Such widespread damage, demonstrated for example in the experimentally induced systemic Schwartzman phenomenon model, is believed to account for various manifestations of septic shock, a situation in which there is excessive formation of TNF. Apparently, TNF has been destined to act primarily as a local inducing agent, but it also enhances defense through some systemic effects, other than those participating in local inflammation. It affects the performance of the heart, causing reduced contractility. Like several other pleiotropic cytokines, e.g. IL-1 and the interferons, it also induces changes in control centers in the brain that allow adjustment to the diseased state. These changes include elevation of the temperature set-point in the hypothalamic thermoregulatory control center (fever), loss of appetite (anorexia), enhanced slow-wave sleep (somnogenic effect) and others. It also contributes to the acute-phase response by affecting, directly as well as through the effects of TNF-induced IL-1 and IL-6, biosynthesis of the acute-phase proteins in the liver.

TNF has a strong inhibitory effect on the *in vitro* expression of several adipocyte proteins associated with the differentiated phenotype of these cells (e.g. glycerophosphate dehydrogenase). It also has a marked catabolic effect when produced by certain implanted tumors in mice. In addition, as mentioned above, it also exerts an anorectic effect. Because of these effects it was thought to contribute to the excessive weight loss observed in certain diseases, and was accordingly dubbed 'cachectin'. It should, however, be noted that there is as yet no information to support the notion that such activities of TNF are relevant to any disease of humans. In fact, it may well turn out that TNF acts in quite the opposite manner in disease. As mentioned above, TNF is produced spontaneously by differentiated adipocytes and therefore may be assumed to be associated with obesity and to contribute to its pathology. It markedly inhibits insulin receptor signaling, apparently via serine phosphorylation in the insulin receptor-associated adaptor protein, IRS1. This effect seems to make an important contribution to the insulin resistance often associated with obesity.

(iv) Effects on 'professional' immune cells. TNF has a wide range of effects on leukocytes. It serves as an autocrine activation and differentiation factor for mononuclear phagocytes and augments its own synthesis by these cells. It also potentiates the activation of granulocytes, as manifested for example by the effectiveness of their respiratory burst, adherence to endothelial cells and degranulation. It acts as an autocrine co-mitogenic signal for T lymphocytes, augmenting the synthesis of interleukin 2 (IL-2), and also as an autocrine co-mitogenic factor for B lymphocytes. It also

increases the expression of the IL-2 receptor in NK and LAK cells. The hematopoietic process manifests both inhibitory and enhancing effects of TNF. The inhibitory effects reflect direct suppression of the growth of bone-marrow progenitors and stem cells. On the other hand, the enhancement of hematopoiesis by TNF occurs in an indirect manner, by induction of the colony-stimulating factors GM-CSF, G-CSF and CSF-1. The nature of the cellular effects of TNFα that underlie its regulation of the development and organization of the splenic follicular architecture is not known.

(b) Functions of LTα and LTβ

As mentioned above, LTα (a secreted soluble protein) and LTβ (known to occur only as a cell-bound protein) associate as cell-bound heterotrimers, LTα1/β2 and LTα2/β1, the former being prevalent in human cells. LTα also occurs as a soluble homotrimer, LTα3. LTβ is apparently transported to the cell surface only when associated with LTα.

Our knowledge of the functions of LTα and LTβ is very limited compared with that of TNFα. The *in vitro* effects observed with trimeric human LTα are qualitatively almost indistinguishable from those of TNFα. Quantitatively, however, there are some clear differences, which seem to stem from differences in the stability of LTα and TNFα as well as in their modes of interaction with CD120a and CD120b.

The LTα2/β1 heterotrimer contains one LTα/LTα interface, which should allow it to bind to CD120a and to CD120b. However, binding to a single receptor molecule is unlikely to result in signaling induction. The heteromeric LT forms indeed exhibit very little pro-inflammatory activity. In addition, LTα2/β1 contains two LTα/LTβ interfaces through which it can bind to the LTβ-R. The LTα1/β2 heterotrimer also contains two such interfaces (as well as one LTβ/LTβ interface, for which no receptor has so far been identified). *In vitro* studies of the LTβ-R disclosed three functional consequences of its triggering: (i) activation of the transcriptional factor NF-κB, (ii) a growth-stimulatory effect on fibroblasts and (iii) a cytocidal effect on some tumor cell lines. Whether these effects or others underlie its *in vivo* role in controlling the morphogenesis of lymphoid organs is not known.

2.4 Structure–function relationships in the TNFs and their receptors

The salient structural features of TNF and its receptors were discussed in part in Section 1. They will be discussed here in somewhat more detail, with particular attention to features that seem to be unique to these specific members of the families.

2.4.1 Structure of the receptor-binding regions in the TNFs and their receptors

Chromatographic characterization performed at an early stage of TNF studies indicated that this molecule occurs and acts in homotrimeric forms. Later crystallographic studies of TNFα and LTα confirmed that these molecules occur as symmetrical trimers. Each of the associated TNFα or LTα monomers forms a structure composed of two packed sheets of eight antiparallel β-strands. This structure bears a strong resemblance to that of coat proteins comprising the icosahedral shells of plant and animal spherical viruses (jelly-roll motif). The three subunits bind edge to face, together forming a triangular conical structure. In crystals, these trimers further associate into higher-order quaternary structures. The residues in the trimer-trimer interfaces of TNF are conserved across a range of mammalian species, suggesting that this higher-order oligomerization plays a functional role, possibly in assisting the clustering of receptor–ligand complexes once they are formed on the cell surface.

2.4.2 The cell-bound forms of TNFα and LTβ

The formation of LTα as a soluble molecule with a typical leader region is exceptional in the TNF family. TNFα, LTα and LTβ, like all other members of this family, are formed as type II transmembrane proteins in which the receptor-binding region constitutes the C-terminal part of the extracellular domain. LTβ is known to exist and act only in this cell-bound form, whereas TNFα also occurs as a soluble molecule which is derived from the transmembrane form by proteolytic cleavage of the spacer region linking the receptor-binding region to the transmembrane region. In the case of LTα such cleavage occurs before exposure of the protein on the cell surface, and the protein is therefore found to act only in a soluble form. As in the case of the proteolytic shedding of the TNF receptors (see below), mutational studies of the spacer region in TNF failed to reveal any sequence motif that dictates its cleavage. The rather short intracellular domains of TNFα (29 residues) and LTβ (16 residues) are highly conserved among different mammalian species, suggesting that they have some functional significance. The nature of this function is still unknown. It is worth noting that the intracellular domain of TNFα is phosphorylated on serine residues, suggesting that the activity of this cytokine is affected by signaling through protein kinases. The presence of a myristyl acylation site in the intracellular domain of TNFα suggests that this domain may help to anchor the protein or to determine its positioning on the cell surface.

Analysis of the cross-linking products of cell-bound TNFα molecules indicated that, like the soluble form of this cytokine, they exist as a trimeric structure. Whether this trimerization involves the interaction of only the receptor-binding moiety in these molecules or also of the intracellular and/or transmembrane domain is unknown.

2.4.3 Mechanism of shedding of TNFα

A metalloproteinase that specifically cleaves the cell-bound form of TNFα and apparently functions as its major processing enzyme was recently isolated and cloned. The enzyme, termed the TNF-alpha-converting enzyme, or TACE, is a transmembrane protein whose extracellular domain has proteolytic activity. It is a member of the adamalysin family, showing notable sequence identity to ADAM, an enzyme implicated in myelin degradation, and to KUZ, a *Drosophila* protein important for neuronal development. Preliminary evidence suggests that TACE may also be involved in the shedding of some other cellular proteins, including the TNF receptors.

2.4.4 Structure of the extracellular domain of the TNF receptors and of the complex it forms with TNFs

The extracellular domains of the TNF receptors are known to be involved in three kinds of molecular interactions: binding to their ligands (which results in triggering of signaling), self-association (so far observed only in crystals of the extracellular domains of CD120a), and interaction with unidentified cell-associated proteases that are responsible for the shedding of these molecules. TNF binding and self-association of the extracellular domain both involve the cysteine-rich module, which in these receptors contain fourfold repetitions of the Cys-motif.

Crystallographic studies of the structure of the cysteine-rich module of CD120a, alone or in a complex with LTα, revealed that the four Cys-repeats in this module are stacked longitudinally to form a bent rod. In each repeat, the cysteine residues form a tandem array of disulfide bonds in the order C1–C2, C3–C5, C4–C6.

Shedding of the receptors involves the spacer region downstream of the cysteine-rich module. In CD120b this region is rather long, comprising 56 residues, and (as in some other receptors of the family) contains clusters of proline, serine and threonine residues that serve as sites of O-linked glycosylation. In contrast, in both CD120a and the LTβ-R the extracellular domains contain rather short spacers of about 15 residues. Detailed mutational analysis of the spacer region in CD120a failed to identify any sequence motif associated with the proteolytic cleavage that results in shedding of the soluble form of the receptor. It did, however, suggest strong dependence of the cleavage process on the conformation of the protein in the vicinity of the cleavage site.

2.4.5 Structural requirements for the interaction of TNF with the TNF receptors and for other interactions of TNF

Mutational and immunological evidence indicates that the 'active site' of the TNF molecule, at least the part involved in cytotoxicity induction, resides near the base of its pyramidal structure, at each side of the groove separating two monomers in the trimeric structure. More detailed mutational analysis of the structural requirement for TNF binding to its receptors revealed, in addition to the region commonly involved in binding to the two receptors, that TNF also contains sites that are specifically involved in its binding to CD120a and CD120b. TNF mutants that preferentially bind to one or other of the receptors could thus be established (see Figure 2). Structural specificity of the binding to the two receptors is also manifested in the inability of human TNF to bind to the mouse CD120b, while it binds to mouse CD120a as effectively as to the human receptor.

Mutational studies of CD120a indicate that its binding to TNF involves all four Cys-motifs. Yet, whereas deletion of the most N-terminal domain abolishes binding, deletion of the most C-terminal domain results only in a decrease in effectiveness of binding, indicating that the role of the latter domain in the binding is secondary. Consistent with this finding, a detailed study of the effect of monoclonal antibodies (mAbs) against different regions in CD120b on its binding to TNF revealed that antibodies which bind to the most C-terminal cysteine loop in this receptor do not decrease its binding to TNFα, and in fact enhance it.

Comparative studies of the ability of different mAbs against CD120a and CD120b to trigger cytotoxicity indicate that the most important determinant of effectiveness is their ability to cross-link the

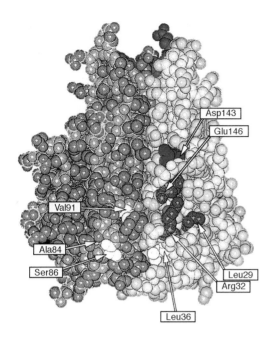

▌ **Figure 2.** Three-dimensional representation of the TNFα molecule. The three subunits are depicted in light, medium and dark gray (mostly hidden by the two subunits in the foreground). Residues involved in receptor binding are marked in white and dark gray; those found to be selectively involved in CD120a (D143) or CD120b (L29, R32 and E146) binding are depicted in dark gray (with the kind permission of Dr W. Fiers).

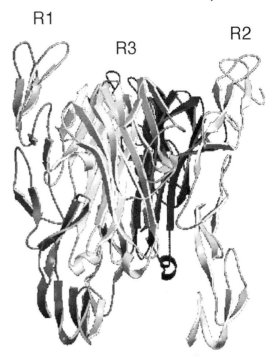

▌ **Figure 3.** Molecular model of the complex of LTα with soluble CD120a (side view as deduced from crystal structure analysis). Shown are the three receptor chains (marked R1, R2 and R3) with the LTα trimer in the center. R3 is almost completely hidden by the three LTα chains. The N-terminal parts of the receptor chains are on the top. The C-termini point towards the bottom. Consequently, the narrow part of the LTα trimer points towards the bottom, in the direction of the cellular membrane (with the kind permission of Dr David W. Banner).

receptor molecules to one other. It was therefore suggested that the TNFs activate their receptors by imposing juxtaposition of several receptor molecules. Consistent with this possibility are the data obtained by analysis of co-crystals of the extracellular domain of CD120a with LTα, showing that each ligand molecule can bind three receptor molecules (Figure 3). As mentioned above, crystallographic studies of the cysteine-rich domain of CD120a alone suggested that this region can also bind to itself. However, whereas the binding of receptor to the ligand results in juxtaposition of the receptors' intracellular domains, the self-association of receptor molecules places their intracellular domains far apart and may thus serve to prevent spontaneous signaling caused by the propensity of the DDs in the intracellular portions of CD120a to self-associate.

Sporadic data suggest that, in addition to the associations described above, TNFα may also participate in other interactions. There is some evidence to suggest that at low pH TNF can interact with phospholipid bilayers, forming pores in them, and that it can also act as a lectin, an activity mapped to a domain which is functionally and spatially distinct from the TNF receptor-binding site. It is doubtful that these interactions play a role in the effect of TNFα on mammalian cells. They seem, however, to contribute to a trypanolytic activity of TNF: the lectin function of TNF facilitates its binding to *Trypanosoma brucei* and its uptake in their digestive organelles, and its direct interaction with the membranes of these organelles (which have an acidic interior) results in lysis of the parasite. There are also some reports showing that certain short peptides derived from the TNF molecule can have similar effects on cells to those of TNF. The relevance of these peptide effects to the *in vivo* function of TNF remains to be clarified.

2.4.6 The intracellular domains of the TNF receptors: functions and structural motifs

(a) Functions of the three receptors

There are wide differences in the amount of knowledge about the functions of the three receptors. Most of the available knowledge concerns the function of CD120a. Triggering of this receptor by antibody cross-linking induces many of the known effects of TNF, including cytotoxicity, prostaglandin synthesis, an antichlamydial effect, co-mitogenic effects in lymphocytes and fibroblasts, an antiviral effect, NO synthase induction and others. In contrast, anti-body cross-linking of CD120b appears to have only limited functional consequences and in most cases the effect observed was the same as that which could also be induced by triggering of CD120a (e.g. cytotoxicity, co-mitogenic effect in lymphocytes, and other). In some cases, the effects triggered by the two receptors were found to be synergic. Apart from its signaling activity, it was suggested that CD120b, which has a higher affinity than CD120a for TNF and a higher dissociation rate from it, might assist in the binding of TNF to CD120a by presenting it to the latter receptor ('ligand passing' mechanism). Least is known about the function of the LTβ-R. The three effects found to be triggered by this receptor, namely cytotoxicity, co-mitogenic effect and NF-κB activation can also be triggered by the other two receptors.

(b) Structure of the intracellular domains of the receptors

The apparent similarity of function between the three receptors cannot be explained by their structure. Their intracellular domains appear completely unrelated. Only two conserved sequence motifs, the DD and the FAN motifs, have been identified in the receptors, both of them within CD120a.

(c) The death domain (DD) motif

This motif, which extends from about amino acid 325 to 410 in the C-terminal part of the receptor's intracellular domain, was initially defined as the region required for induction of death by the receptor, hence its name. A homologous sequence found in CD95 is responsible for death induction by that receptor (see below). The motif was also identified in four adaptor proteins involved in CD120a signaling: TRADD, which interacts directly with the DD of CD120a, MORT1/FADD, RIP, and RIP-associated ICH-1/CED-3-homologous protein (RAIDD/CRADD), whose interactions with each other and with TRADD seem to allow their activation by CD120a. These receptor–adaptor-protein interactions occur by heteroassociation of the DDs in the proteins. Moreover, in several of the proteins that contain DDs, including CD120a, this domain tends to self-associate, a process that apparently contributes to the initiation of signaling. Mutational and NMR studies of the structure of the CD95 DD provided initial information on the structural basis for the function of this motif. The DD motif also occurs in a number of other proteins but with no evident involvement in death induction.

(d) The FAN-binding motif

A sequence motif of about nine amino acids upstream of the DD in the intracellular domain of CD120a binds a Trp–Asp (WD)-repeat protein called FAN, which transmits the signaling for activation of the neutral sphingomyelinase by the receptor. The events further downstream in this signaling pathway are not known.

(e) Other signaling molecules that bind to CD120a

Apart from TRADD and FAN, CD120a reportedly binds a proteasomal component and a protein homologous to the 90 kDa family of heat-shock proteins. The binding in both cases is to a region close to where FAN binds. CD120a was also reported to bind some undefined serine/threonine protein kinases, as shown for CD120b.

(f) The TRAFs

All known CD120b- and LTβ-R-binding proteins belong to the same group of signaling molecules, the TRAFs. This group of proteins is defined by a shared C-terminal domain (the TRAF domain) through which they bind to the receptors and to each other. CD120b binds TRAF2 and the latter associates with TRAF1; it also binds to TRAF5. LTβ-R binds TRAF2, 3 and 5. These proteins serve as adaptors that activate other signaling molecules and are themselves devoid of enzymatic activities; they also participate in cross-talk between different receptors of the TNF receptor family. Thus, TRAF2 can bind to the CD120a-associated DD-containing adaptor protein, TRADD, and is apparently activated by CD120a through this interaction. Upon activation, it triggers the function of the transcriptional factor NF-κB and of the Jun N-terminal kinase. The former function seems to involve activation of NIK, a TRAF2-associated serine/threonine kinase. TRAF2 is also known to bind a protein called A20, which acts as a negative feedback inhibitor of TNF function, with the ability to suppress its NF-κB-activating effect as well as the cytotoxic effect of TNF. Another TRAF2-associated protein, called I-TRAF or TANK, was reported both to inhibit and to potentiate the NF-κB-activating effect. Two proteins, designated c-IAP1 and c-IAP2, which bind to the TRAF2–TRAF1 complex, are related to a baculovirus inhibitor of apoptosis and presumably have an inhibitory effect on cell death induction by TNF.

2.5 Soluble TNF receptors

2.5.1 Occurrence

Soluble forms of both CD120a and CD120b have been detected in the urine and the serum of humans. Serum levels of the soluble TNF receptors are normally in the range of about 1 ng/ml. Although they vary significantly among different individuals, serum levels of the soluble receptors in each individual are fairly constant over long periods. They increase significantly, however, in various disease states, most notably in inflammatory conditions and in various kinds of malignancies. One of the major inducing agents for the soluble receptors, accounting for their increase in inflammatory conditions, appears to be TNF itself.

It is not known whether a soluble form of the LTβ-R exists.

2.5.2 Structure

Initial information on amino acid sequence was obtained first for the soluble TNF receptors and only later for the cell-surface TNF receptors. More detailed studies revealed that the amino acid sequences in the soluble receptors are identical to those of the ligand-binding regions in the cell-surface forms. The urinary soluble CD120a consists of a sequence which in the cell-surface receptor extends from Asp12 to Asn172, and in a minor species to Lys174. Its formation thus seems to involve cleavage at two sites: at its N-terminus by a trypsin-like enzyme and at its C-terminus by an unidentified enzyme (see below). The size of the soluble CD120b is heterogeneous. This is in part because of the heterogeneous length of the protein at the N-terminus, probably reflecting differing extents of proteolytic truncation of the protein following its shedding, and in part due to differences in glycosylation patterns. It has a single C-terminus, corresponding to Val192 in the cell-surface form, which is located 14 amino acids downstream of the C-terminal cysteine.

2.5.3 Mechanisms of formation

Several pieces of evidence indicate that the soluble TNF receptors are formed by proteolytic processing of the cell-surface receptors. Shedding of CD120b can occur via the action of elastase, a soluble protease secreted by granulocytes, as well as via the action of a cell-bound protease, whereas formation of the soluble CD120a seems to involve only a cell-bound protease. Analysis of the effects of various protease inhibitors on the shedding indicates, in the case of both receptors, that the cell-bound protease involved in the shedding is a metalloproteinase. Cell culture studies revealed that a number of different stimulating agents, including TNF and activators of protein kinase C, can trigger the shedding process.

2.5.4 Function

No direct *in vivo* evidence for any functions of the soluble receptors has been presented. However, based on their *in vitro* characterization, these molecules can be assumed to have several kinds of activities. They can inhibit the functions of both TNFα and LTα by competing with them for the cell-surface TNF receptors. The relative effectiveness of the soluble CD120a in blocking LTα binding compared with its effect on TNFα binding is significantly lower than that of the soluble CD120b. The soluble receptors also stabilize TNFα, apparently by preventing the spontaneous dissociation of the trimeric TNF molecules into inactive monomers. Thus, effects of TNFα that require only brief exposure of cells to this cytokine display concentration-dependent inhibition by the soluble receptors, whereas effects that are slowly induced, e.g. the growth-stimulatory effects of TNF on fibroblasts and on B-CLL cells, are enhanced by low concentrations of the soluble receptors and are inhibited only at higher concentrations. In addition, since the soluble receptors bind TNF in a reversible manner, and the TNF molecules dissociating from them are bioactive, the soluble receptors may act as 'carriers', altering the availability of TNF.

2.5.5 TNF viroceptors

The majority of poxviruses for which sequence information is available encode one or more homologs of the TNF receptor family. Most of these 'viroceptors' show particular homology to CD120b and act as specific binding proteins for the TNFs. Such TNF-binding proteins are encoded, for example, by the T2 open reading frame in Shope fibroma virus, the myxoma virus T2 gene and the cowpox virus CrmB and CrmC proteins. Recombinant viruses in which expression of these viroceptors is obliterated by mutation grow normally in tissue culture, yet may be significantly less virulent in their animal host, indicating that these proteins play an important role in the viral resistance to host immune response mechanisms. Some of the viroceptors bind both TNFα and LTα, while others bind only TNFα. The amino acids in the C-terminal portions of these proteins, downstream of the ligand-binding regions, have homologous sequences that are distinct from those found in the cellular TNF receptors, indicating that they serve some virus-specific function(s). These sequences seem to be important for efficient secretion of the viroceptors from infected cells. In some of the viroceptors they associate, forming disulfide-linked dimers that inhibit TNF function more effectively than the monomeric forms do. The TNF viroceptors can also have additional functions, unrelated to their TNF-binding ability. A recent study suggests that the myxoma virus T2 proteins, besides blocking TNF function after their release by infected cells, act intracellularly within infected CD4$^+$ T lymphocytes to inhibit the virus-triggered apoptotic process by as yet unknown mechanisms, thus extending the virus host range for replication in T lymphocytes.

2.5.6 Medical implications

The effects of TNF function have often been described metaphorically as a double-edged sword. Although normally active in immune defense, when TNF is induced either too strongly or for too long its effects may become harmful to the organism. Consequently, TNF — more than any other known regulatory factor — is a major cause of pathological manifestations of both acute and chronic inflammatory disorders. The most compelling evidence for this key pathogenic role comes from the observed beneficial effects of blocking TNF function in various disease states. Such beneficial effects have been noted in some experimental animal models of disease, including acute disease states such as septic shock and cerebral malaria, and chronic inflammatory conditions such as rheumatoid arthritis, cachectic state in cancer, and others. Consequently, intense attempts have been made in recent years to apply TNF blockage as a therapeutic approach in clinical situations. The most extensively tried methods have been injection of antibodies against TNF and of soluble TNF receptors. In addition, some attempts have been made to apply drugs that inhibit the synthesis of TNF, e.g. pentoxifylline or thalidomide, for therapy of 'TNF-related' pathology. Although only recently begun, these trials have already provided compelling evidence for a pathogenic role of TNF and the therapeutic potential of its blockage in various disease states in humans.

Dramatic therapeutic effects of both anti-TNF antibodies and soluble TNF receptors have been reported in acute graft-versus-host disease, rheumatoid arthritis and inflammatory bowel diseases (Crohn's disease and ulcerative colitis).

Blocking of TNF function in animal disease models has also provided compelling evidence for the other edge of the metaphorical sword, namely, the beneficial effect of TNF function in defense against infection and injury. As mentioned above, animals in which TNF function was obliterated by gene knockout or by injection of antibodies or soluble receptors were much more vulnerable to infection by certain pathogens, for example *Listeria* or *Leishmania*, or to intraperitoneal infection resulting from puncture of the intestine. So far, however, no serious attempt has been made to exploit this positive aspect of TNF function for therapy in humans, perhaps because of the risk of deleterious consequences of TNF action.

Another aspect of TNF function of potential therapeutic value is its selective destructive effect on the vasculature within tumors, for which it received the name 'tumor necrosis factor'. Although in the early years following TNF cloning it was thought that this activity could not be applied clinically because of the cytokine's deleterious effects, TNF has recently been used successfully to eradicate melanoma and soft tissue sarcoma in limbs of humans. In these trials, high doses of TNF, together with melphalan, were applied to limbs in which the circulation was temporarily isolated from the main circulation (the limb perfusion approach). Attempts are now being made to use this approach in the application of the anti-tumor activity of TNF in order to eradicate tumors from other organs.

3. HVEM and its ligand, LIGHT

Herpes virus entry mediator (HVEM; also called ATAR and TR2) and its ligand, LIGHT, are a recently described receptor–ligand couple that bind not only to each other but also to ligands and receptors known to participate in TNF function (Figure 1). HVEM also binds to a glycoprotein component of the herpes virus and may thus play a role in the interactions of T lymphocytes with this virus.

3.1 Molecular features

3.1.1 HVEM

HVEM was identified by its ability to mediate the entry of several strains of herpes virus (HSV) into

Chinese hamster ovary (CHO) cells (hence the name HVEM, for herpes virus entry mediator) and also as a cDNA whose encoded protein binds TRAF2 and TRAF5 (the alternative name ATAR stands for another TRAF-associated receptor). It was also cloned during a search of an expressed sequence tag database for proteins showing homology to the Cys-motif. The human and mouse proteins contain 283 and 276 residues, respectively. The extracellular part of the receptor contains three Cys-motifs. The intracellular domain is rather short (46 residues).

3.1.2 LIGHT

LIGHT was cloned from a cDNA library derived from activated peripheral blood lymphocytes by first searching for clones homologous to the Fas ligand (by arbitrary sequencing) and then assessing the ability of the proteins encoded by these cDNAs to bind HVEM. The protein contains 240 amino acid residues, with a C-terminus that contains a TNF homology region, followed by a transmembrane domain of 22 residues and an N-terminal cytosolic domain of 37 residues. Although its binding properties are related to those of lymphotoxin, LIGHT does not form a complex with either LTα or LTβ.

3.2 Occurrence

HVEM is encoded by multiple transcripts, the most abundant of which is 1.7 kb long. The transcripts are highly expressed in peripheral blood mononuclear cells (both T and B lymphocytes, as well as monocytes), spleen, thymus, bone marrow, lung and small intestine, and only to a small extent in brain, skeletal muscle and stomach. No information about the cellular level of the HVEM protein is yet available.

A LIGHT transcript of 2.5 kb was found to be expressed predominantly in the spleen and to a lesser extent, together with a secondary transcript of 3.5 kb, in the brain. Various other tissues exhibit weak expression. Cell-surface expression of the LIGHT protein in T cells is strongly enhanced by combined activation of these cells with 4β-phorbol-12-myristate-13-acetate (PMA) and a calcium ionophore.

3.3 Cellular responses initiated by HVEM and LIGHT

Like other TRAF-associated receptors of the TNF/NGF family, HVEM can activate NF-κB. Presumably this activation can be triggered by

LIGHT. LIGHT also inhibits HSV entry into cells by competing for the binding of the HSV glycoprotein D to HVEM. Antibodies against HVEM effectively block the entry of HSV into T lymphocytes, but not into other cells. Conversely, the HSV glycoprotein D, although showing no homology to the ligands of the TNF family, is capable of blocking the binding of LIGHT to HVEM. These findings suggest that HSV, besides employing HVEM as the receptor facilitating entry into T cells, either mimics or blocks the triggering of this receptor by LIGHT and thus modulates T cell-mediated immune responses.

LIGHT also binds to the LTβ-R and this binding is not inhibited by the HSV glycoprotein D. LTα1/β2, the other known ligand of the LTβ-R, does not bind to HVEM, yet this receptor does bind LTα. These binding patterns may account for differences in the phenotypic consequences of disruption of the LTα, LTβ and LTβ-R genes, such as the presence of cervical and mesenteric lymph nodes in LTβ–/–mice (but not in LTα–/–, nor in LTβ-R–/– mice).

4. CD95 (Fas/Apo-1) and its ligand

4.1 General description

The study of CD95 and its ligand paved the way to recognition of the significance of the death-inducing ability of the TNF/NGF receptor family. Although the cytocidal property of the TNFs was discerned much earlier than that of CD95, we still have no idea of its exact physiological role. In contrast, the physiological and pathological significance of the cell-killing activity of CD95 and its ligand became apparent almost at the same time as these molecules were cloned. CD95 was discovered independently by two research groups through the detection of cytotoxic activities of mAbs against the protein. In both studies, these antibodies were identified in the course of examination of a panel of antibodies raised in mice against intact human cells. The receptor to which their antibody bound was named 'Fas' by one group, for 'FS7-associated surface antigen' (FS7 was the foreskin cell strain used for the immunization), and 'Apo-1' by the other, a reference to the antibody's apoptotic activity. CD95 was cloned by expression cloning, using an anti-CD95 mAb to discern cells that express CD95 among cells transfected with a cDNA library. It was also cloned by first isolating CD95 (by immunoaffinity) and then designing DNA probes based on the amino acid sequence of the protein. The CD95-L was cloned by expression cloning, applying soluble CD95 in order to discern cells that express the ligand among cells transfected with a cDNA library. Emerging evidence indicates that, apart from cell death induction, CD95 may also have other regulatory functions.

4.2 Occurrence of the CD95 ligand and receptor

Our notions of the expression patterns of CD95 and its ligand are in a state of flux. The initial impression was that, as with other members of the TNF ligand and receptor families, expression of CD95-L is transient and quite restricted, occurring predominantly in activated T cells, whereas CD95 is expressed in a variety of cells in a constitutive manner (except for T lymphocytes, in which — as with most other members of the family — its expression is stimulus-dependent). A growing body of evidence indicates, however, that the expression of CD95-L can be induced in a wide range of cells. Thus, apart from its expression in activated T lymphocytes, primarily (though not solely) in Th1 CD4$^+$ cells, it was detected, for example, in B lymphocytes activated by LPS or PMA/ionomycin, in natural killer cells, in hepatocytes during alcoholic liver damage, and in glial cells and macrophages that accumulate in lesions in the brains of multiple sclerosis patients. CD95-L is expressed constitutively by a number of cells, including testicular Sertoli cells, the eye corneal epithelial and endothelial cells, placental cytotrophoblasts, thyrocytes, Paneth cells in the gastrointestinal tract, and monocytes. In addition, various tumor cells express CD95-L ectopically, or upon treatment with chemotherapeutic drugs.

The range of cells known to express CD95 also keeps expanding, and in several of these cells CD95 production is inducible. The beta cells of the Langerhans islets, for example, seem to produce the receptor in response to some T cell cytokines. Likewise, thyrocytes are induced to express it by IL-1, and hepatocytes exhibit a vast increase in CD95 expression during hepatitis B virus-related cirrhosis and in acute liver failure. As described below, both an increase in CD95 expression and induction of its ligand can trigger cell death. It is therefore not surprising that the occurrence of these proteins is highly restricted and hence not always easily discerned. We may still be a long way from knowing the full range of cell types and situations in which their expression occurs.

4.3 Function

(a) The cytotoxic function of CD95 and its physiological significance

The ability to activate a death program in cells is the most prominent and best studied function of CD95. It has been observed in a wide range of cell types, though not all cells which express CD95 are sensitive to its cytocidal effect. Moreover, in those cells that are sensitive, the vulnerability is prone to modulation by a variety of signals, which presumably act to ensure physiological compatibility of the death process.

Our knowledge of the *in vivo* role of CD95-L-induced cytotoxicity comes primarily from studies of three mouse spontaneous mutant strains, which are defective in either the CD95 or the CD95-L gene: (1) *lpr* (lymphoproliferation) mice, whose CD95 gene contains an inserted early-transposable element within the second intron, causing premature termination of the CD95 transcripts and hence greatly reduced expression of CD95; (2) *lpr^{cg}* mice, with a point mutation in the CD95 DD (see below) resulting in a replacement of an isoleucine residue by asparagine. This alteration leads to an extensive conformational change in the DD, with resulting disruption of interactions required for its signaling activity; (3) *gld* (generalized lymphoproliferative disease) mice, with a point mutation in the CD95-L gene resulting in a replacement of a phenylalanine residue by a leucine close to the C-terminus of the molecule. The three mouse strains share the same pathological phenotype, which is similar to that seen in mice with targeted disruption of the CD95 gene as well as in children with inborn defects of CD95.

(i) Role of CD95-induced cytotoxicity in restricting the growth of activated professional immune cells

The most salient feature of the phenotype of the above-mentioned mouse models is excessive lymphoproliferation, manifested in lymph node enlargement and splenomegaly resulting from accumulation of CD4$^-$CD8$^-$ lymphocytes. Depending on other features of their genetic background, mice and humans with such defects may also develop an autoimmune syndrome (glomerulonephritis, vasculitis, arthritis). These abnormalities have been attributed to deficient induction of death by CD95 in at least three kind of cells:

1. *Death of T lymphocytes.* CD95-induced cytotoxicity apparently does not contribute to the selection of T cells in the thymus. It seems, however, to play an essential role in arresting the growth of these lymphocytes in the periphery. Antigenic stimulation of mature CD4[+] lymphocytes results in induction of CD95-L. The physiological role of the cell-killing activity of CD4[+] lymphocytes differs from that of CD8[+] lymphocytes (whose cytotoxic activity is mediated mainly by the combined effect of perforin and granzymes, proteins released by the lymphocytes from stores accumulated within granules). Acting in an MHC type I-restricted manner, CD8[+] cells can destroy pathogen-infected target cells, taking advantage of the ability of the MHC type I molecules to present antigens formed within the infected cells. In contrast, CD4[+] lymphocytes, which recognize their targets in the context of type II MHC molecules, kill mainly immune-competent antigen-presenting cells (APCs). In addition to inducing CD95-L, stimulation of T lymphocytes also leads to enhanced expression of CD95 and increased sensitivity of the cells to the CD95 cytocidal effect. In the absence of co-stimulatory signals, such as those from B7, which can endow the T cells with resistance to the CD95-L effect, these changes result in death of the T cells and consequent restriction of their antigen-induced clonal expansion.

2. *Death of B lymphocytes.* Triggering of CD40 in B cells by T cell-expressed CD40-L sensitizes the B cells to CD95-induced cytotoxicity. Signals provided by the occupied antigen receptor normally protect against this cytotoxicity. However, in autoreactive B lymphocytes, whose chronically engaged antigen receptor is desensitized, these signals may not be generated. The CD95 effect can therefore result in specific elimination of these cells. Failure to eliminate autoreactive B cells in this manner apparently accounts for the autoimmune phenomena resulting from deficient CD95 signaling.

3. *Death of macrophages.* Antigen-stimulated CD4[+] T lymphocytes of the Th1 phenotype kill antigen-presenting activated macrophages by triggering the CD95 that they express. In MRL/*lpr* and MRL/*gld* mice, excessive accumulation of activated macrophages in the absence of this restricting mechanism leads to damage to the arterial wall.

Indirect evidence suggests that CD95-L-mediated cytotoxicity also serves to control the proliferation of eosinophils and perhaps also of neutrophils.

(ii) Role of CD95-induced cytotoxicity in immune privilege

As mentioned above, CD95-L is expressed by cells in the testis and in the eye chamber, both immune-privileged sites. Comparative studies of the immune response of these organs in *gld* and normal mice point to an important role for CD95-L in maintaining the immune privilege. Thus, while grafts of normal testicular tissue survive indefinitely when implanted under the kidney capsule of allogeneic animals, grafts derived from *gld* mice are rejected. Similarly, in normal mice inflammatory cells entering the anterior chamber of the eye in response to viral infection undergo apoptosis and produce no tissue damage, whereas in *gld* mice the inflammatory cells invading the ocular tissue in response to the same infection remain viable and exert local destructive effects. Expression of functional CD95-L on the corneal cells also seems to be a precondition for effective corneal allograft transplantation. Grafts from *gld* mice, or normal grafts transplanted to CD95[-] mice, were found to be fully rejected. In addition to its protective effect on the immune-privileged site, destruction of invading immune cells by the CD95-L expressed within these sites also seems to induce immune tolerance.

CD95-L is expressed during gestation, first in the uterus and then in the placenta, suggesting that, similarly to its function in the testis and in the eye chamber, this molecule acts in the placenta to protect the fetus against the cytolytic action of maternal lymphocytes.

Attempts to apply the presumed immune privilege-related function of CD95-L for preventing graft rejection led to conflicting results. In some of the studies, implantation of CD95-L-expressing cells in the vicinity of the graft resulted in prolonged survival of the foreign tissue. In others it led to enhanced rejection, due to inflammatory reactions, possibly involving the induction of IL-8 by CD95 (see below).

(iii) Role of CD95-induced cytotoxicity in tissue homeostasis

Whereas *lpr* mice (in which CD95 expression is greatly reduced but is not eliminated) do not display any gross abnormalities in non-lymphoid CD95-expressing tissues, mice with targeted disruption of the CD95 gene (and consequently lacking any CD95 expression) show age-dependent hyperplasia in the liver, suggesting a role for CD95 in the removal of senescent hepatocytes. CD95-induced cytotoxicity may well play a similar role in the normal turnover of cells in other tissues.

(b) Non-cytocidal effects of CD95

As in the case of the TNF receptors, cells that do not respond to the cytocidal effect of CD95 may respond to the receptor in other ways. So far, only the following non-cytocidal activities of CD95 have been reported: (i) growth stimulatory effects, observed in lymphocytes and fibroblasts; (ii) induction of synthesis of the cytokines interleukin-8 (IL-8) and interleukin-6. The range of cells in which CD95 expression can induce such effects is much narrower than the range of cells in which these effects can be induced by CD120a.

4.4 Structure–function relationships in CD95 and in its ligand

(a) CD95-L

Most studies of CD95-L have been performed using its cell-bound form. However, some biological fluids have been found to contain a soluble form of the ligand, derived from the cell-surface form by cleavage of the spacer region that links the molecule's receptor-binding portion to its transmembrane domain. As in the case of TNFα, the enzyme that cleaves CD95-L seems to be a metalloproteinase. The soluble form of human CD95-L is biologically active, whereas that of the mouse protein seems to be devoid of activity.

No direct information is available on the three-dimensional structure of CD95-L. Molecular modeling indicates that it attains a trimeric structure similar to that of TNF.

(b) CD95

The ligand-binding region in the extracellular domain of CD95 is comprised of three Cys-repeats. The intracellular domain contains a DD module. The region downstream of this module is rich in serine and threonine residues, as in a number of other DD-containing proteins (though not CD120a). Recent NMR analysis of the fine structure of the DD module in CD95 revealed that it contains six antiparallel, amphipathic α-helices arranged in a novel fold. Molecular modeling suggests that DDs in other proteins have similar structure. Mutational studies indicate that the region in the DD of CD95 involved in self-association of this domain overlaps only partly with the region in the DD that binds MORT1/FADD, the adaptor protein responsible for transmission of the death signal. Mutation of isoleucine at position 225 in the mouse protein, which occurs naturally in *lpr*^{cg} mice, results in gross conformational changes in the DD, eliminating both its self-association and its interaction with MORT1/FADD. A similar change in binding properties is observed following mutation of the DDs in CD120a and TRADD (an adaptor protein

binding to CD120a) at the site corresponding to that of the *lpr*^{cg} mutation.

The sequence Ser-Leu-Val within the serine/threonine-rich region downstream of the DD module in CD95 binds a protein tyrosine phosphatase, FAP1, which inhibits CD95-induced cell killing by an unknown mechanism.

(c) Soluble forms of CD95

Soluble forms of CD95 occur in the serum and have significant inhibitory effects on CD95 signaling, both *in vitro* and *in vivo*. At least part of the soluble CD95 molecules are encoded by distinct transcripts, derived from the CD95 gene by an alternative splicing mechanism.

4.5 Medical implications

Accumulating evidence suggests that both hypoactivity and hyperactivity of CD95 play important roles in the pathology of various diseases.

The pathological consequences of deficient CD95 signaling are most clearly demonstrated in the abovementioned mouse models of inborn mutational obliteration of the signaling, i.e., in *lpr*, *lpr*^{cg} and *gld* mice and in mice with targeted disruption of the CD95 gene. The lymphoproliferative and autoimmune phenotype of humans with inborn defects of CD95 is similar to that observed in these mouse models. Sporadic evidence points to the involvement of more restricted deficiencies of CD95 signaling in a variety of other pathological situations. Sustained overproduction of eosinophils in the idiopathic hyper-eosinophilic syndrome and in some HIV-1-infected individuals has been associated with increased CD4$^-$CD8$^-$ T cells that lack functional CD95 and, in some cases, overexpress soluble CD95 that can block CD95-L function. This probably results in sequential dysregulation of apoptosis: failure of the T cells to die due to the lack of CD95, with consequent excessive formation by the T cells of cytokines that have anti-apoptotic properties for eosinophils. Sera of systemic lupus erythematosus (SLE) patients contain increased concentrations of soluble CD95, and it was suggested that blocking of CD95-mediated killing of activated lymphocytes by these soluble molecules contributes to the pathogenesis of this disease. However, no defect in expression or signaling activity of the cell-bound receptor has been discerned in SLE.

Deficient cell killing by CD95 also seems to play a role in tumor development. Killing of tumor cells by certain chemotherapeutic drugs involves CD95-L up-regulation and cytotoxicity. Mechanisms that obliterate CD95 function may thus contribute to the evolution of tumor resistance to chemotherapy. In line with this possibility, the incidence of complete remission after chemotherapy in patients with acute myeloid leukemia was found to be inversely related to the expression of CD95 on the patients' blast cells.

Escape from cytotoxicity induction by CD95-L (or by TNF) is also a major determinant of the pathogenicity of certain viruses, and various strategies are employed by viruses to this end. The poxvirus CrmA protein and the baculovirus p35 protein block the functions of caspases that are activated in the death pathway. Certain death effector domain (DED)-containing proteins produced by herpes viruses and poxviruses prevent this caspase activation by competing for the binding of CASP 8 and CASP 10, the first caspases in the pathway, to the CD95-associated adaptor protein MORT1/FADD. Herpes simplex virus type 2 possesses mechanisms by which it withholds CD95-L transport to the surface of the infected cell. The protection against immune cytolysis conferred by these mechanisms on the virus-infected cells allows prolonged virus production.

In contrast, the pathology of some other viral infections involves excessive death induction by CD95. Studies both in experimental animal models and in humans indicated that the liver damage sustained in viral hepatitis involves CD95-L-mediated T cell cytotoxicity combined with T cell cytokine-induced up-regulation of CD95 in the hepatocytes. It has also been suggested that the depletion of CD4[+] T cells in acquired immune deficiency syndrome (AIDS) patients involves CD95-induced apoptosis. In this case, increased expression of CD95, as well as stimulation of CD95 by antibodies formed against an epitope that it shares with gp120, were implicated. Moreover, HIV infection has been shown to up-regulate CD95-L expression in normal peripheral blood mononuclear cells. Involvement of excessive CD95-L-mediated lymphocyte-killing in the pathology of HIV infection has, however, been challenged by a recent study showing that monocytes isolated from the blood of HIV-infected patients do not exhibit an increase of CD95-L expression but, on the contrary, a marked decrease relative to the amounts normally expressed on these cells. This finding raised the possibility that, as with some other viruses, progression of the disease in these patients involves not an excess but rather a deficiency of CD95-L-mediated destruction of the infected cells, allowing prolonged virus production.

Up-regulation of CD95 or CD95-L or both was implicated in autoimmune-mediated tissue injury, including graft rejection, the death of beta cells of Langerhans islets in the mouse non-obese diabetic (NOD) model, the destruction of thyrocytes in Hashimoto's thyroiditis, and the death of oligodendrocytes in multiple sclerosis. In Hashimoto's thyroiditis, up-regulation of CD95 by a cytokine (IL-1β) seems to be a limiting factor for destruction of the target cells. The source of the ligand is the target cells themselves. Thyrocytes normally express CD95-L, and their death upon up-regulation of CD95 therefore occurs by fratricide or juxtacrine triggering. Fratricide also appears to be the mechanism for death of hepatocytes in patients with alcohol liver damage. In this case, however, the process seems to result from a change of the inverse nature — up-regulation in these cells of the CD95-L, which in turn triggers their constitutively expressed CD95.

Abnormal regulation of CD95-L function can contribute to pathological manifestations of diseases in other ways. Various tumor cells express CD95-L ectopically. It has been suggested that this enables the tumor cells to evade destruction by T lymphocytes through the effect of their CD95-L on the lymphocytes, a mechanism resembling the presumed mode of action of CD95-L in immune-privileged sites. There are also reports, however, that in some tumor models expression of CD95-L promotes destruction of the tumor cells, as a result of attraction of granulocytes, perhaps mediated by CD95-induced IL-8. Study of a mouse model of acute graft-versus-host disease indicated that expression of CD95-L by the grafted T lymphocytes plays a major role in the disease through two different effects of the CD95-L: mediation of anti-host T cell cytolytic activity and enhancement of the expansion and maturation of the grafted lymphocytes. The latter effect may involve reversed signaling by the CD95-L molecules for growth of the lymphocytes expressing them, or be secondary to effects of the lymphocytes on host cells.

5. Apo-3/WSL-1/DR3/TRAMP/LARD and its ligand TWEAK/Apo3-L

Apo-3/WSL-1/DR3/TRAMP/LARD was identified by screening of DNA databases for expressed sequence tags (EST) showing similarity to CD120a and CD95. Hybridization of cDNA libraries with oligonucleotide probes designed for homologous regions of the intracellular domains of CD120a and CD95, 2-hybrid screening in yeasts using the DD of CD120a as a bait, and data bank screening for sequences homologous to CD95 led to the identification of the same cDNA. The receptor is a 417-amino-acid transmembrane protein with an approximate molecular weight of 47–54 kDa and 2 potential *N*-glycosylation sites at positions 67 and 106. Its extracellular domain contains 4 cys-motifs typical of the TNF/NGF receptor family. The intra-cellular domain contains a DD. The homology between Apo-3 and CD120a is 48% in this region, whereas their overall homology is 29%. The chromosomal localization of the Apo-3 gene was attributed to the short arm of chromosome 1 at position 1p36, which contains the genes of several other TNF/NGF family members, C120b, CD30, 4-1BB and OX40 (see Table 1, p. 55). In contrast to these other genes, however, the gene of Apo-3 resides in a region that is frequently deleted in some of the human neuroblastomas (1p36.3). Apo-3 mRNA is expressed in many lymphoid and nonlymphoid tissues. Altogether, Apo-3 has at least 11 splice variants, whose expression pattern varies in a tissue-specific manner. Native lymphocytes express the full-length receptor at very low levels, though they express the other splice variants and are induced by activation to express the full-length protein. Interestingly, a 1-kb transcript is selectively upregulated in tissues of the *lpr* mice, which exhibit a defect of the CD95 molecule. This finding may suggest involvement of Apo-3 in the pathogenesis of autoimmune abnormalities and/or compensation for the CD95 deficit by another death-inducing receptor. Transfection of HEK-293 or HeLa cells with an expression vector encoding Apo-3 resulted in spontaneous apoptosis of the transfected cells. As in the case of cell killing induced by CD120a and CD95, cotransfection with the cowpox virus gene *crmA* or treatment of cells with another caspase inhibitor, zVAD-fmk, inhibited apoptosis. Overexpression of Apo-3, like overexpression of CD120a, induced an increase in NF-κB DNA-binding activity. Signal transduction by Apo-3 was found to be mediated by a set of molecules (TRADD, TRAF2, NIK, MORT1/FADD and caspase-8) known to interact with CD120a.

The Apo-3 ligand (Apo3-L), TWEAK (for TNF-related ligand with <u>weak</u> ability to induce cell death) was cloned serendipitously in the course of a search for transcripts showing sequence similarity to erythropoietin as well as by searching an expressed sequence tag database for sequences that have similarity to TNF. It is a type II transmembrane protein with a characteristic extracellular TNF-related receptor-binding motif, which is apparently *N*-glycosylated. The 'stalk' region linking the TNF-related motif to the transmembrane domain contains a high proportion of basic residues that are likely to render it vulnerable to proteolysis. Its expression pattern in cultured cells indeed indicates that the protein occurs mainly as a soluble molecule. Unlike LTα, however, and like TNFα, TWEAK is also expressed in a cell-surface form, which is rapidly shed. There is considerable sequence conservation between human and mouse within the TNF-related motif of TWEAK (93%), greater than that observed for any of the other ligand family members.

Unlike the situation with most members of the TNF ligand family, the transcript of TWEAK (about 1.5 kb) is abundant in most tissues. This transcript contains, however, a prominent mRNA destabilizing motif, which may effectively restrict its translation. No information on the tissue expression pattern of TWEAK/Apo3-L protein has yet been presented.

6. TRAIL-R1, -R2, -R3 and -R4, and their ligand, TRAIL

6.1 Molecular features of TRAIL and its receptors

(a) TRAIL

The TRAIL (Apo-2) ligand was cloned by EST database screening using a consensus amino acid sequence of the most conserved region of the TNF ligand family. The full-length cDNA contains an open reading frame encoding a type II transmembrane protein of 281 or 291 amino acids in human or mouse, respectively. There are potential sites of *N*-glycosylation in the C-terminal extracellular domains, at amino acids 109 and 52 in the human and mouse receptor, respectively. Homology of the C-terminal domain of TRAIL to the related family members CD95-L, TNFα and LTα is 28%, 23% and 23%, respectively. Unlike in other members of the family, the N-terminal cytoplasmic domain of TRAIL is not conserved between human and mouse. The soluble form of the protein exists as a homotrimer.

(b) TRAIL-R1, -R2, -R3 and -R4

There are five known receptors to TRAIL. Four, TRAIL-R1, -R2, -R3 and -R4 are apparently specific to TRAIL. The fifth, OPG (see below) also binds to the ligand TRANCE. TRAIL-R1 (DR4) was cloned in a search of EST databases with the CD120a DD, TRAIL-R2, -R3 and -R4 were cloned in searches for sequences homologous to TRAIL-R1. In all four receptors, the ligand-binding region in the extracellular domain comprises one partial and two complete Cys-motifs. The extent of sequence homology between these regions in the four TRAIL receptors is greater than with the corresponding regions in any of the other receptors of the TNF family. TRAIL-R1 and TRAIL-R2 (DR5) also exhibit significant homology in other regions within these molecules, and — most notably — the intracellular domains of both receptors contain DDs of closely related structure. The structures of TRAIL-R3 (DR6/TRID/DcR1) and TRAIL-R4 are, however, quite different. The intracellular domain of TRAIL-R4 contains only part (about one-third) of a DD motif. TRAIL-R3 does not contain an intracellular domain at all, nor does it have a transmembrane domain. It is linked to the membrane through a glycosyl-phosphatidylinositol anchor. Another distinctive feature of the structure of TRAIL-R3 is the presence of five tandem repeats of a 13 amino acid motif (also found in the Son-of-sevenless protein and in the endothelial leukocyte adhesion molecule) at the C-terminus of the protein.

6.2 Occurrence of TRAIL and its receptors

The transcripts of all TRAIL receptors, as well as of TRAIL itself, are widely expressed in a variety of tissues. Only a few distinctive features of their cellular patterns of expression have been noted. TRAIL is expressed in spleen, prostate and lung, but not in brain, liver or testis, nor in freshly isolated T cells from peripheral blood, while TRAIL-R3 is expressed to a much greater extent in peripheral blood leukocytes than in any other tissue examined. The level of TRAIL-R2 transcripts is increased in various tumor lines, while the expression of TRAIL-R3 in cells of a series of tested tumor lines was found to be far lower than in normal tissues. TRAIL-R1 is expressed in the form of three transcripts of 2.6, 4.6 and 7.2 kb, and TRAIL-R3 in the form of five transcripts of about 1.3, 2.5, 3.0, 4.0 and 7.0 kb. TRAIL, TRAIL-R2 and TRAIL-R4; however, all have a single major transcript of about 4.0 kb. No information on the functional significance of these distinct expression features has yet been presented.

6.3 Cellular responses initiated by TRAIL

Only two activities of the soluble and membrane-anchored forms of TRAIL have so far been noted:

induction of cell death and activation of the transcription factor NF-κB. The apoptotic effect is induced in a wide range of cell lines, while normal cells are, in general, resistant to it. TRAIL-R1 and -R2 can both induce apoptosis. They can bind to each other, suggesting that they cooperate in signaling. As in the case of of death induction by CD95 and CD120a, the cytotoxic effect of TRAIL-R1 and -R2 involves activation of caspases. This activation is apparently initiated by the activation of caspase 10 and probably also of caspase 8. There is some evidence that the receptors can bind MORT1/FADD, TRADD and RIP, and that their cytotoxic effect can be blocked by expression of a non-functional MORT1/FADD mutant. Data from other studies, however, suggest that the effectiveness of binding of these adaptor proteins to the receptors is lower than that of their binding to CD95 and CD120a. These findings were interpreted as suggesting that the TRAIL receptors induce death through binding some other, as yet unknown, adaptor protein(s).

TRAIL-R4 lacks death-inducing activity, and its high expression endows cells with resistance to the cytocidal effect induced by TRAIL-R1 and -R2. It does, however, activate NF-κB. Expression of a TRADD dominant-negative mutant blocks NF-κB activation by TRAIL-R4, as well as by TRAIL-R1 and -R2, suggesting that these receptors, like CD120a and DR3, employ a TRADD-dependent pathway to activate NF-κB.

TRAIL-R3 seems to be devoid of signaling activity and to act as a decoy receptor that inhibits TRAIL signaling.

At the time of writing of this review, no information was yet available on the physiological significance of TRAIL function, nor of its role in the pathology of any disease.

7. CAR1

CAR1 is a 368 amino acid chicken protein that recognizes the envelopes of two cytopathic subgroups (B and D) of avian leukosis-sarcoma viruses (ALVs). Its extracellular domain contains one partial and two complete Cys-motifs. Among the known mammalian members of the TNF/NGF family, the TRAIL receptors show highest homology to CAR1. Like these receptors, the intracellular domain of CAR1 also contains a DD motif. The induction of apoptosis by CAR1 may underlie the cytopathic effects of the B and D subgroups of the ALVs. Indeed, application of an immunoadhesin containing a B subgroup-specific envelope protein to CAR1-expressing cells, either in the course of infection or following transfection, induced apoptotic cell death. Interesting information will no doubt be obtained by the future mapping and characterization of the gene encoding CAR1, which is probably an allelic form of the *tv-b* polymorphic gene in chicken. Comparison of this gene with the *tv-b* allelic forms encoding receptors for non-cytopathic subgroups of ALVs should yield deeper insights into the mechanism of cell killing initiated by the cytopathic viruses as well as the mechanism differentiating the two subgroups of ALVs. The physiological function and identity of the ligand of CAR1 are still unknown.

8. Osteoprotegerin (OPG)/Osteoclastogenesis-inhibitory factor (OCIF)

Osteoprotegrin/osteoclastogenesis-inhibitor factor (OPG/OCIF) is a receptor of the TNF family which is apparently expressed only as a soluble, secreted molecule. It is a glycoprotein which contains four cys-motifs and occurs primarily as a cysteine-linked dimer. Its transcript (about 3.0 kb) occurs in a variety of tissues, including liver, lung, heart, kidney and placenta. Notably, in mouse embryo it was found to be expressed mainly in cartilaginous primordia of developing bones. Within the bone, it is expressed both in the bone marrow stromal cells and osteoblastic cells. Its expression in the latter kind of cells is enhanced by various positive regulators of osteoclastogenesis (e.g. TNF, LT, IL1 beta, 1,25-dihydroxyvitamine D) as well by TGF-beta1, that acts as a negative regulator of osteoclastogenesis.

OPG/OCIF apparently serves as a natural inhibitor of the function of two ligands of the TNF family: TRAIL and TRANCE/RANK-L/ODF. These two ligands bind to distinct cell surface receptors, and yet both bind to OPG/OCIF at a high affinity (in the range of 1–5 nM), and this binding blocks the interaction of both ligands with their respective cell surface receptors. By interfering with the function of TRANCE/RANK-L/ODF, OPG/OCIF blocks osteoclast differentiation. Its over-expression in mice causes marked osteopetrosis, whereas targeted disruption of OPG/OCIF gene leads to ostoperosis.

In vivo over-expression of OPG results in splenomegaly too. Whether this effect reflects suppression of TRAIL-induced death of lymphocytes or of some other function(s), mediated either by TRAIL or by TRANCE/RANK-L/ODF remains to be determined.

9. TRANCE/RANK-L/osteoclast differentiation factor (ODF) and its receptor, RANK

9.1 Structure of the receptor and its ligand

Human RANK is a 616-amino-acid type I transmembrane protein with a predicted extracellular domain of 184 amino acids and a cytoplasmic domain of 383 amino acids. The ligand-binding region in the extracellular domain comprises 4 cys-motifs. Among the members of the TNF/NGF family, the highest sequence homology to RANK in its extracellular domain is shown by CD40. Likewise, among the members of the TNF ligand family, the highest sequence homology to TRANCE/RANK-L/ODF extracellular domain is shown by CD40-L, and next to it — to TRAIL. The ligand molecule is a typical type II transmembrane protein of 316 amino acid residues, with a predicted extracellular domain of 247 residues which is apparently *N*-glycosylated, and an intracellular domain of 48 residues.

9.2 Cellular expression and regulation of the receptor and its ligand

The transcript of RANK (of about 4.5 kb) is ubiquitously expressed, with the highest levels in skeletal muscle, thymus, liver, colon, small intestine and adrenal gland. Expression of the RANK protein, however, seems to be more restricted. Its most prominent expression was noted in dendritic cells (DC) of bone marrow, lymph nodes and the spleen origin, where it could be further enhanced by CD40-L. No expression could be detected in freshly isolated T cells, thymocytes, or peritoneal macrophages. Human foreskin fibroblasts also express significant amounts of the receptor, and peripheral blood T cells, although normally expressing very little of the receptor (even when activated), can be induced to express high amounts of it by their treatment (upon activation) with IL-4 and TGF-β.

The TRANCE/RANK-L/ODF transcript (about 2.3 kb) is predominantly expressed in lymph nodes and the thymus and to a much lesser degree in various other organs. It is abundant in lymph node-derived T cells but not in B cells. T cell receptor cross-linking results in extensive upregulation of the TRANCE/RANK-L/OPG-L transcript. TRANCE/RANK-L/OPG-L is also expressed in skeletal mesenchimal cells. The skeletal patterns of expression of this ligand during embryonic development and in the adult are consistent with its suspected role as a major regulator of osteoclast differentiation and activation. Agents known to trigger osteoclast activation by osteoblasts, e.g. 1,25-dihydroxyvitamine D, parathyroid hormone and IL-11, enhance TRANCE/RANK-L/ODF expression in osteoblastic cell lines whereas TGF-beta1 (which acts as a negative regulator of osteoclastogenesis) suppresses it.

9.3 Cellular responses initiated by the receptor and its ligand

Three functions of TRANCE/RANK-L/ODF have so far been described; one concerns the differentiation/activation of the osteoclasts, another the differentiation/activation of dendritic cells (DC), and the third relates to the function of T lymphocytes.

9.3.1 Role of TRANCE/RANK-L/ODF as a regulator of osteoclast differentiation and function.

Soluble TRANCE/RANK-L/ODF induces differentiation of osteoclast precursors and activation of mature osteoclast *in vitro* and, when injected into mice causes massive osteoclast activation. Conversely, soluble RANK as well as OPG/OCIF (which acts as a soluble decoy receptor for TRANCE/RANK-L/ODF) block osteoclast activation by osteoblasts. These findings, and others mentioned above, suggest that bone density is regulated through RANKL/TRANCE/ODF-RANK mediated communication between the osteoblasts and osteoclasts: biological activators of osteoclasts induce RANKL/TRANCE/ODF expression in ostelbasts. This ligand then triggers the signaling activity of RANK in osteoclasts, and the effectivity of triggering is subject to modulation (on the level of ligand availability) through the function of OPG/OCIF.

9.3.2 Role of TRANCE/RANK-L/ODF as a T-cell produced regulator of DC function

Several effects of TRANCE/RANK-L/ODF on cultured DC have been noted: (a) enhancement of survival, an effect which is an outcome of a TRANCE/RANK-L/ODF-induced increase in expression of Bcl-x_L; (b) induction of aggregation, even though, unlike CD40-L, it does not induce increased expression of CD2, CD11a, CD54 or CD58, cell-surface proteins believed to contribute to a similar effect; (c) enhancement of DC-mediated T cell proliferation, probably through its effect on DC survival.

9.3.3 Role of TRANCE/RANK-L/ODF as a regulator of T cell function

TRANCE/RANK-L/ODF was reported to activate the c-Jun N-terminal kinase in T cells. As in the case of the TNF effect on this kinase, the signaling to this effect by TRANCE/RANK-L/ODF appears to involve TRAF2.

10. CD40 and its ligand

CD40 and its ligand, CD40-L, are critical for the development of the humoral immune response. Their interaction triggers the proliferation and regulates the differentiation of B cells and directs the isotype switch. In addition, this receptor–ligand pair exerts important functions during T-cell priming and in inflammation. Therefore, loss-of-function mutations in the genes of CD40 or CD40-L, as found in patients with X-linked hyper-IgM (HIM) syndrome, result in severe immune deficiency. CD40 is expressed in many tissues with high proliferative potential, which may suggest that it serves a more general role in the regulation of tissue homeostasis.

10.1 Molecular features of CD40 and CD40-L

CD40 is an integral transmembrane glycoprotein with an apparent molecular weight of 48–50 kDa on SDS–polyacrylamide gel electophoresis under reducing and non-reducing conditions. In some cell types CD40 may form covalently linked homodimers. While the extracellular part of CD40 displays the architecture typical for the TNF/NGF family, the cytoplasmic part exhibits no obvious homologies to other molecules. Human and murine CD40 molecules are highly conserved.

CD40-L exhibits all the features of a typical TNF ligand family member. It is a 33kDa type II membrane glycoprotein with a highly conserved 22 amino acid cytoplasmic domain, a 24 amino acid transmembrane domain and a 216 amino acid extracellular domain that contains five cysteines and a conserved N-linked glycosylation site. Crystallographic structural analysis revealed that CD40-L exists as a trimer, which is probably composed of membrane-anchored and soluble CD40-L monomers. CD40-L also exists in soluble form; the biological significance of this form is not yet clear. As with the other TNF ligand family members, CD40-L is thought to activate CD40 via trimerization. Notably, some CD40 functions require more intense aggregation than can be achieved by soluble anti-CD40 mAb or soluble CD40-L. Accordingly, in many studies either the antibody is presented via Fc receptor-bearing cells or CD40-L expressing transfectants are used.

10.2 Occurrence of CD40 and CD40-L

10.2.1 CD40 expression

CD40 is expressed mainly on cells with a high proliferative potential and on cells that are able to present antigen. These include B lymphocytes at almost all differentiation stages, dendritic cells, activated monocytes and epithelial cells.

(a) B lymphocytes

CD40 is expressed on almost all B cells irrespective of their origin, stage of maturation or activation status; the exceptions are stem cells and mature plasma cells. Expression of CD40 becomes less pronounced only during differentiation of the B cells into antibody-secreting cells. Consequently, cells from almost all B-cell malignancies express CD40. The receptor has been detected in cells from most chronic B-cell leukemias, non-Hodgkin's and Burkitt's lymphomas and in virtually all Epstein–Barr virus (EBV)-transformed B-cell lines. Cell lines derived from plasmocytoma patients are not always CD40-positive.

(b) Dendritic cells and monocytes

Dendritic cells located in the T-cell rich areas of spleen and tonsils express high levels of CD40. Their precursors, the mucosal Langerhans cells, show low CD40 levels but become strongly positive when cultured *in vitro*. Similarly, the low expression of CD40 in freshly isolated blood monocytes is strongly up-regulated after treatment with GM-CSF, IL-3 or IFNγ.

(c) Other cell types

Immunohistological examination revealed CD40 expression by various cell populations in the thymus, including some stromal cells, cortical and medullary epithelial cells, interdigitating cells and B cells. The CD40 molecule has also been demonstrated on endothelial cells, smooth muscle cells, cardiac myocytes and epithelial cells of various origin. After the initial finding that CD40 is expressed on urinary bladder carcinoma cells, malignancies originating from many other tissues were shown to be CD40-positive. These included colon, prostate, breast, kidney and lung carcinomas and some melanomas.

(d) Regulation of CD40 expression

Various pro-inflammatory cytokines may up-regulate expression of CD40. The most potent of these is IFNγ, but IL-1, TNF and IL-6 may also increase CD40 expression. Interestingly, CD40 often coregulates with MHC II.

10.2.2 CD40-L expression

Compared with CD40 expression, CD40-L expression is much more restricted and dependent on the activation state of the cell. *In vivo* application of thymus-dependent (TD) antigen results in CD40-L expression on CD4[+] T cells that are found in close proximity to B cells producing antibodies to the relevant antigen. This is consistent with the suggested function of CD40-L during T cell/B cell interaction. Another cell type that expresses large amounts of CD40L are activated platelets. *In vitro*, CD40-L is found mainly in activated CD4[+] lymphocytes of all subclasses, i.e. Th0, Th1 and Th2. Activation of CD8[+] and γδ T cells also results in CD40-L expression. Eosinophils and activated basophils also express CD40-L. Dendritic cells start to express CD40-L after being stimulated via CD40. Co-expression of CD40-L with CD40, as found in some malignant melanomas, indicates that this receptor–ligand pair may support tumor growth in an autocrine fashion.

10.3 Function

The interaction between CD40 and CD40-L controls key steps in the development of TD humoral immunity. B-cell proliferation, priming of appropriate T-helper cells, isotype switching, generation and maintenance of germinal centers and development of B-cell memory depend on this receptor–ligand pair. However, CD40/CD40-L interactions also influence a broad range of immune functions that are not mediated through B cells, including the inflammatory activity of macrophages, the development of T-helper cell subtypes and the ability of APCs to prime T-cell responses. Thus loss-of-function mutations, as found naturally in the hyper IgM (HIM) syndrome patients and introduced artificially in CD40 or CD40-L knockout mice, lead not only to a complete loss of TD B-cell responses but also to defects in

T-cell immunity and to weakened inflammatory responses.

10.3.1 CD40/CD40-L functions in B cells

(a) B-cell generation and proliferation

One of the first activities described for CD40 was its co-mitogenic effect on B cells. As deduced from *in vivo* studies with gene-targeted mice, CD40-mediated growth signals appear not to be essential for generation of B cells in the bone marrow. No abnormalities in lymphocyte counts or in the overall composition of B and T lymphocyte subpopulations were observed in either CD40- or CD40-L-deficient mice.

(b) B-cell growth and differentiation

Although CD40/CD40-L interactions are important for B-cell proliferation during the primary immune response, this process seems not to rely exclusively on this receptor–ligand pair. A recent study in mice demonstrated that the TD B-cell proliferation after immunization may occur in the absence of CD40 signals and appears to be mediated through multiple mechanisms that act independently of CD40. Surprisingly, Fas-L seems to contribute significant B-cell growth activity in this situation. *In vitro* resting B cells enter cell cycle and increase their adhesiveness after CD40 ligation. After CD40 stimulation they increase in size and exhibit homotypic aggregation mediated through LFA1–ICAM1 and CD23–CD21 interactions. Increased VLA-4-dependent adhesiveness to endothelial cells was also demonstrated. The mitogenic effect of CD40 usually requires preactivation through surface immunoglobulin or other mitogenic stimuli and is particularly strong when combined with T-cell cytokines. IL-4, IL-10 and IL-13 synergically augment the mitogenic effect of CD40 on B cells. If appropriately combined with a continuous CD40 stimulus, these cytokines allow long-term *in vitro* culturing of resting B cells and the development of various B-cell subpopulations, including those found in the mantle zone of germinal centers. In addition, CD40 may regulate B-cell growth in an autocrine fashion via the induction of IL-6 and IL-10.

(c) Antibody production and isotype switch

Antigen contact of naive B cells in the presence of the appropriate T-helper cells normally results in the production of soluble immunoglobulins. Initially these are predominantly of the IgM isotype, but because of progressive isotype switching during further B-cell differentiation IgG and IgA become the prevalent isotype species at later stages of the response. Both this primary antibody and the secondary antibody response (which also relies on the reactivation of memory B cells) depend essentially on CD40/CD40-L interactions. *In vitro* studies showing that a combination of CD40 ligation with IL-4 treatment results in increased immunoglobulin production and in IgE switching provided the first evidence for this function of CD40. Final proof came from observations made in patients with HIM syndrome where the isotype switching is absent or greatly reduced. These findings were reconfirmed in mice treated with anti-CD40-L antibodies, or exhibiting genetic defects in their CD40 or CD40-L genes. Only antibody responses to thymus-independent antigen, including isotype switching, seem to occur independently of CD40 signals. Further *in vitro* studies established that isotype switching is actually determined by CD40 signals in combination with appropriate cytokine mixtures. Thus IL-4 favors switching to IgG2 and IgE; IL-4 combined with IL-5 to IgG1 and IgE; IL-13 to IgG3 and IgE; and IL-10 to IgG1, IgG2 and IgG3. A combination of TGFβ and IL-10 results in production of IgA and suppression of other isotypes. IFNγ promotes switching to IgG2a and IgG3.

(d) Germinal center formation

The germinal center (GC) is the microenvironment in which B-cell proliferation, somatic hypermutation and affinity maturation occur. Without GC formation, memory B cells cannot develop. CD40-L expression patterns in GCs of mice immunized with TD antigen and suppression of TD humoral responses with anti-CD40-L antibodies, gave the first clues to the importance of CD40/CD40-L interactions during GC formation. Final proof came from the finding that genetic defects in CD40 or CD40-L genes resulted in complete absence of GC formation. Recent studies in CD40–/– mice indicated, however, that GC formation not only depends on CD40-mediated signals but is in fact the result of cross-talk between T and B cells in which CD40 and CD40-L play a leading role. Thus, CD40–/– mice formed a limited number of GCs when they were simultaneously treated with a TD antigen and CD40Fc to cross-link CD40-L. These data were in agreement with results from an earlier *in vitro* study, and strongly supported the notion that the engagement of CD40-L by CD40 triggers a co-stimulatory signal in T cells that contributes to GC formation. Therefore, immigration of appropriately primed T cells appears to be the initial step and an essential requirement for GC formation, even though these cells represent a minority in the GC. GC formation remains incomplete, however, in the CD40Fc-treated CD40-/- mice, indicating that CD40 signaling also plays an important role in B-cell activation for GC formation.

It is well established that GC B cells may be rescued through activation of CD40. It therefore came as no surprise that treatment with antibodies against CD40-L results in disappearance of pre-existing GCs. It seems unlikely, however, that massive apoptosis is the exclusive cause of GC disappearance. An alternative explanation might be terminal differentiation of memory B cells to plasma cells and their subsequent emigration from the GC when IL-2/IL-10-stimulated GC B cells are deprived of CD40-L. The outer zone of GCs indeed contains a subset of T cells with high CD40-L expression and it might be these T cells that maintain the integrity of the GC.

(e) Hypermutation

In view of its dominant role in the regulation of B-cell growth and differentiation, it seemed likely that CD40/CD40-L interaction is also involved in the somatic hypermutation of the variable chain (IgV) genes. However, the studies done so far could not provide any evidence that CD40-L initiates or sustains the hypermutation process.

10.3.2 T-cell priming

Interaction between CD40 and CD40-L does not result in a unidirectional signal to B cells. The activation of CD40 on B cells or other APCs seems rather to be an essential checkpoint for the successful cross-talk between an APC and its T cell. Failure of this cross-talk results in incomplete priming of the participating T cell and hence a defective immune response. CD40/CD40-L interactions may affect T-cell priming both directly and indirectly. As indicated above, the engagement of CD40-L appears to result in retrograde signaling and T-cell activation. At the same time, CD40 signals regulate the expression of important co-stimulatory molecules like B7.1 (CD80) and B7.2 (CD86) on APCs. Thus, blockade of CD40/CD40-L interaction during T-cell priming results in a lack of co-stimulatory signals to the T cell and induction of anergy. Both CD40-deficient and CD40-L-deficient mice therefore show clearly attenuated T-helper functions.

CD40/CD40-L interactions are important for both Th1 and Th2 priming. Differences in CD40-induced co-stimulatory activities on the APCs appear to determine which kind of T-helper cell will develop. In B cells the spectrum of CD40-induced co-stimulatory molecules appears to favor Th2 priming, while in macrophage-type APCs the same signal triggers a combination of co-stimulatory signals that promotes efficient Th1 priming, probably by the induction of IL-12. Consequently, the absence of CD40 signals suppresses allo-responses *in vivo*, results in improved acceptance of allografts, blocks graft-versus-host reactions, and prevents the development of T-cell-mediated inflammatory diseases. In addition, both CD40- and CD40-L-knockout mice are unable to respond to infection with leishmania. Lack of CD40 signals also leads to incomplete activation of macrophage effector mechanisms such as the production of pro-inflammatory cytokines and nitric oxide.

10.3.3 CD40 functions on non-immune cells

(a) Regulation of adhesion

An important step during inflammatory reactions is the extravasation of immune cells. Endothelial cells express increased levels of CD40 during inflammatory reactions and triggering of CD40 on endothelial cells, for example through CD40L, expressed by activated platelets, can lead to up-regulation of vascular cell adhesion molecule (VCAM), ICAM and E-selectin. This may explain why in some autoimmune disease models, treatment with anti CD40-L antibodies leads to reduced T-cell infiltration. These findings indicate that CD40/CD40-L interactions play a role in the formation of inflammatory infiltrates.

(b) Regulation of growth and differentiation

Little is known of the function of CD40 in nonimmune cells with high proliferative potential. Its co-mitogenic activity in B cells, and the finding that CD40 expression is often increased in malignant cells, raises the possibility that the CD40/CD40-L system plays a more general role in the regulation of cell growth. Evidence that some malignant melanomas co-express CD40 and CD40-L also points in this direction. However, there is no direct evidence for mitogenic effects of CD40 on tumor cells. In fact, *in vitro* data support a role for CD40 as a negative growth regulator. On keratinocytes, CD40 ligation inhibits growth and promotes differentiation. In some transformed cells of epithelial and mesenchymal origin, the ligation of CD40 even results in apoptosis. Whether or not CD40 mediates 'veto' signals to restrict unlimited growth in proliferation-competent tissues remains to be elucidated.

10.4 Hyper-IgM syndrome

Much of what is known today about the biology of CD40 and CD40-L was learned from an inborn immunodeficiency syndrome known as hyper-IgM (HIM) syndrome. This disease can be explained by genetic defects in the CD40-L gene. Examination of the CD40-L locus in HIM syndrome patients demonstrated either inactivating point mutations or deletions that lead to the expression of inactive CD40-L. Sequencing of the CD40-L cDNA from 13 patients revealed no mutational hot spot. In one of these patients no alterations were found, indicating the existence of other genetic alterations that lead to HIM-like syndrome. The clinical appearance of HIM syndrome patients is consistent with the postulated biology of CD40 and CD40-L. The condition is characterized by increased susceptibility to bacterial infections. Also typical is a high incidence of opportunistic infections, such as *Pneumocystis* pneumonia and *Cryptosporidium* diarrhea. Because there is no isotype switching these patients have extremely high IgM serum levels and very low levels of IgG. Immunization with TD antigen results in poor or absent humoral responses.

11. CD27, 4-1BB, GITR, CD30, OX-40, TACI and their ligands

The receptors listed in this group are all preferentially expressed in T lymphocytes and act as accessory molecules in lymphocyte activation, proliferation and differentiation. Most of their activities have been determined *in vitro* and their specific biological roles *in vivo* remain to be clarified. It is therefore not yet known whether these receptors are redundant parts of the network involved in the regulation of T-cell activation, or whether they act in unique situations and influence particular steps during the generation of effector T cells, such as cytotoxic T lymphocytes (CTLs) or T-helper cells.

Within this group of receptors, CD27, 4-1BB (ILA) and GITR share significant sequence homology in their intracellular domains, indicating that they operate via related signaling mechanisms. Whether their functional similarity to OX-40, CD30 and the most recently identified TACI, none of which shares this homology, is relevant or just an incidental result of the *in vitro* systems used to determine their activities cannot be judged on the basis of the currently available data. Initial *in vivo* studies indicate, however, that these receptors play specific roles in the generation and activation of particular T-cell subsets.

11.1 CD27 and its ligand, CD70

CD27 was originally identified as a dimeric membrane glycoprotein present on the surface of the majority of human T lymphocytes. More detailed studies revealed its expression on medullary thymocytes, most peripheral blood T cells, a subset of mature B cells, and NK cells. Some studies associate CD27 expression with the T-helper phenotype (CD45RA$^+$), whereas most memory T cells (CD45RA$^-$, CD45RO$^+$) lack this receptor. T-cell activation commonly results in CD27 up-regulation.

The human CD27 ligand (CD27-L, also known as CD70 antigen) has been identified by the use of a particular mAb (Ki-24). The molecule is regularly found on cell lines derived from peripheral T or B lymphomas. In the immune system CD70 expression is usually dependent on activation. It is detectable on phytohemagglutinin (PHA)-stimulated T and B lymphocytes. The expression patterns of CD27 and CD70 suggest an important role for these molecules in interactions between T cells, T cell activation, and the regulation of immunoglobulin synthesis.

Experiments in which CD27 was ligated along with treatment with suboptimal doses of 'classical' T cell proliferation stimuli clearly established that CD27 has a co-stimulatory effect on T cells *in vitro*. In certain lymphomas this activity may contribute to enhanced tumor growth. CD27 ligation also augments NK cell activity. Pokeweed mitogen (PWM)-activated B cells respond to CD27 activation with increased immunoglobulin production, indicating a role for CD27 in the regulation of immunoglobulin synthesis in B cells. This is in contrast to data obtained with a transformed B cell line (Ramos), in which CD27 ligation resulted in apoptosis. The latter effect appears to be mediated by the recently identified CD27-binding protein 'siva' (named after the Indian god of destruction). This signal protein contains a DD homology region as well as ring- and zinc-finger like motifs.

Inital *in vivo* studies suggest that CD27–CD70 interaction is involved in the regulation of thymocyte maturation. Thus, prevention of this interaction with neutralizing antibodies results in reduced expansion and differentiation of the precursor CD4$^-$CD8$^-$CD25$^+$ to CD4$^+$CD8$^+$CD25$^-$ thymocytes.

11.2 4-1BB (ILA) and its ligand

4-1BB was cloned by differential screening in the course of a search for cDNAs that are expressed preferentially in murine cytolytic and helper T cell clones. Its human homolog, initially called ILA, for receptor induced by lymphocyte activation, was cloned from a cDNA library derived from activated human T-cell leukemia virus type 1-transformed human T lymphocytes.

Like CD27, 4-1BB is preferentially expressed on the surface of various activated T-cell subpopulations. Its transcript has also been detected in several other tissues, including B cells, monocytes, fibroblasts, epithelial cells and various cell lines derived from malignancies. Activation with T or B cell stimulatory reagents such as mitogenic lectins, anti-CD3 or anti-μ antibodies up-regulates 4-1BB expression. In epithelial and hepatoma cells, increased expression was also found after IL-1β treatment.

The 4-1BB ligand was first identified on activated macrophages and mature B cells and was later also found to be expressed by mitogen-activated T cells.

The biological functions of 4-1BB observed *in vitro* include co-stimulatory effects on various T-cell subpopulations and, as with GITR, protection of T cells from activation-induced cell death. Paradoxically, the interaction between 4-1BB and its ligand may also lead to T-cell apoptosis. The latter effect appears to ensue from reverse signaling through the 4-1BB ligand. This receptor–ligand pair thus appears to have a rather complex role in the regulation of T cell survival and death.

Recent *in vivo* studies suggest that 4-1BB plays a major role in the control of cytolytic T cell activity. Injection of anti 4-1BB mAbs into tumor-bearing mice was shown to result in eradication of tumors, even rather large ones. This effect was accompanied by marked augmentation of tumor-selective cytolytic T cell activity. In a murine model of acute graft-versus-host disease, injection of such antibodies enhanced the generation of allospecific cytotoxic T cells and thus the rejection of the transplant. It was suggested that this activity involves the tyrosine kinase p56^{lck}, which associates with the intracellular domain of 4-1BB.

11.3 GITR

GITR (for glucocorticoid-induced TNFR family-related gene) is a type I transmembrane receptor whose extracellular domain contains three Cys-motifs. Its intracellular domain shows clear sequence homology to the intracellular domains of CD27 and 4-1BB. Expression of the transcript of GITR has been discerned only in T cells: thymocytes, spleen and lymph node T lymphocytes and cells of a T-cell hybridoma. The levels of constitutive expression of the transcript are rather low, but increase significantly after T-cell activation and in response to dexamethasone. The only activity of GITR reported so far is signaling for resistance to anti-CD3 antibody-induced apoptosis, observed in cells of T-cell hybridomas that over-express the transfected receptor. This effect appears to correspond to a mechanism that acts regularly to restrict the apoptotic process, since expression of antisense of GITR resulted in augmented anti-CD3-induced cell death. Death induced by CD95 triggering, dexamethasone treatment or UV irradiation is not affected by GITR. The identity of the ligand of GITR is not yet known.

11.4 CD30 and its ligand

CD30 was initially identified with the use of antibodies (Ki-1) that were developed to visualize Reed–Sternberg cells in Hodgkin's disease. Subsequent studies demonstrated that this TNF receptor family member is preferentially expressed in tissues containing T lymphocytes. The 3.5 kb transcript for murine CD30 was detected in thymus, mitogen-activated spleen cells and various T-cell lines. The identity of the particular T-cell subpopulation in which CD30 is expressed is currently under debate. CD30 expression in malignancies appears to be common. In addition to Reed–Sternberg cells, CD30 has also been found in various non-Hodgkin lymphomas, leukemias, embryonal carcinomas, malignant melanomas and mesenchymal tumors. Some poxviruses express viral CD30 homologs.

Although CD30-L shares less than 20% identity with other members of the TNF ligand family, it exhibits all the typical features of this gene family. The human and mouse proteins share 72% identity. Unlike other members of the TNF ligand family, CD30-L is extensively glycosylated. CD30-L transcription is triggered by calcium ionophores in tonsillar T cells and by LPS or IL-1β in monocytes. CD30-L expression has also been observed in unstimulated neutrophils, normal B cells and some B-cell lines.

While the murine CD30 contains three cysteine-rich pseudo-repeats in its extracellular domain, the human gene encodes six such domains, indicating that it has evolved from the structure found in the mouse by a replication event. The functional implications of this finding are not yet clear. Like the p75 TNFR, CD40, CD27, 4-1BB and HVEM, CD30 signals through interaction with TRAFs. Direct association with TRAF1, TRAF2 and TRAF3 has been demonstrated. The functional relevance of the cross-species conserved PEST sequences (i.e. sequences rich in proline (P), glutamic acid (E), serine (S) and threonine (T)) and various potential phosphorylation sites in the signaling domain of CD30 remain to be determined.

Most of the biological characterization of CD30 is based on data from *in vitro* studies, thus making it difficult to judge its biological role *in vivo*. CD30 ligation with ligand or agonistic antibodies results in enhanced proliferation of peripheral blood T cells. In some *in vitro* test systems, CD30 enhances development of Th2-type helper cells and mediates activation and proliferation of human $\gamma\delta$ T cells and cytokine secretion in CD8$^+$ CTLs. As expected from its association with TRAFs, CD30 signaling results in NF-κB activation. This activity probably accounts for the ability of CD30 to enhance HIV replication in T cells from HIV-infected individuals. Ligation of CD30 also stimulates B-cell proliferation as well as production and secretion of immunoglobulin. Many studies on CD30 function concern its role in the biology of malignant lymphomas, where it may have mitogenic as well as anti-growth effects. The physiological context in which CD30-mediated activation signals are of importance remains to be determined. CD30 triggers apoptosis *in vitro* by a mechanism which is still unclear. The *in vivo* significance of this finding was substantiated in mice with targeted disruption of the CD30 gene. Thymocytes of these mice were more resistant to activation-induced cell death and the mice showed clear defects in negative thymocyte selection.

Recent studies have indicated that CD30/CD30-L interactions provide bidirectional signals. Thus, it was shown that engagement of CD30-L on freshly isolated neutrophils results in increased IL-8 production and a rapid oxidative burst, and in peripheral blood T cells, activation of CD30L is co-mitogenic and leads to cytokine production.

In view of its high expression in various lymphomas, considerable efforts have been made to use CD30 as a specific disease marker and to develop therapeutic strategies using CD30 as a target structure. Initial clinical trials targeting CD30 with antibodies turned out to be ineffective. Use of antibodies conjugated to various toxins resulted in incidental regressions, but remissions were of short duration. Future clinical trials will be needed in order to demonstrate whether CD30 is a useful target structure for an efficient and specific therapy in CD30-positive lymphomas.

11.5 OX40 and its ligand

The OX40 molecule was first discerned using a mAb that was raised against a 47–51 kDa cell-surface antigen expressed exclusively on CD4$^+$ T lymphoblasts. Expression cloning of the cDNA identified this molecule as a member of the TNF/NGF receptor family with three Cys-repeats in its extracellular domain and a short cytoplasmic domain of 37 amino acids in human and 36 in mouse and rat. There are conflicting data concerning the tissue distribution of OX40. Most reports indicate that the expression of the protein is largely restricted to activated T cells and, in smaller amounts, to the testis. However, there have also been reports describing wider expression of the protein: in lung, spleen and thymus and at lower levels in the heart, placenta and peripheral blood mononuclear leukocytes. Further studies will be required to clarify this matter.

Two lines of research led to the discovery of the OX40 ligand. The search for molecules specifically expressed in HTLV-1-infected T cells led to the discovery and cloning of a Tax-regulated 34 kDa glycoprotein expressed on HTLV-1-infected T-cell clones. This molecule was recognized as the OX40 ligand when its cDNA was isolated by several groups in an expression-cloning approach using OX40Fc constructs as molecular probes.

Despite its relatively low homology (about 20%) with other members of the TNF-ligand family, OX40-L displays all of this family's typical structural features. The OX40-L protein was detected on the surface of activated T cells (at a higher level on CD4$^+$ than on CD8$^+$ lymphocytes), on B cells, on some B-cell lines and on *in vitro* cultured endothelial cells. Analysis of its tissue distribution suggests a broad expression pattern with species differences between rodents and humans. OX40-L mRNA-positive tissues include spleen, thymus, brain, kidney, heart, lung, liver, pancreas, skeletal muscle and gonads. Future studies will be required to determine which cell types in these tissues express the OX40-L protein.

Initial *in vitro* studies established a co-stimulatory effect of OX40 on mitogen or anti-CD3-treated CD4$^+$ T cells. The observation that B cells in the periarteriolar lymphoid sheath of the spleen (PALS) express OX40-L triggered studies that suggested an important role for OX40/OX40-

L interactions in the generation of antibody-producing B cells. The same studies also demonstrated that OX40-L, like CD27-L, CD30-L and CD40-L, triggers an activation signal (when engaged with OX40Fc). The *in vivo* relevance of these findings was supported by the demonstration that anti-OX40-treated mice showed greatly decreased TD dependent antibody responses. This coincided with reduced formation of periarterteriolar lymphoid sheath (PALS)-associated B-cell foci. Germinal center formation under these experimental conditions remained intact, suggesting that OX40 and CD40 regulate distinct aspects of the B-cell response.

Recent studies suggest that OX40⁺ T cells are involved in the development of several autoimmune diseases, including rheumatoid arthritis, graft-versus-host disease and experimental autoimmune encephalomyelitis (EAE; in rats). Thus, a high percentage of CD4⁺ T cells found in the afflicted tissues expressed OX40, and an anti-OX40 immunotoxin effectively suppressed the development of EAE in rats. The exact role of OX40 in these diseases remains to be elucidated. Another function of the OX40/OX40-L system may be the mediation of adhesion between activated T cells and vascular endothelial cells. However, the *in vivo* circumstances under which T-cell adhesion to the endothelium is mediated by this receptor ligand pair remain to be elucidated. Another open question is the relative significance of OX40/OX40-L up-regulation by HTLV-1 infection for the pathology of this virus.

11.6 TACI

TACI (for transmembrane activator and CAML interactor) is a member of the TNF/NGF family whose extracellular domain contains two Cys-motifs and lacks leader sequence (a type III transmembrane protein). Its intracellular domain, which lacks homology to known sequence motifs, binds specifically to CAML (calcium modulator and cyclophilin ligand), an adaptor protein involved in activation of the transcription factor NF-AT. Expression of the TACI transcript (about 1.4 kb) and protein seem to be restricted to B lymphocytes (resting and activated) and T lymphocytes (both CD4 and CD8 cells, though only after activation).

TACI activates NF-AT in a CAML-dependent way, and also activates AP-1 and NF-κB independently of CAML. CAML is located in intracellular vesicles as an integral membrane protein. The way in which TACI gains access to it, as well as the identity of the ligand of TACI, are not yet known.

12. APRIL

APRIL (for **a** **p**roliferation-**i**nducing **l**igand) is a member of the TNF ligand-family that was just recently cloned by screening public data bases for cDNAs that display homology to known members of the family. It shows highest homology to CD95. Human APRIL is a 201-amino-acid type II transmembrane protein with a single N-linked glycosylation site within the receptor-binding motif. APRIL transcripts of several different sizes are expressed in various tissues, though at rather low levels: transcripts of 2.1 and 2.4 kb in the prostate, of 1.8 kb in PBL and (at lower amounts) of 2.1 kb in colon, spleen and pancreas. Rather high levels of the 2.1 transcript are found, however, in cells of various tumor lines as well as in certain tumors.

The only known function of APRIL is growth enhancement. It was found to stimulate the growth of cells of various tumor lines, both of hematopoietic and non-hematopoietic origins, particularly at a low serum concentration. It also enhanced the *in vivo* growth of NIH-3T3 cells in nude mice. The receptor for APRIL is still unknown, nor is there any knowledge of the signaling mechanism by which it affects cell growth.

13. The low-affinity receptor for nerve growth factor (p75 NGF-R)

The low-affinity receptor for nerve growth factor (p75 NGF-R) is expressed predominantly on nerve and glial cells, but some expression is also noted on lymphoid cells. The receptor molecule is of approximately 400 amino acid residues. The extracellular domain has four Cys-motifs typical of the TNF/NGF family of receptors. The cysteine-rich domain forms the ligand-binding site for NGF, and all four of its motifs are required for proper ligand binding. The p75 NGF-R can also bind other neurotrophins, such as the brain-derived neurotrophic factor and neurotrophins 3 and 4. The intracellular domain of the receptor has no intrinsic enzymatic activity. However, association of p75 NGF-R with the ERK1 and ERK2 kinases has been reported, and was implicated in signaling by NGF. Moreover, the C-terminal region of the intracellular domain was found to contain a DD motif. In analysis of the structure of the DD in this receptor it was noted that part of it shows similarity to mastoparan, raising the possibility that it is involved in G protein binding.

Neurotrophic ligands can also bind receptors

belonging to another family, called Trk. These receptors are encoded by the protooncogene *trk* and its relatives, and are characterized by having an immunoglobulin-like extracellular domain as well as by intrinsic activity of the tyrosine kinase which resides in their intracellular domain. It had initially been suggested that neurotrophins signal mainly through the Trk receptors, by inducing the phosphorylation-mediated regulation of the MAP kinase cascade, phospholipase $C\gamma$, and phosphoinositide -3-kinase. The p75 NGF-R had been initially proposed to serve only as an accessory molecule for recruiting the neurotrophic ligand to the receptor encoded by Trk (or its homolog) or participating in the formation of a high-affinity site for NGF binding. However, there is now evidence that the p75 NGF-R also plays an active signaling role, which resembles at least in some respects the functions of other receptors of TNF/NGF family. It can either promote survival or induce apoptosis of nerve and glial cells. The eventual cellular response probably depends on the type of cell or its developmental stage. In addition, stimulation via p75 NGF-R can induce activation of the NF-κB transcription factor and sphingomyelin hydrolysis. In contrast to other receptor–ligand pairs of the TNF family, neither NGF (the major ligand of the p75 NGF-R) nor any other neurotrophin possesses homology to TNFα or any other typical structural features of the TNF-ligand family.

14. Crinkly-4

Crinkly-4 is the first member of the TNF/NGF family to be found in cells that are not of animal origin. It is a maize transmembrane receptor with an as yet unknown ligand that acts to regulate differentiation. Its mutation results in a 'crinkly' shape of the leaves resulting from abnormal differentiation of the leaf epidermis, and in incomplete development of the aleurone layer in the seeds. The extracellular domain of Crinkly-4 contains a cysteine-rich region that closely resembles the second Cys-motif in the Shope fibroma virus TNF receptor. The intracellular domain contains a serine/threonine protein kinase domain characteristic of the serine/threonine innate immune kinases, a family of kinases involved in host defense and embryonic morphogenesis.

Acknowledgements

We would like to thank Drs Marcus Schuchman and Patricia Spear and Mr. Andrei Kovalenko for their help in the preparation of the manuscript, Ms Shirley Smith for editorial assistance and Drs Avi Ashkenazi and Carl Ware for providing us with published and unpublished material. Research in the authors' laboratories was supported in part by grants from Inter-Lab Ltd, Ness Ziona, Israel, from Ares Trading S.A., Switzerland and from the Israeli Ministry of Arts and Sciences (to D. W.), from the State Committee for Scientific Research, Poland (no. 6 P207 121 06, to J.B.), and from the Deutsche Forschungsgemeinschaft (SFB217 and Gerhard Hess-Programm, to H. E.).

References*

The first record of TNF action

Menkin, V. (1948). *Newer Concepts of Inflammation*. Thomas, Springfield.

Earlier reviews on the TNF ligand and receptor families

Beutler, B. and van Huffel, C. (1994). Unraveling function in the TNF ligand and receptor families [comment]. *Science* **264**, 667–8.

Cosman, D. (1994). A family of ligands for the TNF receptor superfamily. *Stem Cells* **12**, 440–55.

Gruss, H. J., and Dower, S. K. (1995). Tumor necrosis factor ligand superfamily: involvement in the pathology of malignant lymphomas. *Blood* **85**, 3378–404.

Meakin, S. O. and Shooter, E. M. (1992). The nerve growth factor family of receptors. *Trends Neurosci.* **15**, 323–31.

Smith, C. A., Farrah, T., and Goodwin, R. G. (1994). The TNF receptor superfamily of cellular and viral proteins: activation, costimulation, and death. *Cell* **76**, 959–62.

Wallach, D. (1996). Suicide by order: some open questions about the cell-killing activities of the TNF ligand and receptor families. *Cytokine Growth Factor Rev.* **7**, 211–23.

Reviews on the mechanisms of action of the ligands and the receptors

Feinstein, E., Wallach, D., Boldin, M., Varfolomeev, E., and Kimchi, A. (1995). The death domain: a module shared by proteins with diverse cellular functions. *Trends Biochem. Sci.* **29**, 342–4.

Van Ostade, X., Tavernier, J., and Fiers, W. (1994). Structure–activity studies of human tumour necrosis factors. *Protein Eng.* **7**, 5–22.

Wallach, D., Boldin, M. P., Goltsev, Y. V., Malinin, N. L., Kovalenko A. V., and Varfolomeev, E. E. (1999). TNF receptor and Fas signaling mechanisms. *Annu. Rev. Immunol.* **17**, in press.

* At the request of the publisher, the detailed reference list of more than 500 publications, which was originally submitted with this review, has been omitted for space reasons. The short list that replaces it refers solely to some previous reviews or, in the case of the most recently discovered family members, to publications describing the initial characterization of the proteins. We apologize to all the scientists whose work is cited in this review for being unable to provide full documentation of their achievements.

Reviews on TNF

Aggarwal, B. B. and Natarajan, K. (1996). Tumor necrosis factors: developments during the last decade. *Eur. Cytokine Netw.* **7**, 93–124.

Aggarwal, B. B. and Vilcek, J. (1992). *Tumor Necrosis Factors.* Marcel Dekker, New York.

Wallach, D. (1996). The sixth international congress on TNF: a decade of accumulated knowledge and emerging answers. *Eur. Cyokine Netw.* **7**, 713–24.

Reviews on Fas/Apo-1

Green, D. R. and Ware, C. F. (1997). Fas ligand: privilege and peril. *Proc. Natl Acad. Sci. USA* **94**, 5986–90.

Nagata, S. and Golstein, P. (1995). The Fas death factor. *Science* **267**, 1449–56.

A review on TRAIL, its receptors and Apo-3

Goldstein, P. (1998). Cell death: TRAIL and its receptors. *Curr. Biol.* **7**, R75–R753.

Reviews on CD40

Van Kooten, C. and Banchereau, J. (1997). Functions of CD40 on B cells, dendritic cells and other cells. *Curr. Opin. Immunol.* **9**, 330–7.

Grewall, I. S. and Flavell, R. A. (1998). CD40 and CD154 in cell-mediated immunity. *Annu. Rev. Immunol.* **16**, 111–35

Initial characterization of the recently discovered members
HVEM and its ligand, LIGHT

Mauri, D. N., Ebner, R., Montgomery, R. I., Kochel, K. D., Cheung, T. C., Yu, G.-L., Ruben, S., Murphy, M., Eisenberg, R. J., Cohen, G. H., Spear, P. G., and Ware, C. F. (1998). LIGHT, a new member of the TNF superfamily, and lymphotoxin are ligands for herpesvirus entry mediator. *Immunity* **8**, 21–30.

Montgomery, R. I., Warner, M. S., Lum, B. J., and Spear, P. G. (1996). Herpes simplex virus-1 entry into cells mediated by a novel member of the TNF/NGF receptor family. *Cell* **87**, 427–36.

TWEAK/Apo3-L

Chicheportiche, Y., Bourdon, P. R., Xu, H., Hsu, Y.-M., Scott, H., Hession, C., Garcia, I., and Browning, J. L. (1997). TWEAK, a new secreted ligand in the tumor necrosis factor family that weakly induce apoptosis. *J. Biol. Chem.* **272**, 32401–10.

Marsters, S. A., Sheridan, J. P., Pitti, R. M. Brush, J., Goddard, A., and Ashkenazi, A. (1998). Identification of a ligand for the death-domain containing receptor Apo3. *Curr. Biol.* **8**, 525–528.

CAR1

Brojatsch, J., Naughton, J., Rolls, M. M., Zingler, K., and Young, J. A. T. (1996). CAR1, a TNFR-related protein is a cellular receptor for cytopathic avian leukosis-sarcoma viruses and mediates apoptosis. *Cell* **87**, 845–55.

TRANCE/RANK-L/ODF and its receptor

Anderson, D. M., Maraskovsky, E., Billingsley, W. L., Dougall, W. C., Tometsko, M. E., Roux, E. R., Teepe, M. C., DuBose, R. F., Cosman, D., and Galibert, L. (1997). A homologue of the TNF receptor and its ligand enhance T-cell growth and dendritic-cell function. *Nature* **390**, 175–9.

Wong, B. R., Rho, J., Arron, J., Robinson, E., Orlinick, J., Chao, M., Kalachikov, S., Cayani, E., Bartlett, F. S. R., Frankel, W. N., Lee, S. Y., and Choi, Y. (1997). TRANCE is a novel ligand of the tumor necrosis factor receptor family that activates c-Jun N-terminal kinase in T cells. *J. Biol. Chem.* **272**, 25190–194.

Lacey, D. L., Timms, E., Tan, H. L., Kelley, M. J., Dunstan, C. R., Burgess, T., elliott, R., Colombero, A., Elliott, G., Scully, S., Hsu, H., Sullivan, J., Hawkins, N., Davy, E., Capparelli, C., Eli, A., Qian, Y. X., Kaufman, S., Sarosi, I., Shalhoub, V., Senaldi, G., Guo, J., Delaney, J., and Boyle, W. J. (1998). Osteoprotegerin ligand is a cytokine that regulates osteoclast differentiation and activation. *Cell 17* **93**, 165–76.

GITR

Nocentini, G., Giunchi, L., Ronchetti, S., Krausz, L. T., Bartoli, A., Moraca, R., Migliorati, G., and Riccardi, C. (1997). A new member of the tumor necrosis factor/nerve growth factor receptor family inhibits T cell receptor-induced apoptosis. *Proc. Natl Acad. Sci. USA* **94**, 6216–21.

TACI

von Bulow, G. U. and Bram, R. J. (1997). NF-AT activation induced by a CAML-interacting member of the tumor necrosis factor receptor superfamily. *Science* **278**, 138–41.

Osteoprotegerin

Simonet, W. S., Lacey, D. L., Dunstan, C. R., Kelley, M., Chang, M. S., Luthy, R., Nguyen, H. Q., Wooden, S., Bennett, L., Boone, T., *et al.* (1997). Osteoprotegerin: a novel secreted protein involved in the regulation of bone density. *Cell* **89**, 309–19.

Emery, J. G., McDonnell, P., Burke, M. B., Deen, K. C., Lyn, S., Silverman, C., Dul, E., Appelbaum, E. R., Eichman, C., DiPrinzio, R., Dodds R, A., James, I. E., Rosenberg, M., Lee, J. C., and Young, P. R. (1998). Osteoprotegerin is a receptor for the cytotoxic ligand TRAIL. *J. Biol. Chem.* **273**, 14363–7.

Crinkly-4

Becraft, P. W., Stinard, P. S., and McCarty, D. R. (1996). CRINKLY4: a TNFR-like receptor kinase involved in maize epidermal differentiation. *Science* **273**, 1406–9.

APRIL

Hahne, M., Kataoka, T., Schroter, M., Hofmann, K., Irmler, M., Bodmer, J. L., Schneider, P., Bornand, T., Holler, N., French, L. E., Sordat, B., Rimoldi, D. and Tschopp, J. (1998). APRIL, a new ligand of the tumor necrosis factor family, stimulates tumor cell growth. *J. Exp. Med.* **188**, 1185–90.

V The interleukin-1 system

Alberto Mantovani

1. Introduction

Interleukin-1 (IL-1) is the term for two polypeptide mediators, IL-1α and IL-1β, which are among the most potent and multifunctional cell activators described in immunology and cell biology. Recently, a cytokine called interferon-γ inducing factor was discovered; it was suggested that it be named IL-18, but the finding that it had the same conformation as IL-1 led to it being called IL-1γ. Since it does not interact with IL-1 receptors, it will not be discussed here.

The spectrum of action of IL-1 encompasses cells of hematopoietic origin, from immature precursors to differentiated leukocytes, vessel wall elements, and cells of mesenchymal, nervous and epithelial origin. Occupancy of a few receptors, or perhaps only one receptor, per cell is sufficient to elicit cellular responses.

The activity of IL-1 overlaps largely with that of tumor necrosis factor (TNF) and other cytokines. This overlap renders it difficult to trace unequivocally the history of the discovery of IL-1 in the 'premolecular' era. Many of the early descriptions of IL-1 activities, for instance that of endogeneous pyrogen, can pertain to other cytokines too. The identification of lymphocyte activating factor (LAF) was a landmark because it provided a reliable, easy and yet fairly specific *in vitro* assay for purifying and, eventually, cloning of IL-1. It is ironic that the actual importance of IL-1 in the generation of T-cell immunity is still not clearly defined, in spite of its role in the discovery.

The production and action of IL-1 are regulated by multiple control pathways, some of which are unique to this cytokine (Figure 1). This complexity and uniqueness is best represented by the term 'IL-1 system'. The IL-1 system consists of the two agonists, IL-1α and IL-1β, a specific activation system (IL-1-converting enzyme, or ICE), a receptor antagonist (IL-1ra) of which there are different isoforms, and two high affinity surface binding molecules. This chapter will focus largely on IL-1 and its receptors, with summaries of the properties of the other elements of the system (IL-1ra, ICE).

2. IL-1 gene structure and regulation

In 1984, a human IL-1β cDNA was identified and sequenced; other sequences for human IL-1 were obtained later. Earlier biochemical studies had identified two forms of IL-1 which differed in their isoelectric points. That two forms of IL-1 indeed existed was confirmed by the identification of two different cDNAs.

The human IL-1α gene is approximately 10 kb long and its mRNA 2000–2200 nucleotides long. The corresponding cDNA contains a 5'-untranslated region (59 bp), followed by a single open reading frame of 813 bp which encodes a precursor protein of 271 amino acids. The sequence does not contain an N-terminal signal peptide of hydrophobic residues, nor does it have a long internal hydrophobic stretch. The coding sequence is followed by a 3'-untranslated region of 1141 bp, and terminated with a poly(A) tail. The 3'-untranslated sequence of IL-1α (and that of IL-1β) contains several copies of the ATTTA motif, which is thought to be involved in reducing transcript stability.

The human IL-1β gene is 7 kb long, and the mRNA 1800 nucleotides long. The corresponding cDNA has a 5'-untranslated region of approximately 70 bp. The coding region is 807 bp long, thus coding for a 269 amino acid precursor protein. The protein encoded by this sequence does not contain a classical leader peptide. Human IL-1α and IL-1β share only 26% homology at the protein level, but

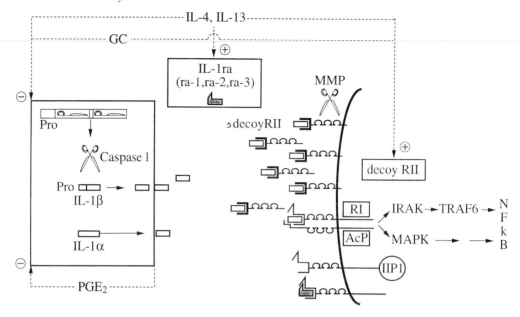

▮ **Figure 1.** An overview of the IL-1 system. Abbreviations: GC, glucocorticoid hormones; ra, receptor antagonist. The plus and minus signs indicate stimulation or inhibition of production, respectively. RI, RII, receptor I, II; AcP, accessory protein; IRAK, IL-1 receptor associated kinase; MAPK, mitogen associated protein kinase; TRAF6, TFN activated factor 6; NFkB, nuclear factor kappa enhanced binding protein.

show significantly more homology (45%) occurs at the nucleotide level, raising the possibility that these genes arose from a duplication event.

The human IL-1 genes are located on chromosome 2, in position 2q13, in the same region as IL-1ra and the type I and type II receptors. Although IL-1α and IL-1β exon sequences differ considerably, the intron–exon structures of these two genes are very similar. Moreover, IL-1α and IL-1β genes show considerable homology in the intron sequences, suggesting a regulatory role for these regions in IL-1 expression (see below).

Both genes have seven exons (Figure 2). The first intron of IL-1β contains a highly conserved homopurine tract. Other sequences in the first intron of the human IL-1β gene may exert either positive or negative regulatory activity in transfection experiments. The fourth intron of the human IL-1α gene and the third intron of the IL-1β gene contain Alu sequences. There is a 46 bp tandem repeat in intron 6 of human IL-1α that could act as an enhancer or a suppressor. Each repeat contains a binding sequence for the transcriptional factor SP-1, an imperfect copy of a viral enhancer element and an inverse complementary copy of the glucocorticoid responsive element (GRE). The number of repeats varies. Intron 5 of IL-1β gene contains a sequence which resembles the consensus GRE (TGTYCT).

The role, if any, of these sequences in mediating glutocorticoid-induced suppression of IL-1 gene transcription has not been determined. Only one variant IL-1β allele has been described.

The IL-1 genes are not expressed in unstimulated blood monocytes, vascular cells (smooth muscle and endothelial cells) or fibroblasts. The IL-1α promoter, in contrast to the IL-1β promoter, lacks a CAT box and has a very poor TATA box. Transient transfection experiments using upstream sequences of the human IL-1β gene demonstrated that CAT expression is detectable in the human promonocytic cell line THP-1 but not in HeLa cells. In contrast, a fusion gene containing also the first intron of the IL-1β gene was expressed in HeLa cells. Moreover, when the first intron is present in the expression construct, only 132 bp of IL-1β promoter are sufficient to promote transcriptional activity.

The human IL-1β gene contains a phorbol myristate acetate (PMA)-responsive enhancer between positions −2982 and −2795 upstream from the transcriptional start site. This enhancer sequence contains a DNA motif similar to the AP-1-responsive element. The human IL-1β gene contains a lipopolysaccharide (LPS)-responsive element between positions −3757 and −2729. The LPS-inducible element appears to mediate PMA

(a)

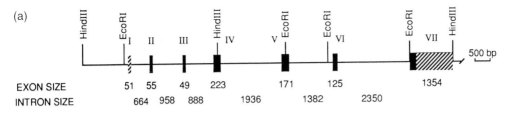

EXON SIZE	51	55	49	223		171	125		1354
INTRON SIZE		664	958	888	1936		1382	2350	

(b)

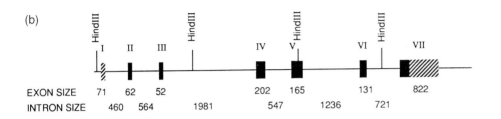

EXON SIZE	71	62	52		202	165		131	822
INTRON SIZE		460	564	1981		547	1236	721	

(c)

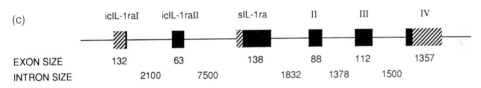

EXON SIZE	132		63		138		88	112		1357
INTRON SIZE		2100		7500		1832		1378	1500	

∎ **Figure 2.** Structures of the IL-1α (a), IL-1β (b) and IL-1ra (c) genes. Exon and intron sizes are indicated, as are selected restriction sites.

and IL-1 responsiveness in monocytes and fibroblasts.

A variety of external stimuli can activate transcription of IL-1 genes, including endotoxins from Gram-negative bacteria, exotoxins from Gram-positve bacteria, phorbol esters, calcium ionophores, UV light, T cells, complement components and adhesion to extracellular matrix molecules. IL-1β induction by LPS does not require intact protein synthesis machinery, suggesting that it involves activation of pre-existing transcriptional factors. The cytokines that induce IL-1 gene transcription are IL-1 itself, TNF and IL-2. Certain cytokines synergize with LPS in inducing IL-1 gene transcription, as demonstrated for interferon gamma IFN-γ and granulocyte–macrophage colony-stimulating factor (GM-CSF). Histamine alone does not induce protein synthesis or production of IL-1β transcripts in human peripheral blood mononuclear cells (PBMC), but is associated with a two- to three-fold increase in IL-1α-induced production of IL-1β transcripts and protein via H2 receptors. In contrast, histamine reduces LPS-

induced IL-1 expression in monocytes isolated by plastic adherence. In mononuclear phagocytes the level of LPS-induced transcripts reaches a peak 4–6 h after stimulation. Early gene products from cytomegalovirus can induce the reporter gene activity of a fusion gene that includes nucleotides −1097 to +14 of the human IL-1β promoter.

IL-1 gene transcription is also under negative control. The T_n2-derived cytokines IL-4, IL-10 and IL-13 suppress IL-1 expression in LPS-treated monocytes. Also IL-6 down-regulates IL-1 expression induced by LPS. IL-4 reduces IL-1 expression also when IL-1 is induced synergically by IFN-γ and LPS. In contrast, IFN-γ down-regulates IL-1β and IL-1α induction by IL-1. Apart from cytokines, the best inhibitors of IL-1 transcription are glucocorticoids, which reduce IL-1 production at both transcriptional and post-transcriptional levels and block IL-1β transcript expression in IL-1-stimulated astrocytoma cells and in LPS activated PBMC and U937 cells. One study reported that glucocorticoids did not suppress IL-1 transcription, but this may have been because the high LPS concentration used ($10\,\mu g/ml$) may have inhibited glucocorticoid activity. Prostaglandins do not affect IL-1 gene expression induced by IL-1 in vascular cells, but block IL-1 release at the post-transcriptional level. The same was found in LPS-stimulated PBMC and murine macrophages. In contrast with these studies, prostaglandin E_2 (PGE_2) and cAMP agonists have been shown to augment IL-1β mRNA in LPS-activated murine macrophages, without any effect on IL-1α transcripts. Also IL-1-induced expression of both IL-1α and IL-1β production has been found to be enhanced by PGE_2 in human PBMC. These discrepancies have yet to be explained. Inducers of heat-shock response also downregulate IL-1 expression. In summary, IL-1 gene transcription is modulated by positive and negative signals and can be modulated by different cytokines, either alone or in combination.

IL-1 expression is also differentially regulated during the maturation of blood monocytes into macrophages. Gene expression studies revealed that, 4 h after LPS stimulation, the level of IL-1β transcripts in macrophages was three-fold lower than in that in monocytes. However, total (i.e. intracellular and secreted) IL-1β protein production was higher in macrophages than in monocytes, suggesting that the efficiency of translation was higher in macrophages. Nevertheless, macrophages secrete only 1–5% of intracellular IL-1β, whereas monocytes release 5–20%, explaining why monocytes secrete more IL-1β than macrophages.

There is only a low level of homology between the IL-1α and IL-1β promoters, so it is not surprising that these two genes are expressed differently in different cell types. Keratinocytes and T cell clones express two to four times more IL-1α than IL-1β transcripts, whereas LPS-stimulated PBMC express predominantly IL-1β mRNA. PMA can induce IL-1β, but not IL-1α, transcripts in U937 cells. IL-1α induces both transcripts in PBMC, but only IL-1β transcripts in vascular cells (smooth muscle and endothelial cells). In these latter cell types, IL-1α transcripts were observed when cells were treated with the protein synthesis inhibitor cycloheximide. Cycloheximide is also known to superinduce LPS-induced IL-1 transcripts, by increasing the stability of IL-1 transcripts. In contrast, PMA-induced IL-1 mRNA is more stable and is not affected by the presence of cycloheximide. Thus, LPS and PMA induce IL-1 gene expression by different mechanisms.

Sequences involved in binding of transcription-regulatory proteins have been identified in the IL-1β gene promoter. Using an electrophoretic mobility shift assay, a DNA-binding activity was found in the promoter region between nucleotides −58 and +11 using nuclear extracts from resting and activated PBMC. This region is highly conserved between human and murine IL-1β promoters, and contains the TATA box sequence. A more detailed analysis revealed that a protein termed NFIL-1βA binds upstream of the TATA box, at nucleotides −49 to −38, suggesting that this nuclear factor may interact with TATA box-binding factors.

IL-1 expression is also controlled at another level: the regulation of its mRNA stability. As mentioned above, IL-1 3'-untranslated regions contain multiple AU-rich regions, which are known to be involved in transcript decay. LPS-activated THP-1 cells transcribe IL-1α and IL-1β genes at similar rates, but IL-1α transcripts are much more unstable than those coding for IL-1β. The ability of IFN-γ to increase LPS-induced IL-1 production is mediated by both enhanced IL-1β transcription and increased mRNA stability. Identical results have been obtained with PMA-induced expression of IL-1β in fibroblasts and THP-1 cells. Since a transfected mutated H-*ras* gene increases both IL-1β transcription and mRNA stability, a role for G-proteins in these processes is likely. It has been reported that the IL-1β mRNA from monocytes is less stable than that from of macrophages (approximate half-lives of 2–3 h and 10 h respectively). Since IL-1β transcripts are more abundant in monocytes than in macrophages, though, this implies that the transcriptional rate of this gene is higher in monocytes. Histamine decreases IL-1β mRNA stability, but augments IL-1β transcripts in IL-1α-treated PBMC, presumably by increasing transcriptional rate. TNF increases the level of IL-1β transcripts in fibrosarcoma cells by making transcripts more stable, while leaving gene transcription unchanged. This effect of TNF is mediated by activating protein kinase C.

Of the negative regulators of IL-1 expression, both IL-4 and the glucocorticoid dexamethasone, in addition to inhibiting IL-1 gene transcription, also increase IL-1 mRNA decay. The destabilizing effect of IL-4 and dexamethasone on IL-1 mRNAs is blocked by the presence of a protein synthesis inhibitor, thus indicating that this effect requires *de novo* protein synthesis.

3. Agonists

The mature human IL-1α (pI = 5) and IL-1β (pI = 7) polypeptides share 26% amino acid identity.

Comparison of mature IL-1β from different animal species indicates that 75–78% of amino acids are conserved; IL-1α sequences are less conserved among species (60–70%). The primary translation products of IL-1α and IL-1β are 271 and 269 amino acids long, respectively, corresponding to molecular weights of 30 606 and 30 749, respectively. The IL-1α propeptide is biologically active whereas the IL-1β precursor is not. Although the sequence contains glycosylation sites, sugars are not important in the biological activity of IL-1. IL-1α may be glycosylated and mannose sites may be important for its association with the cell membrane.

IL-1α and IL-1β lack a signal peptide. IL-1α remains mostly in the cytosol and associated with the plasma membrane. The pathway of secretion of mature IL-1β (amino acids 117–269 of the precursor) is not defined and may be common with that of other leaderless proteins.

Cleavage of pro-IL-1β to mature IL-1β is mediated by ICE, a cysteine protease representative of a novel class of proteolytic enzymes. ICE and related molecules are involved in the regulation of apoptosis. It remains unclear whether and how regulation of cell death and IL-1β processing are related. It is noteworthy that induction of apoptosis induces IL-1 in monocytes.

IL-1α and IL-1β have been crystallized. In spite of limited sequence similarity, IL-1α and IL-1β have similar three-dimensional structure, consisting of a β-barrel with four triangular faces, forming a tetrahedron. A number of site-derived mutagenesis studies have been attempted to identify residues critical for receptor binding and biological activity. The two cysteine residues of IL-1β are essential for activity, whereas exposed lysines are not critical. Mutation studies indicated that Arg4, Leu6, Thr9, Arg11, His30 and Asp146 are important for biological activity. Arg120 is also important in either stabilizing the tertiary structure or the interaction with the receptor. An Arg127→Gly substitution generates a protein with no biological activity, which binds to IL-1R.

4. Interleukin-1 receptor antagonists

Multiple forms of interleukin-1 receptor antagonist (IL-1ra) have been identified:

- soluble IL-1ra (sIL-1ra) is a 152 amino acid protein (pI = 5.2), secreted both in unglycosylated (18 kDa) and glycosylated (22 kDa) forms;

- intracellular IL-1ra type I (icIL-1ra I) has been cloned in keratinocytes and consists of 159 amino acids; it lacks a signal sequence and remains within cells;

- icIL-1ra type II, which contains an extra 21 amino acids at the N-terminus.

The molecule now termed sIL-1ra was discovered as a 22–25 kDa inhibitory activity against IL-1 in supernatants from stimulated human monocytes. The cDNA for this molecule was cloned from a library obtained from IgG-treated monocytes and from the myelomonocytic cell line U937. The cDNA contains a single open reading frame coding for a 177 amino acid protein, preceded by a short (15 bp) 5′-untranslated region and followed by a long (1133 bp) 3′-untranslated region that does not contain any AUUUA motifs (these are involved in mRNA stability). The total length of IL-1ra cDNA is 1.8 kb. Structural analysis of the IL-1ra sequence revealed the presence of a 25 amino acid hydrophobic stretch at the N-terminus resembling a signal peptide. Removal of this leader peptide generates a 152 amino acid mature protein, with a predicted molecular mass of 17 775 Da. The protein sequence contains a potential *N*-glycosylation site.

The icIL-1ra type I and sIL-1ra cDNAs have identical sequences except at the 5′ end: the first 85 bp of sIL-1ra are replaced by a different 130 bp sequence in icIL-1ra I. The icIL-1ra I is generated by an alternative splicing event, in which a different first exon is spliced into an internal acceptor site located in the first exon of sIL-1ra. As a consequence, icIL-1ra I is composed of 159 amino acids, of which 152 are identical to those in sIL-1ra, lacking the signal peptide found in sIL-1ra. Thus this form of IL-1ra remains intracellular. Although icIL-1ra I is also referred as to as keratinocyte or epithelial-type IL-1ra, it is now evident that other types of cell can express icIL-1ra, including monocytes, polymorphonuclear cells (PMN) and fibroblasts. A substantial proportion of total IL-1ra produced by monocytes and PMN remains cell-associated, even after stimulation. The biological function played by the cell-associated fraction of IL-1ra remains largely obscure.

The gene for IL-1ra (Figure 2) is on the long arm of chromosome 2, in the same region as IL-1α and IL-1β. In humans this chromosomal region contains also IL-1RI and IL-1RII genes, whereas in mice the IL-1R genes are on chromosome 1.

The IL-1ra gene contains two alternative first exons followed by three exons. The N-terminal regions of IL-1ra proteins have little homology with the corresponding regions of IL-1α and IL-1β, so it seems that the two alternative first exons are unrelated to IL-1 sequences. The IL-1ra gene is 6.4 kb long, and the first icIL-1ra-specific exon is a further 9.6 kb upstream. Thus, the sIL-1ra promoter is within the first intron of icIL-1ra. This promoter contains a TATAA box and sequences for binding of transcriptional factors, including binding sites for NF-κB, NFIL-1βA and AP-1, and a cyclic AMP responsive element (CRE). In transfection experiments, the sIL-1ra promoter was active only in cell types which naturally can be induced to express IL-1ra. The putative promoter region of icIL-ra apparently lacks any TATAA or CAAT motif, and so it may use alternative mechanisms to start transcription.

The genomic structure of IL-1ra is probably more complex than detailed above, as it has recently been shown that there is another exon between the ones coding for the soluble and intracellular forms of IL-1ra. Usage of this new exon generates a new isoform of icIL-1ra, which we have termed icIL-1ra type II (icIL-1ra II). By RT-PCR it was shown that icIL-1ra II is expressed in keratinocytes and, at lower levels, in monocytes and PMN.

IL-1ra is produced by different cell types, including monocyte–macrophages, PMN and fibroblasts. Keratinocytes and other cells of epithelial origin produce almost exclusively icIL-1ra.

Adherent human monocytes stimulated with LPS produce nearly equivalent amounts of IL-1β and IL-1ra. By contrast, IgG-induced monocytes produce little, if any, IL-1, whereas transcription and translation of IL-1ra is sustained. IL-1ra transcripts induced by LPS have the same half-life as IL-1β (2–4 h), whereas those induced by adherent IgG are extremely stable ($t_{1/2} > 15$ h).

GM-CSF augments IL-1ra production in monocytes. IL-1α and IL-1β are weak inducers of IL-1ra in monocytes, whereas IL-1β enhances the induction by IgG. IL-4, while inhibiting IL-1 expression and production (see above), induces IL-1ra in monocytes and PMN. In the same vein, IL-10 enhances LPS-induced expression of IL-1ra in both monocytes and PMN. IL-13 has also been found to induce IL-1ra in human myelomonocytic cells. It induces transcripts of both the intracellular and soluble forms. We found that transforming growth factor beta (TGFβ) induces transcripts of both soluble and intracellular forms of IL-1ra in human PMN.

Levels of IL-1ra produced by cells of the mononuclear phagocyte lineage depend upon the differentiation state. Differentiation *in vitro* of monocytes into macrophages augments levels of IL-1ra production, which in macrophages appears to be constitutive whereas monocytes require an appropriate activation. IL-1ra production in macrophages is further augmented by GM-CSF but not by LPS and IgG which, as mentioned above, are good inducers of IL-1ra in monocytes.

Transcripts for icIL-ra and sIL-1ra are differentially regulated. Although both transcripts are induced in myelomonocytic cells by IL-13 and TGFβ, in human fibroblasts LPS preferentially induced sIL-1ra, whereas PMA induced icIL-1ra transcripts.

5. Receptors

The first IL-1R was cloned from murine and then human T cells. Soon after identification and cloning of this T-cell IL-1R, the type I receptor (IL-1RI), a second receptor for IL-1 was identified; it is expressed in B lymphocytes and myelomonocytic cells, and is referred to as the type II receptor (IL-1RII) (Figure 3). The IL-1R family is complex and is evolutionarily conserved from plants to humans.

Transcripts of human and mouse IL-1RI are approximately 5 kb in length. A single open reading frame encodes a protein of 552 amino acids. The molecular mass of the fully glycosylated protein is 80–85 kDa and that of the unglycosylated protein is 62 kDa.

On the basis of their structures, IL-1RI and IL-1RII have been assigned to the IgG-like superfamily of receptors, with the extracellular portion containing three IgG-like domains. The extracellular region of IL-1RI is 319 amino acids long and contains seven potential N-glycosylation sites; it is followed by a 20 amino acid transmembrane region, and then by a 213 amino acid cytoplasmic portion. The cytosolic region has no homology with any kinase described so far, and the only protein with which it shows some homology is the Drosophila *Toll* protein, which has recently been suggested to play a role in innate immunity.

IL-1RII transcripts are approximately 1803 bp long. The human transcript encodes a 386 amino acid protein of 68 kDa (treatment with N-glycosidases reduces it 55 kDa). Five potential sites of N-glycosylation have been identified. The extracellular region (332 amino acids) shares only 28% homology with the corresponding region of IL-1RI in humans. A 26 amino acid transmembrane domain is then followed by a very short cytoplasmic domain of 29 residues.

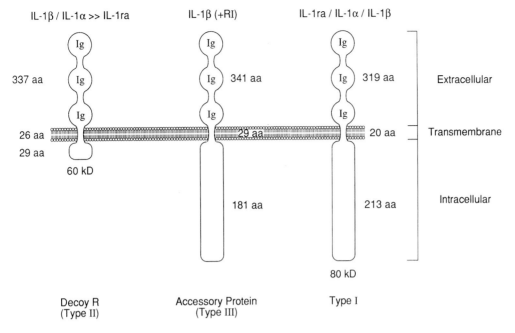

IL-1β / IL-1α >> IL-1ra   IL-1β (+RI)   IL-1ra / IL-1α / IL-1β

337 aa   341 aa   319 aa   Extracellular

26 aa   29 aa   20 aa   Transmembrane
29 aa

60 kD

181 aa   213 aa   Intracellular

80 kD

Decoy R   Accessory Protein   Type I
(Type II)   (Type III)

∎ **Figure 3.** IL-1 receptors. Relative affinity for different ligands are shown.

As already mentioned, the genes of IL-1Rs are located on chromosome 2 (band q12–22) in humans and in the centromeric region of chromosome 1 in mice. The promoter region of IL-1RI has been recently described. IL-1RI and IL-1RII are usually co-expressed. However IL-1RI is expressed as the predominant form in fibroblasts and T cells, whereas B cells, monocytes and PMN express preferentially IL-1RII.

IL-1RI and IL-1RII have different affinities for the three ligands of the IL-1 family. Although results vary between studies, IL-1RI binds IL-1α with higher affinity than IL-1β ($K_\delta = 10^{-10}$ M and 10^{-9} M, respectively). By contrast, IL-1RII binds IL-1β more avidly than IL-1α ($K_\delta = 10^{-9}$–10^{-10} M and 10^{-8} M, respectively). IL-1ra binds to IL-1RI with an affinity similar to that of IL-1α, whereas IL-1RII binds IL-1ra 100-fold less efficiently than IL-1RI.

Given the existence of two distinct forms of IL-1R, a number of studies have investigated the actual role played by each of them in IL-1 signaling. As summarized briefly hereafter, all available evidences indicate that IL-1-induced activities are mediated exclusively via the IL-1RI, whereas IL-1RII has no signaling activity and inhibits IL-1 activities by acting as a decoy for IL-1.

In a number of different cell types, there is circumstantial evidence that different IL-1 activities are mediated by IL-1RI. Human endothelial cells, in which IL-1 regulates functions related to inflammation and thrombosis, express exclusively IL-1RI, indicating at minimum that IL-1RII is dispensable for IL-1 signaling.

Blocking monoclonal antibodies directed against IL-1RI inhibited IL-1 activities in the hepatoma cell line HEPG2, which expresses nearly equal amounts of IL-1RI and RII. IL-1α-induced co-stimulatory activity in CD4+ murine T-cell clones, which express both receptors, was mediated solely by IL-1RI. Similar results were obtained in keratinocytes. Using selective ligands, fever has been shown to be mediated only by IL-1RI.

IL-1RI appears to be the only signaling receptor in cell types expressing predominantly IL-1RII and only minute amounts of IL-1RI. Blocking monoclonal antibodies against IL-1RI totally blocked IL-1-induced expression of cytokines and adhesion molecules in the human monocytic cell line THP-1, human circulating monocytes and PMN. Also IL-1-induced survival of PMN was blocked by anti-type I blocking antibodies. A new IL-1R-related protein (accessory protein, or IL-1 AcP, or IL-1RIII) was recently cloned. It augments the affinity of IL-1RI for IL-1β (but not IL-1ra) and there is evidence that it plays a central role in signal transduction.

Whereas the signaling activity of IL-1RI and IL-1RIII/AcP is well established, unequivocal evidence supporting a signaling function of IL-1RII is still lacking. Blocking monoclonal antibodies against IL-1RII did not inhibit the biological activities of IL-1 in a number of different cell types, including lymphocytes, monocytes, PMN and hepatoma cells. In monocytes, anti-type I antibodies blocked IL-1 activities, whereas anti-type II antibodies did not block the responsiveness of cells to IL-1, rather they augmented it, consistently with a model in which IL-1RII is an inhibitor of IL-1 (see below).

In addition to lack any signaling function, the IL-1RII is shed in a soluble (sIL-1RII) form. sIL-1RII has been found in the supernatants of the B lymphoblastoid cell line Raji and of mitogen-activated mononuclear cells. IL-1RII is also released by cytokine- and dexamethasone-treated PMN and monocytes. sIL-1RII is rapidly shed within minutes after treatment of PMN and monocytes with chemotactic stimuli and oxygen radicals, indicating that release of this receptor represents an aspect of the complex reprogramming of myelomonocytic cells in response to these mediators. An alternatively spliced mRNA encoding a soluble form of IL-1RII was recently cloned.

The findings that IL-1RII has no signaling function and that it is shed in a soluble form, suggested that this molecule could act as an inhibitor of IL-1. We examined this hypothesis in human PMN, in which we found that IL-1 is a potent inducer of PMN survival in culture. Since it had previously been found that IL-4 inhibits IL-1-mediated survival, and IL-4 up-regulates IL-1RII expression and release in these cells, we reasoned that the inhibitory activity of IL-4 on IL-1 activity could be mediated by up-regulation of IL-1RII. The inhibitory activity of IL-4 is totally abrogated by the presence of blocking antibodies directed against IL-1RII, thus demonstrating that IL-1RII inhibits IL-1 activity. We proposed that this inhibition results from IL-RII acting as a 'decoy' target for IL-1, since it binds IL-1 without any signaling function, thus sequestering it and preventing the cytokine from binding to the IL-1RI, the only IL-1R with a cell signaling function.

Consistently with the decoy model of action of IL-1RII, blocking antibodies to IL-1RII augment the activity of suboptimal concentrations of IL-1 on IL-1-induced expression of cytokines and adhesion molecules in human circulating monocytes. The released decoy IL-1RII inhibits processing of the precursor for IL-1β. To validate the decoy model of action of the IL-1RII, we overexpressed this receptor in human fibroblasts or keratinocytes expressing the type I receptor. As expected, IL-1 activity was reduced in fibroblasts expressing high levels of IL-1RII. The inhibitory effect of transfected type II receptors was evident at suboptimal concentrations of IL-1, whereas saturating amounts of IL-1 overcame the IL-1RII-mediated inhibition of IL-1 activity.

Glucocorticoids and T_n2-derived cytokines (IL-4 and IL-13) up-regulate IL-1RII expression and release; this is in keeping with the concept that IL-1RII may represent a physiological pathway of inhibition of IL-1. Induction of expression and release of IL-1RII may contribute to the anti-inflammatory properties of T_n2-derived cytokines and glucocorticoids.

6. Transduction pathways

The mechanism of transduction of the IL-1 signal is a highly controversial and confused area. In certain cellular systems, but not in others, elevations of cAMP have been reported after exposure to IL-1. Levels diacylglycerol, a lipid mediator of the IL-1 signal, increase in the absence of phosphoinositide hydrolysis; the source of diacylglycerol is phosphatidylcholine or phosphatidylethanolamine in different cell types. Recently it has been shown that ceramide, originating from the hydrolysis of sphingomyelin, is involved in IL-1 signaling; this is reminiscent of the TNF signaling system. Cells from patients with Niemann–Pick disease, who have a profound deficiency of acid sphingomyelinase, have full responsiveness to IL-1 and TNF though.

IL-1 causes rapid serine/threonine phosphorylation of diverse proteins, including cytoskeletal components, a membrane receptor and Hsp27. Tyrosine phosphorylation has been detected after exposure to IL-1. IL-1 activates a cascade of protein kinases, eventually resulting in the phosphorylation of Hsp27. An IL-1-activated serine/threonine kinase (IRAK) and a specific transducer (TRAF 6) have recently been cloned. IRAK couples to IL-1 AcP while IRAK-2 interacts with IL-1RI. Via the MyD88 adaptor, TRAF-6 is recruited with activation of NIK, and subsequent NF-κB activation. Thus, elements of a cascade leading to NF-κB activation are beginning to be unraveled.

7. Cellular sources and production

Cells of the monocyte–macrophage lineage are the main cellular source of IL-1, though most cell types have the potential to express this cytokine. In the absence of *in vitro* or *in vivo* stimulation, the IL-1 genes are not expressed. Diverse inducers including bacterial products (e.g. LPS), complement components, cytokines (TNF, IFNγ, GM-CSF, IL-1 itself), induce transcription (see above), but this does not necessarily result in translation. For instance, adhesion causes accumulation of IL-1 mRNA which requires a triggering stimulus (minute amounts of LPS) for translation into protein.

The synthesis of IL-1 is inhibited by endogenous agents, particularly prostaglandins and glucocorticoids. In monocytes or monocytic cell lines *in vitro*, IL-1 production is inhibited by the addition of PGE_2. This inhibition is also to be considered a negative feedback mechanism since when cells are stimulated (with LPS, or phorbol esters) to produce IL-1 they also produce PGE_2 which will limit IL-1 production. Accordingly, addition of inhibitors of prostaglandins up-regulates IL-1 production.

More importantly, glucocorticoids inhibit the synthesis of IL-1 both at the transcriptional and translational levels, like that of most pro-inflammatory cytokines (including TNF, IL-8, IL-6, IL-2 and MCP-1). Also in this case, glucocorticoids act as endogenous inhibitors of IL-1 production as demonstrated by the increase in IL-1 production observed in adrenalectomized animals (for a role of endogeneous glucocorticoids see Section 8.1.3 below).

Stimuli that induce IL-1 also cause production of IL-1ra which may counterbalance the action of IL-1. IL-1ra and icIL-1raI and II are expressed in mononuclear phagocytes and in PMN. icIL-1raI and II are also expressed by fibroblasts and epithelial cells. Signals that induce IL-1 usually also cause production of IL-1ra, which may counteract the action of agonist molecules. However, expression of IL-1 and IL-1ra can be dissociated. Immune complexes and glycans preferentially trigger production of IL-1ra rather than IL-1. Monocytes cultured *in vitro* to resemble mature tissue macrophages, and alveolar macrophages express IL-1ra constitutively with stimulation in response to GM-CSF. Finally, IL-4, IL-13 and IL-10 inhibit IL-1 expression but amplify IL-1ra production.

8. Biological activities

8.1 Spectrum of action of IL-1

IL-1 affects a wide range of cells and organs. Its spectrum of action is similar to that of TNF and, to a lesser degree, that of IL-6. Induction of secondary cytokines, including IL-6, colony stimulating factors and chemokines, is involved in many of the *in vitro* and *in vivo* activities of IL-1. The vast phenomenological literature on the activities of IL-1 is summarized here based on target organs/tissues; the reader is referred to previous reviews for more detailed analysis (see reference list).

8.1.1 Hematopoietic cells

IL-1 affects the hematopoietic system at various levels, from immature precursors to mature myelomonocytic and lymphoid elements. 'Hemopoietin-1' activity was found to be mediated by IL-1α. IL-1 induces production of colony-stimulating factors in a variety of cell types including elements of the bone marrow stroma. It acts synergically with hematopoietic growth factors on various stages of hematopoietic differentiation. IL-1 is also active as an hematopoietic growth factor *in vivo*: it stimulates production of colony-stimulating factors, accelerates bone marrow recovery after cytotoxic chemotherapy or irradiation and has radioprotective activity.

IL-1 acts on T and B lymphocytes. In particular it co-stimulates T-cell proliferation in the classic co-stimulator assay. The AP-1 transcription complex in the promoter of IL-2 is one molecular target for the co-stimulatory activity of IL-1. Although the LAF assay has been invaluable for the identification of IL-1, the actual role of IL-1 co-stimulation in T-cell physiology has not been fully established. Recent data showing that IL-1ra favors the development of Th1-type responses may suggest a role of IL-1 in the generation of Th2-type responses.

IL-1 affects mature myelomonocytic elements. It induces cytokine production in monocytes, while it is a poor inducer in PMN (see above). It prolongs the *in vitro* survival al PMN by blocking apoptosis.

8.1.2 Vascular cells

IL-1 profoundly affects the function of vessel wall elements, particularly endothelial cells. IL-1 activates endothelial cells in a pro-inflammatory, pro-thrombotic way. It induces production of tissue factor and platelet activating factor, down-modulates the protein C-dependent anti-coagulation pathway and induces production of an inhibitor of thrombus dissolution (PAI-1). IL-1 induces gene expression-dependent production of vasodilatory mediators (nitric oxide and prostaglandin I_2), expression of adhesion molecules and production of chemokines in cultured endothelial cells. The concerted action of changes in rheology, adhesion and chemotactic factors underlies leukocyte recruitment at sites of IL-1 production or injection.

8.1.3 Neuroendocrine system

Infection and inflammation induce an elevation of blood corticosteroids through an activation of the so-called hypothalamus–pituitary–adrenal axis (HPAA), very similar to that observed with stress. This is the result of a central action whereby IL-1 stimulates the release of corticotropin-releasing hormone (CRH) by the hypothalamus which will induce adrenocortinotrophic hormone (ACTH) production by the pituitary, ultimately causing a release of corticosteroids in the bloodstream by the adrenals. This release in inhibited by anti-CRH antibodies. The increase in corticosteroids resulting from HPAA activation by IL-1 may have several consequences in view of the wide range of immunosuppressive, antiinflammatory and metabolic actions of glucocorticoids. We would like to stress here that corticosteroids are (as previously mentioned in Section 4) potent inhibitors of the synthesis of IL-1 (and of other cytokines). They also prevent the hemodynamic shock associated with injection of IL-1, TNF or LPS, and protect against IL-1 or LPS toxicity. Therefore, activation of HPAA may be considered a feedback mechanism to control IL-1 production and toxicity. The fact that adrenalectomized animals are extremely susceptible to IL-1 toxicity strongly support this hypothesis.

8.1.4 The acute-phase response

IL-1 is a key mediator of the series of host responses to infection and inflammation known as the acute-phase response. One previous name of IL-1 was leukocytic endogenous mediator (LEM). This was originally identified as a major mediator of the acute phase response, particularly hypoferremia and induction of acute-phase proteins. Hypoferremia seems to be mediated by an effect on neutrophils, that would be stimulated to release lactoferrin to sequester iron in the tissues. It could constitute a 'nutritional immunity' against infection since iron is essential for the growth of many bacteria.

Acute-phase proteins are proteins whose synthesis is increased during inflammation. They include C-reactive protein, serum amyloids, fibrinogen, hemopexin and various proteinase inhibitors. These molecules may have protective, antitoxic and other functions yet to be defined. The synthesis of some of these proteins (e.g. serum amyloid A) can be directly induced in hepatocytes by IL-1. Others, such as fibrinogen, are induced indirectly through IL-6. Therefore both IL-1 and IL-6 (like other cytokines) behave like hepatocyte-stimulating factors (HSFs; this name is applied to cytokines that stimulate liver acute-phase protein synthesis).

The increased synthesis of acute-phase proteins is part of a rearrangement of liver metabolism where the synthesis of 'normal liver proteins' is decreased: one such negative acute-phase reactant is albumin, whose gene expression is decreased by IL-1.

8.1.5 Central nervous system

IL-1 is the main endogenous pyrogen. In 1943, Menkins suggested that leukocytes relase a pyrogenic substance, 'pyrexin', which was subsequently detected in the circulation of febrile rabbits. Human leukocytic pyrogen was purified in 1977 and an immunoassay for it was developed. It is now clear that the main endogenous pyrogen was IL-1, and recombinant IL-1 induces fever in experimental animals, an activity shared with other cytokines including TNF and IL-6 (although these are much less potent than IL-1). The pyrogenic action of IL-1 is due to the increased production of prostaglandins. In fact IL-1 fever is abolished by prostaglandin synthesis inhibitors and IL-1 directly stimulates PGE_2 release by hypothalamic tissue.

In addition to fever, IL-1 has other effects on the central nervous system. These include induction of slow-wave sleep and anorexia, typically associated with infections. It also activates the hypothalamus to produce CRH as described in Section 8.1.3.

8.1.6 Other effects

It is almost impossible to list all the activities of IL-1. Since it has become available as a recombinant protein, a large number of papers have been published reporting extremely diverse effects of this cytokine. IL-1 has a number of local effects that have been termed 'catabolic', and play a role in destructive joint and bone diseases. In particular, IL-1 induces production of collagenase by synovial cells and of metalloproteinases by condrocytes. It

IL-1 also stimulates fibroblast proliferation and collagen synthesis and thus plays a role in fibrosis. It induces profound hypotension, an effect which is inhibited by a cyclooxygenase inhibitor and where IL-1 and TNF act synergically. Another important action of IL-1 is its toxicity for insulin-producing beta cells in the islets of Langerhans, suggesting that it may be involved in the pathogenesis of insulin-dependent type I diabetes.

8.1.7 Genetically modified mice

The development of transgenic or knockout mice for components of the IL-1 system has provided new evidence as to the *in vivo* function of these molecules. IL-1$\beta^{-/-}$ mice show reduced acute-phase response to local irritants, whereas the responses to systemic LPS are normal. Gene targeting and overproduction have revealed a critical role of endogenous IL-1ra in animal growth, responses to infection and inflammation and cytokine regulation. Selective overexpression of IL-1β and the type I receptor in skin result in pathology consistent with the concept that IL-1 is important in skin pathophysiology

9. Clinical implications

IL-1 is a central mediator of local and systemic inflammatory reactions. Blood IL-1 levels reach relatively modest levels in response to septic conditions, but IL-1ra increases to levels orders of magnitude higher. This may represent a feedback control mechanism. Most of the interest in IL-1 is for its pathogenetic role in septic shock syndrome and rheumatoid arthritis. However, *in vitro* studies, animal models and results of studies reporting levels of IL-1 in human diseases have indicated other pathologies where blockade of IL-1 might be beneficial. These include vasculitis, disseminated intravascular coagulation, osteoporosis, neurodegenerative disorders (such as Alzheimer's disease), diabetes, lupus nephritis, immune complex glomerulonephritis and autoimmune diseases in general. Inhibition of IL-1 by IL-1Ra, anti-IL-1 or anti-IL-1R antibodies is protective in various animal models including endotoxin-induced hemodynamic shock and lethality, arthritis, inflammatory bowel disease, spontaneous diabetes in BB rats, graft-versus-host disease in mice, heart allograft rejection and experimental autoimmune encephalomyelitis.

Since TNF is produced concomitantly and IL-1 synergizes with TNF (but not with IL-6) in many systems, it is likely that these two cytokines act in concert in the pathogenesis of these disorders.

IL-1 is a autocrine/paracrine growth factor for acute and chronic myeloid leukemia cells, plasmacytoma (via IL-6) and possibly some solid tumors such as ovarian carcinoma. Interestingly AML blasts usually express IL-1β, but not IL-1ra. Blocking IL-1 using IL-1ra inhibits AML proliferation *in vitro*.

The exploration of the therapeutic potential of IL-1 has provided an opportunity to examine the *in vivo* activity of systemic IL-1 administration in humans. Phase I trials have been conducted with IL-1α and IL-1β in studies ultimately aimed at exploiting the hematopoietic/radioprotective action of these molecules. The results of these studies largely confirm data from animal experimentation. Systemically (intravenously) administered IL-1 (1–10 ng/ml) causes fever, sleepiness, anorexia, myalgia, arthralgia and headache. At doses of 100 ng/ml or higher, a rapid fall in blood pressure occurs.

Therapeutic strategies aimed at blocking IL-1 have received considerable attention in experimental models and in humans. To date, IL-1ra and engineered sIL-1R type I have been tested in the clinic. While initial results with IL-1ra in septic shock syndrome encourage optimism, subsequent phase III study failed to substantiate the phase II data. *A posteriori*, IL-1ra may have been beneficial in patients with more serious disease and organ failure. Initial results with IL-1ra in graft-versushost disease and in rheumatoid arthritis are encouraging. The recent availability of a type IR antagonist may open new vistas in the domain of IL-1 antagonism.

Another possible site for pharmacological action could be the processing of IL-1β by caspase-1. Inhibitors of IL-1β secretion acting by inhibiting ICE/caspase-1 are under study, on the basis that IL-1β is quantitatively more important than IL-1α. While peptides have been the molecules most extensively investigated to date, simple chemicals with anti-caspase activity are currently under study and hold considerable promise.

References

Arend, W. P. (1993). Interleukin-1 receptor antagonist. *Adv. Immunol.*, **54**, 167–227.

Auron, P. E., Webb, A. C., Rosenwasser, L. J., Mucci, S. F., Rich, A., Wolff, S. M., *et al.* (1984). Nucleotide sequence of human monocyte interleukin 1 precursor cDNA. *Proc. Natl Acad. Sci.*

USA, **81**, 7907–11.

Bazan, J. F., Timans, J. C., and Kastelein, R. A. (1996). A newly defined interleukin-1? *Nature*, **379**, 591.

Bertini, R., Bianchi, M., and Ghezzi, P. (1988). Adrenalectomy sensitizes mice to the lethal effects of interleukin 1 and tumor necrosis factor. *J. Exp. Med.*, **167**, 1708–12.

Cao, Z., Henzel, W. J., and Gao, X. (1996). IRAK: a kinase associated with the interleukin-1 receptor. *Science*, **271**, 1128–31.

Cao, Z., Xiong, J., Takeuchi, M., Kurama, T., and Goeddel, D. V. (1996). TRAF6 is a signal transducer for interleukin-1. *Nature*, **383**, 443–6.

Colotta, F., Re, F., Muzio, M., Bertini, R., Polentarutti, N., Sironi, M., *et al.* (1993). Interleukin-1 type II receptor: a decoy target for IL-1 that is regulated by IL-4. *Science*, **261**, 472–5.

Colotta, F., Dower, S. K., Sims, J. E., and Mantovani, A. (1994). The type II 'decoy' receptor: novel regulatory pathway for interleukin-1. *Immunol. Today.*, **15**, 562–6.

Colotta, F., Orlando, S., Fadlon, E. J., Sozzani, S., Matteucci, C., and Mantovani, A. (1995). Chemoattractants induce rapid release of the interleukin 1 type II decoy receptor in human polymorphonuclear cells. *J. Exp. Med.*, **181**, 2181–8.

Dinarello, C. A. (1994). Blocking interleukin-1 receptors. *Int. J. Clin. Lab. Res.*, **24**, 61–79.

Dinarello, C. A. (1996). Biological basis for IL-1 in disease. *Blood*, **87**, 2095–147.

Fantuzzi, G. and Dinarello, C. A. (1996). The inflammatory response in interleukin-1β-deficient mice: comparison with other cytokine-related knock-out mice. *J. Leukocyte Biol.*, **59**, 489–93.

Fenton, M. J. (1990). Transcriptional factors that regulate human IL-1/hemopoietin 1 gene expression. *Hematopoiesis*, **120**, 67–82.

Greenfeder, S. A., Nunes, P., Kwee, L., Labow, M., Chizzonite, R., and Ju, G. (1995). Molecular cloning and characterization of a second subunit of the interleukin 1 receptor complex. *J. Biol. Chem.*, **270**, 13757–65.

Haskill, S., Martin, G., Van Le, L., Morris, J., Peace, A., Bigler, C. F., *et al.* (1991). cDNA cloning of an intracellular form of the human interleukin 1 receptor antagonist associated with epithelium. *Proc. Natl Acad. Sci. USA.*, **88**, 3681–5.

Hirsch, E., Irikura, V. M., Paul, S. M., and Hirsh, D. (1996).

Functions of interleukin-1 receptor antagonist in gene knock-out and overproducing mice. *Proc. Natl Acad. Sci. USA*, **93**, 11008–13.

Mantovani, A. (1994). Tumor-associated macrophages in neoplastic progression: A paradigm for the *in vivo* function of chemokines. *Lab. Invest.*, **71**, 5–16.

Mantovani, A., Bussolino, F., and Dejana, E. (1992). Cytokine regulation of endothelial cell function. *FASEB J.*, **6**, 2591–9.

Mantovani, A., Bussolino, F., and Introna, M. (1997). Cytokine regulation of endothelial cell function: from molecular level to the bed side. *Immunol. Today.*, **18**, 231–9.

McMahan, C. J., Slack, J. L., Mosley, B., Cosman, D., Lupton, S. D., Brunton, L. L., *et al.* (1991). A novel IL-1 receptor, cloned from B cells by mammalian expression, is expressed in many cell types. *EMBO J.*, **10**, 2821–32.

Medzhitov, R., Preston-Hurlburt, P., and Janeway, C. A., Jr (1997). A human homologue of the *Drosophila* Toll protein signals activation of adaptive immunity. *Nature*, **388**, 394–7.

Muzio, M., Polentarutti, N., Sironi, M., Poli, G., De Gioia, L., Introna, M., *et al.* (1995). Cloning and characterization of a new isoform of the interleukin-1 receptor antagonist . *J. Exp. Med.*, **182**, 623–8.

Muzio, M., Ni, J., Feng, P., and Dixit, V. M. (1997). IRAK (Pelle) family member IRAK-2 and MyD88 as proximal mediators of IL-1 signaling. *Science*, **278**, 1612–15.

Oppenheim, J. J. and Gery, I. (1993). From lymphodrek to interleukin 1 (IL-1). *Immunol. Today*, **14**, 232–4.

Re, F., Muzio, M., De Rossi, M., Polentarutti, N., Giri, J. G., Mantovani, A., *et al.* (1994). The type II 'receptor' as a decoy target for IL-1 in polymorphonuclear leukocytes: characterization on of induction by dexamethasone and ligand binding properties of the released decoy receptor. *J. Exp. Med.*, **179**, 739–43.

Re, F., Sironi, M., Muzio, M., Matteucci, C., Introna, M., Orlando, S., *et al.* (1996). Inhibition of interleukin-1 responsiveness by type II receptor gene transfer: a surface 'receptor' with anti-interleukin-1 function. *J. Exp. Med.*, **183**, 1841–50.

Rossi, V., Breviario, F., Ghezzi, P., Dejana, E., and Mantovani, A. (1985). Prostacyclin synthesis induced in vascular cells by interleukin-1. *Science*, **229**, 174–6.

Sims, J. E., March, C. J., Cosman, D., Widmer, M. B., Mac Donald, H. R., McMahan, C. J., *et al.* (1988). cDNA expression cloning of the IL-1 receptor, a member of the immunoglobulin superfamily. *Science*, **241**, 585–9.

VI Interleukin 12

Giorgio Trinchieri

1. Introduction

Interleukin 12 (IL-12) is a heterodimeric cytokine, composed of a heavy chain of 40 kDa (p40) and a light chain of 35 kDa (p35). It was originally called natural killer stimulatory factor (NKSF) or cytotoxic lymphocyte maturation factor (CLMF). It is produced within a few hours of infection, particularly in the case of infections by bacteria and intracellular parasites. It acts as a pro-inflammatory cytokine, activating natural killer (NK) cells, and, through its ability to induce interferon gamma (IFNγ) production, enhancing the phagocytic and bacteriocidal activity of phagocytic cells and their ability to release pro-inflammatory cytokines, including IL-12 itself. Furthermore, IL-12 produced during the early phases of infection and inflammation sets the stage for the ensuing antigen-specific immune response, favoring differentiation and function of T-helper type 1 (Th1) T cells while inhibiting the differentiation of Th2 T cells. Thus, in addition to being a potent pro-inflammatory cytokine, IL-12 is a key immunoregulator molecule in Th1 responses.

2. The IL-12 molecule: its genes and its receptor

The two genes encoding the two chains of IL-12 are separate and unrelated; the gene encoding the light (p35) chain has limited homology with other single-chain cytokines, whereas the gene encoding the heavy (p40) chain is homologous to the extracellular domain of genes of the hematopoietic cytokine receptor family. The p35 and the p40 chains are covalently linked to form a biologically active heterodimer (p70). Thus, the IL-12 heterodimer resembles a cytokine linked to a soluble form of its receptor. An analogous but not identical situation is observed for IL-6, IL-11 and ciliary neurotrophic factor (CNTF), three cytokines which can bind in solution to the soluble form of one chain of their receptor to create complexes that bind to other transmembrane chains of their receptor, including the shared gp130 chain, inducing signal transduction and biological functions. Interestingly, the p40 chain of IL-12 is most homologous to the CNTF receptor and to the IL-6 receptor (α-chain), whereas the p35 chain has some homology with IL-6 itself; thus, it is likely that IL-12 is evolutionarily derived from a primordial cytokine similar to IL-6/CNTF and from a chain of its original receptor.

The p35 subunit has seven cysteine residues, six of which are involved in intrachain disulfide bonds and form the intersubunit disulfide pairing. The p40 subunit has ten cysteine residues, eight of which are involved in intrachain disulfide bonds. All the intramolecular pairs support the homology of IL-12 p35 and p40 to IL-6 and IL-6 receptor, respectively. The p35 gene is located on human chromosome 3p12-q13.2. The Cys252 of the p40 chain does not have a corresponding residue in the IL-6 receptor, and it is not paired with any other cysteine in IL-12, but is cysteinylated or contains thioglycolate paired with sulfur. The gene encoding the p40 chain of IL-12 is located on human chromosome 5q31-q33 within or close to a cluster of genes encoding other cytokines and cytokine receptors, including IL-4.

Two or more IL-12-binding affinities are observed on IL-12-responsive cells, and the receptors with the highest affinity (in the picomolar range) are probably responsible for IL-12 biological activity. Two chains of the IL-12 receptor, IL-12Rβ1 and IL-12Rβ2, have been cloned. They are members of the cytokine receptor superfamily and within that family are most closely related to gp130. Both IL-12Rβ1 and IL-12Rβ2 have a cytoplasmic region that contains the characteristic box I and II

p40

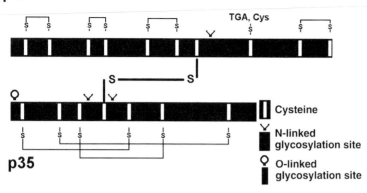

p35

Cysteine

N-linked glycosylation site

O-linked glycosylation site

■ **Figure 1.** Schematic model of the interleukin-12 molecule, a heterodimer formed by a heavy chain (p40, above) and a light chain (p35, below).

motifs found in other cytokine receptors. However, conserved cytoplasmic tyrosine residues are missing from the $\beta1$ subunit, whereas the IL-12 R$\beta2$ subunit contains three cytoplasmic tyrosine residues which are probably involved in signaling processes. Each subunit alone binds IL-12 with only low affinity (K_d = 2–5 nM). Co-expression of both receptor subunits results in both high affinity (K_d = 50 pM) and low affinity (K_d = 5 nM) IL-12-binding sites. IL-12 p40 interacts mostly with IL-12R$\beta1$, whereas IL-12 p35 or possibly an epitope on IL-12 composed of both ligand subunits appears to interact mostly with the IL-12R$\beta2$ of the receptor complex.

The cells producing IL-12 secrete more of the free p40 chain than the biologically active heterodimer, from a few times more (as observed in activated phagocytic cells) to up to 100–1000 times more. Recombinant p40 IL-12, both human and murine, is secreted by transfected cells both as a disulfide-bonded homodimer or as a monomer, and, in the mouse, production of homodimeric p40 has also been observed *in vivo*, although such an observation has not been made in humans. Murine p40 homodimers bind to the IL-12R$\beta1$ chain with an affinity similar to that of the heterodimers and compete with the heterodimers for binding to the IL-12 receptor, effectively blocking the biological functions of IL-12 on murine cells. On human cells, the homodimers bind to the IL-12R with a much lower affinity than the heterodimers and act as antagonists only at much higher concentrations than on murine cells. Thus, in the mouse, but probably not in humans, the IL-12 p40 homodimer may represent a physiological antagonist of IL-12.

Resting T cells and NK cells do not express the IL-12R or express it only at very low levels;

however, resting peripheral blood T cells and NK cells rapidly respond to IL-12 with IFNγ production and enhancement of cytotoxic functions, suggesting that the receptor is present at least in a proportion of the cells and/or it can be rapidly activated in culture. Activation of T cells and NK cells induces up-regulation of IL-12R, as identified by low- and high-affinity binding and up-regulation at least of the IL-12R$\beta1$ gene. It should be noted that certain cell types, e.g. human B lymphoblastoid cell lines, or normal B cells, also express the IL-12R$\beta1$ mRNA, without, in most cases, expressing IL-12-binding sites, suggesting that IL-12R$\beta1$ is essential but not sufficient for expression of functional, high affinity IL-12R and that the IL-12R$\beta2$ subunit may be more restricted in its expression than IL-12R$\beta1$.

Signal transduction through IL-12R induces tyrosine phosphorylation of the Janus family kinases Jak2 and Tyk2 and of the transcription factors STAT3 and STAT4; IL-12 is the only inducer known to activate STAT4. Phosphorylation and activation of the 44 kDa MAP kinase may be responsible for a serine phosphorylation also observed in STAT4 upon stimulation of T cells within IL-12. During developmental commitment of BALB/c CD4+ T cells to the Th2 lineage, the ability of T cells to signal in response to IL-12 is extinct due to down-regulation of IL-12R$\beta2$ expression.

3. Cell types producing IL-12

IL-12 was originally discovered as a product of Epstein–Barr virus (EBV)-transformed B-cell lines, which all constitutively produce at least low levels of IL-12 p40. Although malignant or EBV-transformed B-cell lines produce IL-12, the physi-

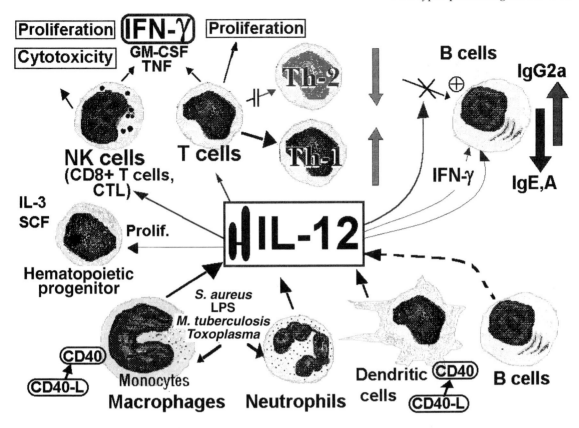

∎ **Figure 2.** Production and biological functions of interleukin-12 (IL-12). IL-12 was originally described as a product of B-cell lines; it is predominantly produced by phagocytic cells and dendritic cells, in response to infectious agents or upon interaction of the CD40 receptor with CD40 ligand on activated T cells. IL-12 is active on T cells and NK cells and induces cytokine production, proliferation, enhanced cell-mediated cytotoxicity, and generation of T-helper type 1 cells. It is also active on hematopoietic progenitor cells in combination with other hematopoietic growth factors and mediates both direct and indirect effects on B-cell functions.

ological relevance of IL-12 production from normal B cells remains to be established and subsequent studies suggested that phagocytic cells are the major physiological producers of IL-12, a conclusion now supported by many *in vitro* and *in vivo* studies. Monocytes produce high levels of IL-12 p40 and p70 when stimulated by bacteria, such as heat-fixed *Staphylococcus aureus* or *Streptococcus* extracts, or bacterial products such as lipopolysaccharide (LPS). In addition to monocytes, polymorphonucleated cells (PMN) also respond to LPS stimulation by producing IL-12.

On both monocytes and PMN, IFNγ has a powerful enhancing effect on IL-12 production, probably potentiating it within inflammatory tissues. The ability of IFNγ to enhance the production of IL-12 by phagocytic cells is of particular interest because

IL-12 is a potent inducer of IFNγ production by T cells and NK cells. Thus, IL-12-induced IFNγ acts as a positive feedback mechanism in inflammation by enhancing IL-12 production. Also, because IL-12 is the major cytokine responsible for the differentiation of Th1 cells which are producers of IFNγ, the enhancing effect of IFNγ on IL-12 production may represent a mechanism by which Th1 responses are maintained *in vivo*. The ability of IFNγ to enhance IL-12 production is particularly evident and required for IL-12 production in the case of certain infectious agents, e.g. mycobacteria, which are rather poor inducers of IL-12 production. However, with many other inducers, such as LPS, *Toxoplasma gondii* and *S. aureus*, IL-12 production *in vivo* and *in vitro* both precedes and is required for IFNγ production.

The positive feedback amplification of IL-12 production mediated by IFNγ represents a potentially dangerous mechanism leading to uncontrolled cytokine production. There are, however, potent mechanisms of down-regulation of IL-12 production and of the ability of T cells and NK cells to respond to IL-12. IL-10 is a potent inhibitor of IL-12 production by phagocytic cells; the ability of IL-10 to suppress production of IFNγ and other Th1 cytokines is primarily due to its inhibition of IL-12 production from antigen-presenting cells (APCs) as well as by inhibition of expression of co-stimulatory surface molecules (e.g. B7) and soluble cytokines (e.g. tumor necrosis factor alpha (TNF-α), IL-1β]. Another powerful inhibitor of IL-12 production is transforming growth factor beta (TGF-β). IL-4 and IL-13 also partially inhibit IL-12 production, suggesting that Th2 cells, by producing cytokines such as IL-10, IL-4 and IL-13, suppress IL-12 production and prevent the emergence of a Th1 response.

Keratinocytes are also reported to express IL-12 mRNAs and possibly secrete minute levels of IL-12 protein. However, the physiological significance of this production is doubtful because, when used as APCs, keratinocytes induce no stimulation of IFNγ production unless exogenous IL-12 is added to the cultures. Also, analysis of IL-12 production by skin cells suggests that Langerhans cells rather than keratinocytes are the major producers.

The production of IL-12 by Langerhans cells raises the issue of the ability of 'professional' APCs such as dendritic cells to produce IL-12 and its role during antigen-presentation and T-cell activation. Definitive evidence that dendritic cells are producers of IL-12 came from studies demonstrating that, when endogenous IL-4 production is blocked, these cells induce a Th1 response, which is prevented by neutralizing anti-IL-12 antibodies. Extensive studies with both human and mouse dendritic cells have now confirmed that dendritic cells are efficient producers of the IL-12 that acts in inducing Th1 responses upon antigen presentation by these APCs.

In addition to the induction of IL-12 observed in response to infectious agents, activated T cells stimulate production of IL-12 by macrophages and dendritic cells. The mechanism of this T-cell-dependent induction of IL-12 is based on the interaction of CD40 ligand (CD40-L) on the surface of activated T cells with CD40 on the APCs and can be mimicked by cross-linking CD40 on the surface of IL-12-producing cells with anti-CD40 antibodies or recombinant CD40-L. The induction of IL-12 by bacterial or other infectious agents and by activated T cells represents two independent pathways of APC activation, as clearly shown by the observation that spleen cells from CD40 knockout mice are completely unable to produce IL-12 in response to activated T cells, but produce normal levels of IL-12 in response to endotoxin or *S. aureus*. However, it is probable that during an infection or an immune response *in vivo* both pathways are activated, the T-cell-independent one during the inflammatory phase of innate resistance and the T-cell-dependent one during the subsequent adaptive immune response: thus, the inflammatory pathway may be responsible for the initiation of the Th1 response and the T-cell-dependent pathway for its maintenance.

4. Molecular control of IL-12 production

Upon activation of phagocytic cells with LPS or *S. aureus*, accumulation of IL-12 p40 mRNA is observed within 2–4 h, slightly later than that of other pro-inflammatory cytokines such as TNF-α; it then subsides after several hours. The induction of p40 expression is largely controlled at the transcriptional level and both the enhancing effect of IFNγ and the inhibitory effect of IL-10 are reflected in changes in the rate of IL-12 gene transcription. The promoter of the p40 gene is constitutively active in EBV-transformed cell lines and inducible in myeloid cell lines, but not in T-cell lines; IFNγ priming of the myeloid cells greatly enhances the activation of the promoter by LPS. A region responsible for promoter induction and activity is between nucleotides −196 and −224, and binds a series of IFNγ- and LPS-induced nuclear proteins, including Ets2 and/or Ets-related factors. Promoter constructs with deletions or mutations in this region display a reduced but still detectable IFNγ/LPS-inducible promoter activity, contributed in part by a site between nucleotides −123 and −99, to which NF-κB binds.

Expression of the p35 gene is also up-regulated upon activation of phagocytic cells, although its ubiquitous constitutive expression has complicated analysis of its expression using non-purified cell preparations. p35 up-regulation is inhibited by IL-10, whereas IFNγ enhances transcription and mRNA accumulation of the p35 gene. In activated phagocytic cells and in B-cell lines, p40 mRNA is approximately 10-fold more abundant than p35 mRNA, explaining the overproduction and secretion of the free p40 chain over the p35-containing heterodimer.

5. *In vitro* activities of IL-12

IL-12 synergizes with other hematopoietic factors in enhancing survival and proliferation of early multipotent hematopoietic progenitor cells and lineage-committed precursor cells. Although IL-12 has prevalently stimulatory effects on hematopoiesis *in vitro*, IL-12 treatment *in vivo* results in decreased bone marrow hematopoiesis and both transient anemia and neutropenia. The toxic effects of IL-12 on hematopoiesis are mostly mediated by IFNγ and, in its absence, treatment with IL-12 results in stimulation of hematopoiesis only.

IL-12 induces T cells and NK cells to produce several cytokines and particularly IFNγ. IL-12-induced IFNγ production requires the presence of low levels of both TNF and IL-1. The importance of IL-12 as an IFNγ inducer rests not only in its high efficiency at low concentrations, but also in its synergy with many other activating stimuli. IL-12 is required for optimal IFNγ production *in vivo* during immune responses, especially in bacterial or parasitic infections. In response to macrophage-produced IL-12, NK cells readily produce IFNγ which activates macrophages and enhances their bactericidal activity, providing a mechanism of T-cell-independent macrophage activation during the early phases of innate resistance.

IL-12 does not induce proliferation of resting peripheral blood T cells or NK cells, although it potentiates the proliferation of T cells induced by various mitogens and has a direct proliferative effect on preactivated T and NK cells. IL-12 is effective at lower concentrations than IL-2, although the levels of proliferation obtained with IL-12 are much lower than those observed with IL-2. However, co-stimulation through the CD28 receptor by the B7 ligand or anti-CD28 antibodies strongly synergizes with IL-12 in inducing both efficient T-cell proliferation and cytokine production. Because B7 is a surface molecule and IL-12 is a secreted product of APCs, their synergic effect on T cells plays an important role in inducing T-cell proliferation and IFNγ production upon antigen presentation to T cells.

IL-12 also enhances the generation of cytotoxic T cells (CTLs) and lymphokine-activated killer (LAK) cells, and potentiates the cytotoxic activity of CTLs and NK cells. Some of the effects of IL-12 on cell-mediated cytotoxicity are due to increased formation of cytoplasmic granules and induction of transcription of genes encoding cytotoxic granule-associated molecules such as perforin and granzymes. The ability of IL-12 to induce expression of adhesion molecules on T cells and NK cells also may affect their cytotoxic activity and their ability to migrate to tissues.

6. IL-12 induction of Th1 responses

IL-12 is required for Th1 cell development during the immune response to pathogens, and the type of T$_h$ cell differentiation is probably determined early after infection by the balance between IL-12 and IL-4 which favor Th1 and Th2 development, respectively. IL-12 is produced by phagocytic cells, other APCs and possibly B cells, whereas IL-4 is produced by subsets of T cells and by mast cells.

The defining characteristic of the Th1 and Th2 cells is that they stably express the ability to produce certain cytokines, but not others. When present at the time of clonal expansion, IL-12 shows a particularly powerful ability to prime T cells, both CD4$^+$ and CD8$^+$, to produce high levels of IFNγ upon restimulation. IL-12, produced during the inflammatory phase of infections or immune responses, induces NK cells and T cells to produce IFNγ; then, IL-12, in cooperation with IFNγ, induces the T-cell clones expanding in response to the specific antigens to differentiate into Th1 cells by priming them for expression of cytokines such as IFNγ and by exerting other positive or negative selective mechanisms, including, for example, deletion of IL-4-producing cells or preferential expansion of cells with Th1-like phenotype. Once a Th1 response has been induced *in vivo*, IL-12 is in most cases not necessary for maintaining it. However, differentiated Th1 cells maintain IL-12 responsiveness and IL-12, produced by APCs during cognate antigen presentation to T cells, appears to be important, at least in autoimmune diseases, for optimal proliferation and cytokine production of the Th1 cells in response to antigens.

Many of the effects of IL-12 on B-cell activation and immunoglobulin isotype production could be interpreted as mediated by either subset of T$_h$ cells or by their products, IFNγ in particular. However, evidence has been provided that IL-12 may directly affect B-cell proliferation and differentiation.

7. Role of IL-12 in infections

The pro-inflammatory functions of IL-12, its ability to stimulate innate resistance and to generate a Th1-type immune response are essential for resistance to different types of infection, particularly bacteria, fungi and intracellular parasites. The most acute instance of IL-12 production resulting in IFNγ induction is observed in the models of endotoxic-induced shock. Similar pathogenetic mechanisms mediated by IL-12 are involved in the toxic shock-like syndromes induced by superantigenic exotoxins produced by Gram-positive bacteria, e.g. *Streptococcus pyogenes* and *S. aureus*. The role of endogenous IL-12 in resistance to infection and the possibility of using IL-12 in therapy of infections have been analysed in many studies. The ability of IL-12 to facilitate Th1 immune responses is at the basis of its successful use as an adjuvant in vaccination in combination with either soluble antigens or killed pathogens, usually poor vaccines, to induce a Th1-biased memory response able to maintain protective immunity. In helminthic infections, Th2 responses are most effective in the elimination of the pathogen and their eggs and in those infections IL-12 treatment decreases the resistance of the host.

Unlike what is observed in bacterial and intracellular parasite infection, IL-12 has a relatively minor role in the resistance to virus infection, and IL-12-independent mechanisms of IFNγ production are operative in virus infection. Even if IL-12 does have a modest role in antiviral resistance, its immunoregulatory activity can be utilized as an adjuvant in vaccine composed of inactivated viruses or isolated viral proteins.

The ability of peripheral blood mononuclear cells from human immunodeficiency virus (HIV)-infected patients to produce IL-12 *in vitro* is profoundly and selectively depressed. Although HIV-infected patients are depressed in their ability to produce IL-12, their NK and T cells respond normally to exogenous IL-12. Furthermore, IL-12 can prevent apoptosis in T cells of HIV patients and correct *in vitro* their defective proliferative response to recall antigens, alloantigens and mitogens. Because IL-12 has immunoregulatory effects favoring generation of Th1 cells and cell-mediated cytotoxicity, and *in vivo* treatment of animals with neutralizing anti-IL-12 antibodies results in a loss of delayed type hypersensitivity responses similar to that observed in HIV-infected patients, a deficient *in vivo* production of IL-12 may be in part responsible for the HIV-induced immunodefi-

ciency. A transient loss of delayed-type hypersensitivity and immunodeficiency is also observed in patients infected or vaccinated with measles virus; *in vitro*, measles virus selectively inhibits production of IL-12 and transcription of the two genes encoding IL-12, without significantly affecting the production of other pro-inflammatory cytokines. This inhibition is also observed using other ligands of the measles receptor (CD46 or membrane cofactor protein) such as anti-CD46 monoclonal antibodies or polymerized complement.

8. Concluding remarks

The immunoregulatory role of IL-12 is of central importance in the immunity against those pathogens and tumors that require immunity based on cell-mediated mechanisms and phagocytic cell activation, supported by Th1 cells. Although IL-12 as a powerful pro-inflammatory cytokine plays a crucial role in the first line of defence against infection, it is also responsible for some of the negative side effects of inflammation, from hematological alterations, tissue damage, e.g. at the level of liver and gastrointestinal tract, up to sensitizing to or acting as an effector molecule in the lethal effects of endotoxic shock. Certain pathogens, e.g. measles virus and, probably, HIV, induce immunosuppression by selectively inhibiting IL-12 production. Because of the role of IL-12 in inducing and maintaining Th1 response, overexpression of this cytokine is likely to play a role in organ-specific autoimmunity, as confirmed by emerging evidence in many of these syndromes. The antitumor effect of IL-12 is indirect and based on several mechanisms: activation of both innate and antigen-specific adaptive immunity against the tumor cells, and ability through IFNγ to inhibit tumor angiogenesis. Based on its ability to block Th2 cell differentiation and in particular IL-4 production, IL-12 has also an anti-allergic activity, clearly shown by its ability to prevent airway hyperresponsiveness and asthma.

References

Biron, C. A. and Gazzinelli, R. T. (1995). Effects of IL-12 on immune responses to microbial infections: a key mediator in regulating disease outcome. *Curr. Opin. Immunol.* **7**, 485–96.

Brunda, M. J. (1994). Interleukin-12. *J. Leuk. Biol.* **55**, 280–8.

Brunda, M. J. and Gately, M. K. (1994). Antitumor activity of interleukin-12. *Clin. Immunol. Immunopathol.* **71**, 253–5.

Chehimi, J. and Trinchieri, G. (1994). Interleukin-12: a bridge between innate resistance and adaptive immunity with a role in infection and acquired immunodeficiency. *J. Clin. Immunol.* **14**, 149–61.

Gately, M. K. (1993). Interleukin-12: a recently discovered cytokine with potential for enhancing cell-mediated immune responses to tumors. *Cancer Invest.* **11**, 500–6.

Germann, T. and Rude, E. (1995). Interleukin-12. *Int. Arch. Allerg. Immunol.* **108**, 103–12.

Presky, D. H., Gubler, U., Chizzonite, R. A., and Gately, M. K. (1995). IL-12 receptors and receptor antagonists. *Res. Immunol.* **146**, 439–45.

Seder, R. A., Kelsall, B. L., and Jankovic, D. (1996). Differential roles for IL-12 in the maintenance of immune responses in infectious versus autoimmune disease. *J. Immunol.* **157**, 2745–8.

Trembleau, S., Germann, T., Gately, M. K., and Adorini, L. (1995). The role of IL-12 in the induction of organ-specific autoimmune diseases. *Immunol. Today* **16**, 383–6.

Trinchieri, G. (1994). Interleukin 12: a cytokine produced by antigen-presenting cells with immunoregulatory functions in the generation of T helper cells type 1 and cytotoxic lymphocytes. *Blood* **84**, 4008–27.

Trinchieri, G. (1995). Interleukin-12: a proinflammatory cytokine with immunoregulatory functions that bridge innate resistance and antigen-specific adaptive immunity. *Annu. Rev. Immunol.* **13**, 251–76.

Trinchieri, G. (1996) Role of interleukin-12 in human Th1 response. In Romagnani, S. (ed.) *Chemical Immunology: Th1 and Th2 Cells in Health and Disease*, Vol. 63, pp. 14–29. Karger, Basel.

Wolf, S. F., Sieburth, D., and Sypek, J. (1994). Interleukin 12: a key modulator of immune function. *Stem Cells* **12**, 154–68.

Zitvogel, L. and Lotze, M. T. (1995). Role of interleukin-12 as an antitumor agent: experimental biology and clinical application. *Res. Immunol.* **146**, 628–37.

VII Transforming growth factor-β

Robert J. Lechleider and Anita B. Roberts*

1. Introduction

Transforming growth factor-β (TGF-β) is the prototype of a large family of secreted, multifunctional growth factors now known to include over 30 different proteins belonging to at least three distinct functional groupings: (i) the TGF-βs, (ii) the inhibins/activins, and (iii) the bone morphogenetic proteins (BMPs), as well as more divergent individual gene products such as Müllerian inhibiting substance and the inhibin α-subunit. Although TGF-β was the first to be described, genetic analysis suggests that members of the family have progressively diverged from *BMP2* and *BMP4*, and their *Drosophila* counterpart, *decapentaplegic* (*dpp*). All family members share structural features based on an almost invariant positioning of seven cysteine residues and signal through a novel pathway involving transmembrane serine/threonine kinases to regulate a broad spectrum of biological activities involved in morphogenesis, development and differentiated function. This review will focus specifically on TGF-β, a molecule with profound effects on immune function.

2. Structural features of the TGF-βs

Mammalian species all express three distinct isoforms of TGF-β, namely TGF-β1, -β2 and -β3, encoded by unique genes located on different chromosomes (human chromosomes 19q13, 1q41 and 14q24, respectively). All three genes share a similar intron/exon structure with a total of seven exons spanning >100 kb of genomic DNA. The mRNAs for each of the isoforms have extensive 5′ and 3′-

untranslated regions (UTRs), resulting in mRNA transcripts of 2.5 and 1.9 kb for TGF-β1, several species ranging from 4.1 to 6.5 kb for TGF-β2, and 3.0 kb for TGF-β3. The longer TGF-β1 transcript is poorly transcribed, often resulting in a lack of correspondence of levels of mRNA and secreted peptide. In contrast, the 1.9 kb TGF-β1 transcript, as seen in mitogen-activated lymphocytes and expressed in other cell types in response to injury or stress, appears to initiate downstream of negative regulatory sequences in the 5′-UTR and be more efficiently transcribed.

The TGF-βs are expressed as approximately 390 amino acid precursor proteins, and, like all family members, processed by a furin-like protease at a tetrabasic site to the mature, C-terminal peptide of 112 amino acids (see Figure 1). Biologically active forms of the TGF-βs exist as 25 kDa homodimers or heterodimers. The amino acid sequences of the active C-terminal domains of the individual TGF-β isoforms are >96% conserved between mammalian species (porcine, simian and bovine TGF-β1 are identical to human TGF-β1, and mouse TGF-β1 differs by only a single amino acid), and the three human isoforms are >72% similar. The TGF-βs and activins are unique in that they contain one extra pair of cysteines not found in any other family member. These two cysteines are disulfide bonded to each other and contribute in an essential way to the structure of the N-terminus of the peptides. Solution NMR and X-ray crystallographic studies suggest that all family members share a common cysteine knot structure consisting of a compact central core or 'knot' containing a conserved sequence motif and three invariant interlooped disulfide bonds. These serve to link together two pairs of antiparallel β-strands in a bow-like structure that is highly resistant to denaturants such as heat or extremes of pH. Because the flat shape of the monomer precludes a hydrophobic core, dimer

*Corresponding author.

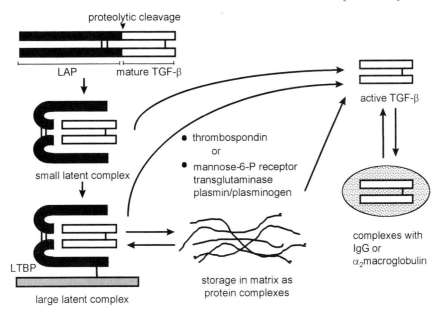

∎ **Figure 1.** The formation of latent TGF-β complexes and their activation are key to regulation of the activity of the TGF-βs *in vivo*. All TGF-βs are expressed as large precursor proteins which are processed at tetrabasic sites to the latency-associated protein (LAP) and the mature, biologically active C-terminal dimer. TGF-β is secreted from cells in non-covalent association with LAP in the form of the 'small' (100 kDa) latent complex; in the 'large' (220 kDa) latent complex the latent TGF-β-binding protein (LTBP) is covalently linked to LAP. LTBP directs the latent complex to the extracellular matrix. Activation of latent TGF-β complexes can occur by LAP-mediated binding to the mannose-6-phosphate/IGF-II receptor followed by proteolysis of LAP by the plasmin/plasminogen pathway, or by binding of thrombospondin to the latent complex in a protease-independent manner. Mature TGF-β is also found in complexes with IgG or α_2-macroglobulin.

formation is favored, stabilized by hydrophobic contacts and an interchain disulfide bond covalently linking the monomeric units. The cysteine knot structure is now known to be a structural feature common to disparate proteins including nerve growth factor, platelet-derived growth factor, vascular endothelial growth factor and human chorionic gonadotrophin, with the different orientation of the monomeric units providing additional structural diversity.

Whereas in other TGF-β family members, the divergent N-terminal region of the unprocessed precursor serves only to facilitate proper folding and secretion of the mature peptides, for the TGF-βs, this portion of the molecule minus the signal peptide (called latency-associated peptide, LAP) serves a unique role, remaining associated non-covalently with the C-terminal fragment to form a 'latent' molecule which is unable to bind to receptors (Figure 1). This unique ability of LAP to confer latency is important in preventing the binding of secreted TGF-β to ubiquitously expressed receptors and to maintenance of an extracellular reser-

voir of TGF-β that can be activated on demand. Even though the LAPs of TGF-β1, -β2 and -β3 are significantly more diverged from one another than are their C-terminal fragments, *in vitro* experiments show that TGF-β1 LAP can preferentially confer latency on TGF-β2 and -β3, suggesting that complex interactions between isoforms may further regulate activation. Most cells secrete TGF-β in the form of large latent complexes containing additional proteins that associate with LAP. Best characterized of these are latent TGF-β-binding proteins (LTBPs) which are covalently linked to LAP by disulfide bonds. Three LTBP genes have been identified thus far. All are characterized by epidermal growth factor (EGF)-like repeats common to many extracellular proteins, and by cysteine repeats, found also in fibrillins, which are major constituents of connective tissue. LTBPs are important both in secretion of TGF-β and in targeting it to extracellular matrix, although >90% of LTBP is not bound to TGF-β and probably serves a structural role in extracellular matrix.

Activation of latent TGF-β has been studied intensively *in vitro*, where extremes of pH or heat and treatment with proteases or chaotropic agents have all been shown to release active TGF-β from the latent complex. However, *in vivo*, where activation of extracellular matrix stores of TGF-β represents a key epigenetic mechanism for regulating the local concentration of active TGF-β, the steps involved in activation are still unclear. In co-cultures of endothelial cells and smooth muscle cells, latent TGF-β is activated by a complex process involving binding of the latent form to the mannose-6-phosphate (IGF-II) receptor through mannose-6-phosphate residues on LAP followed by the action of both transglutaminase and plasmin/plasminogen (Figure 1). Activation of TGF-β by retinoids, and possibly also other members of the steroid hormone family, is also proteolytic and can be blocked by protease inhibitors *in vitro*. However, the physiological significance of the plasmin-dependent mechanism of activation has recently been questioned since mice deficient in various components of this proteolytic pathway show no apparent dysfunction in TGF-β-dependent activities. Another mechanism, possibly of physiological significance, involves activation of both the small and large latent TGF-β complexes by a protease-independent mechanism involving the binding of thrombospondin, a component of platelet alpha granules and of extracellular matrix. Moreover, complexes of TGF-β bound to immunoglobulin have been shown to mediate its uptake by macrophages via Fc receptors; by unknown mechanisms the macrophages can then present active TGF-β directly to interacting lymphocytes. Active TGF-β bound to immunoglobulin has also recently been shown to be made by B cells and plasma cells and suggested to be important in host defense against infection. Non-covalent complexes of TGF-β and α₂-macroglobulin are also latent and can be activated by certain cells. Further study of the mechanisms of activation of TGF-β by proteins such as thrombospondin and immuno globulins is likely to provide insight into important pathways regulating the physiological trafficking and activation of the TGF-βs. Whether unique mechanisms exist for activation of the different TGF-β isoforms *in vivo* is not known.

3. Cellular origins of TGF-β isoforms

Nearly every cell in culture can be stimulated to secrete TGF-β. In most cells and tissues, TGF-β1 is the predominant isoform. The most abundant sources of TGF-β1 are platelets, bone and spleen; TGF-β1 is also present in plasma at levels of about 5 ng/ml, suggesting an endocrine function unique to this isoform. TGF-β2 predominates in fluids such as the aqueous and vitreous humors of the eye and amniotic fluid. In addition, significant quantities of TGF-β isoforms are sequestered in extracellular matrix and are present in the circulation covalently bound to α₂-macroglobulin. Each isoform is expressed in distinct temporally and spatially controlled patterns in development and in various pathologies. The basis of this is clear from studies of the promoters of the three TGF-β isoforms. Expression of TGF-β2 and -β3 is regulated principally by hormone-responsive elements; mice in which these genes are knocked out have serious developmental defects. In contrast, expression of TGF-β1 is induced acutely by a variety of signals related to stress. Because of this, TGF-β1 is the isoform implicated in various pathologies including carcinogenesis, as well as fibroproliferative, parasitic and autoimmune diseases. Moreover, TGF-β1 is the principal, and possibly the only, isoform of TGF-β secreted by hematopoietic cells and immune cells. The importance of this is best illustrated by TGF-β1 null mice, which, although they exhibit no developmental abnormalities, die at about 3 weeks of age from wasting, multifocal inflammation, dysregulated hematopoiesis as evidenced by myeloid hyperplasia, and symptoms of autoimmune-like disease. These observations clearly demonstrate the non-redundant activities of this isoform in regulation of immune cell homeostasis and hematopoiesis.

4. TGF-β receptors and signaling pathways

Like other cytokines, TGF-β mediates its effects on the cell through a set of transmembrane receptors. The mechanisms of TGF-β signaling, however, are unique. Initial cross-linking studies of cell surface proteins demonstrated that three major species of membrane-associated TGF-β-binding proteins exist in most cell types. These were designated the type I, II and III TGF-β receptors, based on their characteristic electrophoretic mobility in denaturing polyacrylamide gels. All three receptor types are glycosylated; the type I and II receptors migrate as cross-linked bands of approximately 70 kDa and 90 kDa respectively, while the type III receptor, which is also known as betaglycan, migrates as a high molecular weight smear of >200 kDa. Studies in chemically mutagenized mink lung cells demonstrated that the type I and II receptors are essential for mediating TGF-β responses, but that the type III receptor plays an accessory role. Analysis of cell mutants lacking either type I or II receptors demonstrated that the type II receptor can directly bind ligand, but requires the presence of the type I receptor to transmit a signal successfully. The type III receptor functions to present TGF-β to the other receptors, but only in the case of TGF-β2 does it appear to be essential.

TGF-β receptors are found on most cell types. Most cells of epithelial origin, including many carcinomas, express all three receptors. Granulocytes, macrophages and lymphocytes all possess functional TGF-β type I and II receptors but lack the type III receptor. Endothelial cells express a unique high molecular weight transmembrane glycoprotein receptor called endoglin, which is analogous to the type III receptor, and appears to play a critical role in mediating TGF-β-induced signals in these cells. Fibroblasts and other mesenchymal cells possess all three receptor types and are major targets of TGF-β biological activity, especially in regard to the regulation of extracellular matrix.

The paradigmatic model for TGF-β signaling at the cell surface suggests that the ligand acts as an adaptor for interaction of the type I and II receptors (see Figure 2) In this scheme, the type II receptor binds TGF-β and then recruits the type I receptor to generate a signaling complex. The accessory receptors (type III and endoglin) serve to present the ligand to the signaling receptors, a critical function when the ligand is TGF-β2. Activin and the BMPs, the two other best studied TGF-β like molecules, signal in a similar manner. In fact, genetic evidence from *Drosophila* and *Caenorhabditis elegans* confirms this model as a highly evolutionarily conserved one.

Our understanding of the mechanism by which TGF-β initiates its intracellular signaling cascade took a giant leap forward with molecular cloning of the transmembrane receptors. Cloning of the TGF-β type I and II receptors revealed characteristic signal peptide sequences, a putative ligand binding domain, a transmembrane domain and a canonical serine/threonine kinase phosphorylation sequence. The serine-threonine kinase motif is shared by receptors for entire TGF-β family (except for the distantly related GDNF receptor), and distinguishes them from other cytokines. The type III receptor is a proteoglycan with a large, highly glycosylated extracellular domain, a transmembrane

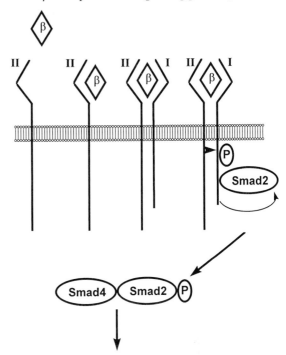

■ **Figure 2.** The mechanism of TGF-β signaling. The type II receptor (II) binds ligand (β) and recruits the type I receptor (I). The type II receptor then transphosphorylates the type I receptor in the GS box. Following type I phosphorylation, Smad2 is recruited to and phosphorylated by the type I receptor. Smad2 phosphorylation releases it from the receptor so that it can interact with Smad4 and transmit a downstream signal.

domain and a short intracellular domain with no known enzymatic motifs. The type II receptor is slightly larger than the type I, with two small inserts in the kinase domain, and a short C-terminal tail.

Following TGF-β binding and recruitment of the type I receptor, the type II receptor transphosphorylates the type I receptor in a region known as the 'GS box', which is a small intracellular region rich in glycine and serine residues which lies between the kinase and transmembrane domains. Phosphorylation in the GS box is necessary for activation of downstream signaling events. Mutation of either the type I or type II receptor in the ATP-binding site, which abrogates all kinase activity, effectively eliminates all signaling. This clearly demonstrates the importance of receptor interaction and sequential activation. Interestingly, a threonine to aspartate mutation, just outside the GS box at position 204 in the type I receptor, yields a constitutively active receptor that can signal in the absence of ligand or type II receptor. This suggests that the major role of the type II receptor is to phosphorylate and activate the type I receptor.

Two approaches have been used to identify signaling intermediates downstream from the type I receptor. The first is a yeast two-hybrid screen using the type I receptor as bait. With this system several proteins have been identified as interacting with the receptor complex. FKBP-12 is a small (12 kDa) *cis–trans* prolyl-isomerase which is well known for its interaction with the immunosuppressants rapamycin and FK506. FKBP-12 appears to function as a negative regulator of TGF-β receptor activity, keeping the receptor in an 'off' state unless ligand is bound. The farnesyltransferase α-subunit also binds to the receptor, but its role is not clear. Likewise, a WD-repeat containing protein, TRIP-1, has been shown to interact with the type II receptor, but its function is unknown.

Developmental model systems have been a much more fruitful area of investigation for identification of downstream signaling intermediates in the TGF-β pathway. In particular, cloning of the *Drosophila* gene *Mothers against decapentaplegic* and the *C. elegans sma* genes provided the first real immediate downstream mediators of TGF-β signaling. The outlines of Smad (for Sma and Mad-related proteins) signaling have rapidly become apparent. Three broad classes of Smad exist. The first class is most similar to *Drosophila* Mad. This includes Smad1 and Smad5, which have been shown to be involved in BMP pathways, and Smad2 and Smad3, which appear to be activin- or TGF-β-regulated. These receptor-activated proteins are phosphorylated in response to ligand, and are translocated to the nucleus following phosphorylation, where they interact with DNA-binding proteins to mediate transcriptional responses. The second class of Smads is most closely related to the product of the *Drosophila* gene *medea*, and has only one human member, Smad4 or DPC4. Smad4 acts as a common mediator, interacting with receptor-activated Smads in response to the appropriate ligand and translocating to the nucleus where it participates in transcriptional complexes. The third class, represented by Smad6 and Smad7, are inhibitory Smads. These proteins, which are rapidly up-regulated by treatment with ligand, block Smad signaling and provide a mechanism for negative feedback regulation of TGF-β signaling.

Loss of TGF-β signaling pathways is probably a key event in the development of many carcinomas and leukemias. Those cells able to escape from TGF-β growth-inhibitory effects would be at a growth advantage compared with neighboring cells still responsive to the cytokine. In human cancers two key signaling elements have been shown to be lost or mutated in a significant proportion of certain cancers. In patients with defects in DNA replication (so-called replication error positive defects), mutations in a long poly(A) tract in the type II receptor coding region are found in a high percentage of right-sided colon cancers. Such mutations cause a frameshift and produce inactive protein. Loss of functional type II receptor renders these cells insensitive to TGF-β growth inhibitory effects. Similar defects have been found in gastric cancers. At least one example of loss of TGF-β signaling pathways contributing to the development of leukemias has also been observed. A dominant inhibitory mutation in the kinase domain of the type II receptor has been isolated from a patient with an aggressive cutaneous T-cell leukemia. This mutation caused loss of TGF-β receptors and signaling. Cells isolated from the same patient at an earlier stage of the disease showed no defects in TGF-β receptors or signaling pathways. This clearly indicates that loss of TGF-β signaling pathways can contribute to worsening of clinical disease. In pancreatic cancer, the *Smad4* or *dpc4* gene has been shown to be mutated or lost with high frequency. Smad4 is a critical intermediate in the TGF-β signaling pathway, and its loss can, like loss of the type II receptor, render the cell unresponsive to TGF-β. Defects in other signaling components, such as the type I receptor or Smad2, have also been demonstrated, but the significance of these is not yet completely understood. The molecular genetic evidence is supported by animal models of TGF-β responsiveness and tumorigenicity. MCF-7 human breast carcinoma cells which express little type II receptor are insensitive to TGF-β and produce tumors in nude mice. When these same cells are transfected with a type II receptor construct, they regain responsiveness and lose their ability to form tumors. Thus the animal model supports the findings from human cancers.

Although the role of TGF-β signaling pathways in mediating responses to other diseases is less well understood, at least one example of such a direct interaction is known. Intact TGF-β receptors and signaling pathways are essential for successful infection of the parasite *Trypansoma cruzi*. Loss of either cell surface receptors or downstream signaling components renders the cells immune to infection. The exact role played by the TGF-β signaling pathway in this infection is not understood, nor is it known if TGF-β or its signaling pathway play a facilitating role in infection by other pathogens. Clearly, the ability to manipulate the TGF-β signal transduction pathway will have beneficial effects in numerous disease processes.

Perhaps because of the essential roles that TGF-β plays in normal development and homeostasis, there has to date been only one example of a hereditary genetic abnormality attributed to TGF-β or its signaling pathways. Hereditary hemorrhagic telangiectasia type I has been shown to be due to a defect in endoglin, which is expressed on endothelial cells. Loss of endoglin causes the cells to be unresponsive to TGF-β2, and leads to the pathological defect seen.

5. Biological activities of the TGF-βs

Given the almost universal ability of cells to respond to TGF-β, including epithelial cells, endothelial cells, mesenchymal cells and bone-marrow-derived cells, the broad spectrum of biological activities attributed to this molecule is not surprising. Major effects include stimulation of chemotaxis, production of extracellular matrix and regulation of cell growth and differentiation. For the purposes of this review, we will focus on the profound effects of this cytokine on nearly every aspect of immune cells from development of the immune system to regulation of the function of mature immune cells. Although the isoforms often function interchangeably *in vitro*, it is typically TGF-β1 which acts as the primary immunomodulator *in vivo*. This is clearly illustrated in TGF-β1 null mice which, despite unimpaired expression of the TGF-β2 and -β3 genes, exhibit pathogenic disruption of immune cell homeostasis and profound alterations in hematopoiesis. In wild-type mice, TGF-β1 produced by thymic epithelial cells has been shown to regulate the production of CD4$^+$CD8$^+$ thymocytes. At femtomolar concentrations, it is a potent chemoattractant for macrophages, neutrophils and lymphocytes. At picomolar concentrations, it can block the production of superoxide and nitric oxide by macrophages and both modulate the production of and antagonize the response of cells to inflammatory cytokines including tumor necrosis factor-α, interferon-γ, and various interleukins. It has profound suppressive effects on the growth of both T and B cells, at the same time inhibiting the differentiated function of mature B cells by suppressing expression of IgG and IgM; it has also been implicated in IgA isotype switching. TGF-β has been shown to play a key role in the immunological aspects of many diseases including

- carcinogenesis, where TGF-β expressed by tumor cells refractory to its autocrine growth inhibitory effects acts in a paracrine fashion to suppress surveillance by immune cells;

- autoimmune disease, where expression of TGF-β1 has been implicated both in suppression of primary disease and in mediating effects of tolerance;

- parasitic infections, where production of TGF-β1 by the infected cell can suppress the host response; and

- chronic inflammatory disease, where its production by inflammatory cells stimulates elaboration of extracellular matrix by mesenchymal cells.

Many of these effects of TGF-β will be discussed in greater detail in subsequent chapters.

References

Barcellos-Hoff, M. H. (1996). Latency and activation in the control of TGF-β. *J. Mammary Gland Biol. Neopl.* **1**, 351–61.

Caver, T. E., O'Sullivan, F. X., Gold, L. I., and Gresham, H. D. (1996). Intracellular demonstration of active TGF-β1 in B cells and plasma cells of autoimmune mice. *J. Clin. Invest.* **98**, 2496–506.

Derynck, R. and Feng, X.-H. (1997). TGF-beta receptor signaling. *Biochim. Biophys. Acta* **1333**, F105–50.

Feige, J. J., Negoescu, A., Keramidas, M., Souchelnitskiy, S., and Chambaz, E. M. (1996). Alpha 2-macroglobulin: a binding protein for transforming growth factor-beta and various cytokines. *Hormone Res.* **45**, 227–32.

Fynan, T. M. and Reiss, M. (1993). Resistance to inhibition of cell growth by transforming growth factor-β and its role in oncogenesis. *Crit. Rev. Oncogen.* **4**, 493–540.

Heldin, C. H., Miyazono, A., and ten Dijke, P. (1997). TGF-β signalling from cell membrane to nucleus through SMAD proteins. *Nature* **390**, 465–71.

Isaacs, N. W. (1995). Cystine knots. *Curr. Opin. Struct. Biol.* **5**, 391–5.

Kingsley, D. M. (1994). The TGF-β superfamily: new members, new receptors, and new genetic tests of function in different organisms. *Genes Dev.* **8**, 133–46.

Knaus, P. I., Lindemann, D., DeCoteau, J. F., Perlman, R., Yankelev, H., Hille, M., Kadin, M. E., and Lodish, H. F. (1996). A dominant inhibitory mutant of the type II transforming growth factor β receptor in the malignant progression of a cutaneous T-cell lymphoma. *Mol. Cell. Biol.* **16**, 3480–89.

Letterio, J. J. and Roberts, A. B. (1998). Regulation of immune responses by TGF-β. *Annu. Rev. Immunol.* **16**, 137–61.

Markowitz, S. D. and Roberts, A. B. (1996). Tumor suppressor activity of the TGF-β pathway in human cancers. *Cytokine Growth Factor Rev.* **7**, 93–102.

Massagué, J. and Weis-Garcia, F. (1996). Serine/threonine kinase receptors: mediators of transforming growth factor beta family signals. *Cancer Surveys* **27**, 41–64.

McCartney, N. L. and Wahl, S. M. (1994). Transforming growth factor *β*: a matter of life and death. *J. Leukocyte Biol.* **55**, 401–9.

Miyazono, K., Ichijo, H., and Heldin, C.-H. (1993). Transforming growth factor-*β*: latent forms, binding proteins and receptors. *Growth Factors* **8**, 11–22.

Nunes, I., Munger, J. S., Harpen, J. G., Nagano, Y., Shapiro, R. L., Gleizes, P. E., and Rifkin, D. B. (1996). Structure and activation of the large latent transforming growth factor-beta complex. *Int. J. Obesity-Related Metabol. Disorders* **3**, 4–8.

Roberts, A. B. and Sporn, M. B. (1990). The transforming growth factors-*β*. In Sporn, M. B. and Roberts, A. B. (eds) *Handbook of Experimental Pharmacology*, Vol 95: Peptide growth factors and their receptors I, pp. 418–72. Springer-Verlag, New York.

Roberts, A. B. and Sporn, M. B. (1992). Differential expression of the TGF-*β* isoforms in embryogenesis suggests specific roles in developing and adult tissues. *Mol. Reprod. Dev.* **32**, 91–8.

Roberts, A. B. and Sporn, M. B. (1993). Physiological actions and clinical applications of transforming growth factor-*β* (TGF-*β*). *Growth Factors* **8**, 1–9.

Shull, M. M., Kier, A. B., Diebold, R. J., Yin, M., and Doetschman, T. (1994). The importance of transforming growth factor *β*1 in immunological homeostasis, as revealed by gene ablation in mice. In *Overexpression and Knockout of Cytokines in Transgenic Mice*, pp. 135–59, Jacon. C. (ed.). Academic Press, Chicago, IL.

Stach, R. M. and Rowley, D. A. (1993). A first or dominant immunization. II. Induced immunoglobulin carries transforming growth factor-beta and suppresses cytolytic T cell responses to unrelated alloantigens. *J. Exp. Med.* **178**, 841–52.

VIII   The interferons: biochemistry and biology

Daniel H. Kaplan and Robert D. Schreiber*

1. Introduction

The term 'interferon' was first coined by Isaacs and Lindenmann in 1957 to describe an activity present in the supernatant of virally infected cells which protected other cells from subsequent infection by a variety of different viruses. Today we know that this antiviral activity is manifested by a family of proteins which can be divided into two general classes on the basis of structural and functional criteria. The first class, termed type I interferons, is the classical type of interferon and is induced by viral infection of cells. The components of this protein class can be further divided according to their cell of origin:

- 21 different type I interferons are made by leukocytes and are collectively termed interferon α (IFNα);
- a single type I interferon is produced by fibroblasts and is termed IFNβ.

The second class of interferons, termed type II interferon or IFNγ, was initially detected by Wheelock as an antiviral activity synthesized by mitogen-stimulated human leukocytes and was identified as an interferon with physicochemical properties distinct from those of the type I interferons by Younger and Slavin shortly thereafter. Importantly, IFNγ is not produced in direct response to viral infection, but rather is generated as a result of stimulation of T lymphocytes and natural killer (NK) cells. Moreover, IFNγ effects a more diverse range of protective cellular responses than type I interferons, including a critical ability to modulate host immune responses. For this reason, the type I interferons have been described as true antiviral agents that

display immunomodulatory activity while IFNγ has been described as an immunomodulator that also displays antiviral activity.

The past 15 years have seen an explosive growth in our understanding of interferon biology. During this time, the genes for the interferon proteins as well as their receptors have been cloned and the polypeptides they encode characterized. Recombinant forms of the proteins together with highly specific monoclonal antibodies (mAbs) are now available in large quantities. In addition, mice with genetic deficiencies of the cytokines or their receptors have been generated and in some cases the corresponding genetic human deficiencies identified as well. Finally, extensive investigation of potential therapeutic effects of these proteins in infectious and neoplastic diseases has been carried out. Taken together, the combination of basic and clinical interferon research efforts has established a relatively clear view of the biochemistry and biology of the interferon system.

The purpose of this chapter is to review the key aspects of the interferon system as we now see them, giving special attention to IFNγ because of its extremely important function in modulating immune system activity and its indisputable role in promoting host resistance to microbial pathogens and tumors.

2. Molecular properties of interferon family members

The interferon family contains three types of proteins, denoted IFNα, IFNβ and IFNγ, which can be differentiated on the basis of structure, function and cell of origin (Table 1). IFNα, also known as leukocyte interferon, is a type I interferon produced by leukocytes in response to viral infection.

* Corresponding author.

◼ **Table 1.**   Properties of the interferons

Property	IFNα	IFNβ	IFNγ
Nomenclature	Type I (leukocyte)	Type I (fibroblast)	Type II (immune)
Major inducer(s)	virus	virus, LPS, double-stranded poly RNA	antigens, mitogens
Physical properties			
M_r of predicted/mature protein (kDa)	19.2–19.7/16–27.5	20/20–26	17/34–50
no. of amino acids	165–166	166	143
N-linked glycosylated?	in some species	yes	two sites
subunit composition	single polypeptide	Single polypeptide	Noncovalent homodimer
pH stability	stable	stable	labile
Gene structure			
number of genes	26	1	1
chromosomal location			
murine	4	4	10
human	9	9	12
no. of introns	none	none	3
Cellular source	T cells, B cells and macrophages	fibroblasts and epithelial cells	T cells and NK cells
Cellular receptors			
(i) α-chain			
M_r of predicted/mature protein (kDa)		60.5/95–100	52.6/85–95
no. of amino acids			
total		530	472
in extracellular domain		409	228
in transmembrane domain		21	23
in intracellular domain		100	221
no. of N-glycosylation sites		12	5
chromosomal location			
murine		16	10
human		21	6
(ii) β-chain			
M_r of predicted/mature protein (kDa)		57.7/95–100	34.8/62
no. of amino acids			
total		489	316
in extracellular domain		217	226
in transmembrane domain		21	24
in intracellular domain		251	66
no. of N-glycosylation sites		3	5
chromosomal location			
murine		16	16
human		21	21

A total of 26 genes for IFNα have been identified, although at least five of these genes are pseudogenes. The purpose of the multiple IFNα gene products remains undefined. The genes encoding IFNα family members are organized in an identical manner, lack intervening sequences (introns) and are thought to have been derived from a common ancestral precursor 100 million years ago. The IFNα gene cluster is located on human chromosome 9 and mouse chromosome 4. IFNα proteins are single-chain polypeptides which contain 165–167 amino acids in their mature form. Some of the forms are glycosylated and thus at least 25 distinct forms of mature human IFNα have been detected with M_r ranging from 16 to 27.5 kDa. Most forms of IFNα are heat- and pH-stable. The different human IFNα gene products display 68% sequence identity with one another and approximately 40% homology with murine IFNα forms. Some types of IFNα display species-specificity in their abilities to bind cell surface receptors and activate cellular responses.

Interferon β is also a type I interferon and is secreted from fibroblasts in response to viral infection. There is only a single IFNβ gene which is located adjacent to the IFNα gene cluster in both human and mouse genomes. The organization of the gene for IFNβ is closely related to that of the IFNα genes, suggesting that IFNβ and IFNα descended from a common ancestor. Mature human and murine IFNβ are glycoproteins and consist of 166 and 161 amino acids, respectively. The natural proteins display M_r of 20 and 26 kDa, respectively. Although the organization of the IFNα and IFNβ genes are closely related, the proteins are only 15–30% identical at the protein level. Nevertheless there is sufficient similarity between key functional domains of the proteins to enable them to bind to the same cell surface receptor and induce an overlapping array of biological responses.

IFNγ, also known as immune interferon, is a type II interferon and is unrelated to IFNα or IFNβ. It is encoded by a single gene of 6 kb which contains four exons and three introns. The gene is located on human chromosome 12 and murine chromosome 10. The human gene encodes a 1.2 kb mRNA that gives rise to a 166 amino acid polypeptide containing a 23 amino acid signal sequence. The mature 143 amino acid polypeptide displays a molecular mass of 17 kDa and contains two potential *N*-glycosylation sites which can be independently glycosylated, thus producing mature polypeptides of M_r 17, 20 and 25 kDa. The murine gene produces a 1.2 kb mRNA and a 134 amino acid mature polypeptide which, depending on its glycosylation state, displays M_r of 15.4, 20 and 25 kDa. In both species the fully glycosylated form of IFNγ predominates. Human and murine IFNγ share only 60% homology at the DNA level and 40% identity at the protein level. The lack of sequence identity explains the strict species-specificity shown by human and murine IFNγ in binding to and activating responses in human and murine cells respec-

tively. Under physiological conditions, the IFNγ polypeptides self-associate to form a homodimer of M_r 50 kDa. The dimer is held together entirely by noncovalent bonds and can therefore be easily denatured by extremes of heat ($>65\,°C$) and pH (<4 and >9). Importantly, only the homodimeric form of IFNγ is able to bind to the IFNγ receptor and activate biological responses in cells. X-ray crystallographic studies have determined that the IFNγ monomers associate in an antiparallel fashion to form a symmetrical dimer. The antiparallel arrangement of the monomers together with data from radioligand binding studies predicts that the IFNγ homodimer contains two identical receptor binding sites. This prediction has recently been confirmed by structural analysis of the complex of IFNγ and the ligand binding chain of the IFNγ receptor.

3. Interferon biosynthesis

Biosynthesis of type I interferons is not a specialized cellular function; most cells are capable of producing some form of type I interferon following exposure to the appropriate inducing stimuli. The most important physiological inducer of IFNα/β is viral infection. However, type I interferon family members can also be induced by double-stranded RNA, inflammatory stimuli such as bacterial endotoxin, interleukin-1, Tumor Necrosis Factor (TNF) and interferon itself. Analysis of the regulatory elements of the type I interferon genes reveals the presence of binding sites for NF-κB and interferon regulatory factor 1 (IRF-1) transcription factors which are activated by the known interferon inducers. Type I interferon induction is rapid and IFNα/β can be detected in the extracellular environment within 6 h after cellular stimulation.

Biosynthesis of IFNγ is more complicated, due to both its cellular restriction and its highly regulated nature. T cells are one of the two major cellular sources of IFNγ in the normal host. All CD8$^+$ T cells and certain subsets of CD4$^+$ T cells (the Th0 and Th1 T-helper cell subsets) secrete IFNγ in response to stimuli that effect T-cell activation (Figure 1). *In vivo*, interaction of the antigen receptor on T cells with cells bearing antigen in the context of the major histocompatibility complex (MHC) and appropriate co-stimulation leads to the production of IFNγ along with other T-cell-derived cytokines. *In vitro*, this process can be mimicked by stimuli that effect T-cell activation such as antibodies to CD3, mitogens (such as concanavalin A and phytohemagglutinin), or pharmacological

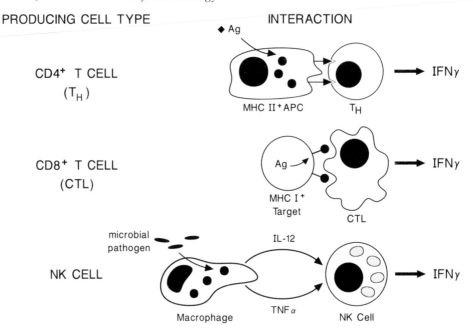

∎ **Figure 1.** Cellular sources of IFNγ. IFNγ can be produced either by the Th0 and Th1 CD4⁺ T cell subsets in response to antigen in the context of MHC class II molecules or by CD8⁺ T cells following recognition of antigen associated with MHC class I. In addition, NK cells elaborate IFNγ through an MHC-independent mechanism involving production of IL-12 and TNFα by macrophages.

stimuli (such as the combination of phorbol myristate acetate and calcium ionophore). The IFNγ gene is not constitutively activated in resting T cells. Following T-cell stimulation, IFNγ transcripts can be detected within 6–8 h, reach maximum levels by 12–24 h and decline to baseline levels thereafter. The IFNγ protein is not stored intracellularly but rather is secreted directly into the extracellular environment where it reaches maximal levels 18–24 h after T-cell stimulation.

The other major cellular source of IFNγ is activated NK cells. In contrast to IFNγ production by T cells, NK cells, a key cellular component of innate immunity, can produce the cytokine in a rapid and MHC-unrestricted manner. NK-dependent IFNγ production provides the host with a first line of cellular defense against infectious agents until such time as specific sterilizing immunity can be generated. Careful examination of mice infected with the gram-positive bacterium, *Listeria monocytogenes*, has revealed an intricate cytokine network linking IFNγ and IL-12. Immunodeficient SCID mice lack functional B and T cells yet are able to contain *Listeria* infection through a mechanism that involves IFNγ generation and macrophage

activation. Upon interaction with bacterial products, resting macrophages produce low amounts of two cytokines (TNFα and IL-12) which in turn stimulate NK cells to secrete low levels of IFNγ. IFNγ produced by the NK cells then activates additional macrophages in the vicinity, leading to enhanced production of TNF and IL-12 and subsequently to enhanced production of IFNγ from NK cells. This generates a positive amplification loop that produces substantial quantities of IFNγ early during the course of infection and facilitates the generation of large numbers of activated macrophages with antimicrobial activity. Thus, the mutual synergic effects that IL-12 and IFNγ have on macrophages and NK cells provides a non-immune mechanism for IFNγ production which in turn promotes the control of infection.

IL-10 is an important negative regulator of IFNγ production from T cells and NK cells. In both cases IL-10 appears to exert its effects by acting indirectly. In the case of NK cells, IL-10 prevents macrophage secretion of TNF and IL-12 thereby preventing NK activation and IFNγ production. In the case of T cells, the data are somewhat less clear. It is known that IL-10 prevents T-cell activation not through a direct effect on the T cell but through its actions at the level of the antigen-presenting cell (APC). It is thought that IL-10 limits the ability of the APC to provide sufficient co-stimulation for T-cell activation and subsequent IFNγ secretion. IL-10 produced late during the course of infection also down-regulates the production of IL-12 and IFNγ and hence may play an important role limiting the progression of inflammatory responses.

4. Interferon receptors and signal transduction

4.1 The IFNγ receptor

The IFNγ receptor system is the best defined of all the cytokine receptors. IFNγ exerts its pleiotropic effects on cells by interacting with a specific receptor that is expressed on nearly all cell surfaces. IFNγ receptors consist of two species-matched polypeptides (Table 1). The first is a 90 kDa receptor α-chain (also known as IFNγR1 or CDw119) encoded by a gene on human chromosome 6 and murine chromosome 10 that is predominantly responsible for ligand binding, ligand trafficking through the cell and signal transduction. The second is a 62 kDa receptor β-chain (also known as IFNγR2 or accessory factor 1 (AF-1)) encoded by a gene on human chromosome 21 and murine chromosome 16 that plays only a minor role in ligand binding but which is obligatorily required for function. IFNγ signal transduction is known to require at least three other components in addition to the receptor polypeptides. Two of these (Jak1 and Jak2) are tyrosine kinases that belong to a four-member family known as the Janus tyrosine kinases (or Jaks). These enzymes are unusual because they contain two C-terminal kinase-like domains, though only the domain at the extreme C-terminus has authentic kinase activity. The third component, Stat1, belongs to the seven-member family of latent cytosolic transcription factors known as 'signal transducers and activators of transcription', or STATs. Members of this family of proteins are unique among transcription factors in that they possess Src homology 2 (SH2) domains capable of binding to phosphotyrosine-containing sequences. Recent experiments using mice with genetic deficiencies in Stat1 or Jak1 have shown that all biological responses induced by either IFNγ or IFNα/β are mediated through the Jak–STAT signaling pathway.

Extensive mutagenesis of both chains of the IFNγ receptor has led to a detailed understanding of the proximal events involved in IFNγ signal transduction. A block of four amino acids (Leu-Pro-Lys-Ser) at positions 266–269 in the receptor α-chain intracellular domain situated 13 residues from the transmembrane domain is required for ligand-induced activation of intracellular tyrosine kinase activity, phosphorylation of the receptor α-chain intracellular domain and induction of IFNγ-mediated cellular responses. This sequence is important for mediating the constitutive binding of Jak1 to the receptor α-chain. Two functionally important regions on the receptor β-chain located near the transmembrane domain have also been identified: Pro-Pro-Ser-Ile-Pro (residues 263–267) and Ile-Glu-Glu-Tyr-Leu (residues 270–274). These two regions are required for the constitutive binding of Jak2 to the receptor β-chain. The third region required for receptor signaling is a membrane-distal sequence on the receptor α-chain that contains the residues Tyr-Asp-Lys-Pro-His at positions 440–444. Structure–function analyses revealed that within this sequence only three amino acids are important: Tyr440, Asp441 and His444. The tyrosine at position 440 was found to be a physiological substrate site for the IFNγ-activated tyrosine kinase activity. Using purified proteins and phosphopeptides, the phosphorylation of the Tyr-Asp-Lys-Pro-His α-chain sequence was shown to form a specific docking site on the receptor for latent Stat1. The residues necessary for Stat1 binding (phosphorylated Tyr440, Asp441 and His444) were precisely the same as the residues required for receptor function.

Based on the large body of data concerning the IFNγ receptor and the Jak–STAT signaling pathway it is possible to construct a relatively comprehensive model of IFNγ signal transduction (Figure 2). In the absence of ligand, the IFNγ receptor α- and β-chains are not associated with each other but are pre-associated with inactive forms of the cytosolic kinases, Jak1 and Jak2, respectively. Signal transduction is initiated when IFNγ, a homodimeric ligand, binds to the IFNγ receptor α-chain, effecting its dimerization. The ligand-induced α-chain dimer now (presumably) expresses two binding sites for receptor β-chains which are then recruited into the complex. Within the ligand-assembled receptor complex the subunit-associated kinases are brought into close apposition and transactivate one another. The activated kinases then effect the phosphorylation of Tyr440 in the receptor α-chain, thereby forming a paired set of docking sites for Stat1. Two Stat1 molecules then bind to the paired docking sites, are brought into close proximity with

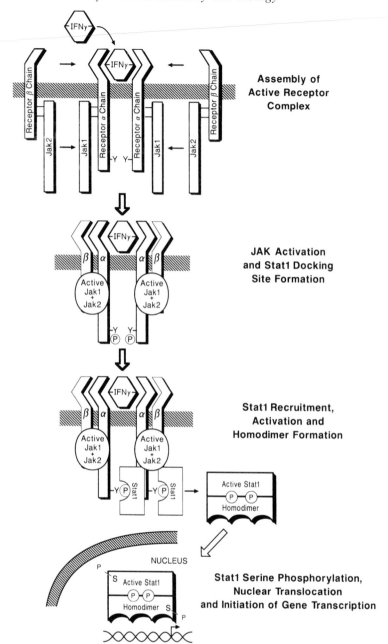

Assembly of
Active Receptor
Complex

JAK Activation
and Stat1 Docking
Site Formation

Stat1 Recruitment,
Activation and
Homodimer Formation

Stat1 Serine Phosphorylation,
Nuclear Translocation
and Initiation of Gene Transcription

∎ **Figure 2.** Proposed signaling mechanism of the IFNγ receptor. The details of this model are described in the text.

the receptor-associated, activated tyrosine kinases and are themselves phosphorylated on Tyr701. Phosphorylated Stat1 proteins then dissociate from their receptor tether, form homodimers, become serine-phosphorylated and translocate to the nucleus. In the nucleus, the Stat1 homodimers bind to specific promoter elements (known as IFNγ-activated sites or GAS elements) and thereby effect transcription of IFNγ-induced genes. Thus, the ligand-induced phosphorylation of the IFNγ receptor α-chain is the event that couples the activated receptor to its signal transduction system.

4.2 The IFNα/β receptor

The IFNα/β receptor system displays considerable homology to the IFNγ receptor system with a few notable exceptions. First, the subunits of the IFNα/β receptor are distinct from those of the IFNγ receptor (Table 1). The IFNα/β receptor is comprised of two polypeptides: a 110 kDa receptor α-chain (also known as IFNAR1 or IFNαR1) and a 102 kDa receptor β-chain (also known as IFNAR2 or IFNαR2). The genes for both these subunits are located together with the gene for the IFNγ receptor β-chain in a gene cluster that is present on human chromosome 21 and murine chromosome 16. Although IFNAR1 is sufficient to confer upon cells the ability to bind one IFNα family member (IFNα8), binding of *all* IFNα family members requires the co-expression of both IFNα receptor subunits. Although a distinct receptor for IFNβ has been proposed, no data are currently available to prove the existence of such a separate receptor protein.

IFNα/β receptor signaling is mediated through the Jak–STAT pathway but is more complicated and less well defined than signaling through the IFNγ receptor. Development of type I interferon biological responses requires the participation of five Jak–STAT pathway proteins. Two of these are the Janus kinases Jak1 and Tyk2, two are the STAT proteins Stat1 and Stat2 and the fifth is a DNA-binding protein known as p48 or ISGF3γ. In an unstimulated cell, Tyk2 is constitutively bound to a membrane-proximal region in the IFNAR1 intracellular domain and Jak1 is preassociated with a Leu-Pro-Lys-Val sequence at residues 274–277 of IFNAR2. Exposure of cells to IFNα/β leads to assembly of the receptor subunits, activation of Jak1 and Tyk2, and presumably phosphorylation of a Tyr-Val-Phe-Phe-Pro sequence at residues 466–470 of IFNAR1, thereby forming a Stat2 docking site on activated receptor. Receptor-associated Stat2 is tyrosine-phosphorylated and then serves as a docking site for Stat1. Stat1 becomes phosphorylated upon binding this complex. On the receptor, the Stat1/Stat2 complex has two fates. Some of the activated Stat1 proteins dissociate from their Stat2 tethers and form Stat1 homodimers which then translocate to the nucleus, bind to GAS elements and activate interferon-inducible genes. To at least some extent this mechanism explains the overlapping pattern of genes induced by IFNα/β and IFNγ. However, the Stat1/Stat2 heterodimer (known as ISGF3α) can dissociate as an intact complex from the IFNα/β receptor. By itself this complex can neither translocate to the nucleus nor bind DNA. These functions are conferred upon the heterocomplex by the DNA-binding protein p48. Association of p48 with ISGF3α forms the active ISGF3 trimolecular complex which can now translocate to the nucleus. ISGF3 then interacts with a distinct promoter element known as the interferon-stimulated response element (ISRE) and promotes transcription of IFNα/β-inducible genes. In this manner, type I interferons can effect activation of genes which are distinct from those induced by IFNγ.

5. Biological activities of interferon

5.1 Antiviral and antiproliferative actions

Both type I and type II interferons can protect cells from viral infection and have profound antiproliferative actions on normal and neoplastic cells. The molecular basis of the antiviral effects of the interferons has been extensively studied over the last 15 years. The interferons promote antiviral responses that are either intrinsic to the infected cell itself or effect recognition and destruction of infected cells by host components which comprise either the innate or specific limbs of the immune response.

Interferon induces several proteins in cells that promote resistance to infection by viruses. Although much work is still needed to elucidate all of interferon's activities in promoting cellular resistance to viral infection, three mechanisms have been studied in relative detail. The first is a family of enzymes known as 2′–5′-oligoadenylate synthetases. These enzymes are induced by both type I and type II interferons and are activated in the presence of double-stranded RNA (which are intermediates or by-products of viral RNA replication). The activated enzymes polymerize ATP into 2′–5′-linked oligomers which in turn activate RNase L, a latent constitutively expressed endoribonuclease. Activated RNase L degrades single-stranded RNA and thereby inhibits protein synthesis in cells. The second is protein kinase R (PKR; also known as double-stranded RNA dependent kinase, P1 kinase, p68 kinase or eIF-2 kinase), which is a serine/threonine kinase that is also induced by interferon and activated by double-stranded RNA. PKR phosphorylates and inactivates the eukaryotic protein synthesis initiation factor (eIF-2) thereby blocking protein synthesis in cells. Thus, these two mechanisms produce their antiviral effects by inhibiting cellular protein synthesis. The third interferon-induced protein is the Mx protein which is induced only by type I and not by type II interferons. The mechanism of action of this protein is poorly understood but is directed primarily at inhibiting replication of orthomyxoviruses such as influenza virus.

Extrinsic antiviral mechanisms induced by the interferons are largely those that direct innate and specific immunity. Type I interferons are known to induce enhanced cytolytic activity in NK cells and thereby promote the capacity of these cells to lyse virally infected target cells. Both type I and type II interferons promote antigen processing and presentation and thereby play a key role in the induction of antiviral cellular and humoral immune responses. These actions will be discussed in more detail later in this chapter.

Both classes of interferon have antiproliferative effects on cells. However, although these biological effects are well documented, their molecular basis is not yet well defined. Recent work has revealed that at least some of the antiproliferative actions of the interferons are due to the induction of proteins that inhibit the enzymes involved in cell cycle progression. For example, interferon has been shown to induce via the Jak–STAT pathway expression of the protein p21$^{WAF1/CIP1/CAP1}$ which is an inhibitor of CDK2. However, this process occurs in a relatively cell-specific manner and it is presently uncertain whether other cell cycle inhibitors may also be involved in the process. Nevertheless, this negative biological response can still be attributed to a positive induction of a particular gene product.

5.2 Interferons, macrophage activation and innate immunity

Of the interferon family members, IFNγ is clearly distinct in its ability to function as the major macrophage activating factor (MAF). As such it is crucial in promoting nonspecific host defense mechanisms against a number of pathogens. Data supporting this come from both *in vitro* and *in vivo* studies which have demonstrated that IFNγ can induce in macrophages the capacity to kill nonspecifically a variety of intracellular and extracellular parasites as well as neoplastic cells. In addition, IFNγ reduces the susceptibility of macrophages to microbial infection and enhances recognition of targets during the early innate phase of immunity through regulation of macrophage cell surface proteins which are to date undefined. The physiological role played by IFNγ in macrophage activation

and host defense against microbial pathogens has been clearly demonstrated in several *in vivo* murine infection models. Mice pretreated with neutralizing antibodies to IFNγ lose the ability to resist a sublethal challenge of a variety of microbial pathogens such as *Listeria monocytogenes*, *Toxoplasma gondii* or *Leishmania major*. In addition, mice with disrupted genes for IFNγ, the IFNγ receptor α-chain, or the interferon signaling protein Stat1, die when challenged with sublethal doses of microbial pathogens such as *Mycobacterium bovis*, *L. monocytogenes* or *L. major*.

Macrophages kill microbial targets using a variety of mechanisms such as reactive oxygen and nitrogen intermediates which are induced by IFNγ. Much of the antimicrobial effect of IFNγ-activated macrophages can be attributed to the actions of nitric oxide (NO) and/or reactive oxygen intermediates. NO is generated in macrophages by the inducible form of nitric oxide synthetase (iNOS). This enzyme is induced following treatment of the cells with IFNγ plus a variety of second signals which activate the transcription factor NF-κB such as TNFα, IL-1 or bacterial endotoxin. The enzyme catalyzes the conversion of L-arginine to L-citrulline, giving rise to large amounts of the toxic gas, NO. NO is thought to kill target cells by one of two mechanisms:

(1) It can form an iron–nitrosyl complex with Fe–S groups of aconitase and complexes I and II, thereby causing the inactivation of the mitochondrial electron transport chain.

(2) It may react with superoxide anion (a reactive oxygen intermediate formed by the IFNγ-induced enzyme NADPH oxidase) to form peroxynitrite, which decays rapidly once protonated to form the highly toxic compound hydroxyl radical.

Evidence that NO is responsible for macrophage killing of intracellular parasites comes from a number of studies with *Listeria* and *Leishmania*. Mice pretreated with the L-arginine analog *N*-monomethyl-L-arginine (NMMA), which is an iNOS inhibitor, were unable to resolve footpad infection with *Leishmania* parasites. Similarly, mice treated with another iNOS inhibitor, aminoguanidine, succumbed to sublethal *Listeria* infection.

5.3 Interferon and antigen processing/presentation

One of the major immunoregulatory roles of the interferons is their ability to promote the inductive phase of immune responses. These cytokines significantly influence the generation and presentation of antigenic peptides on cell surfaces. Among the interferon family members, IFNγ is uniquely capable of regulating expression of MHC class II proteins, thereby promoting enhanced CD4$^+$ T-cell responses. IFNγ induces MHC class II protein expression on many cells such as mononuclear phagocytes, endothelial cells and epithelial cells. Interestingly, IFNγ inhibits IL-4-dependent MHC class II expression on B cells, although the molecular basis for this apparently discordant effect is unknown. Type I interferon cannot induce MHC class II proteins on cells by itself. However it can either inhibit or enhance IFNγ's ability to induce MHC class II. In the case of mononuclear phagocytes, pre-exposure of cells to type I interferons induces a state of unresponsiveness to IFNγ. In contrast, treatment of cells with IFNα/β either concomitant with or subsequent to IFNγ treatment leads to enhanced MHC class II expression. A similar type of modulation of IFNγ's MHC class II inducing activity is also observed with other stimuli such as TNF, bacterial endotoxin and immune complexes. Thus, a cell's ability to express MHC class II in response to IFNγ is influenced by the composition of the microenvironment.

One function shared by all interferon family members is the ability to regulate expression of molecules involved in the MHC class I antigen presentation pathway. Some of these effects are at the level of regulating cell surface molecules. Both type I interferon and IFNγ enhance expression of MHC class I proteins and β_2-microglobulin on a wide variety of cell types, thereby increasing cell surface levels of functional MHC class I molecules. The proteins also enhance expression of several cell surface proteins such as ICAM-1 and B7 responsible for increasing target cell–T cell contact and T-cell co-stimulation, respectively.

The interferons also promote antigen processing by regulating expression of many intracellular proteins required for antigenic peptide generation. IFNγ has been shown to play a key role in modifying the activity of the proteasome, a multisubunit enzyme complex which is responsible for the generation of all peptides that bind to MHC class I proteins. To some extent, IFNγ mediates its effects by modulating the expression of the enzymatic proteasome subunits. It induces increased expression of LMP2, LMP7, and MECL1 and decreased expression of subunits X, Y and Z, thereby altering proteasome composition and specificity. Purified 20S and 26S proteasomes from IFNγ-treated cells show an increased capacity to cleave peptides after hydrophobic and basic residues, thus favoring the production of the types of peptides which most commonly bind to MHC class I. Thus, the proposal was formulated that the IFNγ-mediated alteration of the proteasome composition changed its substrate specificity and thereby altered the types of peptides available to the MHC class I presentation system.

Recently, another IFNγ-regulated proteasome subunit, PA28, also known as the 11S regulator of the proteasome, has been implicated in generation of antigenic peptides. PA28 consists of two subunits, α and β, both of which are IFNγ-inducible. Expression of PA28α to levels normally induced by IFNγ stimulation in cells that express a model antigen leads to enhanced target cell recognition by specific cytotoxic T lymphocytes (CTLs). Interestingly, enhanced recognition was not observed in cells which overexpressed LMP2 and/or LMP7. Moreover, PA28 enhances the ability of the proteasome to generate dual cleavages of protein substrates, thereby increasing the efficiency with which immunodominant, nine-amino-acid peptides are generated. Thus, PA28 may be the molecule that has the most significant effects on determining the repertoire of peptides generated for the MHC class I presentation system.

IFNγ also increases expression of the peptide transporters TAP1 and TAP2, which transfer peptides that have been generated in the cytoplasm by the proteasome into the endoplasmic reticulum where they can bind to nascently produced MHC class I. In addition, IFNγ increases expression of the heat-shock protein gp96 which may play a role in both transferring peptide within the cell from the TAPs to MHC class I and between cells from non-professional APCs to a subset of macrophages. These data thus suggest that IFNγ plays an important role in enhancing immunogenicity by increasing both the quantity and the repertoire of peptide displayed on MHC class I.

5.4 IFNγ and the development of T-helper cell phenotype

Human and murine CD4$^+$ T cells can be divided into two subsets defined by their pattern of cytokine secretion after stimulation. T-helper 1 cells (Th1) promote cell-mediated immunity and delayed-type hypersensitivity (DTH) responses through their production of IFNγ, lymphotoxin and IL-2. In contrast, Th2 cells predominantly produce IL-4, -5 and -10 and thereby provide help for humoral immune responses such as antibody production and isotype switching. IFNγ plays an important role in Th1 development. *In vitro*, antibody neutralization of IFNγ greatly reduces the development of Th1 cells and augments the development of Th2 cells. Importantly, administration of exogenous IFNγ either *in vitro* or *in vivo* does not drive a Th1 response. Thus, IFNγ is necessary, but is not sufficient for Th1 development.

The single most important cytokine that drives T cells to the Th1 pole is IL-12, as shown in both *in vitro* and *in vivo* studies. Bacterial products induce Th1 cell development though the induction of IL-12 secretion from APCs such as macrophages. In addition, mice deficient in the gene for either IL-12 or the IL-12 signaling protein Stat4 are unable to generate Th1 cells and display reduced DTH responses.

The role of IFNγ in this process has recently been shown to be due to its effects both at the level of the macrophage and the CD4⁺ T cell. The effects of IFNγ on macrophages were elucidated in studies that used transgenic mice lacking IFNγ sensitivity specifically in the macrophage compartment. IFNγ-insensitive macrophages were unable to support efficient Th1 development due to a severely reduced capacity to produce IL-12. More recently, IFNγ has been shown to have direct effects on the developing T-helper cells themselves. IFNγ maintains expression of the $\beta 2$ subunit of the IL-12 receptor on developing T cells, thereby preserving sensitivity of these cells to IL-12 and promoting their development into a Th1 phenotype. IFNγ also blocks development of the Th2 phenotype through two mechanisms. First, IFNγ inhibits synthesis of IL-4 from undifferentiated, antigen-stimulated T cells, thereby inhibiting production of the cytokine that is absolutely required for Th2 development. Second, IFNγ inhibits Th2 cell expansion by directly inhibiting proliferation of Th2 cells. The antiproliferative effects of IFNγ are not observed on Th1 cells due to the lack of expression of the IFNγ receptor β subunit on Th1. Thus, IFNγ inhibits the generation of Th2 cells which augment humoral immune responses while simultaneously enhancing the development of Th1 phenotype and DTH responses.

5.5 Interferon and humoral immunity

The interferons play complex and conflicting roles in regulating humoral immunity. Most studies to date have been directed at defining the influence of IFNγ in the process, although more recent observa-

tions suggest that the type I interferons may also cause many of the same biological effects. The interferons exert their effects either indirectly (as described above) by regulating the development of specific T-helper cell subsets or directly at the level of the B cell. In the latter case the interferons are predominantly responsible for regulating three specialized B cell functions: (i) B cell development and proliferation, (ii) immunoglobulin secretion and (iii) immunoglobulin heavy chain switching.

In the case of murine B-cell differentiation and/or proliferation, IFNγ has been shown to inhibit both IL-4-dependent induction of MHC class II protein expression and proliferation of B cells stimulated with anti-immunoglobulin and IL-4. In contrast, IFNγ enhances proliferative responses in human B cells activated with anti-immunoglobulin. IFNγ can also enhance or inhibit immunoglobulin secretion by either murine or human B cells. However, in this process as well, IFNγ's effects are dependent on the differentiation state of the B cell, the timing of IFNγ stimulation and the nature of the activating stimulus.

The best characterized B-cell-directed actions of the interferons are their ability to influence immunoglobulin heavy chain switching. Immunoglobulin class switching is significant because the different immunoglobulin isotypes promote distinct effector functions in the host. IgE is the only isotype which can bind to Fcε receptors on mast cells and basophils and thereby promotes immediate type hypersensitivity and allergic reactions. IgG2a fixes complement and can also, in monomeric form, bind to FcγRI on murine macrophages, a high affinity Fc receptor that is up-regulated during IFNγ-induced macrophage activation. Activated macrophages can efficiently use antibodies of the IgG2a isotype to mediate antibody-dependent cellular cytotoxicity (ADCC). In addition, IgG3 can self-aggregate (which may enhance its opsonic activity) and, along with IgG2a, can bind to the NK cell IgG receptor CD16 and effect NK-mediated ADCC. Thus, by favoring the production of IgG2a and IgG3 while inhibiting the production of IgE isotypes, the interferons can facilitate the interaction between the humoral and cellular effector limbs of the immune response and increase host defense against certain bacteria and viruses.

In vitro, IFNγ is able to direct immunoglobulin class switching from IgM to the IgG2a subtype in LPS-stimulated murine B cells and to IgG2a and IgG3 in murine B cells which have been stimulated with activated T cells. Moreover, IFNγ blocks IL-4-induced immunoglobulin class switching in murine B cells to IgG1 or IgE. The validity of these observations has been stringently tested by injecting mice with polyclonal anti-IgD serum which is a polyclonal activator of B cells. These mice produced large quantities of IgG1 and IgE. However, when IFNγ was administered to the mice before anti-IgD treatment, they produced high levels of IgG2a and decreased levels of IgG1. Thus IFNγ is clearly an important regulator of immunoglobulin class switching *in vivo*.

Recently, a role for the type I interferons in this process has also been identified. Mice deficient in the receptors for either IFNγ, IFNα/β or both type I and type II interferons were infected with live lymphocytic choriomeningitis virus (LCMV) and the profile of the LCMV-specific antibodies generated was determined. Comparable levels of LCMV-specific IgG2a antibodies were observed in the sera of normal mice and of mice which were unresponsive to *either* IFNγ or IFNα/β. In contrast, IgG2a antibodies were not produced in mice with combined unresponsiveness to *both* types of interferon. These results demonstrate that, if induced during the immune response, type I interferons can indeed function in a redundant manner to IFNγ in effecting immunoglobulin class switching.

5.6 The role of interferon family members in tumor immunity

An accumulating body of evidence suggests that interferon family members play important roles in promoting host defense to tumors. The interferons can exert profound antiproliferative effects on a variety of tumor cells. However, IFNα/β is, in general, more active in promoting this response than IFNγ. The interferons have also been proposed to inhibit tumor generation by directly increasing expression of several tumor suppressor genes such as IRF-1 and PKR. However, there are a number of anti-tumor activities that are unique to IFNγ. *In vitro*, IFNγ is able to activate macrophages to lyse certain tumor cells non-specifically. In addition, IFNγ is required for robust IL-12 secretion by macrophages. This cytokine has been shown in several studies to possess potent anti-tumor activities even in mice with pre-established tumors. Injection of recombinant IL-12 into these animals reduces the rate of metastasis, slows tumor growth, and in some cases effects complete tumor regression. IFNγ is required for IL-12-mediated anti-tumor effects. This conclusion is based on the observation that neutralizing IFNγ specific mAbs ablate the effects of IL-12 on tumors *in vivo*.

An additional body of data supports the concept that IFNγ itself is able to effect anti-tumor responses under physiological conditions. Direct injection of IFNγ *in vivo* has been shown to reduce the frequency of spontaneous melanoma and hepatic metastasis as well as tumor growth in several murine tumor models. The actions of IFNγ in these models appear to be mediated through an augmentation of tumor immunity. For example, lymphocytes harvested from draining lymph nodes of mice harboring the fibrosarcoma MCA105 were able to transfer immunity adoptively to naive mice. This effect was abrogated by *in vivo* administration of neutralizing IFNγ antibodies to recipient mice. In addition, several groups have transduced tumors with the gene for IFNγ such that the cytokine is secreted by the tumor cell itself and can achieve locally high concentrations. In general, the growth of these resulting transduced tumors is controlled though host-dependent processes involving induction of specific immunity and CTLs.

Work with UV light-induced murine sarcomas has elegantly demonstrated that CD8$^+$ CTLs can recognize tumor-specific rejection antigens and effect anti-tumor responses. Thus, for a tumor to grow progressively in an immunocompetent host it must evade lysis by CD8$^+$ CTLs. Several examples of tumors with specific deficits in many sites along the MHC class I peptide presentation pathway have been identified. Successfully established tumors have been found which have lost expression of antigenic epitopes, MHC molecules, or proteins involved in MHC class I peptide loading. Since IFNγ serves to increase the expression of many of these proteins, it is likely that a major effect of IFNγ in this process is to increase antigenicity of transformed cells, thereby enhancing tumor-specific immunity. In fact, there are examples where IFNγ treatment is able to restore immunogenicity to non-antigenic, aggressively growing tumors both *in vitro* and *in vivo*. In the B16 murine melanoma model, IFNγ treatment of the tumor cells enhanced *in vitro* lysis of the cells by CD8$^+$ CTL. IFNγ treated B16 melanoma cells when injected into syngeneic mice demonstrated a decreased rate of metastasis and a subsequent increase in host survival when compared with untreated cells.

A clear demonstration of a role for IFNγ in immune-mediated tumor rejection comes from a series of studies using the Meth A tumor cell line. Meth A is a methylcholanthrene (MCA)-induced fibrosarcoma of BALB/c mice that grows

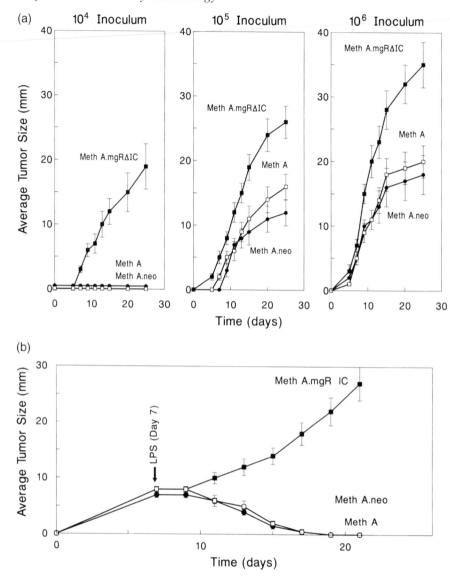

■ **Figure 3.** IFNγ-insensitive Meth A tumor cells demonstrate enhanced growth *in vivo* and cannot be rejected by administration of LPS. (a) Enhanced *in vivo* growth of IFNγ-insensitive tumor cells. Groups of five BALB/c mice were injected on day 0 with different numbers of either wild-type Meth A (open squares), mock transfected controls (Meth A.neo, closed circles), or IFNγ-insensitive Meth A (MethA.mgRΔIC, closed squares). Tumor growth was monitored daily. (b) IFNγ-insensitive tumors are not rejected in LPS-treated syngeneic mice. Groups of five BALB/c mice were injected subcutaneously on day 0 with 5×10^5 of either wild-type Meth A tumor cells (open squares), mock transfected controls (Meth A.neo, closed circles), or IFNγ-insensitive Meth A (MethA.mgRΔIC, closed squares). On day 7, LPS (3 mg/kg) was administered intraperitoneally. Tumor regression was monitored daily.

progressively when transplanted intradermally in syngeneic mice. Although the tumor is highly aggressive and eventually kills the host, a tumor-bearing host can be induced to reject the tumor by administration of LPS. At least one component is involved in this process, namely TNF. However, using neutralizing mAb specific for murine IFNγ, it was recently determined that IFNγ is obligatorily required for LPS-induced tumor rejection. In addition, Meth A tumors grew significantly more rapidly in anti-IFNγ-treated syngeneic mice.

Although these results established the importance of IFNγ in Meth A growth regulation, they did not identify the cellular target of IFNγ's actions. By introducing a mutant non-functional IFNγ receptor α-chain into Meth A tumor cells it was possible specifically to ablate tumor cell responsiveness to the cytokine and demonstrate that the tumor itself was the major target of IFNγ's actions. IFNγ-insensitive Meth A grew more rapidly in naive syngeneic mice than did control tumors and were not rejected when the tumor-bearing mice were treated with LPS (Figure 3). In addition, the IFNγ-insensitive tumors neither primed naive mice for induction of Meth A immunity nor were rejected in mice with pre-established immunity to the wild-type tumor cell. This effect was not due to the antiproliferative actions of IFNγ. Similar studies using different murine fibrosarcomas have yielded similar results. Thus, the ability of the immune system to recognize and reject certain tumors is critically dependent on the tumor's ability to respond to IFNγ.

To expand this finding, inbred IFNγR–/– mice as well as inbred 129/sv controls were treated with MCA to evaluate whether endogenously produced IFNγ is important in controlling growth of nascently forming transformed cells. At every MCA dose examined, IFNγR–/– mice developed tumors significantly more frequently than control mice. Tumor cells derived from the IFNγR–/– mice grew progressively with identical kinetics when introduced in wild-type and IFNγR–/– mice. Moreover, reconsitution of IFNγ-unresponsive tumor cells with the IFNγ receptor α-chain led to the generation of an immunogenic tumor which was quickly rejected in wild-type immunocompetent 129/Sv mice but not lymphocyte-deficient mice. Similarly, IFNγ-unresponsive mice generate tumors much more quickly than controls when bred on to a genetic background that displays a high frequency of spontaneous tumor development. These results thus demonstrate that the interplay between IFNγ and the specific immune system forms the basis of an effective tumor surveillance system in the host. In addition, these observations indicate that the major target of IFNγ's surveillance functions is the tumor cell itself.

6. Concluding remarks

In this chapter we have reviewed the basic biochemical and biological aspects of the interferon family, concentrating on the unique characteristics that distinguish IFNγ from IFNα/β. Recent advances in the field have led to a dramatic increase in our understanding of the molecular mechanisms that underlie interferon receptor signal transduction and in the physiological roles played by the interferons in promoting host defense. It is likely that future research into interferon biology will continue to reveal novel functions of these cytokines in promoting host defense and immunopathological reactions.

References

Abbas, A.K., Murphy, K.M. and Sher, A. (1997). Functional diversity of helper T lymphocytes. *Nature* **383**, 787–93.

Bach, E.A., Aguet, M. and Schreiber, R.D. (1997). The IFNγ receptor: a paradigm for cytokine receptor signaling. *Annu. Rev. Immunol.* **15**, 563–91.

Brunda, M.J. (1994). Interleukin-12. *J. Leukocyte Biol.* **55**, 280–8.

Chin, Y.E., Kitagawa, M., Su, W.S., You, Z., Iwamoto, Y. and Fu, X. (1996). Cell growth arrest and induction of cyclin-dependent kinase inhibitor p21 mediated by Stat1. *Science* **272**, 719–22.

Dalton, D.K.., Pitts-Meek, S., Keshav, S., Figari, I.S., Bradley, A. and Stewart, T.A. (1993). Multiple defects of immune function in mice with disrupted interferon-γ genes. *Science* **259**, 1739–42.

Darnell, J.E., Jr, Kerr, I.M. and Stark, G.R. (1994). Jak-STAT pathways and transcriptional activation in response to IFNs and other extracellular signaling proteins. *Science* **264**, 1415–21.

Dighe, A.S., Richards, E., Old, L.J. and Schreiber, R.D. (1994). Enhanced *in vivo* growth and resistance to rejection of tumor cells expressing dominant negative IFNγ receptors. *Immunity* **1**, 447–56.

Dighe, A.S., Campbell, D., Hsieh, C.-S., Clarke, S., Greaves, D.R., Gordon, S., Murphy, K.M. and Schreiber, R.D. (1995). Tissue specific targeting of cytokine unresponsiveness in transgenic mice. *Immunity* **3**, 657–66.

Farrar, M.A. and Schreiber, R.D. (1993). The molecular cell biology of interferon-γ and its receptor. *Annu. Rev. Immunol.* **11**, 571–611.

Gaczynska, M., Rock, K.L. and Goldberg, A.L. (1993). γ Interferon and expression of MHC genes regulate peptide hydrolysis by proteasomes. *Nature* **365**, 264–7.

Germain, R.N. (1993). Antigen processing and presentation. In Paul, W.E. (ed.) *Fundamental Immunology*, pp. 629–76. Raven Press, New York.

Groettrup, M., Soza, A., Eggers, M., Keuhn, L., Dick, T.P., Schild, H., Rammensee, H.G., Koszinowski, U.H. and Kloetzel, P.M. (1996). A role for the proteasome regulator PA28 alpha in antigen presentation. *Nature* **381**, 166–8.

Huang, S., Hendriks, W., Althage, A., Hemmi, S., Bluethmann, H., Kamijo, R., Vilcek, J., Zinkernagel, R. and Aguet, M. (1993). Immune response in mice that lack the interferon-γ receptor. *Science* **259**, 1742–5.

Meraz, M.A., White, J.M., Sheehan, K.C.F., Bach, E.A., Rodig, S.J., Dighe, A.S., Kaplan, D.H., Riley, J.K., Greenlund, A.C., Campbell, D., *et al.* (1996). Targeted disruption of the STAT1 gene in mice reveals unexpected physiologic specificty in the Jak-STAT signaling pathway. *Cell* **84**, 431–42.

Nastala, C.L., Edington, H.D., McKinney, T.G., Tahara, H., Nalesnik, M.A., Brunda, M.J., Gately, M.K., Wolf, S.F., Schreiber, R.D., Storkus, W.J. and Lotze, M.T. (1994). Recombinant IL-12 administration induces tumor regression in association with IFN-γ production. *J. Immunol.* **153**, 1697–706.

Nathan, C. (1995). Natural resistance and nitric oxide. *Cell* **82**, 873–6.

Pardoll, D.M. (1995). Paracrine cytokine adjuvants in cancer immunotherapy. *Annu. Rev. Immunol.* **13**, 399–415.

Schreiber, H., Ward, P.L., Rowley, D.A. and Strauss, H.J. (1988). Unique tumor-spcific antigens. *Annu. Rev. Immunol.* **6**, 465–83.

Snapper, C.M. (1996). Interferon-gamma. In Snapper, C.M. (ed.), *Cytokine Regulation of Humoral Immunity*, pp. 325–46. Wiley, Chichester.

Unanue, E.R. (1997). Interrelationship among macrophages, NK cells and neutrophils in early stages of *Listeria* resistance. *Curr. Opin. Immunol.* **9**, 35–43.

Uzé, G., Lutfalla, G. and Mogensen, K.E. (1995). Alpha and beta interferon and their receptor and their friends and relations. *J. Interferon Res.* **15**, 3–26.

van den Broek, M.F., Muller, U., Huang, S., Aguet, M. and Zinkernagel, R.M. (1995). Antiviral defense in mice lacking both alpha/beta and gamma interferon receptors. *J. Virol.* **69**, 4792–6.

Vilcek, J. & Sen, G.C. (1996). Interferons and other cytokines. In Fields B.N., Knipe, D.M. and Howley, P.M. (eds), *Virology*, pp. 375–400. Lippincott-Raven Publishers, Philadelphia.

Yan, H., Krishnan, K., Greenlund, A.C., Gupta, S., Lim, J.T.E., Schreiber, R.D., Schindler, C.W. and Krolewski, J.J. (1996). Phosphorylated interferon-α receptor 1 subunit (IFNaR1) acts as a docking site for the latent form of the 113 kDa STAT2 protein. *EMBO J.* **15**, 1064–74.

IX Chemokines

Anja Wuyts, Sofie Struyf, Paul Proost and Jo Van Damme*

1. Introduction

The migration of leukocytes from blood vessels to inflammatory sites is an important phenomenon in host defence. Adhesion molecules and chemotactic factors are key mediators in leukocyte recruitment. Transendothelial migration is a multistep process. A first step is the interaction of leukocytes with vascular endothelial cells mediated by selectins, which are expressed by leukocytes and endothelial cells after the release of inflammatory mediators. The selectins interact with sugar counter-ligands on the endothelial cells and leukocytes, respectively. This interaction is weak; it leads to leukocyte rolling on the endothelium, which slows down their transit through the circulation. In a second step, integrin molecules on leukocytes are activated by chemotactic factors, and interact with immunoglobulin-like cell surface adhesion molecules on endothelial cells, which results in firm leukocyte adhesion to the endothelium. In addition, chemotactic factors induce shedding of selectins on leukocytes. The third step consists in the migration of leukocytes through the vascular endothelium into the tissues. The chemotactic factors, locally produced at the site of inflammation, form a concentration gradient that induces the directional migration of leukocytes from areas of low concentration to the production site of these chemotactic factors. Chemotactic factors can also activate leukocytes to release enzymes which can facilitate migration through the extravascular tissue.

The anaphylatoxin C5a and the bacterial-derived peptide formyl-methionyl-leucyl-phenylalanine (fMLP) are chemoattractants for both neutrophils and mononuclear cells. In the last decade, a family of structurally related cytokines showing chemotactic activity for specific types of leukocytes has been identified. These chemotactic cytokines are now called chemokines. Other terms used in the past to designate these proteins are intercrines, the Scy (small cytokine) family and SIS (small inducible, secreted) cytokines. Chemokines are low molecular weight proteins with a basic nature and with affinity for heparin. They are produced by many different cell types after stimulation with appropriate inducers. It is believed that chemokines are preferentially immobilized through low affinity binding to proteoglycans on the vascular endothelium and to extracellular matrix proteins in the tissues, rather than remaining in solution. Binding to leukocytes and subsequent signaling is mediated by high affinity receptors belonging to the G-protein-coupled serpentine-like receptor family.

The hallmark of the chemokine family is the conservation of four cysteine residues which are important for the tertiary structure of the proteins. Disulfide bridges are formed between cysteines 1 and 3 and between cysteines 2 and 4 (Figure 1). Chemokines can be divided into three groups:

(1) The CXC group, in which the first two cysteines are separated by one amino acid (i.e. Cys-Xaa-Cys); CXC chemokines mainly attract and activate neutrophils.

(2) The CC group, in which the first two cysteine residues are adjacent; CC chemokines attract and activate monocytes, lymphocytes, basophils, eosinophils, natural killer (NK) cells and dendritic cells, but not neutrophils.

(3) The C chemokine group, of which lymphotactin is the only member; it lacks two of the four cysteines and shows chemotactic activity for T cells and NK cells.

Recently, a fourth chemokine type that contains the motif CX_3C (i.e. two cysteine residues with three

*Corresponding author.

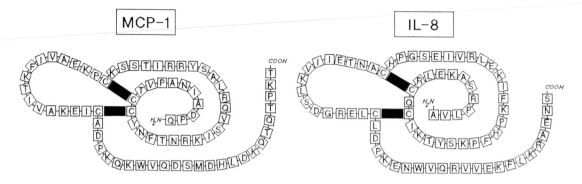

■ Figure 1. Structure of CC and CXC chemokines. The molecular hallmark of chemokines is the conservation of four cysteine residues, which form two disulfide bridges (black bars), essential for biological activity. In CC chemokines (e.g. MCP-1, left), the first two cysteines are adjacent, whereas they are separated by one amino acid in CXC chemokines (e.g. IL-8, right).

IL-8	EGAVLPRSAKELRCQCIKTYSKPFHPKFIKELRVIESGPHCANTEIIVKLSD GRELCLDPKENWVQRVVEKFLKRAENS	100%
NAP-2	AELRCMCIKTTSG IHPKNIQSLEVIGKGTHCNQVEVIATLKD GRKICLDPDAPRIKKIVQKKLAGDESAD	48%
GROα	ASVATELRCQCLQTLQG IHPKNIQSVNVKSPGPHCAQTEVIATLKN GRKACLNPASPIVKKIIEKMLNSDKSN	42%
GROβ	APLATELRCQCLQTLQG IHLKNIQSVKVKSPGPHCAQTEVIATLKN GQKACLNPASPMVKKIIEKMLKNGKSN	41%
GROγ	ASVVTELRCQCLQTLQG IHLKNIQSVNVRSPGPHCAQTEVIATLKN GKKACLNPASPMVQKIIEKILNKGSTN	40%
ENA-78	AGPAAAVLRELRCVCLQTTQG VHPKMISNLQVFAIGPQCSKVEVVASLKN GKEICLDPEAPFLKKVIQKILDGGNKEN	34%
PF-4	EAEEDGDLQCLCVKTTSQ VRPRHITSLEVIKAGPHCPTAQLIATLKN GRKICLDLQAPLYKKIIKKLLES	34%
GCP-2	GPVSAVLTELRCTCLRVTLR VNPKTIGKLQVFPAGPQCSKVEVVASLKN GKQVCLDPEAPFLKKVIQKILDSGNKKN	30%
SDF-1	DGKPVSLSYRCPCRFFESH VARANVKHLK ILNTPNCAL QIVARLKNNNRQVCIDPKLKWIQEYLEKALNKRFKM	30%
Mig	TPVVRKGRCSCISTNQGTIHLQSLKDLKQFAPSPSCEKIEIIATLKN GVQTCLNPDSADVKELIKKWEKQVSQKKK... ...QKNGKKHQKKKVLKVRKSQRSRQKKTT	28%
IP-10	VPLSRTVRCTCISISNQPVNPRSLEKLEIIPASQFCPRVEIIATMKKKGEKRCLNPESKAIKNLLKAVSKEMSKRSP	23%

■ Figure 2. Sequence alignment of human CXC chemokines and percentage homology with IL-8. The conserved cysteines and the ELR motif are underlined.

other residues between them) has been identified. This chemokine, called fractalkine or neurotactin, is active on T cells and monocytes. The polypeptide chain of the CX_3C chemokine is part of a 373 amino acid protein that carries the chemokine domain on top of an extended mucin-like stalk. The protein exists as a membrane-anchored molecule, as well as a 95 kDa shed glycoprotein.

Some chemokines have been identified by purification of their bioactivity and biochemical characterization of the protein by sequence analysis. Other chemokines have been discovered as a result of molecular cloning techniques.

2. CXC chemokines

There are 11 known members of the human CXC chemokine subfamily. Figure 2 shows the amino

acid sequence of the mature proteins and the percentage homology with the most extensively studied CXC chemokine, interleukin 8 (IL-8). The homology between the CXC chemokines ranges from 23% to 88% (Table 1). CXC chemokines can be further subdivided depending on whether they have a Glu-Leu-Arg (ELR) motif just in front of the first cysteine residue:

(1) The ELR$^+$ CXC chemokines include IL-8, GROα, GROβ, GROγ, platelet basic protein (PBP), epithelial cell-derived neutrophil attractant 78 (ENA-78) and granulocyte chemotactic protein 2 (GCP-2). These chemokines all attract neutrophils and use CXC chemokine receptor 2 (CXCR2).

(2) The ELR$^-$ CXC chemokines include platelet factor 4 (PF-4), interferon gamma (IFNγ)-inducible protein 10 (IP-10), monokine induced

∎ **Table 1.** Percentage identical amino acids between CXC chemokines

	IL-8	NAP-2	GROα	GROβ	GROγ	ENA-78	PF-4	GCP-2	SDF-1	Mig	IP-10
IL-8	100										
NAP-2	48	100									
GROα	42	59	100								
GROβ	41	55	88	100							
GROγ	40	52	85	84	100						
ENA-78	34	51	49	48	51	100					
PF-4	34	55	46	42	40	43	100				
GCP-2	30	44	44	41	44	77	41	100			
SDF-1	30	27	27	25	28	26	26	29	100		
Mig	28	33	40	44	42	36	28	31	28	100	
IP-10	23	31	25	27	23	26	32	31	25	38	100

by IFNγ (Mig) and stromal cell-derived factor 1 (SDF-1).

2.1 CXC chemokine gene and protein structure and cellular sources

CXC chemokines are produced as precursor molecules containing a signal sequence of 17–37 amino acids. After cleavage, a mature protein of 70–103 amino acids is secreted. The four cysteines are conserved in the primary structure of all CXC chemokines. The CXC chemokines contain no *N*-glycosylation sites and are probably not glycosylated.

The genes for all known CXC chemokines, except one, are localized on human chromosome 4q12–21. The genes probably arose through gene duplication and subsequent divergence. The gene for SDF-1 is located on chromosome 10. The genes for IL-8, GROα, GROβ, GROγ, ENA-78, GCP-2, IP-10 and SDF-1 consist of four exons and three introns, whereas the genes for PF-4 and PBP contain three exons and two introns. The first and second introns are conserved within this family.

2.1.1 ELR+ CXC chemokines

(a) IL-8

IL-8 was discovered by cloning studies, but it has also been identified as a neutrophil-activating peptide produced by stimulated peripheral blood monocytes (MONAP or monocyte-derived neutrophil activating peptide; MDNCF or monocyte-derived neutrophil chemotactic factor), and as a factor (GCP

or granulocyte chemotactic protein) isolated from stimulated mononuclear cells which induces an early skin reaction in rabbits. IL-8 has also been purified as a T-cell chemotactic peptide from the conditioned medium of peripheral blood mononuclear cells. In addition to mononuclear cells, IL-8 is produced by other leukocyte cell types (myeloid precursors, NK cells, neutrophils, eosinophils and mast cells), various tissue cells (e.g. fibroblasts, endothelial cells and epithelial cells) and tumour cells (Table 2). The production of IL-8 can be induced by a variety of stimuli including cytokines (e.g. IL-1 and tumour necrosis factor α (TNF-α)), bacterial products (e.g. lipopolysaccharide (LPS)), viral products (e.g. double-stranded RNA), plant products (e.g. concanavalin A) and various other inducers.

The 5′ flanking region of the human IL-8 gene contains a TATA box and potential binding sites for several nuclear factors, indicating their role as possible regulatory sequences for IL-8 expression. The stability of IL-8 mRNA may be influenced by RNA instability elements in the 3′ untranslated region. The cDNA encodes a 99 amino acid precursor protein with a signal sequence of 22 amino acids which is cleaved to yield the 77 residue mature protein. A 79 residue form of IL-8 has been detected and originates from a cleavage within the predicted signal peptide. IL-8 is processed further at the N-terminus yielding different truncation analogues (with 77, 72, 71, 70 and 69 amino acids). The occurrence of the N-terminal forms depends on the producer cells and culture conditions. The truncation is due to proteases which are released by the IL-8-secreting cells or by accessory cells. The two major forms are the 77 and the 72 amino acid pro-

∎ **Table 2.** Production of CXC chemokines by different cell types

	IL-8	NAP-2	GROα	GROβ	GROγ	PF-4	ENA-78	GCP-2	IP-10	Mig
myeloid precursors	+									
monocytes/macrophages	+	+	+	+	+		+		+	+
lymphocytes	+	+	+			+			+	
natural killer cells	+									
neutrophils	+	+	+	+	+		+		+	
eosinophils	+									
mast cells	+									
platelets		+				+	+			
endothelial cells	+		+	+	+		+		+	
mesothelial cells	+									
mesangial cells	+									
fibroblasts	+		+	+	+		+	+	+	
chondrocytes	+									
epithelial cells	+	+	+	+	+		+			
keratinocytes	+		+						+	
smooth muscle cells	+									
astrocytes	+									
hepatocytes	+									
synovial cells	+		+						+	
amnion cells	+									
endocrine cells	+									
choriodecidual cells	+									
stromal cells	+								+	
tumour cells	+		+	+	+		+	+	+	

teins. Fibroblasts and endothelial cells predominantly produce intact IL-8, whereas leukocytes mainly secrete N-terminally truncated forms. *In vitro*, the 72 amino acid form of IL-8 is a more potent chemoattractant and activator of neutrophils than the 77 residue form. *In vivo*, both IL-8 forms are equipotent, possibly due to proteolytic cleavage of 77 amino acid IL-8.

IL-8 is stable between pH 2.4 and pH 9.0 as well as under mild oxidizing and reducing conditions. Heating to 100 °C does not alter its activity. IL-8 resists exposure to detergents and organic solvents as well as to plasma peptidases. IL-8 is inactivated only slowly at 37 °C by cathepsin G, elastase and proteinase 3. These properties presumably result from the conformation of the chemokine. IL-8 can be inactivated by a protease found in serosal fluid. Reduction of disulfide bonds leads to unfolding and inactivation of IL-8, but its activity is regained after dialysis. IL-8 can be frozen, stored at 4°C in 0.1% trifluoroacetic acid, pH 2.0 or lyophilized with only little loss of activity. Storage without additional protein in phosphate-buffered saline at 4°C results in a drastic loss of activity, apparently by binding of IL-8 to plastic material.

The three-dimensional structure of IL-8 (72 residues) has been studied by nuclear magnetic resonance (NMR) spectroscopy and by X-ray crystallography. IL-8 occurs as a dimer of two identical subunits in a concentrated solution and on crystallization. The monomer contains a disordered N-terminus, followed by a loop region, three antiparallel β-strands and a prominent α-helix extending from amino acid 57 to the C-terminus. The dimer is stabilized by six hydrogen bonds between the first β-strands of the partner molecules (residues 23–29) and by other side-chain interactions. The dimer consists of two antiparallel α-helices lying on top of a six-stranded antiparallel β-sheet. However, monomeric IL-8 is probably the active form. A chemically synthesized IL-8 analogue, containing a methylated Leu25 to block dimerization, is equivalent to IL-8 for neutrophil activation and receptor binding, indicating that the monomer is a functional form. It has also been shown that the association state of IL-8 is characterized by an equilibrium between monomers and dimers. Dimers predominate at concentrations above 100 μM, but at nanomolar concentrations, which are physiologically relevant and induce maximal biological activity, the chemokine occurs almost exclusively in the monomeric form. Based upon these findings, the three-dimensional structure of monomeric IL-8$_{4-72}$, using the chemically synthesized analogue with methylated Leu25, has been determined by NMR spectroscopy. The structure is well-defined except for N-terminal residues 4–6 and C-terminal residues 67–72. The structure consists of a series of turns and loops in the N-terminal region followed by three β-strands and a C-terminal α-helix. The structure is largely similar to the native dimeric IL-8 structure, but a major difference is that the C-terminal residues 67–72 are disordered in the monomeric structure, whereas they are helical in the dimeric structure.

(b) GROα, β and γ

Melanoma growth-stimulatory activity (MGSA) or GROα has been isolated as an autocrine growth factor from a human melanoma cell line. The protein has been found to be identical to the gene product of a cDNA, called GRO, which is over-expressed in transformed fibroblasts. The cDNA encodes a precursor protein of 107 residues which yields the mature protein after cleavage of 34 amino acids. Two highly related cDNAs, GROβ and GROγ, have been cloned from a leukocyte library. The 5′ untranslated region of the GRO genes contains a TATA box and several binding motifs for nuclear factors. Messenger RNA instability elements in the 3′ untranslated region may influence RNA degradation. The three GRO proteins show 84–88% sequence homology. GROα, β and γ are produced by leukocytes, other tissue cells and tumour cells (Table 2) after stimulation with different inducers. The three-dimensional structure of GROα is similar to that of IL-8, but differences exist in the ELR region and in residues 12–23 and 31–35, regions which are considered critical for function. At physiological concentrations, GROα occurs as a monomer.

(c) PBP, β-thromboglobulin, CTAPIII and NAP-2

In addition to β-thromboglobulin, which is secreted from α-granules of blood platelets, an N-terminally truncated form with chemotactic activity for neutrophils has been isolated from thrombin-stimulated platelets. This truncated protein is more abundantly generated in the conditioned media of blood platelets cultured in the presence of monocytes and is called neutrophil activating protein 2 (NAP-2). β-Thromboglobulin itself is an N-terminally processed form of PBP. The 5′ flanking region of the PBP gene contains a TATA box and

has an alternating purine/pyrimidine tract which can form Z-DNA structures and may be involved in regulation of gene expression. The 3′ untranslated region contains a mRNA instability element. The cDNA for PBP encodes a protein of 128 amino acids corresponding to mature 94 residue PBP after cleavage of the signal peptide. Further processing of PBP at the N-terminus yields at least three distinct molecules with different biological activities: connective tissue activating protein III (CTAPIII; also called low affinity platelet factor 4 (LA-PF-4); 85 residues), β-thromboglobulin (81 residues) and NAP-2 (70 amino acids). Only NAP-2 has neutrophil-activating properties. PBP and CTAPIII are both storage proteins present in α-granules of platelets. CTAPIII is believed to be the primary species in the α-granules, but further processing after degranulation is suspected. Recently, the precursor protein of PBP has been isolated from conditioned media of neutrophils, macrophages and T lymphocytes as a mitogenic factor for fibroblasts and has been called leukocyte-derived growth factor (LDGF). PBP and its derivatives can form dimers and tetramers. Removal of the N-terminal residues from PBP attenuates dimer and tetramer formation. For all forms, the monomeric form is favoured under physiological conditions and is probably the biological active state. The tertiary structure of PBP is similar to that of IL-8. An α-helix and antiparallel β-sheet structure are conserved among PBP and its derivatives. For PBP, there is additional α-helix formation within the elongated N-terminal segment. The N-termini of PBP, CTAPIII and β-thromboglobulin may fold over the NAP-2 structure important for neutrophil activation, blocking normal activity.

(d) ENA-78

ENA-78 is a 78 residue chemokine originally isolated from the supernatant of a lung type II alveolar epithelial cell line stimulated with IL-1 or TNF-α. The cDNA encodes a protein of 114 amino acids. One research group supposes that this precursor protein contains a signal peptide of 31 amino acids, whereas others propose a signal peptide of 17 residues. Both groups suggest that ENA-78 is proteolytically processed after secretion to yield the 78 residue protein. The 5′ flanking region of the ENA-78 gene contains a TATA box and potential binding sites for several nuclear factors. The 3′ untranslated region contains mRNA instability elements. In addition to epithelial cells, monocytes, neutrophils, fibroblasts and endothelial cells, as well as platelets and tumour cells can produce ENA-78 (Table 2).

(e) GCP-2

GCP-2 is a 6 kDa protein (77 residues) isolated from the supernatant of cytokine-stimulated osteosarcoma cells. The chemokine shows high homology with ENA-78 (77%) and contains no *N*-glycosylation sites. Synthetic GCP-2 has the same apparent relative molecular mass as natural GCP-2, indicating that the chemokine is not glycosylated. Different forms of GCP-2, missing two, five or eight N-terminal amino acids of the mature protein, have been identified and can be separated from each other by reversed-phase high performance liquid chromatography. Recently, the GCP-2 cDNA and gene have been cloned. The cDNA encodes a precursor protein of 114 amino acids, yielding mature GCP-2 after cleavage of the signal peptide. The 5′ flanking region of the gene contains potential binding sites for several nuclear factors. The 3′ untranslated region contains mRNA instability elements. In addition to tumour cells, stimulated diploid fibroblasts express GCP-2 (Table 2).

2.1.2 ELR⁻ CXC chemokines

(a) PF-4

PF-4 is released from α-granules upon platelet stimulation together with platelet-derived growth factor, transforming growth factor β and β-thromboglobulin. PF-4 binds with high affinity to heparin and promotes blood coagulation. The amino acid sequence of PF-4 was determined in 1977, 10 years before the discovery of IL-8, and revealed a protein of 70 amino acids containing four cysteines. The PF-4 cDNA was isolated in 1987 from a library derived from a human erythroleukaemic (HEL) cell line. The cDNA encodes a protein of 101 residues yielding mature PF-4 after cleavage of the 31 amino acid signal sequence. The 5′ flanking region of the gene contains a TATA box and an alternating purine/pyrimidine tract which can form Z-DNA structures and may be involved in the regulation of gene expression. For PF-4, a similar three-dimensional structure has been described as for IL-8. PF-4 aggregates into dimers and tetramers but at lower concentrations the equilibrium favours the monomeric state.

(b) IP-10

IP-10 was identified as a gene induced by IFNγ in the histiocytic lymphoma cell line U937. The cDNA encodes a protein of 98 amino acids containing a signal peptide of 21 residues. A DNase I-hypersensitive site, located in a region upstream of the RNA initiation site, is induced by IFNγ and pro-

vides the structural basis for the transcriptional activation of this gene by IFNγ. The 3′ untranslated region of the IP-10 gene contains a mRNA instability element. IP-10 is produced by macrophages, T cells, thymocytes, fibroblasts, endothelial cells, keratinocytes, synovial cells and tumour cells and can be induced by stimuli other than IFNγ. Thymic and splenic stromal cells constitutively express high levels of IP-10, which explains the abundant presence of the protein in lymphoid tissue. High levels of IP-10 have also been observed in the liver, possibly due to the presence of activated macrophages.

(c) Mig

Mig was cloned from cultures of the THP-1 monocytic cell line treated with IFNγ. It can also be induced in peripheral blood mononuclear cells by IFNγ, but not by IFNα or LPS. The gene encodes a 125 residue protein containing a signal peptide of 22 amino acids. The 3′ untranslated region contains mRNA instability elements. Monocytes and THP-1 cells stimulated with IFNγ secrete various C-terminally truncated forms of Mig, which is the result of proteolytic processing.

(d) SDF-1α and -1β

Human SDF-1α and SDF-1β were cloned from a cDNA library of a pro-B cell line. SDF-1β differs from SDF-1α by only four additional amino acids at the C-terminus. The two proteins are encoded by a single gene and arise by alternative splicing. The SDF-1 gene consists of four exons and three introns. SDF-1α and SDF-1β are encoded by three and four exons, respectively. The cDNAs encode proteins of 89 and 93 amino acids, respectively, with a signal peptide of 19 amino acids. The SDF-1 gene is expressed in almost all organs, but not in blood cells. The gene contains no TATA box in the 5′ flanking region but does contain a GC-rich sequence which is associated with ubiquitous expression. In contrast to the other CXC chemokines, the gene for SDF-1 is mapped to chromosome 10q11.1. Human SDF-1α differs in only one amino acid from mouse SDF-1α.

2.2 Biological activities of CXC chemokines

2.2.1 ELR⁺ CXC chemokines

(a) IL-8

IL-8 stimulates neutrophil chemotaxis and haptotactic migration (i.e. migration towards a concentration gradient of immobilized chemokines) and induces shape change, actin polymerization and degranulation (release of β-glucuronidase, elastase, myeloperoxidase, gelatinase B, vitamin B_{12}-binding protein and lactoferrin) of neutrophils. IL-8 increases the intracellular calcium concentration ($[Ca^{2+}]_i$) in neutrophils and induces a weak (in comparison with fMLP) respiratory burst. The chemokine primes neutrophils for an enhanced superoxide anion production in response to fMLP, phorbol 12-myristate 13-acetate (PMA) and platelet-activating factor (PAF) and up-regulates formylpeptide receptor expression. IL-8 enhances intracellular killing of *Mycobacterium fortuitum* by human neutrophils by priming the cells to enhance H_2O_2 production upon stimulation with preopsonized bacteria. Thus, IL-8 facilitates the elimination of microorganisms by increasing the efficiency of bactericidal activity of neutrophils. It stimulates phagocytosis of opsonized particles and enhances the growth-inhibitory activity of neutrophils to *Candida albicans*. IL-8 activates arachidonate-5-lipoxygenase with release of leukotriene B_4 (LTB_4) and 5-hydroxy-eicosatetraenoic acid (5-HETE) in the presence of exogenous arachidonic acid and induces the synthesis of PAF in neutrophils. IL-8 inhibits nitric oxide induction in peritoneal neutrophils, whereas it stimulates nitric oxide production in osteoclast-like cells. IL-8 induces the release of the IL-1 type II decoy receptor from neutrophils. The chemokine stimulates transendothelial migration of neutrophils. The protein induces shedding of the adhesion molecule L-selectin and up-regulation of $\beta2$ integrins (CD11b/CD18 and CD11c/CD18) and complement receptor type 1 (CR1 or CD35) on neutrophils, and alters the avidity of the constitutively expressed integrin molecules. IL-8 promotes adhesion of neutrophils to plastic, extracellular matrix proteins and unstimulated as well as cytokine-stimulated endothelial monolayers. The adhesion is inhibited by antibodies to the $\beta2$ integrin CD11b/CD18. Binding of IL-8 to heparan sulfate enhances its neutrophil chemotactic activity.

IL-8 has also effects on other leukocyte cell types including lymphocytes, basophils and eosinophils (Table 3). However, it shows no chemotactic activity for monocytes but induces an increase in $[Ca^{2+}]_i$ and a weak respiratory burst in monocytes. These effects are much less pronounced than in neutrophils. The chemokine enhances the expression of integrins on monocytes as well as their adherence to endothelial cells. Some controversy exists about the chemotactic activity of IL-8 on T cells. IL-8 has been isolated as a T-cell chemoattractant but this activity could not be confirmed by others. It has been shown that the cell purification procedure can influence the response of T cells to IL-8. Freshly isolated T cells, both CD4+ and CD8+ cells, migrate in response to IL-8, but incubation of the cells at 37 °C reduces the migration response. The chemokine enhances the $[Ca^{2+}]_i$ in a subset of T cells. IL-8 suppresses the spontaneous production of IL-4 and stimulates its own production by CD4+ T cells. The protein causes chemokinesis of IL-2-activated NK cells. IL-8 is a chemoattractant for B cells but selectively inhibits IL-4-induced IgE production and growth of B cells. IL-8 is chemotactic for basophils and stimulates the adhesion of these cells to endothelial cells, mediated by $\beta2$ integrins. IL-8 also augments the $[Ca^{2+}]_i$ in basophils and induces a release of histamine and leukotrienes from IL-3-pretreated cells. However, lower concentrations of IL-8 than those required to release histamine have been shown to inhibit such release induced by IL-8, monocyte chemotactic protein 1 (MCP-1) and IL-3. NAP-2 also inhibits, although weakly, the IL-8-induced histamine release, whereas CTAPIII and PF-4 have no effect. IL-8 shows no chemotactic activity for normal eosinophils. However, the protein induces an increase in $[Ca^{2+}]_i$, shape change and release of eosinophil peroxidase in eosinophils, isolated from patients with hyper-eosinophilic syndrome. Circulating eosinophils from patients with atopic dermatitis and allergic asthma show an increased migratory response towards IL-8. IL-8 also chemoattracts eosinophils pretreated with IL-3, IL-5 or granulocyte–macrophage colony-stimulating factor (GM-CSF). IL-8 suppresses colony formation of immature subsets of myeloid progenitor cells stimulated with GM-CSF plus steel factor and also inhibits megakaryocyte colony formation.

In addition to the effects on leukocytes, IL-8 has activities on several other cell types (Table 3). It:

- inhibits collagen expression in rheumatoid synovial fibroblasts;

- enhances the replication of cytomegalovirus in fibroblasts, possibly through interaction with the IL-8 receptor CXCR1, which is up-regulated on these cells after infection with the virus;

- induces a loss of focal adhesions (structures linking the cytoskeleton to the underlying extracellular matrix) in fibroblasts and promotes chemotaxis and chemokinesis of these cells;

- induces proliferation and haptotactic migration of melanoma cells, as well as chemotaxis and proliferation of endothelial cells;

■ **Table 3.** Target cells for CXC chemokines

chemokine	target cells	
	leukocytes	others
IL-8	neutrophils	fibroblasts
	monocytes	endothelial cells
	T lymphocytes	keratinocytes
	NK cells	smooth muscle cells
	B lymphocytes	melanoma cells
	basophils	
	eosinophils	
	megakaryocytes	
	myeloid progenitor cells	
GROα	neutrophils	fibroblasts
	monocytes	endothelial cells
	T lymphocytes	melanocytes
	B lymphocytes	melanoma cells
	basophils	
NAP-2	neutrophils	
	basophils	
	megakaryocytes	
ENA-78	neutrophils	
GCP-2	neutrophils	
PF-4	neutrophils	fibroblasts
	monocytes	endothelial cells
	B lymphocytes	chondrocytes
	mast cells	tumour cells
	eosinophils	
	myeloid progenitor cells	
IP-10	monocytes	endothelial cells
	T lymphocytes	
	NK cells	
	B lymphocytes	
	bone marrow progenitor cells	
Mig	T lymphocytes	
SDF-1	monocytes	
	T lymphocytes	
	neutrophils	
	haematopoietic progenitor cells	

- is chemotactic for keratinocytes, increases $[Ca^{2+}]_i$ and promotes their proliferation;
- is a mitogen and chemoattractant for smooth muscle cells and also stimulates these cells to produce prostaglandin E_2, which has a negative feedback on the proliferation and migration.

At high doses ($3\,\mu g$), IL-8 provokes an early (3h) skin reaction in rabbits characterized by swelling, redness and neutrophil infiltration. Upon co-injection with a vasodilator, e.g. prostaglandin E_2, IL-8 (at doses of ≥ 0.2 ng) induces local oedema and neutrophil accumulation in rabbit skin, with a maximum effect after 30 min. The plasma leakage is neutrophil-dependent and the effects can be inhibited by antibodies against leukocyte integrins. Histology of IL-8-induced lesions has also revealed intravascular neutrophil accumulation, cell aggregate formation and venular wall damage. IL-8, injected subcutaneously into the lymphatic drainage areas of rat lymph nodes, causes accelerated emigration of only lymphocytes from high endothelial venules. However, injection of IL-8 in rat ear skin results within 3 h in accumulation of both lymphocytes and neutrophils in the connective tissue. Lower doses of IL-8 (20 pg) selectively attract lymphocytes, whereas higher doses (2 ng) predominantly recruit neutrophils. Subcutaneous injection of IL-8 (100 ng) in mice with severe combined immunodeficiency engrafted with human peripheral blood lymphocytes (hPBL/SCID), induces neutrophil accumulation 4 h after administration, whereas a T-cell infiltrate is only detected after 72 h. The T-cell accumulation may be mediated by chemoattractants, released by neutrophils after stimulation with IL-8. Four hours after a dose of $10\,\mu g$, the chemokine induces an influx of neutrophils and lymphocytes into the skin of sheep with a neutrophil:lymphocyte ratio of 45:1. Intradermal injection of IL-8 into human subjects causes perivascular neutrophil infiltration, detectable after 30 min and increasing between 30 min and 3 hours; no basophils or eosinophils are found and the number of lymphocytes is not increased. At moderate doses (400 ng), no weals, flare, itching or pain is detected, suggesting that IL-8 does not elicit histamine release from local mast cells. The observed differences in lymphocyte accumulation after intradermal injection may be due to differences in animal species and chemokine dose.

Intravenous injection of IL-8 into animals results in immediate leukopenia followed by profound neutrophilia. The neutrophilia is accompanied by the release of non-segmented neutrophils from the bone marrow reservoir. Intravenous injection of IL-8 can inhibit emigration of neutrophils to inflammatory sites. After intraperitoneal injection of IL-8 in mice, rapid mobilization of progenitor cells and pluripotent stem cells occurs. These mobilized cells are able to rescue lethally irradiated mice and to repopulate host haematopoietic tissues completely and permanently. Intracranial injection of IL-8 in mice provokes neutrophil recruitment in the central nervous system, which is associated with breaching of the blood–brain barrier. Intracerebroventricular administration of IL-8 in rats induces fever by a prostaglandin-independent mechanism and suppresses food-intake.

Knockout mice that lack the murine IL-8 receptor homologue look outwardly healthy but show lymphadenopathy (due to an increase in B cells) and splenomegaly (due to an increase in metamyelocytes and mature neutrophils). This indicates that the receptor may participate in the expansion and development of neutrophils and B cells. As expected, these mice also show less neutrophil migration to sites of inflammation.

Finally, IL-8 is angiogenic when implanted in the rat cornea and induces neovascularization in a rabbit corneal pocket model. IL-8 is also a primary mediator of angiogenesis in human bronchogenic carcinoma. In SCID mice, IL-8 promotes the tumour growth of human non-small cell lung cancer through its angiogenic properties.

(b) GROα, β and γ

GROα, β and γ induce chemotaxis, shape change, degranulation, rise in $[Ca^{2+}]_i$ and a weak respiratory burst in neutrophils. GROα up-regulates the expression of the $\beta 2$ integrin CD11b/CD18 and of CR1 and decreases the expression of L-selectin on neutrophils. The protein also enhances the avidity of CD11b/CD18 as well as actin polymerization and phagocytosis of opsonized particles. GROα primes the superoxide anion production by fMLP-stimulated neutrophils and up-regulates the formylpeptide receptor. Pretreatment of neutrophils with a non-stimulatory dose of GROα reduces the degranulation response to higher doses of GROα, NAP-2 or IL-8. GROα is not chemotactic for monocytes, but elicits a weak increase in $[Ca^{2+}]_i$ and a weak respiratory burst in these cells. The GRO family plays a role in monocyte adhesion to endothelium, stimulated with minimally modified low-density lipoprotein. GROα is also chemotactic for T and B cells. GROα, β and γ are chemotactic for basophils and augment the $[Ca^{2+}]_i$, whereas no chemotactic activity for eosinophils has been detected. GROβ suppresses colony formation of immature subsets of myeloid progenitor cells stimulated with GM-CSF plus steel factor.

GROα, β and γ are also active on cells other than leukocytes. GROα has melanoma growth-stimulatory activity (MGSA) and is mitogenic for melanocytes. The protein stimulates fibroblast growth and decreases collagen expression in rheumatoid synovial fibroblasts. GROα and GROβ inhibit basic fibroblast growth factor (bFGF)-stimulated proliferation of capillary endothelial cells, whereas GROγ does not.

Intradermal injection of GROα induces neutrophil accumulation in rabbit skin. GROβ has angiostatic properties since it inhibits blood vessel formation in the chicken chorioallantoic membrane assay and can suppress bFGF-induced corneal neovascularization after systemic administration in mice. GROβ inhibits the growth of murine Lewis lung carcinoma in mice by suppression of neovascularization. These findings are in contrast with another study, which shows that all ELR$^+$ CXC chemokines are angiogenic, whereas ELR$^-$ CXC chemokines are angiostatic.

Transgenic mice expressing KC, the murine homologue of GROα, in the thymus or epidermis show a marked infiltration of neutrophils to the sites of transgene expression without morphological evidence of injury. Thus, KC expression results in neutrophil recruitment without inflammatory reaction. Central nervous system-specific KC expression induces remarkable neutrophil infiltration into perivascular, meningeal and parenchymal sites. Unexpectedly, these mice develop a neurological syndrome of pronounced postural instability and rigidity. The major neuropathological alterations are florid microglial activation and blood–brain barrier disruption. Intracranial injection of macrophage inflammatory protein 2 (MIP-2), the murine counterpart of GROβ/γ, induces neutrophil recruitment in the central nervous system which is associated with a breaching of the blood–brain barrier.

(c) PBP, β-thromboglobulin, CTAPIII and NAP-2

LDGF, the precursor of PBP, is a potent mitogenic factor for connective tissue cells. PBP is biologically inactive and yields CTAPIII after removal of nine amino acids. CTAPIII stimulates the proliferation of connective tissue cells. Additional removal of four amino acids from CTAPIII yields β-thromboglobulin, which is chemotactic for fibroblasts, but not for neutrophils or monocytes. Further removal of 11 amino acids yields NAP-2, which induces chemotaxis, a rise in $[Ca^{2+}]_i$, degranulation and a weak respiratory burst in neutrophils. NAP-

2 enhances the expression of the adhesion molecule CD11b/CD18 and of CR1, as well as the binding activity of CD11b/CD18. The protein induces actin polymerization and phagocytosis of opsonized particles by neutrophils. Priming of neutrophils with a non-stimulatory dose of NAP-2 leads to a reduced degranulation response to higher doses of NAP-2, GROα or IL-8. This reduced response is correlated with a down-regulation and internalization of high affinity NAP-2- and IL-8-binding sites. NAP-2 also augments the $[Ca^{2+}]_i$ in basophils and releases histamine from IL-3-pretreated basophils. No effects on monocytes have been observed. NAP-2 inhibits megakaryocytopoiesis.

NAP-2 variants containing three to five additional amino acids at the N-terminus are less potent neutrophil activators than 70 residue NAP-2. A C-terminally truncated variant of NAP-2, lacking the four C-terminal residues, has also been isolated. This truncated protein shows a three- to four-fold enhanced potency for neutrophil degranulation as well as three-fold enhanced receptor binding affinity.

Intradermal injection of NAP-2 together with a vasodilator results in neutrophil accumulation and plasma extravasation in rabbit skin. Intravenous injection provokes rapid granulocytosis in rabbits. Although the kinetics of neutrophilia are similar for NAP-2 and IL-8, the effect with IL-8 is more pronounced.

(d) ENA-78

ENA-78 stimulates neutrophil chemotaxis, a rise in $[Ca^{2+}]_i$, exocytosis and up-regulates CD11b/CD18. Pre-incubation of neutrophils with ENA-78 enhances their ability to generate superoxide anion in response to fMLP. ENA-78 is capable to desensitize the neutrophil response to human GCP-2, indicating a common receptor usage.

(e) GCP-2

GCP-2 occurs in several N-terminally truncated forms, which are equally potent at chemoattracting neutrophils. GCP-2 also induces degranulation of neutrophils and augments $[Ca^{2+}]_i$. Intradermal injection in rabbit skin results in neutrophil accumulation and plasma extravasation. The murine counterpart of GCP-2 has been isolated from stimulated fibroblasts and epithelial cells. Eleven different N-terminally truncated forms have been identified. In contrast to human GCP-2, the more truncated forms of mouse GCP-2 have a higher specific activity in neutrophil chemotaxis and activation assays than the longer forms. Mouse GCP-2 is

a better chemoattractant than human GCP-2 on human neutrophils and more active than murine KC (GROα) and MIP-2 (GROβ/γ) on mouse neutrophils. Human GCP-2 is not chemotactic for monocytes, lymphocytes or eosinophils.

2.2.2 ELR⁻ CXC chemokines

(a) PF-4

Reports of the neutrophil-activating properties of PF-4 are rather controversial. Earlier studies report chemotactic migration, degranulation and enhanced adhesion of neutrophils to plastic surfaces and endothelial cells, whereas others failed to demonstrate any neutrophil-activating capacity. Studies with highly purified PF-4 revealed that only weak degranulation of specific secondary neutrophil granules and no exocytosis of azurophilic granules occurs in response to this chemokine. The degranulation of secondary granules can be significantly enhanced by pre- or co-incubation of the cells with TNF-α. Thus, PF-4 has no neutrophil chemotactic activity and has no effect on the $[Ca^{2+}]_i$, nor does it compete for IL-8 binding to neutrophils in the absence or presence of TNF-α. However, PF-4 may be weakly chemotactic for monocytes, IgE-stimulated mast cells, B lymphocytes and normal eosinophils, but the migration response of the latter is markedly potentiated by IL-5. Eosinophils from patients with atopic dermatitis have an increased migratory response to PF-4. PF-4 suppresses colony formation of immature subsets of myeloid progenitor cells stimulated with GM-CSF plus steel factor and inhibits megakaryocytopoiesis.

PF-4 is chemotactic for fibroblasts and promotes their proliferation. It stimulates glycosaminoglycan synthesis by chondrocytes, is a growth inhibitor of various tumour cell lines (including human erythroid leukaemia and osteosarcoma), and inhibits endothelial cell proliferation and migration, as well as angiogenesis in the chicken chorioallantoic membrane system. These effects may result from the ability of PF-4 to block binding of bFGF to its receptor on endothelial cells. A PF-4 derivative originating from cleavage of the peptide bond between Thr16 and Ser17 has been isolated from activated human leukocytes. This cleavage leads to structural changes and a 30- to 50-fold greater growth inhibitory activity on endothelial cells than for intact PF-4, suggesting that the CXC motif and the intramolecular disulfide bridges are not necessary for the growth-inhibitory effect of PF-4. The growth of melanoma and human colon carcinoma in the mouse is inhibited by administration of

PF-4. The *in vivo* anti-tumour effects of PF-4 appear to coincide largely with the angiostatic activity. Finally, PF-4 alleviates concanavalin A-induced immunosuppression in mice.

(b) IP-10

IP-10 is not chemotactic for neutrophils, but attracts monocytes, B cells and T cells. It promotes T-cell adhesion to endothelial cells, to the adhesion molecules ICAM-1 and VCAM-1, and to extracellular matrix proteins via $\beta1$ and $\beta2$ integrins. It also induces NK cell migration and degranulation and enhances NK cell-mediated killing of tumour cells. In the presence of growth factors, IP-10 suppresses *in vitro* colony formation by early human bone marrow progenitor cells.

After subcutaneous injection into mice, mainly monocytes, but little or no lymphocytes or neutrophils are recruited within 4 h. Subcutaneous injection of IP-10 into hPBL/SCID mice induces infiltration of murine monocytes and human CD3$^+$ T cells within 24–48 h. IP-10 inhibits endothelial cell proliferation *in vitro* and is a potent inhibitor of angiogenesis *in vivo*. *In vitro*, the protein has no effect on the growth of plasmacytoma cells or mammary adenocarcinoma cells, genetically engineered to secrete high levels of murine IP-10. However, *in vivo*, these engineered tumour cells fail to grow. This anti-tumour response is T-cell dependent, is also observed after injection of a 1:1 mixture of non-transfected and transfected tumour cells and appears to be mediated by the recruitment of an inflammatory infiltrate composed of lymphocytes, monocytes and neutrophils. However, IP-10 has also been shown to be an anti-tumour agent in athymic mice. It causes damage in established tumour vasculature and tissue necrosis in nude mice but no complete regression of human Burkitt's lymphomas is observed. Constitutive expression of IP-10 in Burkitt's lymphoma cells reduces their ability to grow and induces visible tumour necrosis.

(c) Mig

Mig augments [Ca^{2+}]$_i$ and is chemotactic for tumour-infiltrating lymphocytes and activated T cells. Its biological activity is diminished by C-terminal truncation. In hPBL/SCID mice, Mig induces infiltration of monocytes within 4 h after injection, whereas CD3$^+$ T-cell recruitment is observed between 4 and 48 h.

(d) SDF-1α and -1β

Mouse SDF-1α has been cloned from a bone marrow stromal cell line as a factor that supports proliferation of a pre-B cell clone and that augments the growth of bone marrow B cell progenitor cells in the presence of IL-7. A highly efficient lymphocyte chemoattractant, purified from the supernatant of a murine bone marrow stromal cell line, has been identified as SDF-1α lacking two and seven N-terminal residues and with the C-terminal lysine removed by proteolytic processing (giving 67 and 62 residue forms, respectively). *In vitro*, this SDF-1α mixture also induces monocyte chemotaxis. Subcutaneous injection of SDF-1α in mice results in a mononuclear cell infiltrate. SDF-1 augments [Ca^{2+}]$_i$ in CD34$^+$ cells and chemoattracts haematopoietic progenitor cells *in vitro* and *in vivo*. This SDF-1-induced chemotaxis is increased by IL-3. CD34$^+$ progenitors from peripheral blood are less responsive than CD34$^+$ progenitors from the bone marrow. SDF-1 is responsible for B-cell lymphopoiesis and bone marrow myelopoiesis. Mutant mice with a targeted disruption of the gene encoding SDF-1 die perinatally. The mutants have a cardiac ventricular septal defect.

The human homologues of the two isolated forms of murine SDF-1α have been synthesized and tested. The 67 residue form of SDF-1α is chemotactic for and increases [Ca^{2+}]$_i$ in monocytes, phytohaemagglutinin (PHA)-activated peripheral blood lymphocytes and freshly purified peripheral blood T cells but also in neutrophils. The 62 residue form of human SDF-1α is inactive. SDF-1 inhibits infection of CD4, CXCR4 transfectants and peripheral blood mononuclear cells by T-cell line tropic (T-tropic) HIV-1 strains, but does not affect infection by macrophage-tropic (M-tropic) or dual-tropic primary HIV-1 strains.

2.3 CXC chemokine receptors

2.3.1 Identification of CXC chemokine receptors

Using binding studies, high affinity receptors (20–90 $\times 10^3$ per cell) for IL-8 have been identified on the surface of neutrophils, with a K_d ranging from 0.2 to 4 nM. The IL-8 receptors do not bind unrelated chemoattractants such as LTB$_4$, fMLP, C5a or PAF. After binding, IL-8 is rapidly internalized and degraded. More than 90% of the ligand-bound receptors are endocytosed within 10 min. The receptors rapidly reappear on the cell surface, probably through recycling. The addition of lysosomotropic agents does not inhibit ligand binding or internalization, but partially inhibits reappearance of the receptors and also neutrophil chemotaxis. This indicates that recycling of the receptors may

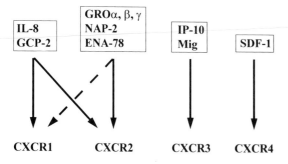

■ **Figure 3.** Receptor usage by human CXC chemokines.

be essential for the chemotactic response of neutrophils.

The existence of two classes of IL-8 receptors has been indicated by binding competition assays and cross-desensitization experiments: one receptor with high affinity for IL-8, GROα, GROβ, GROγ, NAP-2 and ENA-78 and a second receptor with high affinity for IL-8 and low affinity for the other tested chemokines. By cross-linking experiments with iodinated IL-8, two membrane proteins (44–59 kDa and 67–70 kDa) can be demonstrated. The 44–59 kDa receptor (CXCR1) corresponds to the receptor with high affinity for IL-8 and low affinity for the other ELR$^+$ CXC chemokines (Figure 3). On basophils, a single class of high-affinity receptors (K_d = 0.15 nM, 9600 per cell) for IL-8 is present, which also weakly bind NAP-2. A receptor for GROα that does not bind IL-8 has been reported on a human melanoma cell line, as well as on rheumatoid synovial fibroblasts.

2.3.2 Cloning and expression of CXC chemokine receptors

Four different receptors for human CXC chemokines have been cloned. These chemokine receptors are members of the rhodopsin or serpentine receptor superfamily. They contain seven hydrophobic transmembrane-spanning regions and couple to guanine nucleotide binding proteins (G-proteins). The N-terminus and C-terminus are located extracellularly and intracellularly, respectively (Figure 4). Chemokine receptors are small, relative to other rhodopsin-like receptors, due to a short third intracellular loop and short N- and C-termini. The receptors contain *N*-glycosylation sites in the N-terminus and in extracellular regions. The intracellular C-terminus contains several serine and threonine residues which may function as

phosphorylation sites, important for desensitization of the receptors.

In addition to the receptors which specifically bind CXC chemokines, a promiscuous receptor, with seven transmembrane spanning domains, which shows high affinity for both CXC and CC chemokines has been identified. This receptor is expressed on erythrocytes and on endothelial cells of postcapillary venules and is identical to the Duffy antigen, the receptor for the malarial parasite *Plasmodium vivax*. This receptor is now called the Duffy antigen receptor for chemokines (DARC).

(a) IL-8 receptors

For IL-8, two receptors have been cloned. The cDNA for IL-8 receptor A (IL-8RA) or CXCR1 has been identified from human neutrophils and encodes a 350 amino acid protein. The cDNA for IL-8RB or CXCR2 has been cloned from HL-60 cells differentiated into neutrophils and encodes a 355 or 360 residue protein. CXCR1 and CXCR2 show 77% homology in their amino acid sequences. The calculated molecular mass is approximately 40 kDa for both receptors. Glycosylation may explain the difference between the theoretical molecular mass of the cloned receptors and the molecular mass of the native receptors on neutrophils. Both receptors have an extracellular N-terminus which is rich in acidic residues. The genes for the two IL-8 receptors as well as an IL-8RB pseudogene have been assigned to human chromosome 2q34–35. The open reading frame of both CXCR1 and CXCR2 corresponds to a single exon. The gene for CXCR1 consists of two exons, whereas for CXCR2 seven distinct neutrophil mRNAs are formed by alternative splicing of 11 exons.

Expression of CXCR1 or CXCR2 in mammalian cells shows that IL-8 binds to both receptors with high affinity and induces chemotaxis and calcium in both transfected cell types. GROα, NAP-2 and IL-8 are equally potent as attractants for CXCR2-transfected cells, but IL-8 is 300–1000 times more potent than GROα or NAP-2 for chemotaxis of CXCR1-transfectants. The ELR$^+$ CXC chemokines GROα, GROβ, GROγ, NAP-2 and ENA-78 are agonists that potently increase the $[Ca^{2+}]_i$ in CXCR2-transfected cells, but not in CXCR1-transfectants. These chemokines have high affinity for CXCR2 whereas they compete weakly for the high-affinity IL-8 binding site on CXCR1 (Figure 3). Similar to IL-8 but in contrast to the other ELR$^+$ CXC chemokines, GCP-2 is equipotent at augmenting the $[Ca^{2+}]_i$ in both CXCR1- and CXCR2-

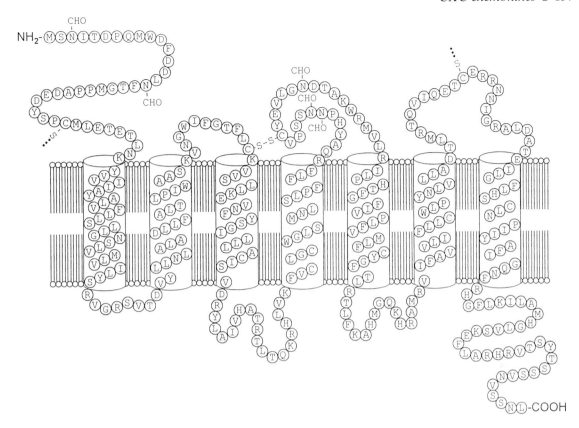

∎ **Figure 4.** Proposed membrane topography of the CXC chemokine receptor CXCR1. Potential *N*-glycosylation sites are indicated by 'CHO' and disulfide bonds by 'S–S'.

transfected cells. The calcium rise in neutrophils after stimulation with GCP-2 can be inhibited by prestimulation of the cells with IL-8. Similarly, the IL-8 response can be abolished by pretreatment with GCP-2. These data indicate that both chemokines use the same receptors. The ELR⁻ CXC chemokines Mig or IP-10 do not induce responses in CXCR1- or CXCR2-transfected cells.

Messenger RNA for CXCR1 has been detected in neutrophils, PHA-activated T-cell blasts, CD4⁺ T cells, monocytes, platelets, megakaryocytes and leukocytic cell lines, as well as in melanocytes, fibroblasts, keratinocytes, endothelial cells and melanoma cells. The gene for CXCR2 is expressed in neutrophils, monocytes, platelets, megakaryocytes, HL-60 cells, AML 193 cells, melanocytes, keratinocytes and melanoma cells.

With specific antibodies against CXCR1 or CXCR2, IL-8 receptors can be detected on neutrophils, monocytes, a subset of CD8⁺ T cells and of

CD56⁺ NK cells. Neutrophils express the highest numbers of both CXCR1 and CXCR2 at an approximately equal ratio, whereas monocytes and IL-8R⁺ lymphocytes express higher levels of surface protein for CXCR2 than of CXCR1. Recently, it has been shown that about 15% of freshly isolated B cells functionally express IL-8 receptors. Also astrocytes, microglia and the human astrocyte cell line HSC2 contain specific mRNA and express surface protein for CXCR2.

The expression of CXCR1 and CXCR2 can be regulated by cytokines. Granulocyte colony-stimulating factor (G-CSF) increases the levels of specific mRNA for both IL-8 receptors in neutrophils, and this correlates with increased binding of IL-8 and subsequent neutrophil chemotaxis. In contrast, a dramatic reduction in IL-8 receptor mRNA is observed after treatment of neutrophils with LPS, which is correlated with decreased IL-8 binding and a reduced chemotactic response. A less marked IL-8 receptor down-regulation occurs after treatment with TNF-α. The G-CSF effect is dependent upon enhanced transcriptional activity of the IL-8 receptor genes. LPS inhibits IL-8 receptor expression on neutrophils by a combination of transcriptional inhibition and a decrease in mRNA stability. In contrast to these findings, it has also been reported that LPS and especially serum-activated LPS, and fMLP augment IL-8 receptor expression on neutrophils. IFNγ up-regulates both CXCR1 and CXCR2 on freshly isolated CD4$^+$ and CD8$^+$ T lymphocytes. TNF-α and IL-2 increase both IL-8 receptors on CD4$^+$ T lymphocytes. On B cells, the IL-8 receptors are down-regulated by IFNγ, IL-2 and TNF-α (only CXCR1) and up-regulated by IL-4 and IL-13.

The two IL-8 receptors are functionally different. GROα stimulates neutrophil chemotaxis exclusively through CXCR2, whereas chemotaxis to IL-8 is mediated predominantly by CXCR1. In contrast, other researchers postulate that CXCR2 may play an active role in the initiation of neutrophil migration distant from the site of inflammation, where the IL-8 concentration is at the picomolar level, whereas CXCR1 may be more important for mediating the IL-8 signal at the site of inflammation where the concentration of IL-8 is high. Elastase release by neutrophils in response to IL-8 is decreased by anti-CXCR1 and even more substantially by anti-CXCR1 plus anti-CXCR2, whereas the elastase release induced by GROα is only inhibited by anti-CXCR2. The superoxide anion production by IL-8 is affected only by anti-CXCR1 and not by anti-CXCR2. The priming effect of IL-8 on fMLP-induced superoxide anion production is predominantly mediated by CXCR1, whereas the same effect of GROα and ENA-78 is mediated by CXCR2. T- and B-cell chemotaxis in response to IL-8 and GROα is probably mediated by CXCR2. Binding of GROα to melanoma cells is partially blocked by anti-CXCR2. The interaction of GROα with CXCR2 is required for melanoma cell growth.

(b) IP-10/Mig receptor

A seven-transmembrane-domain receptor selective for IP-10 and Mig has been cloned from a CD4$^+$ T-cell library. The receptor cDNA encodes a protein of 368 amino acids with a molecular mass of 40 kDa. The gene has been localized on chromosome 8p11.2–12. The receptor shows about 40% homology with the two IL-8 receptors. The IP-10/Mig receptor or CXCR3 is highly expressed in IL-2-activated T cells, but is not detectable in resting T cells, B cells, monocytes or granulocytes.

The receptor mediates Ca^{2+} mobilization and chemotaxis in response to IP-10 and Mig, but does not recognize the other CXC or CC chemokines, nor the C chemokine lymphotactin.

(c) SDF-1 receptor

A cDNA clone, called LESTR, encoding a protein of 352 amino acids and corresponding to a seven-transmembrane-domain, G-protein-coupled receptor has been isolated from a human blood monocyte cDNA library and shows about 34% homology to CXCR1 and CXCR2. There is no binding of the CXC chemokines IL-8, NAP-2, GROα, PF-4 and IP-10 or of the CC chemokines MCP-1, MCP-2, MCP-3, MIP-1α, I-309 and RANTES ('regulated on activation, normally T cell expressed and secreted') nor of C3a or LTB$_4$ to this receptor. A high level of LESTR mRNA is found in neutrophils, PBL, PHA-activated T-cell blasts and HL-60 cells induced to differentiate by $1\alpha,25(OH)_2$-vitamin D$_3$; moderate levels are found in monocytes and undifferentiated HL-60 cells but only low levels are detectable in U937 cells, Jurkat cells and DMSO-differentiated HL-60 cells. This receptor has also been cloned from fetal brain, lung, spleen and peripheral blood mononuclear cell libraries. Recently, it has been shown that SDF-1 is the ligand for LESTR, which is now called CXCR4. SDF-1 induces calcium and chemotaxis in CXCR4-transfected cells. CXCR4, also designated fusin, has been identified as the cofactor for T-tropic HIV-1 infection. The gene for CXCR4 is located on chromosome 2q21.

2.4 Structure–function relationship for chemokines and their receptors

2.4.1 Structural features important for chemokine function

Some studies have investigated which regions of ELR$^+$ CXC chemokines are important for binding to and activation of neutrophils. Removal of the entire C-terminal sequence after the fourth cysteine residue decreases, but does not abrogate, the binding and the biological effect of IL-8 on neutrophils. In contrast, receptor binding and activity of IL-8 are abrogated by removal of the ELR motif. The ELR motif is essential for neutrophil activation by NAP-2, but the remainder of the protein also participates in the interaction with the receptor. For GROα it has been shown that the ELR motif, His19 and Lys49 are important for binding to CXCR2. The importance of the ELR motif is also shown by the differences in activity and IL-8 receptor binding between ELR⁺ and ELR⁻ CXC chemokines. ELR⁺ CXC chemokines all bind to IL-8 receptors and attract and activate neutrophils. PF-4 and IP-10, which do not contain the ELR motif, do not bind to IL-8 receptors and are much less potent or inactive on neutrophils. PF-4 that has been modified to contain the ELR motif in the N-terminus binds to IL-8 receptors and activates neutrophils. Modified IL-8, with ELQ (Glu-Leu-Gln) instead of ELR (Glu-Leu-Arg), is inactive. Modification of the N-terminal region of IP-10 or the CC chemokine MCP-1 to contain the ELR motif does not result in IL-8 receptor binding, indicating that other parts of the structure are involved in receptor binding. Considering the importance of the N-terminal region in receptor binding, IL-8 antagonists have been synthesized. The modified proteins IL-8$_{6-72}$, starting with Arg, and Ala-Ala-IL-8$_{6-72}$ are potent antagonists, both inhibiting IL-8 binding and activity. These antagonists have a much lower receptor affinity than IL-8, indicating that the ELR motif is both a binding and activating motif. Besides the ELR motif, the two disulfide bridges, and residues 10–22 and 30–35 are important for the binding and activity of IL-8. It is supposed that the disulfide bridges and the 30–35 turn, which connects the first two β-strands, provide a scaffold for the N-terminal region which includes the primary receptor binding site (namely the ELR motif) as well as secondary binding and conformational determinants between residues 10 and 22.

The abovementioned structural features of IL-8 that are important for receptor binding and activation have been determined using neutrophils as test cells. Neutrophils express both IL-8 receptors, thus the obtained results show the combined requirements for both CXCR1 and CXCR2. Using cells transfected with either CXCR1 or CXCR2, the important regions of IL-8 for selective receptor binding and activation have been determined separately by testing mutants and chimeric molecules of IL-8 and GROα. The ELR motif is important for IL-8 binding to the two IL-8 receptors, but additional regions are necessary for high-affinity binding. The C-terminal α-helix is important for binding to CXCR2, but does not mediate high-affinity binding to CXCR1. Recently, it has been shown that Tyr13 and Lys15 of human IL-8 are important residues for interaction with CXCR1.

2.4.2 Structural features important for receptor function

ELR⁺ CXC chemokines show different affinities for CXCR1 and CXCR2. Several studies indicate that the N-terminal segment of CXCR1 is a dominant determinant of receptor subtype selectivity, whereas others report that the N-terminal segment of CXCR2 is dominant. Determinants of GROα selectivity have been found both in the N-terminal segment, before transmembrane domain (TMD) 1 beyond the first 15 residues, and in the region from TMD4 to the end of the second extracellular loop of CXCR2.

The N-terminal amino acid sequences of both IL-8 receptors are rich in acidic residues and have been suggested to play a major role in binding IL-8. Residues 10–15 (Trp-Asp-Phe-Asp-Asp-Leu) of CXCR1 and residues 6–10 (Phe-Glu-Asp-Phe-Trp) of CXCR2 play an important role in ligand binding and functionalities of the receptors. The cysteines in the extracellular domains of CXCR1, which may be part of a binding site or may be critical for the overall folding of the receptor, are also important for IL-8 binding and IL-8-mediated signalling. Besides the extracellular domain cysteines, the three residues Arg199, Arg203 and Asp265 of CXCR1 are involved in IL-8 binding and signal transduction. These residues are found in extracellular loops 2 and 3 and are conserved in CXCR2. Substitution of several other residues in the extracellular domains of CXCR1 yields mutants with a reduced affinity for IL-8, but which can still increase [Ca²⁺]ᵢ in response to IL-8. These amino acids might provide the initial contacts that facilitate formation of successively more stable interactions or possibly influence the tertiary structure of the receptor itself. The finding that an antibody against CXCR1 can block IL-8-induced activity without inhibiting binding confirms this model. It is possible that the receptor cannot undergo IL-8-induced conformational changes to elicit biological responses or that the antibody blocks the access of a signalling domain of IL-8 to the appropriate site of the receptor. Thus, IL-8 binding to and signalling through CXCR1 occur in at least two discrete steps involving distinct domains of the receptor. A conformational change of the receptor or ligand after IL-8 binding may play an important role in signalling.

Amino acids Tyr136, Leu137, Ile139 and Val140 in the second intracellular loop and Met241 in the third intracellular loop are essential for binding of the G-protein α-subunit to CXCR1 and for mediating the calcium signalling in response to IL-8. These residues are conserved in CXCR2 and are also here important for signal transduction. Also amino acids 317–324 of the C-terminus of CXCR2 are essential for signalling.

```
MIP-1α              ASLAADTPTACCFSYTSRQ  IPQNFIAD  YFETSSQ  CSKPGVIFLTKRSRQV  CADPSEEWVQKYVSDLELSA
MIP-1β              APMGSDPPTACCFSYTARKL  PRNFVVD  YYETSSL  CSQPAVVFQTKRSKQV  CADPSESWVQEYVYDLELN
RANTES              SPYSSDTTPCCFAYIARPL  PRAHIKE  YFYTSGK  CSNPAVVFVTRKNRQV  CANPEKKWVREYINSLEMS
I-309               KSMQVPFSRCCFSFAEQE  IPLRAILC  YRNTSSI  CSNEGLIFKLKRGKEA  CALDTVGWVQRHRKMLRHCPSKRK
MCP-1               QPDAINAPVTCCYNFTNRK  ISVQRLAS  YRRITSSK  CPKEAVIFKTIVAKEI  CADPKQKWVQDSMDHLDKQTQTPKT
MCP-2               QPDSVSIPITCCFNVINRK  IPIQRLES  YTRITNIQ  CPKEAVIFKTKRGKEI  CADPKERWVRDSMKHLDQIFQNLKP
MCP-3               QPVGINTSTTCCYRFINKK  IPKQRLES  YRRTTSSH  CPREAVIFKTKLDKEI  CADPTQKWVQDFMKHLDKKTQTPKL
MCP-4               QPDALNVPSTCCFTFSSKK  ISLQRLKS  YVITTSR  CPQKAVIFRTKLGKEI  CADPKEKWVQNYMKHLGRKAHTLKT
Eotaxin             GPASVPTTCCFNLANRK  IPLQRLES  YRRITSGK  CPQKAVIFKTKLAKDI  CADPKKKWVQDSMKYLDQKSPTPKP
Eotaxin-2/MPIF-2    VVIPSPCCMFFVSKR  IPENRVVS  YQLSSRST  CLKAGVIFTTKKGQQS  CGDPKQEWVQRYMKNLDAKQKKASPRARAVA
MPIF-1   RVTKDAETEFMMSKLPLENPVLLDRFHATSADCCISYTPRS  IPCSLLES  YFETNSE  CSKPGVIFLTKKGRRF  CANPSDKQVQVCMRMLKLDTRIKTRKN
MIP-3               DRFHATSADCCISYTPRS  IPCSLLES  YFETNSE  CSKPGVIFLTKKGRRF  CANPSDKQVQVCMRMLKLDTRIKTRKN
HCC-1               TKTESSSRGPYHPSECCFTYTTYK  IPRQRIMD  YYETNSQ  CSKPGIVFITKRGHSV  CTNPSDKWVQDYIKDMKEN
HCC-2/MIP-5         SFHFAADCCTSYISQS  IPCSLMKS  YFETSSE  CSKPGVIFLTKKGRQV  CAKPSGPGVQDCMKKLKPYSI
HCC-3   TKTESSSQTGGKPKVVKIQLKLVGGPYHPSECCFTYTTYK  IPRQRIMD  YYETNSQ  CSKPGIVFITKRGHSV  CTNPSDKWVQDYIKDMKEN
TARC               ARGTNVGRECCLEYFKGA  IPLRKLKTW  YQ TSED  CSRDAIVFVTVQGRAI  CSDPNNKRVKNAVKYLQSLERS
LARC/exodus/MIP-3α  ASNFD CCLGYTDRILHPKFIVGFTRQLANEG  CDINAIIFHTKK KLSV CANPKQTWVKYIVRLLSKKVKNM
PARC/MIP-4/DC-CK1   AQVGTNKELCCLVYTSWQ  IPQKFIVD  YSETSPQ  CPKPGVILLTKRGRQI  CADPNKKWVQKYISDLKLNA
Exodus-2/SLC        SDGGAQDCCLKYSQRK  IPAKVVRS  YRKQEPSLGCSIPAILFLPRKRSQAELCADPKELWVQQLMQHLDKTPSPQKPAQG...
             ...CRKDRGASKSGKRKGSKGCKRTERSQTPKGP

ELC/MIP-3β          GTNDAEDCCLSVTQKP  IPGYIVRNFHYLLIKDG  CRVPAVVFTTLRGRQL  CAPPDQPWVERIIQRLQRTSAKMKRRSS
MDC                 GPYGANMEDSVCCRDYVRYRL  PLRVVKHF  YWTSDS  CPRPGVVLLTFRDKEI  CADPRVPWVKMILNKLSQ
TECK               QGVFEDCCLAYHYPIGWAVLRRAWT  YRIQEVSGSCNLPAAIFYLPK  RHRKVCGNPKSREVQRAMKLLDARNKVFAKLHH...
             ...NMQTFQAGPHAVKKLSSGNSKLSSSKFSNPISSSKRNVSLLISANSG
```

■ **Figure 5.** Sequence alignment of human CC chemokines. The conserved cysteines are underlined.

3. CC chemokines

The human CC chemokine subfamily is rapidly expanding and already contains more than 20 structurally identified proteins (Figure 5). The CC chemokine genes for MCP-1, -2, -3 and -4, eotaxin, MIP-1α and -1β, HCC-1 (haemofiltrate CC chemokine 1), HCC-2/MIP-5, I-309, RANTES and PARC (pulmonary and activation-regulated chemokine) are localized on chromosome 17q11–21. However, the genes for some of the recently identified CC chemokines have been mapped to other chromosomes. The genes for TARC (thymus and activation-regulated chemokine) and MDC (macrophage-derived chemokine) are both located on chromosome 16q13. The LARC (liver and activation-regulated chemokine) gene has been mapped to chromosome 2q33–37, whereas the genes for ELC (EBI1-ligand chemokine) and Exodus-2 are localized on chromosome 9p13. All identified CC chemokine genes contain three exons and two introns. Their cDNA encodes a precursor protein of 89–151 amino acids containing a signal peptide, which is cleaved to yield the mature protein (Figure 5). The three-dimensional structure of the MIP-1β dimer has been reported to differ from that of the CXC chemokine IL-8. Whereas the IL-8 dimer is globular, the MIP-1β dimer is elongated and cylindrical. CC chemokines are produced by different cell types (Table 4) after appropriate stimulation with bacterial, viral or plant products or with several

other exogenous as well as endogenous inducers (cytokines are examples of the latter). Since several extensive reviews covering CC chemokines have recently been published, in this chapter only a short overview of these chemokines will be given.

3.1 CC chemokine protein structure, cellular sources and biological activities

3.1.1 MIP-1α, MIP-1β and RANTES

MIP-1α (LD78) was identified as a cDNA isolated from human tonsillar lymphocytes. The closely related chemokine MIP-1β (Act-2) was cloned from a T-cell library. Although MIP-1α and -1β are acidic proteins (in contrast to other chemokines), they do bind to heparin. MIP-1α tends to polymerize and aggregate but the active form is the monomer. MIP-1α and -1β are produced by stimulated leukocytes as well as other tissue cells and tumour cells (Table 4). RANTES was originally identified by molecular cloning as a transcript expressed in T cells. It is constitutively expressed by unstimulated T cells but has also been isolated from platelets, as well as other cell types (Table 4).

The effects of MIP-1α, MIP-1β and RANTES on monocytes and T cells include chemotaxis, a rise in $[Ca^{2+}]_i$, expression of integrins and increased adhesion to endothelial cells. Apart from the induction of a modest increase in $[Ca^{2+}]_i$ and shape change, MIP-1α has no activities on neutrophils. MIP-1α,

∎ **Table 4.** Production of CC chemokines by different cell types

	MCP-1	MCP-2	MCP-3	MCP-4	MIP-1α	MIP-1β	I-309	RANTES	eotaxin
monocytes/macrophages	+	+	+	+	+	+		+	
lymphocytes					+	+	+	+	
neutrophils					+				
eosinophils					+			+	+
platelets								+	
endothelial cells	+		+	+				+	+
mesothelial cells	+								
mesangial cells	+								
fibroblasts	+	+	+		+	+		+	+
chondrocytes	+								
epithelial cells	+	+		+				+	+
keratinocytes	+								
smooth muscle cells	+								
astrocytes	+				+				
synovial cells	+					+			
lipocytes	+								
melanocytes	+								
osteoblasts	+								
tumour cells	+	+	+	+	+	+		+	+

MIP-1β and RANTES provide an important signal for T-cell activation resulting in enhanced proliferation, IL-2 secretion and cell surface IL-2 receptor expression. These chemokines chemoattract and degranulate NK cells and enhance NK cell- as well as CTL-mediated cytolysis. MIP-1α, MIP-1β and RANTES induce chemotaxis and [Ca^{2+}]$_i$ increase in dendritic cells. MIP-1α is also chemotactic for B cells. Finally, in contrast to MIP-1β, MIP-1α and RANTES enhance IgE and IgG4 production by IL-4-stimulated B cells, and chemoattract and activate basophils, IgE-stimulated mast cells and eosinophils.

MIP-1α inhibits colony formation of immature subsets of myeloid progenitor cells stimulated with GM-CSF plus steel factor. This inhibitory effect is antagonized by MIP-1β. The MIP-1 proteins are pyrogenic in rabbits. Injection of RANTES into dog skin induces a profound infiltration of eosinophils and monocytes. Subcutaneous injection of MIP-1α, MIP-1β or RANTES in hPBL/SCID mice induces accumulation of murine monocytes and human CD3$^+$ T cells.

3.1.2 I-309

I-309 cDNA was cloned from activated T lymphocytes. In addition to the characteristic four cysteines, the mature chemokine has an extra pair of cysteine residues. I-309 induces chemotaxis and an [Ca^{2+}]$_i$ rise in monocytes. Recently, I-309 was identified as a factor that protects lymphoma cells from dexamethasone-induced apoptosis. TCA-3, which is thought to be the mouse equivalent of I-309, is chemotactic and mitogenic for mesangial cells and increases mesangial adhesiveness to fibronectin. *In vivo* administration of TCA-3 in mice causes local infiltration of neutrophils and monocytes.

3.1.3 MCP-1, -2, -3 and -4

MCP-1 has been isolated from the conditioned medium of human peripheral blood mononuclear cells, fibroblasts and tumour cells. MCP-1 is produced by almost all cell types upon stimulation with a variety of inducers (Table 4). MCP-2 and -3 were isolated from stimulated human osteosarcoma cells, and MCP-4 was cloned from a human fetal library. Like MCP-1, MCP-2, -3 and -4 contain a pyroglutamate at the N-terminus and show 62%, 71% and 61% amino acid homology with MCP-1, respectively. Although MCP-2 and -3 are often co-produced with MCP-1, the amounts of secreted MCP-2 and -3 are at least 10 times lower than for MCP-1. MCP-1 and -3, but not MCP-2 and -4, contain one *N*-glycosylation site. However, *N*-

glycosylation has not been detected for MCP-1. MCP-1 occurs in several molecular forms (from 9 kDa to 17 kDa) due to the presence of O-linked sugar and sialic acid residues. No glycosylation is expected for 7.5 kDa MCP-2. Although natural 11 kDa MCP-3 is probably not glycosylated, multiple forms of recombinant MCP-3 (11, 13, 17 and 18 kDa, after expression in COS cells) have been detected, due to N- and O-glycosylation. MCP-1 and -2 (but not MCP-3) form dimers at high concentrations, but all MCPs probably function as monomers.

MCP-1, -2, -3 and -4 chemoattract and cause enzyme release by monocytes. MCP-1, -3 and -4 and, to a much lesser extent, MCP-2, augment $[Ca^{2+}]_i$ in these cells. MCP-1 has been shown to induce the respiratory burst, the production of IL-1 and -6 and the expression of $\beta2$ integrins by monocytes. All MCPs are chemotactic for T cells and increase $[Ca^{2+}]_i$ in T-cell clones. As mentioned for MIP-1α, MIP-1β and RANTES, MCP-1 co-stimulates T-cell proliferation and lymphokine production. MCP-1 induces NK cell migration and degranulation and enhances NK cell- and CTL-mediated cytolysis. MCP-2 and -3 also chemoattract NK cells. On basophils, MCP-1, -2 and -3 induce chemotaxis, $[Ca^{2+}]_i$ increase and histamine release. After pretreatment of the basophils with IL-3, IL-5 or GM-CSF, the histamine release by these chemokines is further enhanced, and leukotrienes are also released. MCP-4 has been shown to induce histamine release from pretreated basophils. The MCP-1-induced histamine release can be partially inhibited by preincubation of the cells with IL-8 or RANTES. MCP-1 also chemoattracts mast cells. In contrast to MCP-1, MCP-2, -3 and -4 are chemotactic for eosinophils. MCP-3, but not MCP-1 and -2, attracts dendritic cells and increases $[Ca^{2+}]_i$ in these cells.

MCP-1 suppresses the colony formation of immature subsets of myeloid progenitors stimulated with GM-CSF plus steel factor. The growth of several tumour cell lines cultured in the presence of monocytes is inhibited by the addition of MCP-1. Sarcoma cell clones expressing MCP-1 and MCP-1-transfected melanoma cells show a reduced growth capacity *in vivo* and an enhanced number of tumour-associated macrophages, which is correlated with the MCP-1 production, is observed. Chinese hamster ovary (CHO) cells are not tumorigenic in nude mice if they are transfected to produce MCP-1, and MCP-1 transfectants also inhibit the growth of coinjected, non-transfected HeLa or CHO cells. MCP-1, -2 and -3 induce selective infiltration of monocytes after intradermal injection into rabbits. Subcutaneous injection of MCP-1 into hPBL/SCID mice induces significant infiltration of murine monocytes and human CD3$^+$ T cells. Surprisingly, in transgenic mice expressing MCP-1 in the basal layer of the epidermis, dendritic cells and not monocytes are recruited to the skin. This observation is in contrast with the absence of chemotactic activity of MCP-1 for dendritic cells *in vitro*. Finally, transgenic mice expressing MCP-1 in the thymus and brain have an increased presence of monocytes and macrophages, but not lymphocytes, in these tissues without eliciting inflammation.

3.1.4 Eotaxin, eotaxin-2 (MPIF-2), MPIF-1 and MIP-3

Human eotaxin is a potent chemoattractant for eosinophils. In these cells, eotaxin induces a calcium rise, actin polymerization and the respiratory burst and up-regulates $\beta2$ integrins; it also promotes eosinophil adhesion to endothelial cells. The recently identified chemokine eotaxin-2 also chemoattracts eosinophils. Both eotaxin and eotaxin-2 induce chemotaxis of basophils and stimulate the release of histamine and leukotriene C_4 from IL-3-primed basophils. Intradermal injection of human eotaxin or eotaxin-2 into Rhesus monkeys induces accumulation of eosinophils in the skin. Although eotaxin-2 is functionally very similar to eotaxin, their primary structures are rather different (36% identical amino acids).

Eotaxin-2 is identical to myeloid progenitor inhibitory factor 2 (MPIF-2), except for Ala35 and Ser47 which correspond to Gly35 and Phe47 in MPIF-2. MPIF-2 has been shown to increase the $[Ca^{2+}]_i$ in eosinophils, to attract resting T cells but not monocytes, and to suppress colony formation by a subset of haematopoietic progenitor cells. Eotaxin-2 is expressed in activated T cells and monocytes.

MPIF-1, a CC chemokine with six cysteine residues, attracts resting T cells and monocytes, and suppresses a different subset of myeloid progenitors from MPIF-2. This chemokine is mainly expressed by macrophages and is detectable in liver, lung, bone marrow and placenta. Both MPIF-1 and MPIF-2 show weak chemotactic activity for neutrophils. Structurally, MPIF-1 and MPIF-2 are rather different (29% identical amino acids).

MIP-3 is a chemokine of which only the nucleotide sequence is known. The predicted mature protein sequence is identical to that of residues 24–99 of MPIF-1.

3.1.5 HCC-1, HCC-2 (MIP-5) and HCC-3

HCC-1 (haemofiltrate CC chemokine 1) has been isolated from the haemofiltrate of patients with chronic renal failure. Its cDNA has been cloned from human bone marrow. In contrast to other chemokines, HCC-1 is present at high concentrations in human plasma. The protein weakly activates monocytes to increase the $[Ca^{2+}]_i$ and to release enzymes, but it does not chemoattract these cells. HCC-1 is not active on T cells, neutrophils or eosinophils. HCC-3 is a spliced variant of HCC-1 and has not yet been biologically characterized.

HCC-2 (MIP-5) is expressed in liver, intestine and lung leukocytes and induces chemotaxis of monocytes, T cells and, to a lesser degree, eosinophils, but not that of neutrophils. In contrast to most of other CC chemokines, HCC-2 contains a third disulfide bridge.

3.1.6 TARC, LARC, PARC, Exodus-2 and ELC

TARC is constitutively expressed in the thymus and is up-regulated in PHA-stimulated peripheral blood mononuclear cells. This chemokine induces chemotaxis of T-cell lines but not of monocytes or granulocytes.

LARC is identical to Exodus and MIP-3α (except for Asn69, which is an Asp in MIP-3α). This chemokine is expressed in the liver, lung, lymph nodes, appendix and peripheral blood mononuclear cells. It is induced in U937 cells, THP-1 cells and Bowes melanoma cells by PMA and the expression in lymphocytes, monocytes and dendritic cells is up-regulated by LPS. Endothelial cells express LARC after stimulation with TNF-α. LARC is chemotactic for lymphocytes and weakly chemoattracts neutrophils, but is not active on monocytes. The chemokine inhibits proliferation of myeloid progenitors.

Exodus-2 or SLC (secondary lymphoid-tissue chemokine) (identical except for Ser89 and Arg93 which are replaced in SLC by Thr and Gly, respectively) is expressed in lymph nodes, appendix and spleen. This chemokine has, compared with most other CC chemokines, a C-terminal extension which contains two cysteines, implying that there may be an extra disulfide bridge. Exodus-2/SLC inhibits colony formation by haematopoietic progenitor cells and stimulates chemotaxis of T cells and, to a lesser extent, B cells. It is not chemotactic for monocytes or neutrophils.

ELC is identical to MIP-3β. ELC is expressed strongly in thymus, lymph nodes and appendix and induces chemotaxis of T-cell lines.

PARC is identical to DC-CK1 (dendritic cell-derived CC chemokine 1) and MIP-4 (except for Pro37, which corresponds to Leu37 in MIP-4). This chemokine can be detected in the lung and in lymphoid tissue. Alveolar macrophages, dendritic cells, peripheral blood monocytes stimulated with LPS and PMA-treated U937 cells express this chemokine. The protein is chemotactic for T lymphocytes, but not for monocytes or granulocytes.

3.1.7 MDC and TECK

MDC is highly expressed in the thymus and is produced by macrophages and dendritic cells. It chemoattracts dendritic cells, NK cells and, to a lesser extent, monocytes. TECK (thymus-expressed chemokine) is expressed in the thymus and the small intestine. It is chemotactic for thymocytes, macrophages and dendritic cells.

3.2 CC chemokine receptors

Initial binding studies on monocytic cells using the radiolabelled chemokines MCP-1, MIP-1α and -1β, and RANTES provided evidence for the presence of at least three different CC chemokine receptors (CCRs) on monocytic cells: one receptor specific for MCP-1, a second receptor able to bind both MIP-1α and -1β, and a third receptor shared by MIP-1α, RANTES and MCP-1. Since then, nine CCRs have been cloned. The first to be cloned, CCR1, responds to MIP-1α, RANTES, MCP-2, MCP-3 and HCC-2 (Table 5). The MCP-1 receptor CCR2 also recognizes MCP-2, -3 and -4. CCR3, originally reported to be expressed selectively in eosinophils, responds to eotaxin, eotaxin-2, MCP-2, -3, and -4, RANTES and HCC-2, but not to MIP-1α. The MIP-1α effect on eosinophils must therefore be mediated by another receptor. Expression of the MIP-1α high-affinity receptor, CCR1, has been detected in eosinophils although at low levels. CCR4, originally reported to recognize MIP-1α, MCP-1 and RANTES, has been demonstrated to be the functional receptor for MDC and TARC. The combination of the agonists for CCR5 (RANTES, MIP-1α and MIP-1β) has been shown to inhibit HIV-1 infection of human peripheral blood leukocytes. Like for CCR1, MIP-1α is a stronger agonist than RANTES. The ranking orders are MIP-1α > MCP-3 > RANTES for CCR1, and MIP-1α > RANTES $\geq$ MIP-1β for CCR5. MIP-1β also binds to CCR1 but does not mobilize calcium in CCR1-transfected cells. So far the recently identified CCR6, CCR7 and CCR8 each bind only a single chemokine, the ligands being LARC/Exodus/MIP-

■ **Table 5.** CC chemokine receptors, their ligands and expression

CC chemokine receptor	ligands	expression in leukocytes	expression in tissues
CCR1	MCP-2, MCP-3, MIP-1α, RANTES, HCC-2	monocytes, T cells, eosinophils, basophils, CD34$^+$ cells	placenta, lung, liver
CCR2	MCP-1, MCP-2, MCP-3, MCP-4	monocytes, T cells, basophils	lung, heart, bone marrow, liver, pancreas
CCR3	MCP-2, MCP-3, MCP-4, RANTES, eotaxin, eotaxin-2, HCC-2	Th2 cells, eosinophils, basophils	
CCR4	(MCP-1, MIP-1α, RANTES), TARC, MDC	CD4$^+$ T cells, basophils, platelets	thymus, spleen
CCR5	MIP-1α, MIP-1β, RANTES	adherent monocytes, CD4$^+$ and CD8$^+$ T cells	thymus, spleen
CCR6	LARC	B cells, CD4$^+$ and CD8$^+$ T cells, cord blood derived dendritic cells	spleen, lymph nodes, appendix, fetal liver, thymus, pancreas,' testis and small intestine
CCR7	ELC	B cells, T cells	spleen, tonsils
CCR8	I-309	adherent monocytes, B cells, CD4$^+$ and CD8$^+$ T cells, granulocytes	spleen, lymph nodes, thymus
CCR10	MCP-1, MCP-3, MCP-4, RANTES		placenta, fetal liver, lymph nodes

3α, ELC/MIP-3β and I-309, respectively. CCR10 recognizes MCP-1, -3 and -4 and RANTES. Desensitization experiments have been performed with several receptors but the findings are rather complex and contradictory. This is possibly due to the use of different sources of ligands, cells or methods as well as differences in receptor expression levels.

In addition to the expression in haematopoietic tissues and leukocytes, Northern blot or PCR analysis has revealed the presence of CCR mRNAs, except CCR3 mRNA, in non-haematopoietic tissues such as lung, heart, kidney and liver (Table 5). The expression pattern of CCRs in various peripheral blood leukocyte types correlates well with the selective activity of the corresponding ligands. CCR1 mRNA has been detected in monocytes, IL-2-stimulated T cells, CD34$^+$ cells, basophils and eosinophils. CCR2 is expressed in monocytes, basophils, anti-CD3- and IL-2-stimulated T cells. A more detailed analysis has shown that CCR3 is not selectively expressed in eosinophils as originally reported, since CCR3 also mediates basophil and Th2 lymphocyte chemotaxis. CCR4 transcripts have been detected in CD4$^+$ T cells, IL-5-stimulated basophils and platelets, and CCR5 transcripts have

been reported in adherent monocytes and in CD4$^+$ and CD8$^+$ T cells. CCR6 is expressed in lymphocytes and dendritic cells derived from CD34$^+$ cord blood cells. Infection of lymphocytes with Epstein–Barr virus or human herpes virus 6 and 7, or stimulation with PHA, induces CCR7 gene transcription. CCR8 expression has been detected in B and T cells, adherent monocytes and granulocytes. The presence of CCR10 RNAs has been demonstrated in placenta, fetal liver and lymph nodes, but the occurrence in leukocyte subtypes has not yet been analysed. Studies with monoclonal antibodies against CCR1 and CCR3 confirmed the expression in leukocytes, except for the presence of CCR1 on the cellular surface of B lymphocytes and neutrophils.

The genes for CCR1 to CCR5, CCR8 and CCR10 have been mapped to chromosome 3 (p21.3–p24 region). Within this same region is the GPR13/V28 gene located, coding for CX$_3$CR1. The CCR6 and CCR7 genes have been mapped to chromosome 6 (6q27) and 17 (17q12–17q21.2), respectively.

CCR5 is now well known as the fusion cofactor required for infection by M-tropic HIV-1, HIV-2 and SIV strains. T-tropic HIV-1 strains use fusin or

CXCR4 as cofactor and a dual-tropic primary HIV-1 isolate, 89.6 can utilize fusin, CCR2, CCR3 and CCR5 as entry cofactors. In addition to CCR5, at least some SIV strains can utilize the orphan chemokine receptors STRL33/Bonzo and GPR15/BOB as co-receptors. Interestingly, STRL33 could function as a cofactor for fusion mediated by envelope proteins from both T-tropic and M-tropic HIV-1 strains. HIV-2 isolates have also been shown to use CCR5, CXCR4 or both.

As the number of new CCRs increases, the number of orphan receptors highly homologous to chemokine receptors decreases. However, there are still some orphan receptors for which the ligands remain to be revealed. These receptors may be recognized by a chemokine that remains to be discovered, or by one of the recently cloned CC chemokines Exodus-2/SLC, MIP-3, PARC/MIP-4/DC-CK1, HCC-1, HCC-3 or TECK.

References

Baggiolini, M., Dewald, B., and Moser, B. (1994). Interleukin-8 and related chemotactic cytokines-CXC and CC chemokines. *Adv. Immunol.* **55**, 97–179.

Ben-Baruch, A., Michiel, D. F., and Oppenheim, J. J. (1995). Signals and receptors involved in recruitment of inflammatory cells. *J. Biol. Chem.* **270**, 11703–6.

Clark-Lewis, I., Kim, K.-S., Rajarathnam, K., Gong, J.-H., Dewald, B., Moser, B., Baggiolini, M., and Sykes, B. D. (1995). Structure–activity relationships of chemokines. *J. Leukocyte Biol.* **57**, 703–11.

D'Souza, M. P. and Harden, V. A. (1996). Chemokines and HIV-1 second receptors. *Nature Med.* **2**, 1293–300.

Farber, J. M. (1997). Mig and IP-10: CXC chemokines that target lymphocytes. *J. Leukocyte Biol.* **61**, 246–57.

Horuk, R. (1994). The interleukin-8-receptor family: from chemokines to malaria. *Immunol. Today* **15**, 169–74.

Horuk, R. (1994). Molecular properties of the chemokine receptor family. *Trends Pharmacol. Sci.* **15**, 159–65.

Kelvin, D. J., Michiel, D. F., Johnston, J. A., Lloyd, A. R., Sprenger, H., Oppenheim, J. J., and Wang, J.-M. (1993). Chemokines and serpentines: the molecular biology of chemokine receptors. *J. Leukocyte Biol.* **54**, 604–12.

Murphy, P. M. (1996). Chemokine receptors: structure, function and role in microbial pathogenesis. *Cytokine Growth Factor Rev.* **7**, 47–64.

Oppenheim, J. J., Zachariae, C. O. C., Mukaida, N., and Matsushima, K. (1991). Properties of the novel proinflammatory supergene 'intercrine' cytokine family. *Annu. Rev. Immunol.* **9**, 617–48.

Power, C. A. and Wells, T. N. C. (1996). Cloning and characterization of human chemokine receptors. *Trends Pharmacol. Sci.* **17**, 209–13.

Premack, B. A. and Schall, T. J. (1996). Chemokine receptors: gateways to inflammation and infection. *Nature Med.* **2**, 1174–8.

Proost, P., Wuyts, A., and Van Damme, J. (1996). The role of chemokines in inflammation. *Int. J. Clin. Lab. Res.* **26**, 211–23.

Proost, P., Wuyts, A., and Van Damme, J. (1996). Human monocyte chemotactic proteins-2 and -3: structural and functional comparison with MCP-1. *J. Leukocyte Biol.* **59**, 67–74.

Schall, T. J. (1994). The chemokines. In Thomson, A. W. (ed.), *The Cytokine Handbook*, pp. 419–60. Academic Press, London.

Sozzani, S., Locati, M., Allavena, P., Van Damme, J., and Mantovani, A. (1996). Chemokines: a superfamily of chemotactic cytokines. *Int. J. Clin. Lab. Res.* **26**, 69–82.

Taub, D. D. (1996). Chemokine–leukocyte interactions. The voodoo that they do so well. *Cytokine Growth Factor Rev.* **7**, 355–76.

Taub, D. D. and Oppenheim, J. J. (1994). Chemokines, inflammation and the immune system. *Ther. Immunol.* **1**, 229–46.

Van Damme, J. (1994). Interleukin-8 and related chemotactic cytokines. In Thomson, A. W. (ed.), *The Cytokine Handbook*, pp. 185–208. Academic Press, London.

Van Damme, J., Wuyts, A., Froyen, G., Van Coillie, E., Struyf, S., Billiau, A., Proost, P., Wang, J. M., and Opdenakker, G. (1997) Granulocyte chemotactic protein-2 and related CXC chemokines: from gene regulation to receptor usage. *J. Leukocyte Biol.* **62**, 563–9.

X Cytokines and haematopoiesis

Elke Schneider and Michel Dy

1. Introduction

A limited number of pluripotent stem cells, mainly located in the bone marrow, give rise to all blood cell lineages (Figure 1). Because of their relatively short lifespan, circulating cells must be continually replaced throughout adult life. This task is performed by haematopoietic stem cells which share two central features: the capacity for self-renewal, which prevents their own depletion, and the ability to preserve blood homeostasis. The mechanisms behind the critical choice between lineage-commitment and maintenance of the stem-cell pool remain speculative. They involve a number of complex interactions between haematopoietic progenitor cells at different stages of maturation, stromal cells and their extracellular matrix, as well as a variety of stimulatory or inhibitory cytokines provided by the microenvironment.

Haematopoietic growth factors were first identified in the 1960s when soluble agents produced by tissues like spleen, uterus or lung were found to support the formation of differentiated colonies from haematopoietic progenitor cells in semi-solid culture systems developed by Bradly and Metcalf, as well as Pluznik and Sachs. Hence they were named colony-stimulating factors (CSFs). Since then, most of these molecules have been purified, and their genes sequenced. They are currently available in recombinant form and have been used with success in clinical trials.

Haematopoietic growth factors (or CSFs) can be divided into two categories, according to their target cell specificity. One group comprises the factors whose activity is relatively restricted to the differentiation of a particular cell type, such as macrophages for macrophage colony-stimulating factor (M-CSF, or CSF-1), neutrophils for granulo-cyte colony-stimulating factor (G-CSF), eosinophils for interleukin 5 (IL-5, or Eo-CSF) and megakary-

ocytes for thrombopoietin (Tpo). Erythropoietin (Epo) is also included in this group because of its specific action on the erythroid lineage, though in some ways it resembles a hormone rather than a cytokine.

The second category of growth factors has a wider spectrum of activities, and comprises IL-3 (or multi-CSF) and granulocyte–macrophage colony-stimulating factor (GM-CSF) which target a heterogeneous population of cells, including both primitive and lineage-committed progenitors. There is a certain functional redundancy between these two molecules, whose action can be modulated by a number of cytokines which are not growth factors *per se*. Among these, IL-1, IL-6, IL-9, IL-11 and leukaemia inhibitory factor (LIF) should be mentioned. Stem cell factor (SCF; also called c-Kit ligand or KL) and Flt3 ligand play a particularly important role in the amplification of early stem cell commitment. IL-7 is also noteworthy in this context, with respect to its crucial role in lymphopoiesis, as evidenced by the strong lymphopenia in IL-7-deficient mice. Finally, haematopoiesis can be regulated negatively by a heterogeneous set of molecules, such as interferons, tumor necrosis factor alpha (TNFα), transforming growth factor beta (TGFβ) and compounds like prostaglandins, ferritin and lactoferrin.

The precise function of cytokines during constitutive haematopoiesis in a healthy organism is unclear, though the study of genetically modified mice has given some clues. However, the purpose of haematopoiesis is not only the maintenance of homeostasis, but also a rapid and controled riposte to stress situations. Thus, during the immune response induced by severe infection, the number of circulating white blood cells can increase more than 10-fold. In this case, the cytokines generated by sensitized lymphocytes and activated cells of the immune system play a crucial role in the

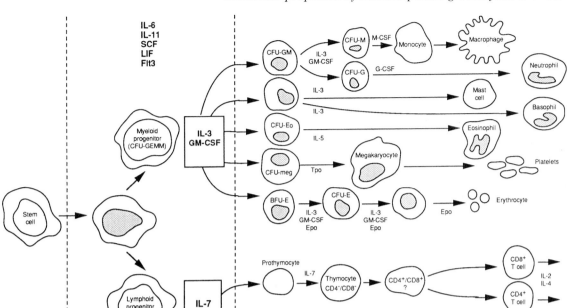

■ **Figure 1.** Simplified haematopoietic differentiation scheme.

recruitment and the differentiation of haematopoietic cells.

2. Molecular properties of haematopoietic growth factors

Haematopoietic growth factors are glycoproteins of variable size with low sequence homology (Table 1). Glycosylation is not required for their biological activity, but enhances their stability, solubility and resistance to proteolysis. All cytokines involved in haematopoiesis contain disulfide bonds which are critical for the maintenance of their three-dimensional structure. In spite of their low sequence homology, these molecules have a similar conformation conferred by a four antiparallel α-helix-bundle structure. This feature explains how these proteins, whose primary sequence is quite distinct, can nevertheless bind to structurally related receptors.

All haematopoietic growth factors exist as precursor molecules that must undergo proteolytic cleavage before secretion. This maturation process is particularly complex in the case of M-CSF which occurs in several isoforms generated by alternative splicing. Three biologically active forms of M-CSF can be distinguished:

(1) soluble glycoproteins, occurring mainly as 85 kDa molecules, rapidly cleaved from the membrane at a specific proteolytic site, and a minor 44 kDa isoform slowly released from the membrane without proteolytic cleavage;

(2) a 68 kDa membrane-bound M-CSF; and

(3) M-CSF proteoglycans, cleaved from their membrane-bound form and anchored in the extracellular matrix by their heavy chondroitin sulfate side-chains.

These mature M-CSF isoforms are probably generated by distinct producer cells and differ in their physiological functions.

Haematopoietic growth factors are encoded by single gene copies with various chromosomal locations (Table 1). IL-3, IL-5 and GM-CSF genes are clustered on human chromosome 5 and on murine chromosome 11 which also contains the gene for G-CSF. These molecules have similar genomic structures, with AT-rich regions at the 3' end, a feature common to a variety of genes that must be turned on and off quickly. A decanucleotide consensus sequence called CK1 or conserved lymphokine

■ **Table 1.** Biochemical properties of haematopoietic growth factors

	Amino acids		S-S bonds	Chromosomal location	Exons	mRNA (kbp)
	precursor	signal peptide				
IL-3	152	19	1	5	5	1
GM-CSF	144	17	2	5	4	2.5
IL-5	134	22	1	5	4	0.9
G-CSF	204	30	2	17	5	1.6
	207					
	256					1.6
M-CSF	554	32	3	1	10	2.3
	438					2.5
Epo	193	27	2	7	5	1.6
Tpo	353	21	2	3	5	1.8

element (CLE1) is repeated twice in the upstream flanking region of GM-CSF, IL-3 and G-CSF genes. A homologous element, CLE0, is present in the IL-5 promoter. This region can bind nuclear proteins in gel-shift assays and confers responsiveness to CD28. A number of recognition sites for transcription factors such as NF-κB, NFAT, Sp1 and AP-1 are found in the upstream region of these genes. A repeated sequence, CATT(A/T), is present in the GM-CSF and IL-5 promoters but not in the IL-3 promoter. Both IL-3 and GM-CSF flanking regions contain consensus binding sites for Ets and Octamer-1, as well as a CK-2 element and a TGTGGT sequence recognized by the core binding protein. Negative regulatory elements have been characterized in the human GM-CSF gene and a binding site for nuclear inhibitory protein (NIP) is present in the IL-3 gene.

The regulation of erythropoietin transcription by hypoxia and certain metal ions is a paradigm of gene regulation. It appears that this system reacts to hypoxia itself, rather than to some secondary event such as cellular damage or impaired cellular respiration. The characterization of a 50 bp control sequence in the 3′ flanking region of the Epo gene, termed the erythropoietin 3′ enhancer, and the subsequent identification of the hypoxia-inducible transcriptional complexes HIF-1 (hypoxia-inducible factor 1) and HNF-4 (hypoxia nuclear factor 4) have contributed greatly to the understanding of the mechanisms underlying the production of Epo. HIF-1 is composed of two basic helix–loop–helix, proline–alanine–serine (PAS) proteins

termed HIF-1α and HIF-1β, the latter being identical to aryl hydrocarbon nuclear receptor translocator (ARNT), an essential molecule for the transcriptional response to certain environmental hydrocarbons. The involvement of identical regulatory elements in such completely different physiological situations raises the question of the origin of this mechanism and suggests that it belongs to a more widely operating system of gene control.

The regulation of the low basal Epo expression in normoxia is less well understood, but seems to involve a highly conserved GATA motif in the –30 region of the Epo promoter acting as a negative response element. The regulatory elements of the more recently characterized Tpo gene are less well known. This is also true for the M-CSF gene whose transcription appears to be controlled by a particularly complex system involving multiple *trans*-acting nuclear factors.

3. Biosynthesis of haematopoietic growth factors

A common feature of haematopoietic growth factors is their generation by multiple cellular sources. These highly inducible molecules are synthesized in response to a variety of stimuli provided by other cytokines or inflammatory compounds, such as endotoxins. Detectible amounts of M-CSF are generated spontaneously by the stroma, which is also a source of constitutive G-CSF. Biosynthesis of Epo and Tpo occurs mainly outside the haematopoietic environment, in the kidney and in the liver, respectively, and is maintained at basal levels in the absence of stimulation.

Activated T lymphocytes, belonging essentially to the CD4[+] subset, are the main IL-3 producers in human and mice. The two types of murine T-helper cells, defined by their cytokine profile, namely Th1 (IL-2 and IFNγ) and Th2 (IL-4, IL-5 and IL-6), produce IL-3 equally well, while GM-CSF originates preferentially from the Th1 subpopulation and IL-5 from the Th2 subpopulation. Stimulated mast cells or basophils constitute another major source of IL-3, in particular after FcεRI cross-linking which promotes the synthesis of GM-CSF and IL-5. The latter cytokine is also produced by eosinophils, its main target cell population. Expression of IL-3 in non-haematopoietic tissues, e.g. astrocytes, epidermal cells, keratinocytes and endothelial cells, has been reported. However, the poor specificity of the techniques employed in these studies could not provide absolute proof for this assertion.

Like most other haematopoietic growth factors, IL-3 is synthesized by several tumor cells, especially by T-cell lymphomas, such as EL-4 or LBRM 33, in response to various stimuli. This cytokine was originally purified from supernatants of the WEHI-3 cell line, which produces IL-3 constitutively as the result of a retroviral insertion into the 5′ region, adjacent to the promoter of the IL-3 gene.

In contrast with other CSFs, IL-3 is not easily detected in serum, even after infection or injection of lipopolysaccharide (LPS). The only experimental situations known so far in which it appears in the circulation, are graft-versus-host disease, anti-CD3 injection and specific antigen challenge after immunization. It is therefore plausible that the effect of IL-3 on haematopoiesis is normally restricted to the immediate neighbourhood of activated T cells.

Like IL-3, GM-CSF is mainly an inducible molecule, but its cellular sources and the number of effective stimuli are more abundant than those of IL-3. Though stimulated T cells are again the predominant source of GM-CSF, its production is more complex than in the case of IL-3. Indeed, the former growth factor is produced both directly and indirectly, via the release of cytokines like IL-1 and TNFα, which induce its synthesis by monocytes/macrophages, fibroblasts and endothelial cells. GM-CSF can be released by T cells in response to IL-2, but T-cell receptor (TCR) cross-linking is a much better stimulus. Both CD4[+] and CD8[+] subsets syn-

thesize the factor, although T-helper cells, mainly those of the Th1 subset, are more effective.

The two other CSFs, G-CSF and M-CSF, are produced both constitutively and in response to a variety of stimuli. M-CSF is easily detected in serum, though the stromal cells responsible for its constitutive *in vivo* production have not been exactly defined. G-CSF is generated spontaneously by the placenta and, to a lesser degree, by human stromal cells. Both factors are potentially significant during embryogenesis since G-CSF is synthesized by cells of the trophoblast and M-CSF is detected in the uterus during gestation. Monocytes/macrophages stimulated by LPS are an important source of both M-CSF and G-CSF. The latter is also induced by a variety of cytokines, including IFNγ, GM-CSF, IL-3, M-CSF, IL-4 and IL-2. Fibroblasts and endothelial cells depend on the presence of IL-1 or TNFα for the generation of G-CSF. Activated T cells certainly produce M-CSF, while their production of G-CSF still remains controversial.

Epo and Tpo share significant homology in their molecular structure. Both cytokines are produced chiefly outside the haematopoietic environment and may therefore be classified as hormones rather than cytokines. In contrast to Epo, whose production seems to be restricted to the kidney, the cellular sources for Tpo are more diverse, comprising kidney, medullary stroma and hepatocytes, the latter being by far the most important producers. The expression of Epo and Tpo is apparently not induced by other cytokines, but their expression depends, respectively, on oxygen and platelet levels in the circulation.

4. Cytokine receptors

The lack of sequence homology between the different cytokines cloned to date, making it impossible to classify them based on sequence. However, once the corresponding receptors had been discovered and the existence of different receptor families established, it at last became possible to divide these molecules into different subgroups. Haematopoietic growth factors bind to receptors of the haematopoietin or class I cytokine family, with the exception of M-CSF which is recognized by receptors with intrinsic protein tyrosine kinase (or receptor tyrosine kinase, RTK) activity belonging to the immunoglobulin superfamily. The haematopoietin receptor family is the largest family and has

■ **Table 2.** Biochemical properties of human haematopoietic growth factor receptors

Receptor		Mol. weight (kDa)	Amino acids			Chromosomal location	K_d (M) ($\times 10^{-9}$)
			extracellular	membrane	intracellular		
IL-3	α	70	287	20	53	pseudoautosomal (X,Y)	120
	βc(HK97)	120	420	25	431	22	0
GM-CSF	α	80	298	26	54	pseudoautosomal (X,Y)	2–8
	βc(KH97)	120	420	25	431	22	0
IL-5	α	60	344	21	55	3	1
	βc(KH97)	120	420	25	431	22	0
G-CSF	G-CSFR	150	603	26	183	1	0.1–0.5
M-CSF	*cfms*	150	512	25	435	5	400
Epo	EpoR	66	226	22	236	19	100–1000
Tpo	*cmpl*	—	472	22	122/66	1	—

been subdivided into smaller groups according to structural similarities.

Receptors for Epo, Tpo and G-CSF are composed of a single chain, while IL-3, GM-CSF and IL-5 receptors are heterodimers sharing a common β-chain (Table 2). In both cases, ligand binding induces dimerization, resulting in the activation of intracellular signalling cascades. All haematopoietin receptors are transmembrane glycoproteins composed of a single extracellular ligand-binding domain, a hydrophobic transmembrane region and a cytoplasmic domain. They share an extracellular cytokine receptor homology (CRH) sequence of about 210 amino acid residues, which may be present in single or duplicated form. Its characteristic features are four spatially conserved cysteine residues in the N-terminus and a tryptophan–serine–X–tryptophan-serine motif (or WSXWS motif, where X stands for any amino acid) between two fibronectin type III repeats arranged in tandem, close to the transmembrane region. The cysteine residues contribute to the tertiary stucture of the receptor by forming disulfide bridges. The fibronectin repeat is a module of about 90 amino acids which forms a barrel-like structure consisting of seven antiparallel β-strands.

Crystallographic studies have shown that the ligand-binding site is located in the hinge region between two fibronectin repeats. Single amino acid mutations in this sequence and in the WSXWS motif, which is part of it, compromise the ligand

binding to the receptor. The cytoplasmic domains of haematopoietin receptors are less conserved than the extracellular part. They lack catalytic activity, unlike the M-CSF receptor. Functionally important domains with limited homology have been identified in the membrane-proximal cytoplasmic region of haematopoietin receptors and termed Box 1 and Box 2 motifs. Box 1 is part of the 20 first intracellular amino acids, and the hallmark of this motif is the Al-Ar-Pro-X-Al-Pro-X-Pro or Ar-X-X-X-Al-Pro-X-Pro sequence, where Al stands for aliphatic, Ar for aromatic and X for any amino acid residue. The region between Boxes 1 and 2, termed variable Box (V-Box) contains a positionally conserved tryptophan residue, mutation of which completely abrogates receptor signalling. Box 2 is composed of a cluster of hydrophobic residues and one or two positively charged amino acids in the C-terminal part of the domain. Mutations or deletions in Box 1, Box 2 or V-Box severely compromise or abrogate biological responses mediated by the receptor.

4.1 The βc family

IL-3, IL-5 and GM-CSF are tightly linked on chromosome 5 in humans and on chromosome 11 in mice, and their similar genomic structure suggests that their genes have evolved from a single ancestor. The amino acid sequences of these cytokines do not display significant homology, but they have similar four α-helix-bundle structure. The cross-competition of IL-3, GM-CSF and IL-5 for receptor binding led to the suggestion that the three factors might share common receptor subunits. This hypothesis was confirmed by the cloning of a specific ligand-binding α-chain and a shared signal-transducing β-chain, forming the βc receptor complex. An additional, unique IL-3 signalling subunit has been demonstrated in mice, which explains why IL-3 does not cross-compete with GM-CSF or IL-5 for murine receptor binding, while it does in human.

The first component of the mouse IL-3 receptor (IL-3R β chain specifically associated with the murine IL-3R) to be molecularly characterized was AIC2A. It was cloned using an anti-AIC2 antibody which partially blocks IL-3 binding. The cDNA of AIC2A encodes a 878 amino acid 120 kDa glycoprotein, with an extracellular region of 417 residues that contains two cytokine type I (CRH) domains. Like all other cytokine receptor chains of this group, its cytoplasmic region is devoid of known catalytic signalling motifs, but it is tyrosine-phosphorylated after ligand stimulation. AIC2A alone binds IL-3 only with low affinity ($K_d = 10^{-8}$ M). The functional high-affinity IL-3R ($K_d = 3 \times 10^{-10}$ M) contains an additional subunit, cloned as SUT-1 cDNA. It encodes 396 amino acids with a single CRH domain in the extracellular portion and a short cytoplasmic region. This molecule is presently known as IL-3Rα, a 70 kDa glycoprotein that binds IL-3 with low affinity ($K_d = 4 \times 10^{-8}$ M), but is unable to mediate a biological response, even at high concentrations of growth factor. After binding of IL-3 it associates not only with AIC2A, the β-chain exclusive to the murine IL-3R, but also with the subsequently cloned AIC2B protein. This subunit has been termed 'common β-chain' (βc) because it is shared by the high-affinity receptors for IL-3, GM-CSF and IL-5 without binding any of these ligands by itself.

There is only one human molecule (KH97) homologous to AIC2, a 130 kDa protein which shows 56% identity with the AIC2A and AIC2B proteins. It was initially identified as a second subunit for the human GM-CSFR, but turned out to be a component of both IL-3R and IL-5R. In terms of its function, human β-chain resembles AIC2B, inasmuch as it cannot bind any of its ligands in the absence of the specific α-chain. Human cells express only the high-affinity IL-3R, suggesting that the α-chain has either weak or no binding capacity for its ligand. Kitamura *et al.* have cloned the human IL-3Rα chain from COS cells transfected with KH97 cDNA, together with cDNA library pools constructed from IL-3R-expressing cells. The cDNA that reconstituted the high-affinity IL-3-binding activity encoded a 378 amino acid protein containing motifs typical of class I cytokine receptors in the extracellular domain. This 70 kDa molecule is structurally similar to its murine counterpart, but there is only 30% homology at the amino acid level. Both human and murine IL-3Rα contain a proline-rich Box 1 domain, which is required for signal transduction. The affinity of human IL-3Rα for its ligand is very low ($K_d = 10^{-7}$ M), but co-expression with the β-subunit reconstitutes the high-affinity receptor ($K_d = 10^{-10}$ M).

The GM-CSFR α-chain was cloned from human placental membrane by expression cloning. Its cDNA encodes a 400 amino acid class I cytokine receptor structurally related to the IL-3Rα. The mature protein, of about 80 kDa, binds GM-CSF with low affinity ($K_d = 5 \times 10^{-9}$ M) but does not mediate any biological response. The structure of murine GM-CSFRα is similar to its human counterpart, although only 35% of the amino acid sequence is identical. Soluble forms of the receptor have been identified. The high-affinity GM-CSFR ($K_d = 10^{-11}$ M) has a unique βc chain, namely AIC2B in murine and KH97 in human.

The murine low-affinity component of the IL-5R was obtained by expression cloning, using a monoclonal antibody (mAb) against the receptor. It is composed of 415 amino acids and contains short extracellular and cytoplasmic domains that are highly homologous to the α-chains of IL-3 and GM-CSF receptors. The proline-rich Box 1 is particularly well conserved. AIC2B associates with the low-affinity subunit to form the signal-transducing complex.

The human IL-5Rα exhibits 70% identity with its murine counterpart. It seems that the high-affinity receptor uses the βc chain mainly as a functional subunit since it apparently does not contribute to the ligand binding by the human receptor. Indeed, co-expression of the α- and β-chains enhances the binding affinity for IL-5 only a few times, compared with the 50- to 100-fold increase observed under similar circumstances for IL-3 or GM-CSF.

Both immature progenitors and committed differentiated haematopoietic cells express the βc receptor subunit. In contrast, it is not present on T cells and most B cells. Cytokines, such as IL-1, TNFα and IFNγ, induce βc expression. The cellular distribution of the specific α-chains reflects the biological activities of each growth factor. The IL-

$3R\alpha$ is expressed most widely on both immature and lineage-restricted progenitors, while the GM CSFRα is associated more particularly with macrophage/monocyte and granulocyte lineages and IL-5Rα expression is limited to eosinophils, basophils and B cells. The βc subunit is not present on the most primitive stem cells or early blast cells, but the α components of IL-3 and GM-CSF receptors are already displayed on early embryonic stem cells before blood formation.

Though the detailed mechanisms regulating receptor expression are not known, it is evident that the control is exerted essentially on transcriptional levels in a differentiation stage specific manner. This does not exclude the possibility that post-transcriptional events are also involved in modulating this process. The promoter regions of the βc, IL-3Rα and GM-CSFRα chains are quite distinct. The only shared feature is the presence of PU.1 and GATA motifs characteristic of haematopoietic cells. PU.1 is a member of the Ets family of transcription factors, and is critical for the development of lymphoid and myeloid cells. Its disruption results in major multilineage defects in haematopoiesis. PU.1 is not required for the commitment to myeloid differentiation, as assessed by the expression of early myeloid genes like G-CSFR and myeloperoxidase in Pu.1$^{-/-}$ cells. It seems, however, to be regulated by genes associated with terminal myeloid differentiation, like the genes encoding M-CSFR and CD11b.

4.2 Single-chain cytokine receptors

This family which comprises the receptors for Epo, Tpo and G-CSF is not composed of functionally or structurally conserved molecules. However, there is some homology, especially between the EpoR and the TpoR. One of the typical features of this group is a restricted expression, as well as a high functional specificity within a particular cell lineage.

The murine G-CSFR cDNA encodes a 812 amino acid transmembrane glycoprotein, whose extracellular region is composed of 601 amino acids. The N-terminus contains an immunoglobulin domain, followed by the CRH domain with four conserved cysteines and the WSXWS motif typical of the class I cytokine receptor family. The cytoplasmic region of the G-CSFR contains the characteristic conserved Box 1 and Box 2 domains, as well as a Box 3 domain requisite for G-CSF-induced differentiation signals.

The human G-CSFR shows 63% homology with its murine counterpart. In addition to its predominant transcript *t*. Three alternatively spliced forms

have been detected. They encode a soluble receptor, a variant with a 27 amino acid insertion in the cytoplasmic region and a molecule with an altered C-terminus, lacking the domain required for the differentiation signal.

Mutational analyses of the G-CSFR have provided some interesting clues to the mechanisms enabling a unique receptor to transduce both proliferation and differentiation signals. They have revealed that the membrane-proximal Box 1 and Box 2 domains are required for G-CSF target cells to grow, while the C-terminal 120 amino acids are responsible for ligand-induced neutrophilic differentiation, as judged by the expression of myeloperoxidase and elastase genes.

The receptors for Tpo and Epo are strongly homologous to each other and to the IL-3Rβ. The murine EpoR was obtained by expression cloning from eythroleukaemia MEL cells. It encodes a 507 amino acid protein that is 82% identical to its human counterpart and is therefore not species-specific. The membrane-proximal 127 amino acids are required for mitogenic signalling and β-globin synthesis in response to Epo. The last 40 amino acids of the C-terminus constitute a negative regulatory domain, as assessed by the enhanced Epo-induced mitogenesis observed after deletion of this region. This notion is further corroborated by the finding that humans with dominantly inherited erythrocytosis (characterized by increased haematocrit) have a truncated Epo receptor, whose C-terminus lacks a 70 amino acid sequence.

The cellular responses to Epo are determined by intracellular regulators whose presence depends on the differentiation stage of the cell. It is not currently known whether these molecules are transcription factors, signalling protein(s) or additional receptor subunits. The data available so far support the notion that the EpoR does not signal commitment to the erythroid lineage, but regulates late stages of maturation and proliferation in cells already engaged in this differentiation scheme.

Like several other cytokine receptors, the EpoR has different well defined splice isoforms. Human bone marrow erythroid precursor cells express, in addition to the wild-type receptor, a truncated molecule without the cytoplasmic regions required for a mitogenic response. The latter acts in a dominant-negative fashion, inhibiting normal receptor function. It is chiefly expressed in the most primitive cells and is replaced by the wild-type receptor during differentiation. The differential expression of alternatively spliced variants may represent a physiological means of regulating cytokine responses in target cells. The mechanisms that regulate the expression of the different EpoR isoforms have not yet been elucidated.

In contrast with the G-CSFR (whose membrane-proximal domain provides mitogenic signals, while the C-terminus is required for differentiation), growth and erythroid-specific gene expression seem not to be determined by distinct portions of the EpoR. It has been shown that chimeric receptors, constructed from the extracellular portion of the epidermal growth factor receptor (EGFR) and a truncated cytoplasmic EpoR including only the membrane-proximal 127 amino acids, can mediate both proliferation and differentiation signals. Others have reported that receptor constructs from the extracellular portion of the EpoR and the cytoplasmic region of IL-3Rβ or IL-2Rβ can signal proliferation and erythroid-specific gene expression, suggesting that the specificity is determined by the extracellular domain of the receptor. One possible explanation for this discrepancy is that the EpoR contains additional subunit(s) interacting with the extracellular region of the receptor.

The hypothesis that the EpoR comprises more than one chain is consistent with the evidence for Epo-binding proteins of distinct molecular weights (66, 85–95 and 100–110 kDa) after cross-linking with radiolabelled Epo. Transfection of the EpoR into fibroblasts and COS cells results only in the expression of the 66 kDa protein which confers Epo responsiveness. The nature of the other two putative receptor subunits is not known.

The receptor for Tpo was identified, and called c-Mpl, before its specific ligand was known. c-Mpl is the cellular counterpart of the retroviral protein v-Mpl, encoded by the oncogene v-*mpl* in the genome of myeloproliferative leukaemia virus (MPLV). It is a truncated molecule lacking a large portion of the extracellular domain of the receptor which is reduced to a 43 amino acid sequence adjacent to the membrane. Infection of adult mice with MPLV results in severe haematopoietic disorders, with a striking expansion of the myeloid compartment. Only haematopoietic cells transform in response to the oncogene, suggesting that the transforming activity depends on the presence of haematopoietic cell-specific proteins.

The TpoR is most homologous to the EpoR and to the IL-3Rβ subunit. Its CRH domain is present as a tandem repeat. The cytoplasmic domain of the receptor contains the Box 1 and Box 2 sequences characteristic of class I cytokine receptors. They are necessary for the cellular proliferation and transforming activity induced by v-*mpl*. The 20 C-terminal amino acids of the cytoplasmic receptor domain are required for c-Fos expression.

The M-CSFR belongs to the immunoglobulin receptor superfamily (or RTK family), which comprises c-Kit, Flk2/Flt3 and the receptors for EGF, platelet-derived growth factor (PDGF), insulin, and insulin-like growth factor (IGF). The cloning of the M-CSFR gene was facilitated by the identification of its viral homologue, v-*fms*, which encodes the transforming protein of the SM and HZ5 feline sarcoma viruses. Its unique gene encodes a 512 amino acid transmembrane protein with five extracellular immunoglobulin domains. The binding site for M-CSF involves the first three of these domains (D1–D3), while the fourth is required for receptor dimerization which, in turn, activates the tyrosine kinase present in the cytoplasmic portion of the receptor.

5. Signal transduction by haematopoietic growth factor receptors

This chapter will not provide an exhaustive account of the complex mechanisms involved in cytokine signalling (see Overview and Chapter XII on mechanisms of signal transduction for more information), but convey some general notions applying to haematopoietic growth factors. In all cases, ligand-induced receptor dimerization is the initial event triggering the signalling process. It enables the cytoplasmic receptor domains to interact, thus initiating downstream signalling cascades. The first step in this process is the rapid induction of tyrosine phosphorylation of several cellular proteins. In the case of M-CSF the cytoplasmic domain of the receptor has its own catalytic activity and undergoes intermolecular autophosphorylation upon dimerization. All other haematopoietic growth factor receptors are associated with cytoplasmic tyrosine kinases which couple ligand-induced receptor activation to intracellular signalling events.

Haematopoietin receptors are associated with the Jak (Janus kinase) family of protein kinases which have also been claimed to be involved in the signalling of M-CSF, as assessed by its capacity to activate Tyk2 or Jak1. Each haematopoietin receptor activates one or more Jak(s) specifically (G-CSF, Epo and Tpo activate Jak2, while IL-3, IL-5 and GM-CSF activate both Jak1 and Jak2). The exact mechanism by which cytokine receptor oligomerization activates Jaks is not known, but it is plausible that the associated conformational changes bring these molecules into sufficiently close contact to enable them to autophosphorylate and activate their respective kinase(s). The signal is then propagated by recruiting other signalling proteins through phosphorylation.

Association and activation of Jaks require the membrane-proximal region of haematopoietin receptors. Jak2 associates with single-chain receptors and with the βc chain of IL-3R, IL-5R and GM-CSFR through Box 1 motifs. It is not yet known whether the proline residues in this sequence are necessary for the interaction itself or for the correct folding of the binding site. The structural Jak domains responsible for this association are not precisely defined, but tyrosine phosphorylation seems not to be required. Indeed, Jaks are constitutively, though weakly, associated with the receptor in the absence of ligand binding. They do not contain Src homolgy 2 (SH2) or phosphotyrosine interaction (PI) domains and there is no phosphorylation site in the Box 1 motif.

In haematopoietin receptor signalling, Jaks appear to be the principal activators of a family of transcription factors termed signal transducers and activators of transcription (Stats). These molecules were first identified in studies of IFN-regulated gene transcription, and their relationship with Jaks was demonstrated by the use of Jak-deficient cell lines and of receptor mutants with defective Stat activation. It has also been shown that Jak kinases are capable of phosphorylating Stats in the correct site and activate their DNA binding activity *in vitro*.

Experiments with knockout mice suggest that each Stat has unique regulatory functions. Furthermore, the patterns of Stat activation used by different haematopoietin receptors are not identical and they change according to the differentiation stage of the cell the receptor is expressed in. For example, in immature myeloid cells IL-3 and GM-CSF activate Stat5a and Stat5b, which are replaced by related 94 and 96 kDa polypeptides during maturation.

It is not entirely clear what determines the specificity of Stats. It is certainly not conferred by Jaks, because there is no correlation between the activation of one Jak and a particular set of Stats. The hypothesis that tyrosine-phosphorylated sequence elements in the receptor are responsible for specific Stat recruitment is most plausible. It is consistent with the requirement of SH2 domains for Stat activation and the identification of distinct tyrosine-containing docking sites for Stats on the receptor.

The Jak–Stat pathway provides haematopoietic cells with a fast phosphorylation route communicating with the nucleus. However, cytokine-induced specific gene regulation is far more complex and requires the coordinated action of several transcription factors. The response to Stats probably occurs in a biphasic manner: an initial phase of propagation in the nucleus, through recognition and activation of promoter sites and a second phase entailing interactions with other transcription factors in the control of specific cytokine-induced late response genes.

Although members of the Jak family are undoubtedly central mediators for haematopoietin receptor signalling, several other tyrosine kinases become activated upon stimulation with haematopoietic growth factors, namely the proteins of the Src family. The receptors themselves become tyrosine-phosphorylated once they have bound their ligand, and most of their downstream signalling proteins contain SH2 or PI domains allowing them to recognize specific phosphotyrosine motifs in the receptor.

Several cellular proteins and signalling pathways are triggered by haematopoietin receptor stimulation, including phosphatidyl inositol 3-kinase, insulin receptor substrates (IRS) -1 and -2, Vav, phospholipase Cγ, several serine/threonine kinases and the Ras–MAP kinase pathway. These pathways are likewise used by other receptor families, such as the immunoglobulin family to which M-CSF belongs. They are discussed in some detail in the Overview (p. 1).

It is important to note that signalling cascades in haematopoietic cells are not linear but communicate with each other. This cross-talk is made up of positive and negative elements. The latter comprise a large number of phosphotyrosine phosphatases whose complex regulatory functions on the cytokine signalling cascade have only just begun to be explored. The importance of this type of molecules is illustrated by the naturally occurring *motheaten* mutation due to haematopoietic cell phosphatase (HCP) deficiency. This mutation is lethal in its homozygous form and results in overproliferation and activation of several haematopoietic lineages in viable heterozygous mice.

More recently, the search for cytokine-induced genes led to the identification of a new family of suppressors of cytokine signalling. These SH2-containing mediators have been cloned, and their products named CIS (cytokine-inducible SH2 molecule), SOCS (suppressor of cytokine signal), SSI-1 (Stat-induced Stat inhibitor-1) and JAB (Jak2-binding protein), and might be partially identical. They represent a new family of factors whose principal function seems to consist in turning off cytokine signalling by blocking the action of specific protein kinases (certain Jaks or Src proteins) or of some Stats.

6. Biological activities of haematopoietic growth factors

All haematopoietic growth factors share at least two complementary functions: they are required for growth and differentiation of all types of blood cells, and they modulate the biological activities of their mature progeny. They are further characterized by a certain degree of functional redundancy and pleiotropy, two features that have complicated the terminology in this research field. To exemplify this point, IL-3 has turned out to be identical to about 20 other previously described factors or activities, including mast cell growth factor, persisting cell-stimulating factor, histamine-producing cell stimulating factor, burst promoting activity, CFU-S-stimulating activity, Thy-1-inducing factor and multi-CSF.

The availability of genetically modified mice with one or more specifically disrupted haematopoietic growth factor or receptor gene(s) has considerably changed our understanding of the physiological significance of haematopoietic growth factors. It has underscored their functional redundancy, leading to a re-evaluation of their actual role during constitutive and inducible haematopoiesis (Table 3).

6.1 IL-3

IL-3 was initially identified as a T-cell-derived factor that stimulates the production of 20α-hydroxy-steroid dehydrogenase in lymphocytes from nude mice. This feature led to the assumption that it was chiefly a T-cell differentiation factor, till it became evident that this enzymatic activity was not T-cell-specific. In semi-solid culture conditions, IL-3 promotes the formation of colonies composed of all myeloid cell lines (CFU-Mix, CFU-GM, CFU-G, CFU-M, CFU-Eo, CFU-Baso, CFU-Mast and CFU-Meg). It can act on even earlier pluripotent progenitors when combined with cytokines like IL-6, IL-11, LIF, G-CSF, M-CSF or the ligands for c-Kit (KL, or SCF) or Flt-3, whose effect is probably mediated through a shortening of the G_0 phase of the cell cycle or IL-3R induction. IL-3 acts also on early erythroid progenitors, rendering them responsive to Epo for complete maturation.

Furthermore, it promotes mast cell and basophil differentiation, respectively, from murine and human progenitors in liquid culture. This latter activity, as well as the capacity to trigger quiescent multipotent progenitors (CFU-S) into cell cycle, is IL-3-specific.

In addition to its effect on 20α-hydroxysteroid dehydrogenase, IL-3 stimulates several other enzymatic activities in haematopoietic precursor cells, such as arginase, ornithine decarboxylase and histidine decarboxylase, and induces the synthesis of IL-4 and IL-6. It modulates the functional capacities of macrophages and their expression of surface antigens, like CD11a/CD18 and MHC class II molecules and prolongs the survival of haematopoietic cells.

The characterization of IL-3 as a predominantly myeloid growth factor has been confirmed *in vivo*. When injected into mice, it elicits a striking increase in extramedullary myelopoiesis and promotes the mobilization of bone marrow progenitors. Yet, the increase in peripheral blood white cell counts remains relatively discrete, as compared with that promoted by more lineage-restricted growth factors, like GM-CSF and G-CSF. This observation suggests that the chief function of IL-3 consists in stimulating early haematopoietic progenitor cells, rendering them responsive to more specific growth factors for terminal differentiation.

In spite of its striking effects *in vivo*, IL-3 does not appear to play an essential role during constitutive haematopoiesis. Indeed, the factor seems to be exclusively produced by activated T cells and is normally detected neither in the circulation nor in fetal or stromal tissues. No significant haematopoietic changes could be demonstrated in naturally occurring murine strains with low IL-3α expression and in mice transgenic for antisense IL-3 (Table 3). It is likely that multi-CSF comes into play exclusively during the immune response, when the expansion of haematopoietic progenitor cells is required for generating enough cells to guarantee the host defence.

6.2 GM-CSF

The biological activities GM-CSF and IL-3 overlapping to some extent, at least as far as their effect on survival, proliferation and differentiation of granulocyte/macrophage precursors is concerned. The action of GM-CSF is more restricted to these particular myeloid lineages, though it also affects early erythroid progenitors and megakaryocytopoiesis. It determines the functional charac-

■ **Table 3.** Main consequences of haematopoietic growth factor or receptor gene inactivation

	Mode of inactivation	Haematopoietic changes and other prominent effects
IL-3	Transgenesis (IL-3 anti-sense)	— Partial IL-3 deficiency associated with pre-B lymphoproliferation, evidence for a cerebellar syndrom
	Gene mutation resulting in low IL-3Rα expression	— Normal haematopoietic progenitor frequencies
GM-CSF	GM-CSF–/–	— Normal haematopoiesis
		— Normal progenitor frequencies in haematopoietic organs
		— Lung pathology with alveolar proteinosis, accumulation of surfactant and peribronchal lymphoid infiltration
	βc(AlC2B)–/–	— Lung pathology, same as GM-CSF–/– mice
G-CSF	G-CSF–/–	— Chronic neutropenia
		— Circulating neutrophils: 20–30% of normal
		— Decreased myeloid progenitor frequencies in bone marrow and spleen
		— Increased susceptibility to bacterial infection
M-CSF	Spontaneous *op/op* mutation	— Decreased haematopoietic progenitor frequencies in the bone marrow, compensated by extramedullary haematopoiesis
		— Osteopetrosis
		— 86–95% decrease in peripheral macrophages
		— 65–70% decrease in spleen, thymus and liver macrophages
		— Impaired fertility in females
		— Age-dependent compensation of deficiencies
IL-5	βc(AlC2B)–/–	— Decreased basal and infection-induced circulating eosinophil levels
Epo	Epo–/– or EpoR–/–	— Lack of erythrocytes, normal erythroid progenitor frequencies
Tpo	Tpo–/– or TpoR–/–	— Important decrease in megakaryocytes and circulating platelets (10–15% of normal)
		— Decreased haematopoietic progenitor frequencies

teristics of monocytes and macrophages by promoting their cytotoxicity against tumour cells, microorganisms and parasites, by increasing their antigen-presenting capacities, as well as adherence and phagocytosis. GM-CSF stimulates antibody-dependent cytotoxicity (ADCC) in both monocytes/macrophages and neutrophils, and induces the production of IL-1, G-CSF and M-CSF by these two populations. Furthermore, it is a potent stimulus for prostaglandin, IL-6 and TNF production by monocytes/macrophages and is a chemotactic agent for neutrophils.

Outside the neutrophil and monocyte/macrophage compartment, GM-CSF is an activator of eosinophils and synergizes with IL-1 and IL-7 on thymocyte proliferation. Together with IL-4, it promotes dendritic cell differentiation from haematopoietic progenitor cells and monocytes *in vitro* and increases the antigen-presenting capacities of Langerhans cells. It has also been reported that the latter population can be induced from monocytes,

provided that TGFβ is added during culture with GM-CSF and IL-4.

In vivo, GM-CSF shows the biological properties expected from *in vitro* studies. Yet, in spite of the increased granulocyte and macrophage production after GM-CSF injection, this growth factor appears to play no irreplacable role in regulating basal granulocyte or monocyte levels. Indeed, genetically modified mice with a disrupted GM-CSF gene exhibit no overt haematopoietic abnormalities (Table 3). They do, however, suffer from alveolar proteinosis with surfactant accumulation, due to a defective functioning of alveolar macrophages. This finding suggests that GM-CSF plays effectively a unique role as a functional regulator of phagocytes, while its growth- and differentiation-promoting activity on granulocyte/macrophage progenitors is redundant. The identity of the compensatory factor(s) remains obscure, and there is no evidence for a rise of IL-3, G-CSF or M-CSF synthesis in GM-CSF knockout mice.

6.3 IL-5

IL-5 was first identified as T-cell-replacing factor (TRF) in antibody responses to thymus-dependent antigens. It is also involved in murine B-cell differentiation, but its effect in humans is a subject of controversy. Its main physiological function is most likely related to inflammatory and allergic reactions, mediated by basophils and eosinophils. IL-5 selectively stimulates eosinophil colony formation, increases the survival and activates the functional capacities of mature eosinophils. It is a chemotactic agent and promotes eosinophil degranulation, as assessed by the release of specific proteins, such as eosinophil-derived neurotoxin (EDN).

As expected from *in vitro* and *in vivo* experiments, constitutive IL-5 expression in transgenic mice results in a marked eosinophilia without other haematopoietic modifications. Disruption of the βc gene decreases the number of basal and parasite-induced circulating eosinophils.

6.4 G-CSF

G-CSF functions specifically to induce proliferation, differentiation and activation of neutrophilic granulocytes. It is critical both for maintaining basal levels of circulating neutrophils and for recruiting this population during infection. Like all other haematopoietic growth factors, G-CSF increases the survival as well as the functional capacities of its target cells, as estimated by enhanced ADCC, phagocytosis and oxidative burst. This activation is concomitant with increased expression of adhesion molecules, immunoglobulin receptors, alkaline phosphatase and arachidonic acid secretion and response to formyl-methionyl-leucyl-phenylalanine (fMLP).

G-CSF modulates the effect of other growth factors, e.g. IL-3 and GM-CSF, on early progenitor cells. *In vivo* treatment of humans and mice results in a striking mobilization of haematopoietic precursors and neutrophils into the circulation. Homozygous inactivation of the G-CSF gene is compatible with normal fetal and post-natal development, but adult mice show profound neutropenia, a reduced capacity to mobilize neutrophils on challenge, and increased susceptibility to infection. Haematopoietic progenitor frequencies are also diminished in these mice.

6.5 M-CSF

M-CSF is a major regulator for normal formation and function of most macrophage subsets, but it is not essential for all members of this lineage since in mice deficient for both M-CSF and GM-CSF, macrophages are still present in some tissues, such as the lung. They are, however, functionally deficient. The spontaneously occurring *op/op* mutation has been very useful for evaluating the biological functions of M-CSF. These osteopetrotic mice have skeletal abnormalities and low body weight. They are toothless, their fertility is impaired, and they exhibit a severe deficiency in macrophages and osteoclasts. *In vivo* administration of M-CSF corrects osteopetrosis and toothlessness, and restores normal numbers of spleen and bone marrow macrophages. It does not cure the deficiencies of pleural and peritoneal macrophages or the partial female sterility. *In vitro*, soluble M-CSF increases survival, migration, chemotaxis and spreading of macrophages, but reduces the resorbing activity of isolated osteoclasts, suggesting that other forms of M-CSF, or other cytokines acting in combination, are required for the physiological regulation of osteoclast activity.

6.6 Epo and Tpo

These two growth factors share several structural and functional characteristics and resemble hormones in their specificity and in that they are produced mainly away from their actual site of action. They express a certain degree of cross-reactivity, at least *in vitro*: the erythroid differentiation factor Epo affects megakaryocytopoiesis, while the megakaryocyte differentiation factor Tpo can modulate erythropoiesis.

Epo acts specifically on lineage-committed erythroid progenitors from the late BFU-E (burst-forming unit–erythroblast) differentiation stage onwards. These cells still depend on IL-3, IL-9 and GM-CSF for their differentiaton and/or proliferation and are relatively insensitive to Epo, but once they have become CFU-E (colony-forming unit–erythroblast), their final maturation is entirely controlled by Epo. The growth factor is responsible for the synthesis of haem and haemoglobin, and induces cell surface and cytoskeletal modifications, including all proteins required for the formation of red blood cells. Furthermore, Epo can accelerate the emigration of medullary reticulocytes into the periphery. The specific effect of Epo on lineage-restricted erythroid progenitors has been confirmed in genetically modified mice in which the Epo or EpoR gene has been disrupted. These mice have no erythrocytes, but normal erythroid progenitor frequencies.

Tpo is the primary regulator of megakaryocy-topoiesis. In its recombinant form it promotes the formation of very large mature polyploid mega-karyocytes from highly purified murine or human progenitor cells in liquid culture. These cells undergo a terminal maturation process, including the development of pro-platelets and shedding of platelet-like structures into the medium. It has been shown that the platelets derived from such cultures do not differ morphologically and functionally from their plasma-derived counterpart.

In addition to its direct effect on haematopoietic progenitors to give rise to megakaryocytes, Tpo interacts synergically or additively with early-and late-acting haematopoietic factors. In murine megakaryocytopoiesis, IL-11, SCF (KL) and Epo synergize with Tpo, while IL-3 and IL-6 have an additive effect. Tpo affects other haematopoietic lineages as well. It has been shown to enhance BFU-E formation in human CD34$^+$ colony assays. This burst-promoting activity is comparable to that of GM-CSF and SCF. It increases both the number of colonies and their size, suggesting that both early and late erythroid progenitors are affected. Tpo stimulates erythropoiesis only in the presence of Epo and has no effect on normal myeloid colony growth *in vitro*.

Pharmacological doses of rTpo cause as much as a 10-fold increase in platelet levels when adminis-tered to mice or monkeys. This effect is preceded by increased CFU-Meg frequencies in bone marrow and spleen, suggesting that Tpo acts pri-marily on haematopoietic progenitors, rather than stimulating platelet formation from mature cells. The megakaryocytes from treated animals are larger than those from controls and show higher ploidy. Red or white blood cell counts are not modified by Tpo administration to mice, but BFU-E and CFU-GM are expanded and CFU-E are redistributed.

Despite its striking effect on platelet production, Tpo has only a modest effect on their functional capacities. *In vitro* studies have demonstrated that it can enhance agonist-induced platelet aggrega-tion, though on its own it has no effect on this activity. However, such a prothrombotic response has never been observed *in vivo*, and thrombotic episodes never occurred in these experiments, even when platelet levels were 4–10 times higher than normal.

Studies of mice rendered deficient for Tpo or its receptor by homologous recombination have confirmed the data resumed above. In both cases, the platelet count drops dramatically, to 10–15% of normal, while a similar decrease of megakaryocytes occurs in spleen and bone marrow. Furthermore, the megakaryocytes still present in knockout mice are smaller and have lower ploidy than those of control mice. The few remaining platelets appear structurally and functionally normal, and are sufficient to prevent haemorrhage. The genes and factors responsible for maintaining these basal megakaryocyte and platelet levels have not been identified. IL-6, IL-11 and SCF induce a modest increase in platelet formation when injected into Tpo- or TpoR-deficient mice and may thus be involved in the maintenance of baseline production.

In addition to its dramatic effect on megakary-ocyte and platelet formation, Tpo or TpoR gene disruption leads to a decreased incidence of haematopoietic progenitors of both erythroid and myeloid lineages. This finding suggests that Tpo acts on a very early pluripotent progenitor cell, in accor-dance with the early expression of c-Mpl on the primitive AA4$^+$Sca1$^+$ murine stem-cell population.

7. Clinical use of haematopoietic growth factors

The discovery of the important regulatory functions of cytokines in general and haematopoietic growth factors in particular has radically changed the way of looking at certain diseases. It has led quite rapidly to therapeutic assays; some haematopoietic growth factors are already in use and others are at various stages of clinical trials. Concomitantly with the administration of cytokines themselves, other strategies have been applied to modulate their biological activity. This approach has resulted in the development of specific antibodies, agonists or antagonists of cytokine receptors or fusion proteins combining the specific activities of several factors.

The growth factors with relatively specific bio-logical activities, like Epo and G-CSF, are the ones that have been most successfully employed in the clinical setting. Epo has found a major application in the treatment of patients whose Epo production is severely impaired by chronic kidney disease. Administration of Epo has greatly improved the quality of life of these patients who were previously dependent on blood transfusions. The most impor-tant complication encountered so far is the appear-ance or the aggravation of hypertension.

In uraemic patients on chronic maintenance haemodialysis, administration of rEpo improves platelet adhesion and aggregation, in addition to its effect on the haematocrit. The growth factor has also been used with success in non-renal forms of anaemia, consecutive to chronic infection, radiation therapy or treatment with cytostatic drugs. Presurgical activation of erythropoiesis, allowing the collection of autologous donor blood, represents another potential use of Epo.

Though Tpo was only discovered quite recently, it may soon be used therapeutically to correct defective megakaryocyte and platelet production. Early results from human clinical trials corroborate the data obtained in animal models. It has been shown that daily injections of a pegylated and truncated form of rTpo (MGDF) (over 10 days) to cancer patients before chemotherapy, increases circulating platelet levels up to four-fold. A similar increase occurred in patients given a single dose of full-length rTpo. Drug-related toxicities have not been reported and no thrombotic events were observed. In myelosuppressed patients, MGDF given after chemotherapy has been shown to reduce the extent of the platelet nadir. As seen in animal studies, bone marrow progenitors were expanded by administration of Tpo, suggesting that it may be useful as a priming agent, like IL-3.

The most important therapeutic application of G-CSF is probably the treatment of neutropenia following chemotherapy and/or radiotherapy. Continuous administration of this growth factor, for several days, increases peripheral neutrophil blood counts 10- to 15-fold, without affecting erythrocytes and platelets. This effect allows the doses of various anti-tumour drugs to be increased. The duration of neutropenia consecutive to chemotherapy or autologous bone marrow transplantation, is markedly decreased by G-CSF, provided that it is given daily at optimal doses. G-CSF is also useful for its capacity to increase and mobilize myeloid progenitors before harvest of bone marrow or blood cells for autologous transplantation.

Marked improvements in haematological parameters have been achieved in patients with congenital neutropenia, neutropenia caused by diffuse infiltration of the bone marrow by malignant non-Hodgkin lymphomas and idiopathic neutropenia. In all cases G-CSF is remarkably effective in reducing the episodes of severe infection and fever inherent to these diseases, warranting a lower dosage of antibiotics.

Like G-CSF, GM-CSF has been used to accelerate haematopoietic recovery in diseases characterized by a reduction in circulating neutrophils. It has been beneficial in attenuating the consequences of immunosuppression following HIV infection, chemotherapy, bone marrow transplantation and myelodysplasic syndrome. Administration of GM-CSF results in a marked increase of circulating granulocytes and monocytes that is maintained for some time after cessation of treatment. It offers the double advantage of accelerating haematopoietic recovery and enhancing the sensitivity to high-dose chemotherapy in acute myeloid leukaemia when given as a part of the anti-tumour regimen before and during administration of cytostatic drugs.

A number of phase III clinical trials evaluating the usefulness of GM-CSF in autologous bone marrow transplantation have been reported. They have confirmed the capacity of the growth factor to increase circulating neutrophil levels, thus shortening the period of hospitalization. Even in long-term treatments no severe side-effects have been observed. However, GM-CSF does not diminish the phase of absolute neutropenia and cannot substitute for platelet transfusion. It also has no effect on the survival rate, but seems to promote a significant decrease in the susceptibility to infections during the period of absolute neutropenia. In conclusion, GM-CSF already constitutes an important therapeutic tool in a number of clinical settings. Its range of applications is expected to widen even further, given its potential application as a vaccine adjuvant and the possibility of using it in conjunction with other cytokines and haematopoietic growth factors, such as IL-3.

The therapeutic possibilities of IL-3 lie chiefly in its capacity to stimulate early progenitors, while its effect on circulating granulocyte levels is less pronounced than that of GM-CSF and G-CSF. It might therefore be useful in a variety of haematopoietic insufficiencies, essentially as a priming agent, rendering primitive cells responsive to more lineage-restricted growth factors given in a combined treatment.

There are fewer data on the clinical applications of IL-5 and M-CSF. Because of the predominant effect of IL-5 on eosinophilopoiesis, treatment of nematode-induced eosinophilia by anti-IL-5 injection seems the most promising application. As far as M-CSF is concerned, its obvious clinical relevance is the correction of impaired haematopoiesis, where it could be employed most efficiently in combination with other haematopoietic growth factors. An interesting therapeutic prospect is also provided by the capacity of macrophages to recognize and kill tumour cells transfected with the membrane form of M-CSF. The examples of clinical applications mentioned are of course not exhaustive, and there are probably many other clinical settings in which these factors will be useful in future, once their biological functions and interactions are better understood.

8. Concluding remarks

In this chapter we have reviewed the principal biochemical and biological features of haematopoietic growth factors. We have summarized the structural and functional characteristics of their specific receptors which generate the signalling cascade that leads from the membrane to the nucleus. Our knowledge of the complex cross-talk between signal-transducing elements and nuclear transcription factors that gives rise to growth and differentiation, remains fragmentary. Nevertheless, haematopoietic growth factors have already been successfully employed in the clinical setting, with few deleterious side-effects as compared with other cytokines. The scope of therapeutic applications is expected to broaden in the near future, and will probably include treatments in which the advantages of several cytokines are combined for a more efficient effect.

References

Adamson, J.W. and Eschbach, J.W. (1990). Treatment of the anemia of chronic renal failure with recombinant human erythropoietin. *Annu. Rev. Med.* **41**, 349–60.

Bartley, T.D., Bogenberger, J., Hunt, P., Li, Y.S., Martin, F., Chang, M.S., Samal, B., Nichol, J.L., Swift, S. *et al.* (1994). Identification and cloning of a megacaryocyte growth and development factor that is a ligand for the cytokine receptor Mpl. *Cell* **77**, 1117–24.

Danzig, M. and Cuss, F. (1997). Inhibition of interleukin-5 with a monoclonal antibody attenuates allergic inflammation. *Allergy* **52**, 787–94.

Dent, L.A., Strath, M., Mellor, A.L. and Sanderson, C.J. (1990). Eosinophilia in transgenic mice expressing interleukin 5. *J. Exp. Med.* **172**, 1425–31.

Estey, E. (1998). Haematopoietic growth factors in the treatment of acute leukemia. *Curr. Opin. Oncol.* **10**, 23–30.

Fixe, P. and Praloran, V. (1997). Macrophage colony-stimulating factor (M-CSF or CSF-1) and its receptor: structure–function relationships. *Eur. Cytokine Netw.* **8**, 125–36.

Fukunaga, R., Ishisaka-Ikeda, E. and Nagata, S. (1993). Growth and differentiation signals mediated by different regions in the cytoplasmic domain of granulocyte colony-stimulating factor receptor. *Cell* **74**, 1079–87.

Goodnought, L.T., Monk, T.G. and Andriole, G.L. (1997). Erythropoietin therapy. *New Engl. J. Med.* **336**, 933–8.

Harrington, M., Konicek, B.W., Xia, X.L. and Song, A. (1997). Transcriptional regulation of the mouse CSF-1 gene. *Mol. Reprod. Dev.* **46**, 39–44.

Ihle, J.N., Keller, J., Orozlan, S., Henderson, L.E., Copeland, T.D., Fitch, F., Prystovsky, M.B., Goldwasser, E., Schrader, J.W., Palaszynski, E. *et al.* (1983). Biologic properties of homogeneous interleukin 3. I: Demonstration of WEHI-3 growth factor activity, mast cell growth factor activity, P cell-stimulating activity, colony-stimulating factor activity and histamine-producing cell stimulating factor activity. *J. Immunol.* **131**, 282–7.

Ihle, J.N., Witthuhn, B.A., Quelle, F.W., Yamamoto, K. and Silvennoien, O. (1995). Signaling through the haematopoietic cytokine receptors. *Annu. Rev. Immunol.* **13**, 369–98.

Jacobs, K., Shoemaker, C., Rudersdorf, R., Neill, S.D., Kaufman, R.J., Mufson, A., Seehra, J., Jones, S.S., Hewick, R., Fritsch, E.F. *et al.* (1985). Isolation and characterization of genomic and cDNA clones of human erythropoietin. *Nature* **313**, 806–10.

Kaushansky, K., Lok, S., Holly, R.D., Broudy, V.C., Lin, N., Bailey, M.C., Forstrom, J.W., Buddle, M.M., Oort, P.J., Hagen, F.S. *et al.* (1994) Promotion of megakaryocyte progenitor expansion and differentiation by the c-Mpl ligand thrombopoietin. *Nature* **369**, 568–71.

Kopf, M., Le Gros, G. Coyle, A.J., Kosco-Vilbois, M. and Brombacher, F. (1995). Immune responses of IL-4, IL-5, IL-6 deficient mice. *Immunol. Rev.* **148**, 45–69.

Leary, A.G., Yang, Y.C., Clark, S.C., Gasson, J.C., Golde, D.W and Ogawa, M. (1987). Recombinant gibbon interleukin-3 supports formation of human multilineage colonies and blast cell colonies in culture: comparison with recombinant human granulocyte-macrophage colony stimulating factor. *Blood* **70**, 1343–8.

Lieschke, G.J., Grail, D., Hodgson, G., Metcalf, D., Stanley, E., Cheers, C., Fowler, K.J., Basu, S., Zhan, Y.F. and Dunn, A.R. (1994). Mice lacking granulocyte colony-stimulating factor have chronic neutropenia, granulocyte and macrophage progenitor cell deficiency, and impaired neutrophil mobilization. *Blood* **84**, 1737–46.

Mach, N., Lantz, C.S., Galli, S.J., Reznikoff, G., Mihm, M., Small, C., Granstein, R., Beissert, S., Sadelain, M., Mulligan, R.C and Drnoff, G. (1998). Involvement of interleukin-3 in delayed-type hypersensitivity. *Blood* **91**, 778–83.

Maxwell, P.H. and Ratcliffe, P.J. (1996). The erythropoietin-producing cells. *Exp. Nephrol.* **4**, 309–13.

Metcalf, D. (1995). The granulocyte–macrophage regulators: reappraisal by gene inactivation. *Exp. Haematol.* **23**, 569–72.

Miyatake, S., Otsuka, T., Yokota, F., Lee, F. and Arai, K. (1985). Structure of the chromosomal gene for granulocyte macrophage stimulating factor: comparison of the mouse and human genes. *EMBO J.* **4**, 2561–8.

Murone, M.M., Carpenter, D.A. and de Sauvage, F.J. (1998). Haematopoietic deficiencies in c-mpl and TPO knock-out mice. *Stem Cells* **16**, 1–6.

Nagata, S., Tsuchiya, M., Asano, S., Yamamoto, O., Hirata, Y., Kubota, N., Oheda, M., Nomura, H. and Yamazaki, T. (1986). The chromosomal gene structure and two mRNAs for human granulocyte colony stimulating factor. *EMBO J.* **5**, 575–81.

Nemumaitis, J. (1997). A comparative review of colony-stimulating factors. *Drugs* **54**, 709–29.

Nicolas, N.A. and Metcalf, D. (1991). Subunit promiscuity among haematopoietic growth factor receptors. *Cell* **67**, 1–4.

Nimer, S.D. and Uchida, H. (1995). Regulation of granulocyte–macrophage colony-stimulating factor and interleukin 3 expression. *Stem Cells* **13**, 324–35.

Pitrak, D.L. (1997). Effects of granulocyte colony-stimulating factor and granulocyte–macrophage colony-stimulating factor on the bacterial function of neutrophils. *Curr. Opin. Haematol.* **4**, 183–90.

Ratcliffe, P.J., Ebert, B.L., Firth, J.D., Gleadle, M., Maxwell, P.H., Nagao, M., O'Rourke, J.F., Pugh, C.W. and Wood, S.M. (1997). Oxygen regulated gene expression: erythropoietin as a model system. *Kidney Int.* **51**, 514–26.

Robb, L., Drinkwater, C.C., Metcalf, D., Li, R., Kontgen, F., Nicola, N.A. and Begley, C.G. (1995). Haematopoietic and lung abnormalities in mice with null mutation of the common β subunit of the receptors for granulocyte-macrophage colony-stimulating factor and interleukins 3 and 5. *Proc. Natl Acad. Sci. USA* **92**, 9565–9.

Schneider, E., Pollard, H., Lepault, F., Guy-Grand, D., Minkowski, M. and Dy, M. (1987). Interleukin 3 and granulocyte–macrophage colony-stimulating factor induce *de novo* synthesis of histidine decarboxylase in haematopoietic progenitor cells. *J. Immunol.* **139**, 3710–17.

Souza, L.M., Boone, T.C., Gabrilove, J., Lai, P.H., Zsebo, K.M., Murdock, D.C., Chazin, V.R., Bruszewski, J., Lu, H., Chen, K.K. *et al.* (1986). Recombinant human granulocyte colony stimulating factor: effects on normal and leukemic myeloid cells. *Science* **232**, 61–5.

Spyridonidis, A., Mertelsmann, R. and Henschler, R. (1996). Haematopoietic cell proliferation and differentiation. *Curr. Opin. Haematol.* **3**, 321–8.

Stanley, E.R., Berg, K.L., Einstein, D.B., Lee, P.S., Pixley, F.J., Wang, Y. and Yeung, Y.G. (1997). Biology and action of colony-stimulating factor-1. *Mol. Reprod. Dev.* **46**, 4–10.

Takatsu, K., Takaki, S. and Hitoshi, Y. (1994). Interleukin-5 and its receptor system: implications in the immune system and inflammation. *Adv. Immunol.* **57**, 145–90.

Wendling, F. and Vainchenker, W. (1995). Thrombopoietin and its receptor, the proto-oncogene c-mpl. *Curr. Opin. Haematol.* **2**, 331–8.

Wendling, F., Maraskovsky, E., Debili, N., Florindo, C., Teepe, M., Titeux, M., Methia, N., Breton-Gorius, J., Cosman, D. and Vainchenker, W. (1994). c-Mpl ligand is a humoral regulator of megakaryocytopoiesis. *Nature* **369**, 571–4.

Wong, G.G., Witek, J.S., Temple, P.A., Wilkens, K.M., Leary, A.C., Luxenberg, D.P., Jones, S.S., Brown, E.L., Kay, R.M., Orr, B.C. *et al.* (1985). Human GM-CSF: molecular cloning of the complementary DNA and purification of the natural and recombinant proteins. *Science* **228**, 810–15.

Wu, H., Liu, X., Jaenisch, R. and Lodish, H.F. (1995). Generation of committed erythroid BFU-E and CFU-E progenitors does not require erythropoietin or erythropoietin receptor. *Cell* **83**, 59–67.

Yang, Y.C., Ciarletta, A.B., Temple, P.A., Chung, M.P., Kovacic, S., Witek-Giannotti J.S., Leary, A.C., Kriz, R., Donahue, R.E., Wong, G.G. *et al.* (1986). Human IL-3 (multi-CSF): Identification by expression cloning of a novel haematopoietic growth factor related to murine IL-3. *Cell* **47**, 3–10.

Yoshida, H., Hayashi, S.I., Kunisada, T., Ogawa, M., Nishikawa, S., Okamura, H., Sudo, T., Shultz, L.D. and Nishikawa, S. (1990). The murine mutation osteopetrosis is in the coding region of the macrophage colony stimulating factor gene. *Nature* **345**, 442–4.

XI Interleukins 16, 17 and 18

S. Chouaib, J.-Y Blay, J. Chehimi and J. D. Marshall

1. Introduction

Recently, new cytokines involved in immune responses have been identified and characterized. This chapter brings together the latest knowledge about interleukins 16, 17 and 18.

2. Interleukin-16

Interleukin 16 (IL-16) was originally called lymphocyte chemoattractant factor (LCF) because of its ability to promote the migration of CD4+ T cells. Its lymphocyte chemoattractant activity was first described in bronchoalveolar lavage fluid obtained from asthmatics after antigen challenge. IL-16 was first identified as a product of CD8+ T cells but CD4+ T cells, bronchial epithelial cells and eosinophils have been reported to be additional sources. Eosinophils constitutively express mRNA transcripts for both IL-16 and RANTES detectable by reverse transcription–PCR (RT-PCR), and contain preformed IL-16 and RANTES demonstrable by enzyme-linked immunosorbent assay (ELISA) of cell lysates of fresly isolated eosinophils. This suggests that these cells could contribute cytokines to enhance the recruitment of additional populations of CD4+ lymphocytes and eosinophils.

IL-16 appears in culture supernatants as a biologically active, non-covalently linked tetramer with an Mr of →56 000 Da. Monomeric peptides are inactive but reaggregate to M_r 56 000, regaining biological activity. The biological activity of IL-16 resides in the C-terminal 114 residues which are completely within the 130 residue sequence. There is no consensus signal sequence in pro-IL-16 and the secretion pathway for pro-IL-16 or its processed products is unknown. Zhang *et al.* have recently provided evidence that pro-IL-16 cleavage is mediated by caspase 3 and that cleavage by this enzyme releases biologically active IL-16 from its inactive precursor.

This cytokine has a unique cDNA which codes for a 130 amino-acid protein which is unrelated to previously described chemokines or other chemotactic/growth factor cytokines. The gene is located on chromosome 15.

In contrast to other interleukins, such as IL-15, IL-16 mRNA expression is almost exclusively limited to lymphatic tissues, underlining the potential of IL-16 as an immune-regulatory molecule.

It is well established that CD4 serves as receptor for the secreted form of IL-16. IL-16 binding to CD4 induces signal transduction, which affects the activation state of the cells. The natural CD4 ligand, IL-16, induces CD4+ T-cell unresponsiveness by a mechanism distinct from that of other CD4 ligands in that expression of CD4 and interleukin-2 receptor (IL-2R) was unaffected. The failure of IL-2 to restore proliferation suggests that the decrease in T-cell responsiveness induced by IL-16 may result from an interruption in the IL-2R signaling mechanism. Furthermore, it has been reported that the binding of IL-16 to CD4 results in activation of p56[lck] and reduces the magnitude of the anti-CD3-induced intracellular Ca^{2+} increase. It has been suggested that while the interaction of CD4 with IL-16 results in antigen-independent chemotaxis and IL-2R expression, this pro-inflammatory state is associated with subsequent transient inhibition of responsiveness via the CD3–T-cell receptor (TCR) complex.

There is evidence that IL-16 inhibits the replication of human immunodeficiency virus type 1 (HIV-1) and simian immunodeficiency virus (SIV) in infected peripheral blood mononuclear cells (PBMC). The binding of IL-16 to CD4 may indicate a role for IL-16 in suppression of HIV replication. It is also likely that IL-16 inhibits viral replication by competing for binding to CD4, but it may induce

CD4-dependent signal transduction events mediating repression of HIV promoter activity. In this context, it has been reported that IL-16 represses HIV promoter activity, prevents Tat activation and plays a role in HIV replication. Recently Bisset *et al.* have reported that, following initiation of indinavir-based therapy, significant decreases in plasma HIV-RNA levels were observed. Concomitantly, significant increase in IL-16 levels occurred. This further confirms the existence of an association between diminishing plasma HIV-RNA levels and the emergence of IL-16 which is capable of inhibiting HIV replication. Furthermore, it has been shown that human CD4⁺ cells transfected with IL-16 cDNA are resistant to HIV-1 infection and that such transfection results in a decrease in the level of HIV-1 transcripts, suggesting that IL-16-mediated inhibition of HIV-1 is not at the level of viral entry or reverse transcription, but at the level of mRNA expression.

To summarize, besides being a chemoattractant factor, IL-16 clearly functions as a modulator of T-cell activation and as an inhibitor of immunodeficiency virus replication.

3. Interleukin 17

Murine interleukin 17 (IL-17) has been identified, and initially termed CTLA-8, as a gene expressed in an activated T hybridoma, using a subtraction cDNA library. There is significant homology between the CTLA-8 molecule and an open reading frame (ORF13) of herpes virus Saimiri (HVS), suggesting that ORF13 could represent a gene captured by this virus. It has been proposed that the protein encoded by ORF13 be termed viral IL-17, in view of its shared biological activities with IL-17.

Only a few papers have been published on IL-17 since its initial characterization, and its present biological function remains unclear. However, it is clear that IL-17 has specific features that distinguish it from other lymphokines.

3.1 Biochemistry

The human IL-17 gene produces a 1.2 kb transcript in PBMC stimulated with phorbol myristate acetate (PMA) and ionomycin. This mRNA contains in its 3′ untranslated region eight ATTTA motifs, which are involved in the regulation of mRNA stability and are present in the mRNAs of several cytokines and oncogenes. The open reading frame of IL-17 mRNA encodes a peptide of 155 amino acids. IL-17 has no sequence similarity with other cytokines. The IL-17 protein contains an N-terminal hydrophobic leader sequence, six cysteine residues which are highly conserved in mouse and rat IL-17 as well as in viral IL-17, and a single *N*-glycosylation site. Cells transfected with human IL-17 cDNA secrete two proteins of 28 and 33 kDa corresponding to homodimers of unglycosylated or *N*-glycosylated human IL-17. IL-17 is highly conserved between human and mouse with 63% conservation at the amino acid level and 72% conservation at the nucleotide level. Human IL-17 is even more closely related to viral IL-17, with 72% homology at the protein level and 75% homology at the nucleotide level.

3.2 Viral IL-17

Herpes virus Saimiri, a *α2* herpes virus, is a a non-pathogenic virus in its natural host but induces fulminant acute leukemia and lymphomas in primates. It can induce a stable transformation of human T lymphocytes, and triggers CD2-dependent autocrine growth of transformed T lymphocytes in an IL-2-dependent manner. ORF13 of HVS, an immediate early gene, is highly homologous to CTLA-8/IL-17, suggesting that HVS has captured the mammalian IL-17 gene. As stated above, it has been proposed that HVS ORF13 be called viral IL-17. This situation may therefore be similar to that of the viral IL-10 gene within the Epstein–Barr virus genome.

Although it is likely that the proposed capture of the mammalian IL-17 gene by the virus may have given a selective advantage to the virus, the biological role of viral IL-17 in the pathophysiology of HVS infection is not known. Viral IL-17 is produced by virus-infected cells and shares several of the biological activities of IL-17, in particular the capacity to induce IL-6 and IL-8 production. The biological activities of the two related genes can be blocked with the same antibodies. Conceivably, viral IL-17 could interfere with the immune response directed against virus-infected cells or protect infected cells against virus-induced death.

3.3 IL-17 receptor

A cDNA encoding an IL-17-binding protein at the cell surface was recently identified and found to be expressed on a wide range of cell types. It encodes a 864 amino acid, type I transmembrane receptor. This receptor does not show significant homology with any of the cytokine receptor family. IL-17R is detectable by Northern blotting in all tissues tested. A soluble form of this receptor inhibits the biological activity of IL-17. Whether other receptor chains are involved in the IL-17 receptor is not known.

3.4 IL-17 expression

Whereas IL-17R expression appears to be ubiquitous, the expression of human IL-17 has so far been demonstrated only in the T-cell lineage. IL-17 mRNA and protein are not detectable in resting PBMC, T or B cells, monocytes, or in any of the other cell types tested. However, IL-17 mRNA is detectable in peripheral blood T cells stimulated with phorbol esters and ionomycin. IL-17 protein is also detectable in these conditions as well as in T cells stimulated with anti-CD3 antibodies or phytohemagglutinin (PHA). CD4$^+$ memory T cells appear to be the major producers of IL-17 although IL-17 may also be produced by a subset of CD8$^+$ cells.

3.5 Biological properties of IL-17

As far as is known at present, IL-17 has a narrow spectrum of biological activity. It has been reported to induce the production of IL-6, IL-8, granulocyte colony-stimulating factor (G-CSF) and prostaglandin E_2 (PGE$_2$) by bone marrow stromal cells, as well as by renal carcinoma cells, endothelial cells, and fibroblasts. IL-17 does not induce tumour necrosis factor (TNF) nor IL-1β production by these cells. Importantly, IL-17 strongly enhances fibroblast-dependent proliferation of marrow CD34$^+$ progenitors, and can trigger their differentiation into mature neutrophils, suggesting that it may play an important role in normal hematopoiesis. In addition, the combination of IL-17 and TNF induces GM-CSF production by fibroblasts, although neither cytokine on its own induced detectable GM-CSF production. IL-17 induces the expression of intercellular adhesion molecule 1 (ICAM-1) but does not affect HLA ABC, DR or LFA-1 expression on fibroblasts.

Despite its potent capacity to induce stromal cells to produce cytokines, IL-17 does not affect production of cytokines (IFNγ, IL-4, IL-6 and IL-10) by PBMC, T cells or monocytes, regardless of the stimulus. IL-17 does not affect (i) immunoglobulin production, (ii) proliferation of B lymphocytes activated through antigen receptor or CD40, or (iii) T-cell proliferation or cytotoxic activity. Although it does not affect the cytotoxic activity of PBMC or T cells stimulated with PHA or IL-2, it is reported to act as a co-stimulator of T-cell proliferation.

The soluble form of the IL-17 receptor can inhibit T-cell proliferation and IL-2 production induced by PHA, concanavalin A (ConA) or anti-TCR antibodies.

In conclusion, although the current body of knowledge regarding IL-17 is still limited, this cytokine is clearly distinct from interleukins in terms of its structure and that of its receptor structure, and its restricted pattern of expression, which contrasts with wide receptor expression. It is also the second known interleukin whose gene seems to have been captured by a herpes virus. This cytokine may play an important role during normal hematopoiesis, T-cell activation, and probably in the pathophysiology of HVS infection.

4. Interleukin 18

As with the discovery of each new cytokine, the introduction of interferon gamma inducing factor (IGIF) has attracted widespread attention even though comparatively little is known about its role (if any) in the immune response. Its activities are startlingly similar to those of IL-12, and it has intriguing structural connections with the IL-1 group of cytokines, so it is hoped that further characterization of IGIF may contribute to filling in some of the many gaps still remaining in our understanding of how the cytokine network is regulated.

4.1 Discovery of IGIF

Injection of Gram-negative bacteria or of bacterial lipopolysaccharide (LPS) into mice causes endotoxic shock, or sepsis, characterized by an uncontrolled burst of pro-inflammatory cytokines, initially led by TNF-α and succeeded by IL-1β; IL-6, IL-12, and IFNγ. Blood sera from mice that have first been primed with heat-inactivated *Propionibacterium acnes* and then challenged with LPS demonstrated significant IFNγ-inducing activity when used to stimulate resting splenocytes *in vitro*. Column chromatography and polyacrylamide gel electrophoretic analysis have shown that this IFNγ-inducing activity was found in a 75–80 kDa factor which acted as a cofactor with Con A, IL-2 or anti-CD3 in inducing IFN-γ secretion as well as mitogenic activity of non-adherent splenocytes. Since *P. acnes*-induced sepsis causes severe hepatitis, indicating that the liver might be a site of elevated expression of this factor, a similar purification process from liver extracts was performed, yielding a 18–19 kDa protein that colocalized with the previously observed 75–80 kDa complex. This factor demonstrated the same cofactor activity for IFNγ induction previously recorded and hence was named IFNγ-inducing factor (IGIF).

Murine IGIF was cloned via a multistep process beginning with purification and trypsin-digestion of the factor responsible for the IFNγ- enhancing activity. The resulting peptides were sequenced to obtain primers for generating cDNA fragments from liver mRNA extracted from a mouse primed with *P. acnes* and challenged with LPS. Screening a liver cell cDNA library from a *P. acnes*/LPS-stimulated mouse with these fragments yielded the full-length clone. Subsequently, murine IGIF cDNA probes were used to screen a human liver cDNA (phage library to obtain clones positive for human IGIF expression. The murine IGIF protein is a 157 amino acid polypeptide, derived from a 192 amino acid precursor, with an M_r of 18.3 kDa, while human IGIF is composed of 193 amino acids and has a similar M_r. Murine and human IGIF share 65% sequence homology.

4.2 Function of IGIF

The best characterized function of IGIF thus far is its cofactor activity for the induction of IFNγ production by T cells, an activity so far observed using 1–1000 ng doses of IGIF. Murine IGIF enhances IFNγ production by non-adherent splenocytes and by established type 1 T-helper cell (Th1) clones in the presence of anti-CD3, Con A, IL-2, IL-12, or antigen plus antigen-presenting cells (APCs). Similarly, recombinant human IGIF enhances IFNγ production by PBMCs stmulated with anti-CD3, Con A or IL-2. However, IGIF has not demonstrated IFNγ-inducing activity to date when used as the sole source of stimulation.

Murine Th2 clones appear to be resistant to modulation by IGIF and do not produce IFNγ in response to this factor. This IFNγ-promoting activity of IGIF can also be indirectly respnsible for influencing the isotype profile of the immunoglobulin response, increasing the IgG2a:IgE ratio in *N. brasiliensis*-challenged mice. GM-CSF can also be induced by recombinant human IGIF in conjunction with anti-CD3 or ConA from human T cells. IGIF has variously been reported to reduce IL-4 and IL-10 protein expression slightly, or to have no effect.

Additionally, murine IGIF appears to have a moderate mitogenic effect on non-adherent splenocytes, but, as with IFNγ-induction, serves only as a cofactor and does not induce significant proliferation without the presence of a mitogenic factor such as anti-CD3, ConA or recombinant IL-2. The proliferation of enriched human T cells stimulated with anti-CD3 can also be further enhanced with the addition of recombinant human IGIF, in an IL-2-dependent fashion, indicating that IGIF may act to increase mitogenesis primarily by the positive regulation of IL-2 expression or function.

There is some evidence that IGIF has a regulatory role in the enhancement of cytolytic activity in both human and murine systems. Murine IGIF induces impressive killing of YAC-1 cells by BDF1 splenocytes at only 10 pg/ml. The lysis of K562 target cells by IL-2-activated PBMC is augmented in a dose-dependent manner by recombinant human IGIF, while recombinant murine IGIF enhances the killing ability of NK clones, apparently by directly up-regulating the expression of FasL on the surface of the NK cells. Murine Th1 clones also exhibit cytolytic activity in a FasL-dependent manner, and this activity is boosted in the presence of IGIF in an IFNγ-independent fashion. It is unclear from this study whether IGIF specifically enhanced the expression levels or enhanced the effective function of Fas.

Finally, administration of anti-IGIF antiserum to mice treated with *P. acnes* and challenged with LPS completely abrogates subsequent hepatic injury, indicating that IGIF has a critical role in promoting the inflammatory response during bacterial sepsis.

4.3 Producer and target cells of IGIF

Several reports point to a constitutive pattern of IGIF mRNA expression in some cell types. IGIF mRNA is constitutively and abundantly expressed in Kuppfer cells, resident macrophage-like APCs of the liver, to an extent that further modulation by LPS or *P. acnes* stimulation is not detectable. Constitutive mRNA expression is also observed in PBMC populations via RT-PCR, the levels of which are also independent of LPS or PHA. Tissues found to be positive for constitutive IGIF expression include pancreas, kidney, skeletal muscle, lung and liver.

Activated macrophages from *P. acnes*-elicited peritoneal exudate also express IGIF mRNA, but only upon LPS activation. Additionally, some osteoblastic stromal cell lines produce IGIF which appears to make the cell lines unable to support osteoclast formation *in vitro* through a mechanism which relies on GM-CSF but is independent of IFNγ. Splenocytes and pancreatic infiltrating mononuclear cells from NOD mice express higher levels of IGIF mRNA than BALB/c control mice when treated with cyclophosphamide, which accelerates the onset of insulin-dependent diabetes mellitus in NOD mice. This wave of IGIF expression precedes the insulitis-causing expansion of IFNγ-producing T cells observed in these mice. Thus far, the accumulated data point to an APC-type of cell, primarily the macrophage, as the major producer of IGIF.

Most investigators have concentrated on the study of the IFNγ-inducing property of IGIF, which appears to target primarily T cells. Recently, highly purified splenic murine B cells that had been stimulated with a synergic combination of anti-CD40, IL-12 and IGIF were found to increase IFNγ production several-fold, both *in vitro* and *in vivo*. Additionally, NK cells can respond to IGIF with an up-regulation in lytic activity, and it seems reasonable to hypothesize that NK cells, which produce IFNγ in response to IL-12 as do T cells, may also react to IGIF with increased IFNγ production.

4.4 Comparison of the functions of IGIF and IL-12

IL-12 and IGIF have no detectable similarities in sequence or structure, yet there are striking parallels between their biological activities. The IFNγ-inducing activity of IGIF has been reported to be equivalent to that of IL-12 in the human system or significantly more potent than IL-12 with murine cells. IL-12 may exert activity at a lower optimal dose than IGIF. When added in combination, IL-12 and IGIF act synergically in augmenting IFNγ production by murine or human freshly isolated T cells or murine Th1 clones. When administered simultaneously, IL-12 and IGIF also show marked synergy with regards to inducing IFNγ production by murine B cells; the presence of IL-12 appears to be necessary for any significant induction of IFNγ by IGIF. In addition, IL-12 and IGIF have an additive enhancing effect on FasL-mediated cytolytic activity, which appears to be independent of the upregulation of IFNγ. Neutralizing antibody to murine IL-12 or murine IGIF *in vitro* fails to curb the effects of IGIF or IL-12, respectively, on IFNγ induction by Th1 clones. Given this observation, and the fact that in many instances an additive and even synergic effect is noted from simultaneous stimulation by both, it may be that IL-12 and IGIF do not exert their effects through the same signaling pathway, and if the activity of one is blocked, the other may compensate. Further evidence pointing in this direction comes from the data described below comparing IGIF with the IL-1 family of molecules, which apparently do not signal through the Jak–Stat system of phosphorylation, recently observed to be the primary mechanism of signaling by IL-12.

4.5 IGIF processing by caspase-1

The amino acid sequences of human and murine IGIF show no significant homology with sequences in databases. However, comparison of protein fold recognition patterns revealed the existence of the IL-1-like signature sequence Phe-X_{12}-Phe-X-Ser-X_6-Phe-Leu within a larger region containing 12 β-strands composing a β-trefoil fold. The β-trefoil fold is an unusual molecular structure composed of 62-stranded hairpins which is known to be shared by several proteins with dissimilar functions such as fibroblast growth factors, the Kunitz family of protease inhibitors, and IL-1α, IL-1β and IL-1 receptor agonist. Despite the structural similarity, human IGIF shares only 15–18% sequence homology with the IL-1 family of molecules.

No putative N-terminal hydrophobic signal peptides have been found in the sequences of either human or murine IGIF, suggesting that the mechanism of secretion of these proteins is not by transfer into the endoplasmic reticulum by the signal recognition particle. A recombinant 24 kDa precursor form of human IGIF (pro-IGIF) composed of the entire 193 amino acid sequence demonstrated very little IFNγ-inducing activity. pro-IGIF is cleaved at Asp35–Asn36 to generate the biologically active mature form, which suggests the involvement of a member of the caspase family, a novel group of cysteinyl aspartate-specific proteases. IL-1β, which is also detectable in a 31–33 kDa inactive precursor form (pro-IL-1β), is converted to the active 17.5 kDa mature form by cleavage at Asp116–Ala117. This catalysis is performed near the plasma membrane by an endoprotease named interleukin-1β-converting enzyme (ICE) or caspase-1, the first described member of the caspase family. Inhibition of ICE function by several methods severely impairs expression of mature IL-1α and IL-1β, although ICE may not be directly responsible for their secretion, as pro-IL-1β is released in large quantities in ICE-deficient mice. This mechanism of secretion has not been described to date. ICE–/– mice are also highly resistant to LPS-induced endotoxic shock and do not release detectable mature IL-1α or IL-1β during such treatment.

Such similarities in precursor processing and lack of signal peptide suggested that an ICE-like enzyme may be involved in the activation of IGIF. Cotransfection of COS cells with murine or human caspase-1 and with human or murine pro-IGIF yielded cytosolic expression of the mature 18 kDa protein complete with IFNγ-enhancing properties. In confirming studies, recombinant human caspase-1 was reported to cleave pro-IGIF *in vitro*, and this cleavage increased the extracellular release of nature IGIF.

In conclusion, these preliminary experiments suggest that IGIF is produced by monocytes, perhaps in a constitutive manner, and that it can target most if not all types of lymphocytes, including T cells, B cells and NK cells. Its primary function appears to be the up-regulation of IFNγ, but how it contributes to the priming, expansion or maintenance of the IFN-producing population, including Th1 cells, alongside the well-known positive regulatory activity of IL-12 is unknown. It is intriguing to note that several of the effects of IGIF are not dependent upon IFNγ production (including enchancement of T and B cell proliferation, cytolytic killing and osteoclast formation) which may indicate a critical role for IGIF even in the presence of other IFNγ-inducing agents such as IL-12.

References

Albrecht, J.C., Nicholas, J., Biller, D., Cameron, K., Biesinger, B., Newman, C., Wittmann, ••., Craxton, M.A., Coleman, H., Fleckenstein, B. *et al.* (1992) Primary structure of the herpesvirus Saimiri gerome. *J. Virol.* **66**, 5047–58.

Bacon, C.M., Petricoin, E.F., III, Ortaldo, J.R., Rees, R.C., Larner, A.C., Johnston, J.A. and O'Shea, J.J. (1995) Interleukin 12 induces tyrosine phosphorylation and activation of STAT4 in human lymphocytes. *Proc. Natl Acad. Sci. USA* **92**, 7307–11.

Bazan, J.F, Timans, J.C. and Kastelein, R.A. (1997) A newly defined interleukin. *Nature* **379**, 591.

Biesinger, B., Muller-Fleckenstein, I., Simmer, B., Lang, G., Wittmann, S., Platzer, E., Desrosiers, R.C. and Fleckenstein, B. (1992) Stable transformation of human T lymphocytes by herpes virus. *Proc. Natl Acad. Sci. USA* **89**, 3116–19.

Bisset, L.R., Rothen, M., Joller-Jemelka, H.I., Dubs, R.W., Grob, P.J. and Opravil, M. (1997) Change in circulating levels of the chemokines macrophage inflammatory proteins I alpha and II beta, RANTES, monocyte chemotactic protein-1 and interleukin-16 following treatment of severely immunodeficient HIV-infected individuals with indinavir. *AIDS* **11**, 485–91.

Cruikshank, W.W., Center, D.M., Nisar, N., Wu, N., Natke, B., Theodore, A.C. and Kornfeld, H. (1994) Molecular and functional analysis of a lymphocyte chemoattractant factor: association of biologic function with CD4 expression. *Proc. Natl Acad. Sci. USA* **91**, 5109–113.

Cruikshank, W.W., Lim, K., Theodore, A.C., Cook, J., Fine, G., Weller, P.F. and Center, D.M. (1996) IL-16 inhibition of CD3-dependent lymphocyte activation and proliferation. *J. Immunol.* **157**, 5240–8.

Cruikshank, W.W., Long, A., Tarpy, R.E., Kornfeld, H., Caroll, M.P., Teran, L., Holgate, S.T. and Center, D.M. (1995) Early identification of IL-16 and MIP1 alpha in bronchoalveolar lavage fluid of antigen-challenged asthmatics. *Am. J. Resp. Cell Mol. Biol.* **13**, 738–47.

Dao, T., Ohashi, K., Kayano, T., Kurimoto, M. and Okamura, H. (1996) Interferon-γ-inducing factor, a novel cytokine, enhances Fas ligand-mediated cytotoxicity of murine T helper 1 cells. *Cell. Immunol.* **173**, 230–5.

Fossiez, F., Djossou, O., Chomarat, P., Flores-Romo, L., Ait-Yahia, S., Maat, C., Pin, J.J., Garrone, P., Garcia, E., Saeland, S. *et al.*. (1996) T cell interleukin 17 induces stromal cells to produce proinflammatory and hematopoietic cytokines. *J. Exp. Med.* **183**, 2593–603.

Ghayur, T., Banerjee, S., Hugumin, M., Buttler, D., Herzog, L., Carter, A., Quintal, L., Sekut, L., Talanian, R., Paskind, M. *et al.* (1997) Caspase-1 processes IFN-γ-inducing factor and regulates LPS-induced IFN-γ production. *Nature* **386**, 619–23.

Gu, Y., Kuida, K., Tsutsui, H., Ku, G., Hsiao, K., Fleming, M.A., Hayashi, N., Higashino, K., Okamura, H., Nakanishi, K. *et al.* (1997). Activation of interferon-γ inducing factor mediated by interleukin-1β-converting enzyme. *Science* **275**, 206–9.

Kohno, K., Kataoka, J., Ohtsuki, T., Suemoto, Y., Okamoto, I., Usui, M., Ikeda, M. and Kurimoto, M. (1997) IFN-γ-inducing factor (IGIF) is a costimulatory factor on the activation of Th1 but not Th2 cells and exerts its effect independently of IL-12. *J. Immunol.* **158**, 1541–50.

Li, P., Allen, H., Banerjee, S., Franklin, S., Herzog, L., Johnston, C., McDowell, J., Paskind, M., Rodman, L., Salfeld, J. *et al.* (1995) Mice deficient in IL-1β-converting enzyme are defective in production of mature IL-1β and resistant to endotoxic shock. *Cell* **80**, 401–11.

Lim, K.G., Wan, H.C., Bozza, P.T., Resnick, M.B., Wong, D.T., Crukshank, W.W., Kornfeld, H., Center, D.M. and Weller, P.P. (1996) Human eosinophils elaborate the lymphocyte chemoattractants. IL-16 and RANTES. *J. Immunol.* **156**, 2566–70.

Mackewicz, C.E., Levy, J.A., Cruikshank, W.W., Kornfeld, H., and Center, D.M. (1996) Role of IL-16 in HIV replication. *Nature* **383**, 488–9.

Micallef, M.J., Ohtsuki, ••., Kohno, K., Tanabe, F., Ushio, S., Namba, M., Tanimoto, T., Torigoe, K., Fujii, M., Ikeda, M., Fukuda, S. and Kurimoto, M. (1996) Interferon-γ inducing factor enhances T helper 1 cytokine production by stimulated human T cells: synergism with interleukin-12 for interferon-γ production. *Eur. J. Immunol.* **26**, 1657–51.

Miller, D.K. (1997) The role of the caspase family of cysteine proteases in apoptosis. *Sem. Immunol.* **9**, 35–49.

Mittrucker, H.-W., Muller-Fleckenstein, I., Fleckenstein, B. and Fleischer, B. (1992) CD2 mediated autocrine growth of herpes virus Saimiri transformed human T lymphocytes. *J. Exp. Med.* **176**, 909–13.

Nakamura, K., Okamura, H., Nagat, K., Komatsu, T. and Tamura, T. (1993) Purification of a factor which provides a costimulatory signal for gamma interferon production. *Infect. Immun.* **61**, 64–70.

Nicholas, J., Smith, E.P., Coles, L. and Honess, R. (1990) Gene expression in cells infected with gammaherpes virus Saimiri: properties of transcripts from two intermediate early genes. *Virology* **179**, 189–200.

Okamura, H. , Nagata, K., Komatsu, T., Tanimoto, T., Nukata, Y., Tanabe, F., Akita; K., Torigoe, K., Okura, Y., Fukuda, S. and Kurimoto, M. (1995) A novel costimulatory factor for gamma interferon induction found in the livers of mice causes endotoxic shock. *Infect. Immun.* **63**, 3966–72.

Okamura, H., Tsutsui, H., Komatsu, T., Yutsudo, M., Hakura, A., Tanimoto, T., Torigoe, K., Okura, T., Nukuda, Y., Hattori, K. *et al.* (1995) Cloning of a new cytokine that induces IFN-γ production by T cells. *Nature* **378**, 88–91.

Rothe, H., Jenkins, N.A., Copeland, N.G. and Kolh, H. (1997) Active stage of autoimmune diabetes is associated with the expression of a novel cytokine, IGIF, which is located near Idd2. *J. Clin. Invest.* **99**, 469–74.

Rouvier, E., Luciani, M.F., Mattei, M.G., Denizot, F. and Golstein, P. (1993) CTLA-8, cloned from an activated T cell, bearing AU-rich messenger RNA instability sequences, and homologous to a herpesvirus Saimiri gene. *J. Immunol.* **150**, 5445–56.

Theodore, A.C., Center, D.M., Nicoll, J., Fine, G., Kornfeld, H. and Cruikshank, W.W. (1996) CD4 ligand IL-16 inhibits the mixed lymphocyte reaction. *J. Immunol.* **157**, 1958–64.

Thornberry, N.A., Bull, H.G., Calaycay, J.R., Chapman, K.T., Howard, A.D., Kostura, M.J., Miller, D.K., Molineaux, S.M., Weidner, J.R., Aunins, J. *et al.* (1992) A novel heterodimeric cysteine protease is required for interleukin-1β processing in monocytes. *Nature* **356**, 768–74.

Tsuitsui, H., Nakanishi, K., Matsui, K., Higashino, K., Okamura, H., Miyazawa, Y. and Kaneda, K. (1996) IFN-γ-inducing factor up-regulates Fas ligand-mediated cytotoxic activity of murine natural killer cell clones. *J. Immunol.* **157**, 3967–73.

Udagwa, N., Horwood, N.J., Elliot, J., Mackay, A., Owens, J., Okamura, H., Kurimoto, M., Chambers, T.J., Martin, T.J. and Gillespie, M.T. (1997) Interleukin 18 (interferon-γ-inducing factor) is produced by osteoclasts and acts via granulocyte/macrophage colony-stimulating factor and not via interferon-γ to inhibit osteoclast formation. *J. Exp. Med.* **185**, 1005–12.

Ushio, S., Namba, M., Okura, T., Hattori, K., Nukada, Y., Akita, K., Tanabe, F., Konishi, K., Micallef, M., Fujii, M. *et al.* (1996). Cloning of the cDNA for human IFN-γ-inducing factor, expression in *Escherichia coli*, and studies on the biologic activities of the protein. *J. Immunol.* **156**, 4274–9.

Wysocka, M., Kubin, M., Vieira, L.Q., Ozmen, L., Garotta, G., Scott, P. and Trinchieri, G. (1995) Interleukin-12 is required for interferon gamma production and limality in lipopolysacharide induced shock in mice. *Eur. J. Immunol.* **25**, 672–6.

Yao, Z., Fanslow, W.C., Seldin, M.F., Rousseau, A.M., Painter, S.L., Comeau, M.R., Cohen, J.I. and Spriggs, M.K. (1995) Herpesvirus Saimiri encodes a new cytokine, IL-17, which binds to a novel cytokine receptor. *Immunity* **3**, 811–21.

Yao, Z., Painter, S.L., Fanslow, W.C., Ulrich, D., MacDuff, B.M., Spriggs, M.K. and Armitage, R.J. (1995) Human IL-17: a novel cytokine derived from T cell. *J. Immunol.* **155**, 5483–6.

Yoshimoto, T., Okamura, H., Tagawa, Y.-I, Iwakura, Y. and Nakanishi, K. (1997) Interleukin 18 together with interleukin 12 inhibits IgE production by induction of interferon-γ production from activated B cells. *Proc. Natl Acad. Sci. USA* **9**, 3948–53.

Zhang, Y., Center, D.M., Wu, D.M.H., Cruikshank, W.W., Yuan, J., Andrews, D.W. and Kornfeld, H. (1998) Processing and activation of pro-interleukin-16 by caspase-3. *J. Biol. Chem.* **273**, 1144–9.

Zhou, P., Goldstein, S., Devadas, K., Tewari, D. and Notkins, A.L. (1997) Human CD4+ cells transfected with IL-16 cDNA are resistant to HIV-1 infection: inhibition of mRNA expression. *Nature Med.* **3**, 659–64.

XII Mechanisms of signal transduction

Tohru Itoh and Ken-ichi Arai*

1. General introduction

Cytokines act as communicators between cells, but they are not effector molecules in their own right; they function by binding to specific receptors that are expressed on surfaces of target cells and coupled to intracellular signal transduction pathways. Thus, the response of cells to a given cytokine is determined by expression of the cytokine receptor as well as by the nature of the signaling pathways that link the ligand-bound, activated form of receptors to various downstream effectors regulating cellular functions. Intracellular signaling pathways consist mainly of protein–protein interactions and the generation of second messenger molecules.

An important feature of the signaling pathways is that some of their components have intrinsic mechanisms that enable them to oscillate between two distinct states, i.e. 'active' and 'inactive' states; thereby functioning as a 'molecular switch' which confers unidirectionality and stimulation-dependency upon signal transduction pathways and regulates the intensity and duration of signals. As an example, second messengers are generated or released and accumulate in response to extracellular stimuli, and levels of these messengers eventually decrease to a basal level as a result of functions of particular proteins, such as enzymes or ion channels.

In cytokine receptor signaling, two known mechanisms play major roles as molecular switches: guanine nucleotide-binding proteins (G proteins) and protein phosphorylation. G proteins usually exist in an inactive, GDP-bound form, and become activated by releasing GDP and subsequent loading of GTP; a signal is terminated when G proteins hydrolyze GTP with their intrinsic GTPase activity,

reverting to the GDP-bound state in the process. Low-molecular-weight G proteins such as Ras are commonly involved in many cytokine receptor systems. In contrast, heterotrimeric G proteins have a role only in signaling of interleukin 8 (IL-8) and related chemokines, the receptors of which belong to the seven-transmembrane-domain receptor family.

Protein phosphorylation can be classified into two types: serine/threonine and tyrosine phosphorylation. In both cases, the phosphorylation state of a particular molecule is determined by the activity of protein kinases as well as by protein phosphatases, the functions of which are antagonistic. Phosphorylation on various signaling proteins regulates their enzymatic activities or creates docking sites for other molecules.

As described in detail in the preceding chapters, cytokine receptors are classified into distinct families, based on their structural features. The major subset of cytokines, including most interleukins and colony-stimulating factors (CSFs), comprise the so-called α-helical cytokines, and receptors for these cytokines belong to the class I cytokine receptor superfamily (or the hematopoietin receptor family). This family also includes receptors for growth hormone and prolactin. Receptors for interferons as well as IL-10 are classified as the class II cytokine receptor family (or the interferon receptor family), and are closely related to the class I receptors. Their characteristic features are that they have no known enzymatic activity, such as kinase or phosphatase activity. Recent studies have revealed that cytoplasmic tyrosine kinases, especially those in the Jak (Janus kinase) family, associate with the cytoplasmic portion of these families of receptors and play critical roles in signaling. Thus, activation of receptor-associated tyrosine kinase(s) and subsequent phosphorylation of the receptor and multiple signaling molecules are central events in signal

*Corresponding author.

transduction. As a downstream target of Jaks, the STAT (signal transducer and activator of transcription) family of latent cytoplasmic transcription factors mediate part of the signals directly from the cell membrane to the nucleus.

The tumor necrosis factor (TNF) receptor family is unique in that some members of this family can transduce signals for cell survival, while others can transduce signals for cell death, and some can even transmit both survival and death signals, depending on the circumstances. These receptors contain no known enzymatic activities, but upon ligand stimulation, interact with various molecules through protein–protein interactions. The molecules recruited on to the receptor cytoplasmic domains in turn transduce signals to downstream of the protease (caspase) cascade, which leads to initiation of cell death by apoptosis, or to activation of the transcription factor NF-κB through a kinase cascade, which renders the cells resistant to apoptosis.

The macrophage colony-stimulating factor (M-CSF) and stem cell factor (SCF) receptors, encoded by proto-oncogenes c-*fms* and c-*kit*, respectively, are members of class III of the receptor-type tyrosine kinase (RTK) family which also includes the platelet-derived growth factor (PDGF) receptor and the vascular endothelial cell growth factor (VEGF) receptor. Structural features of this class of receptors are five tandem arrays of immunoglobulin-like loops in their extracellular domains and a cytoplasmic tyrosine-kinase domain split by a kinase insert (KI) sequence. Signals mediated by RTKs involve ligand-induced dimerization of the receptor followed by activation of intrinsic tyrosine kinase. Subsequently, autophosphorylation of RTKs occurs, and signals are initiated through phosphorylation of other substrates which recognize specific phosphotyrosine residues and are recruited on to the receptors.

Signals of transforming growth factor-β (TGF-β) family ligands are mediated through heteromeric complexes between type I and type II receptors, both of which belong to the TGF-β receptor family and contain the intrinsic serine/threonine kinase domain within their cytoplasmic portions. Ligand binding results in formation of functional receptor complexes and induces phosphorylation of cellular substrates, among which are members of recently expanding Smad family proteins.

Unlike other cytokines with receptors composed of an extracellular and an intracellular region separated by a single transmembrane segment, IL-8 transmits signals through receptors with seven membrane-spanning regions coupled to heterotrimeric G proteins. Although the regulatory mechanisms and downstream pathways of heterotrimeric G proteins are generally known, IL-8 receptor signaling has not been analysed extensively.

The components of signal transduction pathways are not simply arranged in linear cascades, but rather form complicated networks. There is often functional cross-talk and synergy between signaling pathways which are likely to play critical roles in precise regulation of cellular responses. Since activation of signaling pathways do not occur in an 'all-or-none' manner while a signal over the threshold level is required to trigger a biological response, the intensity and duration of signals are also pertinent factors. The function and relative importance of a signaling molecule vary depending on cell types; *in vitro* studies using cultured cell lines and *in vivo* studies using transgenic or knockout animals can give apparently different results. Thus when interpreting results one must take into consideration the experimental system used to analyze signaling molecules and for establishing physiological functions and regulatory mechanisms of signal transduction *in vivo*.

2. Receptors with no direct tyrosine kinase activity: Jak- and STAT-mediated signal transduction of the class I and II cytokine receptors

2.1 Subunit sharing, multiple receptor domains and signaling pathways with distinct functions: molecular basis for pleiotropic and redundant functions of cytokines

It was initially assumed that each cytokine has a unique biological function towards a specific cell type. Later though, after cDNA clones and production of recombinant proteins became available, it was revealed that the actions of cytokines were not as simple as had first been thought, in that cytokines characteristically shows pleiotropy (they can act on various target cells with different responses) and redundancy (several different cytokines can target the same cells and exhibit similar functions).

Many class I cytokine receptors are composed of multiple distinct subunits. In most cases, one (or two) of these subunits are shared by multiple different receptor complexes. This 'subunit sharing' is likely to provide the molecular basis for redundancy in cytokine function. Three groups are known to share common receptor subunits:

(1) Receptors for granulocyte–macrophage CSF (GM-CSF), IL-3 and IL-5. These complexes share the 'common β-subunit' or βc-subunit. This βc-subunit by itself has no apparent affinity for any cytokine, but does form high-affinity receptors together with each ligand-specific α-subunit which alone can bind to the cognate ligand with low affinity.

(2) Receptors for IL-6, IL-11, leukemia inhibitory factor (LIF), oncostatin M (OSM), ciliary neurotrophic factor (CNTF) and cardiotrophin 1 (CT-1). In this group, gp130 (which was initially identified as a signal-transducing component of the IL-6 receptor complex) and, in some cases, the LIF receptor are shared. Homodimerization of gp130, or heterodimerization between gp130 and LIF receptor, is necessary for signal transduction of this group of receptors.

(3) Members of the third group share the 'common γ-subunit', or the γc-subunit, a component of the IL-2 receptor. Receptors for IL-4, IL-7, IL-9 and IL-15 also make use of this γc-subunit. The IL-15 receptor complex shares with the IL-2 receptor not only the γc-subunit but also the β-subunit. The situation is more complicated in the case of the IL-4 receptor: the IL-4 receptor α-subunit is a component of the IL-13 receptor complex, and this complex, in addition to the one composed of the IL-4 receptor α-subunit and the γc-subunit, also binds IL-4. Thus two different combinations of receptor subunits exist for IL-4.

In contrast to the receptors in the three groups described above, those for erythropoietin, granulocyte CSF (G-CSF), growth hormone and prolactin are composed of a single species of the receptor subunit. In these cases, homodimerization (or homo-oligomerization) of each of these subunits is required for activation.

Functional receptors for interferons, including IL-10, are also composed of multiple components which belong to the class II cytokine receptor family. The IFNγ receptor contains two polypeptide subunits; the α-chain (also termed IFNγR1) and the β-chain (IFNγR2, or accessory factor-1 (AF-1)). IFNα is a family of related proteins, and IFNα subtypes and IFNβ share some receptor components.

Several subunits of the IFNα/β receptors, such as IFNAR, have been identified, and reconstitution studies suggest that their functions depend on interactions with additional components endogenously expressed in each line of cell used. The number of subunits and their functions in binding and signal transduction by the IFNα/β subtypes remain to be elucidated. An accessory chain for the IL-10 receptor complex was also identified recently. The previously identified subunit (IL-10Rα or IL-10R1) of human origin, when expressed in hamster cells, is sufficient for human IL-10 binding, but not for inducing signal transduction. Co-expression of CRFB4, an orphan receptor of the class II cytokine receptor family, results in reconstitution of the functional human IL-10 receptor. At the time of writing, a subunit sharing among the IFNα/β, IFNγ, and IL-10 receptor systems had not been reported.

The pleiotropic function of cytokines can be explained, at least in part, by activation of multiple signaling pathways; each signaling pathway leads to a unique set of cellular targets and elicits specific biological functions. Thus, the nature of signaling molecules in each cell plays a key role in defining the cell's response to a particular cytokine. Even when qualitatively the same signaling pathways are apparently activated, quantitative differences — for example in strength and temporal pattern of activation — may lead to distinct final outcomes. Functional cross-talk and synergy with other signaling pathways caused by additional extracellular stimuli also have a considerable effect.

Mutational analyses of the cytoplasmic regions of various class I and II receptors initially revealed that these regions in general can be divided into several domains each possessing distinct functional properties. Subsequent studies have extended this notion and determined the existence of two kinds of fundamental constituents, or motifs, in the receptor cytoplasmic portions that play key roles in receptor signaling: a membrane-proximal domain which recruits Jak, and tyrosine residues which are phosphorylated upon ligand stimulation and then function as binding sites for various signaling molecules. Most class I and II receptors have multiple cytoplasmic tyrosine residues, and these tyrosines each have distinct properties in recruiting signaling molecules, thereby contributing to the initiation of multiple downstream signaling pathways. Notably, the membrane-proximal regions of many receptors alone can transmit some signals without any involvement of the tyrosine residues; for example, a mutant βc-subunit of human GM-CSF receptor with all of the cytoplasmic tyrosines replaced by (a)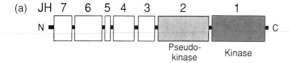

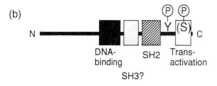

(b)

■ **Figure 1.** Structural features of (a) Jak family tyrosine kinases and (b) STAT family transcription factors. Ⓟ, phosphorylation site; Y, conserved tyrosine residue; (S), serine residue phosphorylated in some STATs.

phenylalanines has severe defects in many aspects of signal transduction including cell proliferation, yet transduces signals sufficient to maintain cell viability. Thus, receptors like this can propagate their signals through both the receptor tyrosine-dependent and -independent manners.

2.2 Jaks as key tyrosine kinases in initiating cytokine receptor signaling

Despite the lack of tyrosine-kinase domains within the cytoplasmic regions of class I and II cytokine receptors, stimulation by their ligands induces rapid and reversible tyrosine phosphorylation of various proteins. Studies using tyrosine kinase inhibitors also support the notion that tyrosine phosphorylation is a key event in cytokine receptor signaling. Therefore, it has been assumed that cytoplasmic tyrosine kinases associate with these receptors and function to initiate signaling. The Src family tyrosine kinases were the initial candidates and, as exemplified by binding of Lck to the IL-2 receptor β-subunit, their interaction was actually found in several receptor systems. However, further studies could not prove that these kinases have a major role in cytokine signaling. Even though Src family kinases may play some role, it is now generally accepted that another family of tyrosine kinases, the Jaks, play a pivotal role in signaling of all class I and II receptors.

Jak-family tyrosine kinases were initially identified, based on sequence homology, to be novel tyrosine kinases, the functions of which remained to be determined. The significance of these molecules in

cytokine receptor signaling was firstly noted in the interferon receptor system in an elegant study involving use of somatic cell genetics. Subsequently, Jak2, one member of this family, was shown to be involved in signaling from the erythropoietin and growth hormone receptors.

The Jak family contains four known members, Jak1, Jak2, Jak3 and Tyk2. These members are about 120–140 kDa in size and do not have Src homology 2 (SH2), SH3 or pleckstrin homology (PH) domains, but do share seven homologous regions termed Jak homology (JH) domains 1 to 7 (numbered starting from the C-terminus (Figure 1)). The JH1 region in the C-terminus is a kinase domain, while the JH2 region, N-terminally adjacent to JH1, is a pseudo-kinase domain without obvious kinase activity. The Jaks were named after Janus, a Roman god of gates who had two faces in order to keep watch in opposite directions, because they contain two kinase(-like) domains. JH3 to JH7 in the N-terminal region function as binding sites for receptors, and probably for other signaling molecules. Jak1, Jak2, and Tyk2 are expressed ubiquitously, while Jak3 expression is limited mainly to lymphoid cells.

Each cytokine activates a defined set of Jaks by virtue of specific interactions between a receptor subunit and a particular Jak molecule (Table 1). The N-terminal region of a Jak is thought to interact with the receptor cytoplasmic region through the membrane-proximal Box 1 motif. Box 1 and Box 2 motifs are modestly conserved among several, if not all, class I cytokine receptors. The Box 1 motifs, especially the characteristic Pro-X-Pro sequence, are necessary for activation of Jaks and for subsequent signaling events. The Box 2 motifs of some receptors are also suggested to be important for signaling but their significance remains to be clarified. These motifs are not completely conserved among the class I receptors and, moreover, cannot be found in the γc-subunit or in class II receptors. Possible involvement of adaptor molecules between Jaks and the receptors thus cannot be ruled out.

Jak3 specifically associates with the γc-chain, and is essential for signaling of cytokines whose receptors uses the γc-chain. Since mutations in the γc gene result in X-linked severe combined immune deficiency (X-SCID), it is reasonable to assume that lack of Jak3 function also leads to an impaired immune function. In fact, mutations in the Jak3 gene have been noted in patients with autosomal SCID and mice defective in this gene also showed defects in lymphoid development and a SCID phenotype.

∎ **Table 1.** Jaks and STATs activated by various cytokines

Group	Cytokine	Jak	STAT
βc	GM-CSF, IL-3, IL-5	Jak1, Jak2	STAT5
gp130 (+ LIFR)	IL-6, IL-11, LIF, CNTF, OSM	Jak1, Jak2, (Tyk2)	STAT3, (1)
γc	IL-2, IL-7, IL-9, IL-15	Jak1, Jak3	STAT3, 5
IL-4Rα + γc	IL-4	Jak1, Jak3	STAT6
IL-4Rα + IL-13R	IL-13	Jak1	STAT6
	IL-12	Jak2, Tyk2	STAT3, 4
	G-CSF	Jak1, Jak2	STAT1, 3
	EPO	Jak2	STAT5
	TPO	Jak2	STAT1, 3, 5
	GH	Jak2	STAT5
	PRL	Jak1, Jak2	STAT5
	Leptin	?	STAT3, 5, 6
IFNR (class II R)	IFNα/β	Jak1, Tyk2	STAT1, 2, 3
	IFNγ	Jak1, Jak2	STAT1
	IL-10	Jak1, Tyk2	STAT1, 3, (5)
RTK	EGF	Jak1	STAT1, 3
	PDGF	Jak1, Jak2, Tyk2	STAT1, 3
	M-CSF	Jak1, Tyk2	STAT1, 3, 5

2.3 Mechanism of receptor activation

Many of the class I and II receptor complexes consist of two or more subunits, so that ligand binding causes heterodimerization or oligomerization of these components (Figure 2). In the case of cytokines that have only one type of receptor subunit, such as erythropoietin and G-CSF, homodimerization is induced. Because Jaks bind to each receptor subunit either constitutively or upon ligand binding, these formations of complexes between receptor subunits mean that two or more Jak molecules can be located within the same receptor complex and in close proximity. These Jaks are activated, probably by reciprocal *trans*-phosphorylation, and activated Jaks in turn phosphorylate various substrates, which include receptor subunits. Phosphorylation on tyrosine residues of the receptors creates docking sites for several signaling molecules with SH2 and/or phosphotyrosine-binding (PTB) domains which recognize and bind phosphotyrosines surrounded by specific motifs. These signaling molecules, either by themselves or via activation of downstream cascades, transduce signals to the nucleus, cytoskeleton, and so on, to modulate cellular functions.

It should be noted that, although the term 'Jak–STAT pathway' is often used, Jaks are required not only for STAT activation; they play a critical role for all receptor functions. For instance, the dominant-negative type of Jak2 suppresses all the signaling events of GM-CSF stimulation, including receptor phosphorylation, activation of signaling molecules, cell proliferation, and gene expression. One of the most critical role of Jaks is to phosphorylate receptors; whether this can be achieved by direct phosphorylation or through other kinases remains to be determined.

2.4 Downstream signaling pathways

2.4.1 c-*myc* expression and DNA replication

Expression of the c-*myc* gene is tightly correlated with, and is thought to be essential for, the induction of DNA synthesis by cytokines such as IL-2, IL-3 and GM-CSF. While the role of c-*myc* in DNA synthesis remains largely unknown, it is known to encode a transcription factor, Myc, the target genes of which involve Cdc25, a cell cycle-related phosphatase.

The signaling pathway from the receptors to c-

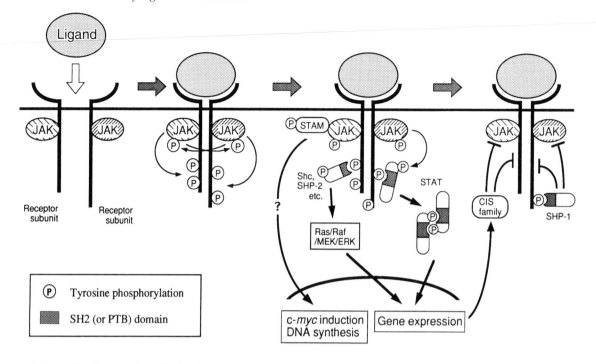

■ **Figure 2.** Proposed model for the activation and inactivation of the class I and class II cytokine receptors.

myc transcription also has to be identified. c-*myc* induction by GM-CSF/IL-3 requires the box 1 motif of the receptor βc-subunit as well as Jak2 activity. Since the dominant-negative STAT5 did not suppress this event, Jak2 is likely to phosphorylate other substrates to transduce signals to activate c-*myc*. Recently, a novel adaptor molecule called STAM (signal transducing adaptor molecule) was identified; it contains an SH3 domain and an immunoreceptor tyrosine-based activation motif (ITAM). STAM is associated with and phosphorylated by Jak2 and Jak3 upon stimulation with GM-CSF and IL-2, respectively. Transient transfection analyses on GM-CSF- or IL-2-stimulated signals have shown that an SH3 deletion mutant of STAM exhibits a dominant-negative effect on DNA synthesis, while the wild-type STAM enhances c-*myc* promoter activation.

In the case of the IL-2 signaling, the Syk tyrosine kinase also associates with the IL-2 receptor β-subunit and has been implicated in c-*myc* transcription. Activation of Syk achieved by antibody-mediated cross-linking of a Syk-containing chimeric molecule can induce the expression of c-*myc* but not of the c-*fos* gene in a Ba/F3 subline, but it is still not clear whether this indicates that

Syk has a physiologically relevant role in IL-2-induced c-*myc* expression.

2.4.2 Ras and its downstream targets

Ligand binding to many of the class I cytokine receptors induces activation of Ras, a well-known low-molecular-weight G protein, and its downstream signaling pathways. Activation of Ras by GTP loading is achieved by the GDP/GTP exchange reaction catalyzed by Sos protein. Sos forms a stable complex with an adaptor protein, Grb2, and it is now generally accepted that recruitment of this Grb2/Sos complex by activated receptor results in localization of Sos to the plasma membrane, where its substrate Ras is anchored and hence activated. Cytokine receptors usually utilize Shc and/or SHP-2 to activate Ras. Shc is another adaptor protein with an SH2 domain and a PTB domain, which bind to phosphotyrosines of activated receptors, and several tyrosine residues to be phosphorylated, which serve as Grb2 binding sites. SHP-2 (previously termed SH-PTP2, PTP1D, or Syp) is a protein tyrosine phosphatase with two SH2 domains, and is also tyrosine-phosphorylated after stimulation with particular cytokines. It may also function as a kind of adaptor protein linking the receptor to Grb2, but whether and how its phosphatase activity is involved in the Ras activation remain to be determined.

Activated Ras has several effectors, of which the serine/threonine kinase Raf and its downstream MEK (MAPK–ERK kinase; MAPK stands for mitogen-activated protein (MAP) kinase, and ERK for extracellular signal-regulated kinase)/ERK cascade are well established. Following activation, ERKs translocate into the nucleus and then phosphorylate and activate several transcription factors. GM-CSF/IL-3-induced activation of this pathway leads to transcription of c-*fos* and *egr*-1 genes through the SRE (serum response element) site, where a transcription factor called Elk-1, one of the targets of ERKs, binds. In the case of the IL-6 receptor system, a target of the Ras/Raf/MEK/ERK pathway is NF-IL6, a transcription factor of the C/EBP (CCAAT/enhancer-binding protein) family.

Another downstream target of Ras is phosphoinositide 3′-OH kinase (PI3K). Multiple forms of PI3K, with distinct mechanisms of regulation and different substrate specificities, function in mammalian cells, and catalyze phosphorylation of the inositol 3′-hydroxyl group of phosphoinositides. Various growth factors and hematopoietic cytokines activate an isoform of PI3K composed of a $p85\alpha$ regulatory subunit and a $p110\alpha$ catalytic subunit. The $p85\alpha$ subunit has two SH2 domains and an SH3 domain, through which it interacts with various signaling molecules. These interactions of p85 with other molecules give the kinase an increased activity in many cases, and may also play an important role in localizing the kinase complex to the cell membrane where its substrate lipids are present. The GTP-bound form of Ras binds directly to the $p110\alpha$ catalytic subunit and can increase the lipid kinase activity of PI3K. The products of PI3K can bind to and then regulate functions of multiple downstream effector molecules.

Recent studies suggested a critical role for PI3K and one of its downstream targets, the protein serine/threonine kinase Akt (also termed protein kinase B, or PKB), in survival signals from various cell-surface receptors, including receptors for cytokines such as IL-3. Phosphatidylinositol-3,4-bisphosphate, a product of PI3K, binds directly to the PH domain of Akt and causes dimerization of this kinase, which could be a part of the mechanism of activation. Akt activity is also regulated by phosphorylation on serine and threonine residues, by other kinases; phosphoinositide-dependent kinase-1 (PDK-1), the activity of which is also regulated by binding of phosphoinositide products to its PH domain, has been purified and cloned as a kinase that phosphorylates and activates Akt. Although Akt also is likely to have several downstream effectors, it has recently become evident that at least one of the critical target molecules of Akt is the death-promoting protein Bad. Bad is a member of the Bcl-2 family of proteins, and heterodimerizes with and hence inactivates the activity of Bcl-x_L, a cell survival factor found in mitochondrial membrane. Activated Akt phosphorylates a critical serine residue on Bad, which results in cytosolic retention through association with 14-3-3 protein, and thus its dissociation from Bcl-x_L. Released Bcl-x_L then acts as a suppressor of cell death, probably by blocking the release of cytochrome *c* from the mitochondria and subsequent activation of death protease (caspase; see Section 3 of this chapter) cascade. Thus, the PI3K–Akt pathway mediates an anti-apoptotic signal through, at least partly, phosphorylating and thereby inactivating the Bad protein. However, there are additional targets of Akt related to cell survival, and the PI3K/Akt-independent pathways also play roles in mediating survival signals. Thus, multiple pathways are likely to act in concert to maintain cell survival, and their relative contribution may vary depending on cell type and stimulus.

2.4.3 STATs

The STATs constitute a family of transcription factors with important roles in class I and II cytokine receptor signaling. To date, seven STAT members, namely STATs 1, 2, 3, 4, 5A, 5B and 6, have been identified in mammals. Studies on the interferon-inducible gene expression led to the discovery of STAT1 and STAT2. In response to IFNα or IFNβ stimulation, a DNA-binding complex is formed consisting of STAT1, STAT2, and another protein, p48, which is not a STAT protein and binds the IFNα-stimulated gene response element (ISRE). In contrast, the IFNγ-induced DNA-binding complex which binds to the IFNγ-activated sequences (GAS) was found to consist of STAT1 homodimers. These findings prompted a search for another STAT members, and two additional ones, STAT3 and STAT4, were identified by homology screening. STAT3 was independently purified and cloned as the acute-phase response factor (APRF), an IL-6-responsive DNA-binding protein. STAT5 was identified as the sheep mammary gland factor (MGF) which is involved in activation of genes encoding milk proteins in response to prolactin. Subsequent cloning of mouse and human counterparts revealed the existence of two highly homologous molecules, STAT5A and STAT5B, encoded by separate genes. STAT6 was identified as an IL-4-induced DNA-binding factor.

STAT family members share conserved structures, with a central DNA-binding domain unique to STATs (Figure 1). All STAT binding sites match the consensus sequence of TTCCXGGAA. Within the C-terminal half is an SH2 domain with critical roles for the function of this protein, as stated below. An SH3-like domain is present between the DNA-binding and SH2 domains, yet is structurally not well conserved and its functional significance is not known. Another important feature is a conserved tyrosine residue located on the C-terminal side of the SH2 domain, which is a site to be phosphorylated upon activation. The most C-terminal end is a transcriptional activation domain.

In the course of STAT activation, its SH2 domain plays two known distinct roles, in other words, it has two distinct phosphotyrosine motifs as its binding target. Upon cytokine stimulation, one or more of the receptor subunits become phosphorylated on their tyrosine residues, which serve as the primary binding site for the SH2 domain of the cognate STAT molecule. This SH2–phosphotyrosine interaction is the major determinant of cytokine specificity in STAT activation. For example, the SH2 domain of STAT3 binds to Tyr-X-X-Gln motifs in receptors such as gp130 and G-CSF receptor, while that of STAT6 recognizes Tyr-X-X-Phe motifs present in the IL-4 receptor α. Introduction of the Tyr-X-X-Gln motif into the erythropoietin receptor, which normally does not activate STAT3, renders the receptor capable of activating STAT3 in response to erythropoietin stimulation. Furthermore, swapping of SH2 domains between the STATs can alter receptor recognition. Interaction with receptor brings STAT to close proximity to the receptor-associated Jak (or other tyrosine kinases), leading to phosphorylation of STAT on its conserved tyrosine. Following STAT phosphorylation, the SH2 domain now serves a second function, by recognizing a phosphotyrosine residue of another STAT molecule; this results in dimer formation between two STAT molecules through reciprocal SH2–phosphotyrosine binding. STATs can form homodimers, and the following heterodimers: STAT1–STAT2, STAT1–STAT3 and STAT5A–STAT5B. The dimerized STATs now translocate into the nucleus, bind to response elements in target genes and activate transcription.

The mechanism of nuclear translocation of STATs is less well understood, but it has been shown that IFNγ-stimulated, tyrosine-phosphorylated STAT1 associates with components of the nuclear pore-targeting complex, including NPI-1, and requires a small G protein, Ran, and its GTPase activity for nuclear translocation. NPI-1 is known to function as a receptor for nuclear localization signals (NLSs) consisting of a cluster of basic amino acid residues located in the primary structure. Interestingly, the STAT1-binding domain of NPI-1 is located in the C-terminal region, which is distinct from the conventional NLS-binding domain. In accordance with this, STAT1 does not have an NLS as defined by the basic amino acid cluster. Since dimerization of STAT1 is shown to be essential for interaction with NPI-1, it is likely that the NLS of STAT1 is not solely defined by its primary sequence but rather involves the tertiary structure of the dimerized complex.

Although dimerization is a prerequisite for nuclear localization and DNA-binding of STATs, it is not necessarily sufficient for their transactivation. There are two STAT1 isoforms produced by alternative splicing, STAT1α and STAT1β, with the latter lacking C-terminal 38 amino acids of the former. Interestingly, STAT1α, but not STAT1β, can support the IFNγ-induced transcriptional activation, suggesting a possible requirement of this C-terminal region for STAT1-mediated transactivation. This notion is supported by the finding that ligand-induced phosphorylation of a serine residue present in this region of STAT1α is required for maximal transcriptional activation by IFNγ stimulation. The requirement of serine phosphorylation for transactivation has also been noted for STAT3, and may also be applicable for STAT4 and STAT5. STAT2 is not subjected to serine phosphorylation, but its C-terminal region is rich in acidic residues and can also serve as the transcriptional activation domain.

Evidence has accumulated showing physical and functional interactions between STATs and other transcription factors: for instance, STAT1 and the specificity protein 1 (Sp1) in the intercellular adhesion molecule (ICAM) gene promoter, or STAT5 and the glucocorticoid receptor in the β-casein gene promoter. Interaction of STAT1 or STAT2 with the transcriptional co-activator CREB-binding protein (CBP) has also been demonstrated. In some cases, multiple STAT dimers interact through their N-terminal regions and cooperatively bind to tandem STAT-recognition sites in a single promoter. Although STATs were initially thought to constitute one of the quickest and simplest pathways from the plasma membrane to the nucleus, functional cross-talk clearly exists between STATs and other signaling pathways, at least at the level of serine phosphorylation and in promoter elements.

2.5 Negative regulation of signaling

2.5.1 Phosphatases

For tight control of cell growth and differentiation, a mechanism for *negative* regulation and termination of signals is as important as a mechanism for receptor activation; for example, unregulated proliferation of hematopoietic cells may cause leukemia. Activation of tyrosine kinases and increasing phosphorylation initiate receptor signaling, hence it is reasonable to assume that dephosphorylation mediated by protein tyrosine phosphatases plays a role in turning off receptor signaling. SHP-1 (previously termed HCP, SH-PTP1, PTP1C or SHP), a tyrosine phosphatase expressed specifically in hematopoietic cells, is structurally related to SHP-2, with two tandem SH2 domains in the N-terminus and a phosphatase domain in the C-terminus. While SHP-2 positively regulates the Ras pathway, SHP-1 functions predominantly as a negative regulator of receptor signaling. Tyrosine 429 of the erythropoietin receptor is required for binding of SHP-1, and cells expressing erythropoietin receptors with a mutation at this position are hypersensitive to erythropoietin and display prolonged activation of Jak2 after erythropoietin stimulation. Interestingly, a C-terminally truncated erythropoietin receptor lacking the same tyrosine causes dominantly inherited benign human erythrocytosis. Likewise, SHP-1 is also required for down-modulation of the IFNα-stimulated Jak1 activity. SHP-1 also binds to the βc-subunit of the GM-CSF, IL-3 and IL-5 receptors and to SCF receptor, and seems to play a general role in negative regulation of cytokine receptor signals. Mice carrying an SHP-1 null mutation (*motheaten*) or a deletion in the phosphatase domain of SHP-1 (*motheaten viable*) have severe defects in their hematopoietic and immune systems, most prominently hyperproliferation and inappropriate activation of myeloid cells, indicating the obligatory function of this phosphatase *in vivo*.

2.5.2 CIS family proteins

Another potential mechanism for down-regulating cytokine signaling involves a molecule named cytokine-inducible SH2 (CIS) and related proteins, which are inducibly expressed by cytokine stimulation and in turn inhibit cytokine signaling, thus constituting a negative feedback loop of cytokine signaling. CIS is a relatively small molecule with a central SH2 domain, and is encoded by *cis*, a gene that was initially identified as a cytokine-induced immediate early response gene. Various cytokines, including IL-2, GM-CSF/IL-3 and erythropoietin, induces *cis* expression through the STAT5-mediated pathway. Overexpression of *cis* inhibits STAT5 activation in response to IL-3 or erythropoietin, as well as resulting in decreased IL-3-induced proliferation of Ba/F3 cells. Thus, this molecule probably negatively modulates IL-3 and erythropoietin signaling. Since CIS associates with tyrosine-phosphorylated IL-3 and erythropoietin receptors, perhaps through its SH2 domain, it is possible that CIS down-regulates receptor signaling by masking phosphotyrosine residues of the receptor and thus competing with other SH2-containing molecules. Alternatively, CIS may recruit other molecules, such as phosphatase or protease, which indeed inhibit receptor function.

Making use of entirely different approaches, three groups of investigators recently identified a novel, CIS-related molecule, variously named SOCS-1, JAB and SSI-1. SOCS-1 (suppressor of cytokine signaling-1) was identified based on its ability to inhibit IL-6-induced macrophage differentiation and growth arrest of murine monocytic leukemic M1 cells; JAB (Jak-binding protein) was found using a yeast two-hybrid screen of molecules that interact with the JH1 (kinase) domain of Jak2, while SSI-1 (STAT-induced STAT inhibitor 1) because its SH2 domain was recognized by a monoclonal antibody raised against the SH2 domain of STAT3. Studies of SOCS-1/JAB/SSI-1 indicate that this molecule, like CIS, is inducibly expressed upon cytokine stimulations and inhibits various aspects of cytokine signaling. SOCS-1/JAB/SSI-1 is not likely to bind to cytokine receptors but instead associates with all four JAKs and inhibits their catalytic activity. When overexpressed, SOCS-1/JAB/SSI-1 inhibits IL-6-induced gp130 phosphorylation and STAT3 activation, erythropoietin-induced STAT5 activity, IL-2- and IL-3-induced c-*fos* promoter activation, and IL-6- and LIF-induced differentiation and growth arrest of M1 cells.

Based on an expression sequence tag (EST) database search, three additional CIS-related genes, SOCS-2/CIS2/SSI-2, SOCS-3/CIS3/SSI-3, and CIS4 were cloned and at least two additional candidates (CIS5, CIS6) of this gene family were found in the database. In this CIS family, the central SH2 domain and a C-terminal 40–50 amino acid region, designated as the SOCS box or the CIS homology (CH) domain, are highly conserved, while the N-terminal regions of these proteins share little similarity. Among these, SOCS-3/CIS3/SSI-3 apparently functions as a negative regulator; it inhibits LIF-induced tyrosine phosphorylation of STAT3 as well as LIF-induced growth arrest of M1 cells. It is controversial whether or not SOCS-2/CIS2/SSI-2 exhibits an inhibitory effect and functions of others remain unknown. Since CIS family members exhibit differences in tissue distributions and cytokine inducibilities, it is conceivable that each of these proteins has distinct functions, particularly *in vivo*. It is also tempting to speculate that CIS family proteins constitute, at least in part, the molecular basis of cytokine cross-talk, where a particular CIS member induced by a cytokine then suppresses the signaling of another cytokine.

2.6 How is the cytokine-specificity in signaling achieved?

2.6.1 Activation of cytokine-specific signaling molecules

The specific functions of each cytokine can be explained, at least in part, by the specific signaling pathway(s) activated by that cytokine. For example, STAT4 and STAT6 are activated specifically by IL-12 and IL-4, respectively. This suggests specific roles for STAT4 and STAT6 in IL-12 and IL-4 signaling, respectively, and studies using knockout mice clearly support this idea. In the case of STAT1, not only interferons but also IL-6, IL-10, growth hormone, and epidermal growth factor (EGF) activate this STAT in cultured cells. Interestingly, however, in STAT1 mutant mice, only the response to interferons was lost and no marked deficiency was observed in the response to other cytokines. Therefore, STAT1 seems to play a specific role in IFN signaling *in vivo*. In the case of other cytokines, however, functions of their cognate STAT molecules do not necessarily correlate with their biological activities; it is likely that the specificity of each cytokine cannot be explained by STATs alone. The same STAT molecule may even activate a different sets of target genes, depending on cell types and/or stimuli. Various factors may contribute to induce a different outcome, including the nature of other signaling pathways and/or transcription factors activated simultaneously, and possibly the chromatin structure.

2.6.2 Lineage-specific expression of receptor subunits: lessons from transgenic mice

Another important factor which also defines the specificity in cytokine action is the cell type- or lineage-specific expression of the receptor subunit(s). In the case of the GM-CSF/IL-3/IL-5 receptors, for example, the receptor βc-subunit is expressed in a broad range of cells. In contrast, the expression of their cognate α-subunits are lineage-specific, as reflected in their different biological functions.

This notion was further supported in a series of experiments using transgenic mice carrying a particular cytokine receptor gene ectopically expressed in various lineages of hematopoietic cells. For example, a transgenic line was established in which human GM-CSF receptors (both the α- and βc-subunits) are expressed constitutively in almost all lineages. Bone marrow cells were derived from these mice, and their responses to cytokines were analyzed by *in vitro* colony formation assay. Bone marrow cells derived from normal mice did not respond to human GM-CSF but formed granulocyte/macrophage colonies in response to mouse GM-CSF. Strikingly, when cells from human GM-CSF receptor transgenic mice were used, human GM-CSF stimulated colony formation not only of the granulocyte/macrophage progenitors but also of the other lineages of cells. Thus, signals from the ectopically expressed human GM-CSF receptor did not affect the cells to differentiate specifically into granulocyte/macrophage lineages, but rather solely supported proliferation and expansion of all the lineages of progenitors whose fate is likely to be determined by intrinsic mechanisms. These results clearly show that response of cells to a particular cytokine is determined, at least in part, by receptor expression. Similar results have been obtained in studies using IL-5 receptor transgenic mice or hematopoietic stem cells with ectopic erythropoietin receptors introduced by viral infection.

3. Receptors of the TNF family

3.1 Mechanism of receptor activation

The cytoplasmic parts of receptors in the TNF receptor family, like class I and II cytokine receptors, have no known enzymatic activities. A characteristic feature for this family of receptors is that activation involves receptor trimerization. TNF family ligands, such as TNF-α, TNF-β (lymphotoxin), or Fas ligand (Fas-L), have a trimeric structure. Analysis using X-ray crystallography indicated that a TNF-β trimer makes a complex with three molecules of the extracellular region of the TNF receptor, suggesting that TNF induces

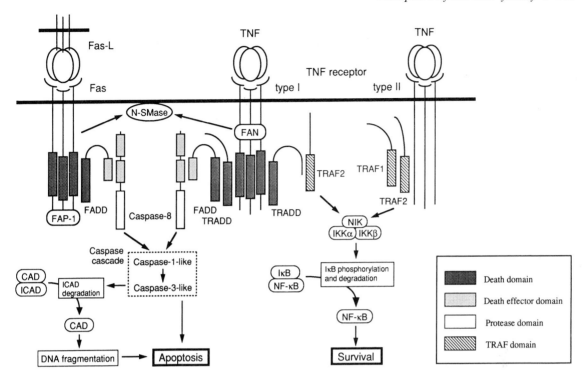

∎ **Figure 3.** Molecular components and possible interactions in TNF receptors and Fas signaling complexes leading to apoptosis and/or cell survival.

trimerization of the receptor to transduce the signal. Dimerization with a divalent anti-Fas or anti-TNF receptor monoclonal antibody does not activate these receptors, whereas antibodies that have a tendency to aggregate can function as potent agonists. In accordance with this, G-CSF-induced dimerization of a chimeric molecule consisting of the G-CSF receptor extracellular and Fas cytoplasmic regions is not sufficient to trigger cell death signals; upon receptor cross-linking using a polyclonal antibody against the G-CSF receptor, cells expressing this chimeric molecule undergo apoptosis. These results together indicate that dimerization of the Fas or TNF receptor type I (TNFRI) cytoplasmic region is not sufficient to transduce signals, and that the active form of these receptors involves oligomerization (probably trimerization). Dimerization of the class I and II cytokine receptors or of the receptor tyrosine kinases leads to activation of the associated or intrinsic tyrosine kinases by cross-phosphorylation, which initiates downstream signaling events. In case of the TNF receptor family, however, it remains to be determined how receptor trimerization leads to recruitment

and activation of signaling molecules associated with the receptor cytoplasmic region.

3.2 Death signaling

Among members of the cytokine receptor superfamily, TNF receptor type I (p55) and Fas are unique in that they can induce apoptosis of the cells in which they are expressed. The induction of apoptosis by TNF or Fas-L can occur in cells whose RNA or protein synthesis is blocked, suggesting that components of the death signaling pathway(s) pre-exist in a latent form within the cells and are modulated following ligand binding. Recent findings revealed a critical function of members of the interleukin-1β-converting enzyme (ICE)/CED-3 family of cysteine proteases, collectively termed 'caspases'. It appears that a basic concept has been established concerning the death signaling pathway from Fas and the TNFRI: ligand binding induces recruitment, and subsequent activation, of a caspase to the receptor complex through specific protein–protein interactions of some adaptor proteins, and the activated caspase in turn triggers downstream 'caspase cascade' where a number of caspases are sequentially activated by protein processing (Figure 3).

3.2.1 Death domain- and death effector domain-mediated protein interactions

TNFRI and Fas possess within their cytoplasmic portions a characteristic domain of about 80 amino acids, called the 'death domain', which is critical for transducing a death signal. Notably, the death domain of the TNFRI is essential not only for death signaling but also for other signaling events, such as NF-κB activation (see below). Several additional transmembrane proteins with death domains have been identified; these include DR3 (also called WSL-1, Apo3, LARD and TRAMP), DR4, and DR5 (also called Trick2, TRAIL-R2 and KILLER/DR5), and all belong to the TNF receptor family. Among them, DR3 is of particular interest in that it is only expressed in the lymphoid organs. The ligand for DR4 and DR5 is the TNF-related apoptosis-inducing ligand (TRAIL)/Apo-2 ligand, another TNF-like protein that can induce apoptosis.

The death domains function as an interface of protein–protein interaction through homophilic binding. These motifs are also found in a variety of other cellular proteins, including FADD (Fas-associated death domain), TRADD (TNFRI-associated death domain protein), RIP (receptor-interacting protein) and RAIDD (RIP-associated ICH-1/CED-3-homologous protein with a death domain), and mediate association of these molecules with the receptor and/or with each other, following ligand binding to the receptor. FADD, also known as MORT1 (mediator of receptor-induced toxicity 1), is a cytoplasmic protein that has a C-terminal death domain which interacts with the Fas death domain. FADD does not interact with TNFRI directly, but does so via an adaptor protein termed TRADD. TRADD has a C-terminal death domain which interacts with death domains of both FADD and TNFRI. RIP is a serine/threonine kinase with a C-terminal death domain, which binds to the death domains of both FADD and TRADD. The kinase domain of RIP is not required for transduction of the death signal and its function remains unknown, while the death domain of this molecule interacts with another death domain-containing molecule, termed RAIDD.

Some of these death domain-containing adaptor molecules in turn recruit members of caspases, and this is likely to trigger activation of cell death pathways. The best characterized is the interaction of FADD with caspase-8, originally termed MACH (MORT1-associated CED-3 homolog) or FLICE (FADD-like IL-1β-converting enzyme). Binding of FADD to caspase-8 involves a shared sequence motif, called the death effector domain (DED). FADD possesses a DED in the N-terminal region, while caspase-8 contains two tandem DEDs in the region upstream of the proteolytic moiety, or prodomain. Association between these molecules occurs via a homotypic interaction between their DEDs, only when the ligand is bound to the TNFRI or Fas. Therefore, binding of FADD to Fas and TNF receptor complexes, by direct interaction and through TRADD, respectively, leads to recruitment of caspase-8, in the same complexes. This event results in activation of caspase-8, probably mediated by its cleavage, leading to a sequential activation of other caspases. Another recently characterized caspase, caspase-10 (or Mch4), also contains the DED motif and binds FADD. Caspase-10 is likely to play a similar but not identical role as caspase-8 in Fas signaling. RAIDD can bind caspase-2 (ICH-1) and recruit it to the TNFRI; their contribution to TNFRI- or Fas-mediated apoptosis remains unknown.

3.2.2 Protease (caspase) cascade

Involvement of caspases in apoptosis was initially considered based on the finding that *ced-3*, the *Caenorhabditis elegans* gene essential for programmed cell death, encodes a protein that has sequence homology to the mammalian cysteine protease ICE (now called caspase-1). This notion has been given further support by subsequent studies, especially those involving use of specific inhibitors such as baculovirus p35 protein, poxvirus CrmA protein, and aldehydes or fluoromethyl ketone-derivatized synthetic peptides. To date, more than 10 caspases have been identified in mammals. Each of these enzymes is synthesized as a pro-enzyme (or zymogen), which is activated after being proteolytically cleaved at specific sites to form a heterodimeric catalytic domain. These sites conform to the substrate consensus for the caspase family proteases, suggesting that activation of these proteases is achieved by auto- and trans-cleavage and is likely to form a cascade. In fact, the sequential activation of caspases in Fas-induced apoptosis has been implicated, where the caspase-3 (CPP32, Yama, or apopain)-like activity is dependent on the preceding activation of a caspase-1-like protease(s). Although the exact role of each caspase in Fas- and TNFRI-mediated apoptosis remains largely unknown, caspase-8 clearly plays a key role by establishing direct linkage between receptor complexes and the downstream caspases.

Activated caspases themselves may induce disassembly of cellular structures and functions by directly cleaving essential components, and/or by activating other effector molecules, through protein processing. A number of proteins, in addition to caspases, have been identified as substrates to be cleaved by caspases, including poly(ADP-ribose) polymerase (PARP) and the catalytic subunit of DNA-dependent protein kinase, both of which are involved in DNA damage sensing and repair. Caspase-6 (Mch2) cleaves nuclear lamins, which may lead to deregulation of nuclear structures. Other identified targets of caspases include the 70 kDa protein component of the U1 small nuclear ribonucleoprotein, topoisomerase I, protein kinase Cδ, α-fodrin, β-actin, and so forth. However, the significance of these substrates in apoptosis remains to be established. Multiple substrates, rather than just a single molecule, probably need to be cleaved to induce efficient apoptosis.

Cleavage of chromatin at internucleosomal sites occurs in most apoptotic cells, resulting in the characteristic 'ladder' detected by gel electrophoresis of chromosomal DNA. Very recently, key components of molecular machinery which mediate this process have been identified, and subsequently proved to be a direct target of the caspase cascade: a caspase-activated deoxyribonuclease (CAD), and its inhibitor (ICAD). These molecules were purified and cloned from mouse cells, and ICAD was found to be highly homologous to the 45 kDa subunit of human DFF (DNA fragmentation factor), a protein that induces DNA fragmentation after it has been activated by caspase-3. In living cells, CAD exists in an inactive form complexed with ICAD; binding of ICAD inhibits the DNase activity of CAD and masks its NLS to retain CAD in the cytoplasm. Importantly, there are two caspase-3 cleavage sites in ICAD, and caspase-3 cleaves ICAD and inactivates its CAD-inhibitory effect. Thus, caspases activated by apoptotic stimuli cleave ICAD, releasing CAD to enter the nucleus where it degrades chromosomal DNA. Intriguingly, overexpression of ICAD in human Jurkat cells completely blocks chromatin degradation upon apoptotic stimuli, yet cannot prevent other characteristic features of apoptosis such as activation of caspase-3, degradation of PARP, surface-membrane expression of phosphatidylserine, and loss of integrity of mitochondria, thereby indicating that the cell can undergo apoptosis without DNA degradation, probably as a result of cleavage of other critical cellular substrates by caspases.

3.3 NF-κB activation and cell survival

Stimulation of the Fas antigen or TNFRI alone can induce cell death in only a limited number of cell types. Treatment of resistant cells with agents that block protein synthesis renders the cells sensitive to induction of apoptosis. This suggests that a mechanism may exist to protect against apoptosis and which depends on protein synthesis. Several lines of evidence revealed that the resistance of cells against TNF-induced cell death results from activation of NF-κB and expression of its target genes by TNF itself. Thus, cells that are normally resistant to TNF-induced death become sensitive to TNF stimulation and undergo apoptosis when NF-κB activation is blocked. Therefore, TNF transmits signals for both apoptosis and survival, and the final outcome of the target cells is likely to be determined by a balance between these two opposing effects.

TNF can induce activation of NF-κB through both type I and II receptors. Induction of NF-κB activation by TNFRI, like that of apoptosis, is mediated by the receptor death domain as well as its binding partner, TRADD. Downstream effector of TRADD in the NF-κB-inducing signaling cascade, however, is not FADD, but another molecule, TRAF2, a member of the TRAF (TNF receptor-associated factor) family of proteins. The TRAF family now includes six members, TRAFs 1 to 6, and shares a structural feature with a C-terminal TRAF domain which is conserved in this family. TRAF2 binds to the N-terminal half of TRADD through the TRAF domain, thereby forming a complex with the TNFRI. Therefore, TRADD functions as a branching point and mediates two distinct aspects of TNFRI signaling: apoptosis through FADD, and NF-κB activation and cell survival through TRAF2. In contrast, the type II TNF receptor (TNFRII; p75) interacts directly with the TRAF domain of TRAF2. Another member of the TRAF family, TRAF1, is also recruited to the TNFRII receptor complex through binding to TRAF2. The cytoplasmic region of CD40 also interacts directly with TRAF2. These receptors can also stimulate NF-κB activation through TRAF2. In addition, activation of NF-κB by the IL-1 receptor type I, which belongs to a different receptor family, involves TRAF6, a member of the TRAF family which interacts with the receptor through an intermediate protein, IRAK.

The NF-κB-inducing kinase (NIK), a member of the MAP kinase kinase kinase family, was identified as a molecule which interacts with TRAF2, and is a critical component of an NF-κB-activating cascade downstream of the TNF receptor as well as of the IL-1 receptor. NIK activates NF-κB when overexpressed, whereas kinase-deficient mutants of NIK function as dominant-negative inhibitors that suppress NF-κB activation mediated by TNF and IL-1 stimulation and by overexpression of TRADD, RIP, TRAF2 and TRAF6.

A protein serine/threonine kinase previously known as CHUK was identified in a yeast two-hybrid screen for NIK-interacting proteins, and subsequent functional studies showed that CHUK functions downstream of NIK and mediates NF-κB activation by phosphorylating IκB-α on two critical serine residues, modifications that are required for targeted degradation of IκB-α via the ubiquitin–proteasome pathway. At the same time, an attempt to purify a TNF-α-induced activity which specifically phosphorylates IκB-α at the two serine residues also led to identification of two protein kinase components of the IκB kinase (IKK) complex; IKKα, which is identical to CHUK, and IKKβ, a novel, IKKα-related kinase. Both IKKα (CHUK) and IKKβ phosphorylate various IκB proteins, although there are differences in efficiency toward a particular IκB. IKKα and IKKβ can each form homodimers and together form heterodimers, and make a ternary complex with NIK. It remains to be determined how NIK is activated by upstream signaling complexes and in turn stimulates activities of IKKs.

Identification of IKKs as critical targets of NIK revealed a contiguous pathway from the TNF (and IL-1) receptor at the cell membrane to activation of cytoplasmic NF-κB, through phosphorylation and degradation of IκB. However, several lines of evidence indicate that the IKK complex, under physiological conditions, contains several additional components, which have yet to be characterized. It is also important to determine the critical target gene(s) of NF-κB which is/are responsible for TNF-induced cell survival.

3.4 Sphingomyelinase and ceramide production

TNFRI as well as Fas initiate sequential activation of a phosphatidylcholine-specific phospholipase C (PC-PLC) and an acidic sphingomyelinase, leading to ceramide production. However, the significance of ceramide in the TNF receptor or Fas signaling as well as its effect on cellular functions remains controversial. It is possible that ceramide triggers downstream signaling events which initiate proteolytic degradation of IκB, leading to NF-κB activation. For activation of this acidic sphingomyelinase pathway, death domains of those receptors are responsible. In contrast, cells treated with membrane-permeable ceramide undergo apoptosis, suggesting the possible involvement of ceramide in the TNF receptor and Fas-mediated apoptosis.

Another signaling pathway initiated by TNFRI involves a membrane-bound neutral sphingomyelinase. This enzyme catalyzes hydrolysis of plasma membrane sphingomyelin to produce ceramide. Ceramide production by neutral sphingomyelinase leads to a sequential activation of ceramide-activated protein kinase, Raf-1 kinase and MAPKs. This neutral sphingomyelinase-mediated activation of MAPK is likely to result in growth promotion and/or phosphorylation and activation of phospholipase A$_2$ (PLA$_2$) in particular cells. PLA$_2$ produces arachidonic acid and hence plays an important role in pro-inflammatory responses. A domain of about 10 amino acids, just N-terminal of the death domain, in the cytoplasmic region of the TNFRI is apparently responsible for activation of this pathway. A novel WD-repeat protein, FAN ('factor associated with neutral sphingomyelinase activation') has been identified; it specifically binds to this domain and regulates ceramide production by neutral sphingomyelinase, but not by acidic sphingomyelinase. Fas also signals for a neutral sphingomyelinase pathway, leading to activation of MAPK and PLA$_2$. Because FAN does not interact with Fas, another molecule may be responsible for Fas-mediated activation of this pathway.

3.5 Other signaling pathways

It has been suggested that members of the MAPK family, including c-Jun N-terminal kinases (JNKs)/stress-activated protein kinases (SAPKs) and p38/RK/Mpk2, are involved in apoptosis. TNF stimulation results in activation of both JNK and p38^{MAPK}, but it is less likely that these molecules play a significant role in TNF-induced apoptosis. TNF-induced activation of JNK, like that of NF-κB, is mediated by TNFRI through TRAF2 and RIP, but not FADD, which are associated with receptor-bound TRADD. Dominant-negative FADD inhibits TNF-induced apoptosis but not JNK or NF-κB activation, while dominant-negative TRAF2 blocks both JNK and NF-κB activation by TNF but not apoptosis, thereby clearly dissociating JNK activation from the induction of apoptosis. Since inhibition of NF-κB activation, but not JNK, potentiates the sensitivity of cells to TNF-induced apoptosis, anti-apoptotic effects of TNF are most likely mediated by NF-κB rather than by JNK. Fas can also activate the JNK pathway, but its significance remains unclear.

At the C-terminus, Fas has a negative regulatory domain, or a 'salvation domain'. This domain of 15 amino acids interacts with FAP-1, a cytoplasmic protein tyrosine phosphatase previously identified as PTP-BAS. Expression of FAP-1 results in inhibition of Fas-induced apoptosis in Jurkat T cells, suggesting a possible role for tyrosine phosphorylation in positive regulation of apoptosis. In accordance with this, tyrosine kinase inhibitors, genistein or herbimycin A, suppress the Fas-induced apoptosis in Jurkat cells and neutrophils. However, it is unclear if tyrosine phosphorylation has a significant role in Fas and TNFRI-mediated death signaling.

4. Receptors with tyrosine kinase activity: M-CSF receptor and SCF receptor

4.1 Mechanism of receptor activation

The mechanism of activation and signaling of RTKs has been extensively analyzed using receptors for growth factors such as EGF and PDGF, and is shared by the M-CSF and SCF receptors. A prominent feature of signal transduction by RTKs is ligand-induced dimerization of the receptor, and autophosphorylation.

M-CSF and SCF, the ligands for these receptors, exist as non-covalently linked homodimers. Thus, the binding of these ligands causes dimerization of their receptors, which, in the cytoplasm, brings two kinase domains into close proximity. The kinases are activated by cross-phosphorylation, then phosphorylate multiple tyrosine residues on their own. The resultant phosphotyrosine residues on cytoplasmic domains of these receptors serve as binding sites for SH2 or PTB domains of other proteins involved in downstream signaling pathways.

Because the genes encoding the M-CSF and SCF receptors were initially identified as cellular counterparts of viral oncogenes, it is possible that mutation in these genes renders the receptors constitutively activated, thereby leading to tumorigenesis. There are two types of gain-of-function mutants of the SCF receptor:

(1) A point mutation leading to replacement of a specific aspartic acid residue within its catalytic domain by tyrosine or valine. Receptors with these mutations, by unknown mechanisms, exist in a hyperphosphorylated state without forming a stable dimer.

(2) A point mutation, or a deletion, affecting the membrane-proximal region of the cytoplasmic

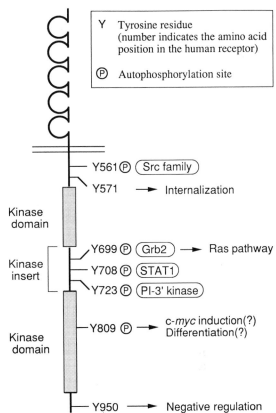

∎ **Figure 4.** Functional architecture of the M-CSF receptor and its downstream signaling pathways.

domain and leading to constitutive dimer formation of the receptor, without binding to SCF.

Receptors with mutations of either type are capable of inducing factor-independent growth of hematopoietic cell lines, and transplantation of hematopoietic stem cells expressing either of them can cause leukemia in mice.

4.2 Downstream signaling

4.2.1 Recruitment of multiple signaling proteins through specific phosphotyrosine–SH2 domain interactions

As mentioned above, ligand binding induces autophosphorylation on tyrosine residues of these receptors and this in turn creates docking sites for multiple signaling molecules with SH2 and/or PTB domains. In the case of the M-CSF receptor, ligand-induced phosphorylation sites as well as their functions are well defined (Figure 4).

At least five tyrosine autophosphorylation sites have been identified within the cytoplasmic region of the M-CSF receptor. M-CSF stimulation results in recruitment to the receptor complex and/or activation of various signaling molecules, including the Src family tyrosine kinases Src, Fyn and Yes, PI3K, SHP-1 and 2, Shc, Grb2, Sos, Cbl and SHIP. Only a few of these proteins have been shown to bind directly to the receptor. Tyrosine 561 in the membrane-proximal region of the human receptor (equivalent to residue 559 in the murine receptor) is a potential binding site for Src family tyrosine kinases. There are three autophosphorylation sites located in the kinase insert domain of the murine M-CSF receptor, at tyrosines 697, 706 and 721 (699, 708 and 723 in the case of the human receptor), which are important for binding of SH2 domain-containing proteins: the SH2 domain of Grb2 has been shown to bind to Tyr697. Grb2 binding to the receptor is believed to result in translocation of the guanine nucleotide exchange factor Sos to the membrane and subsequent Ras activation. The p85 regulatory subunit of PI3K binds to Tyr721, thereby activating PI3K-mediated signaling events. Although deletion of the kinase insert from v-Fms does not affect its ability to transform NIH3T3 cells, either the same deletion or mutations of Tyr697 and/or Tyr721 of the murine M-CSF receptor decreases its potential to induce changes in morphology and to increase cell growth rate in response to M-CSF in Rat-2 fibroblasts. Tyrosine 809 within the kinase catalytic domain of the human M-CSF receptor (corresponding to Tyr807 in the murine receptor) neither binds any known signaling proteins nor is required for induction of the c-*fos* and *jun*B genes, but is important for c-*myc* induction in NIH3T3 cells. A mutant receptor lacking this tyrosine is unable to support M-CSF-dependent proliferation of NIH3T3 cells in serum-free medium and colony formation in semi-solid medium, which are rescued by forced expression of c-*myc*. In contrast, mutation of Tyr807 of the murine receptor was reported to abrogate differentiation of the FDC-P1 clone and conversely increase the rate of M-CSF-dependent proliferation. In any event, this tyrosine is a critical site for the regulation of growth and differentiation induced by M-CSF.

The cytoplasmic region of the SCF receptor associates with PI3K and phospholipase C (PLC)-γ1. Although the site(s) of interaction of PLC-γ1 is unknown, it has been reported that the kinase insert region is required for association with the p85 subunit of PI3K.

4.2.2 Comparison with the receptors with no intrinsic tyrosine kinase activity

Evidence indicates that activation of Jaks and STATs is not restricted to signaling of class I and II cytokine receptors lacking intrinsic tyrosine kinase activity, but is also induced by several RTKs. The most characterized one is activation of Jak1 and STAT1 and STAT3 by the EGF receptor. PDGF has also been shown to activate several Jaks and STATs. In the case of hematopoietic cytokines, M-CSF induces activation of Tyk2 in macrophages and Tyk2 and Jak1 in fibroblasts, as well as STAT1, 3 and 5 in both. Tyrosine 706 of the murine M-CSF receptor (or Tyr708 of the human receptor) is required for efficient activation of STAT1, but not of STAT3, in FDC-P1 cells, and probably binds directly to STAT1 when phosphorylated. It remains to be determined, however, whether activation (phosphorylation) of STATs is mediated by the receptor tyrosine kinase itself, Jaks, Src family members, or another yet-to-be identified kinase. Needless to say, the roles of these molecules in RTK signaling, especially *in vivo*, is an important issue.

As Jaks and STATs are also activated by RTKs, it appears that no clear-cut difference exists between signaling of receptors with and without intrinsic tyrosine kinase activity. It is somewhat enigmatic why two distinct receptor systems have evolved for cytokines, if they can both transmit functionally equivalent signals. In general, the class I and II cytokine receptors are relatively divergent and even the mouse and human counterparts are not perfectly conserved, in contrast to growth factor receptors which are evolutionary well preserved from nematodes to mammals. It is thus tempting to speculate that these cytokine receptors are not essential for development and survival of animals so that they can possess a degree of flexibility and be suited to function in environmental changes and stress, such as in cases of immune and inflammatory systems.

Notably, functional and physical interactions of the erythropoietin and SCF receptors, belonging to two distinct receptor families, have been reported. The SCF receptor physically associates with the erythropoietin receptor through their cytoplasmic domains, and SCF induces tyrosine phosphorylation of the erythropoietin receptor. This phosphorylation of the erythropoietin receptor is likely to be involved in downstream signaling, since SCF can support the proliferation of 32D cells that express the SCF receptor but only if they also express the erythropoietin receptor. Since mutations in the SCF receptor gene affect erythropoiesis, SCF is likely to play a significant role in supporting proliferation and/or differentiation of erythroid progenitors probably by activating the erythropoietin receptor through direct interaction with its receptor. In this case, the class I receptor functions downstream of and is required to mediate signaling from RTK. On the contrary, a recent report presented an example where the class I receptor transmits its signals through an RTK; growth hormone stimulates tyrosine phosphorylation of the EGF receptor and its association with Grb2, leading to activation of ERK and expression of c-*fos*. These signaling events induced by growth hormone require the tyrosine-phosphorylation site of the EGF receptor to which Grb2 binds, but not its intrinsic kinase activity; instead, growth hormone-stimulated Jak2 phosphorylates this tyrosine residue, leading to creation of a docking site for Grb2. Further studies are needed to examine whether these kinds of interactions occur between other class I receptors and RTKs, as well as their functional significance *in vivo*.

4.3 Negative regulation of receptor signaling

The C-terminal tail of about 24 amino acids of the M-CSF receptor contains sequences which negatively regulate tyrosine kinase activity of the catalytic domain. This portion contains a tyrosine residue at position 969 (human), which is analogous to the inhibitory Tyr527 of c-Src tyrosine kinase. Since Tyr969→Phe mutation potentiates transforming activity, this tyrosine may be responsible for negative regulation of the M-CSF receptor signaling through the C-terminus. In the v-Fms oncoprotein, deletion of this C-terminal portion, together with several other mutations which render the receptor kinase M-CSF-independent, is required for its potential to transform efficiently.

As pointed out in the section on class I and II cytokine receptor signaling (see Section 2), tyrosine phosphatase SHP-1 plays critical roles in hematopoiesis and immune function by negatively regulating receptor signals. This is also the case for cytokines which transduce signals through RTKs. In bone marrow-derived macrophages from homozygous *motheaten* (me) mice lacking functional SHP-1, the M-CSF receptor as well as several cellular proteins are hyperphosphorylated upon M-CSF stimulation, suggesting that SHP-1 dephosphorylates these proteins. Furthermore, these cells hyperproliferate in response to M-CSF. These results indicate that SHP-1 plays a role in negative regula-

tion of M-CSF signaling. However, physical association of SHP-1 with the M-CSF receptor has not been observed. In the case of the SCF receptor, SHP-1 binds to this receptor upon ligand stimulation through its SH2 domains, although the responsible tyrosine residue(s) of the receptor remain(s) to be determined. The physiological relevance of this interaction in regulating the SCF receptor function has been demonstrated by genetic analyses using mutant mice: mutations in either the SCF receptor (*dominant white spotting*, or W) or SHP-1 (me) result in lethality and multiple hematopoietic defects. Mice homozygous for both the W^v (*white viable*: a kinase-defective, weak allele of W) and *me* loci show increased viability, suggesting that the lethal effects of these two mutations are mutually compensatory. In addition, several phenotypic abnormalities of both W and *me* single homozygotes are also suppressed in the double homozygous mutant mice, including the lung disease and mast cell deficiency associated with *me* and W mutations, respectively. These results show a strong intergenic complementation between these two loci, and establish that SHP-1 negatively regulates SCF receptor function *in vivo* and that excess signaling of the SCF receptor contributes substantially to the pathology of the *me* phenotype.

After ligand-induced activation, cell surface receptors are internalized by endocytosis via clathrin-coated vesicles and are eventually transferred to secondary lysosomes for degradation of both receptor and ligand. This step may be involved in down-modulating the signal, but a potential role in the signaling process has not been ruled out. In the case of the M-CSF receptor, a four-amino-acid sequence in the juxtamembrane domain, containing Tyr569, is implicated as part of the internalization signal for endocytosis, although phosphorylation of this tyrosine is not a prerequisite. This sequence is entirely conserved in the M-CSF, SCF, and α- and β-PDGF receptors, suggesting a common regulatory mechanism for this family of receptors.

5. Receptors with serine/threonine kinase activity: the TGF-β receptor family

5.1 Mechanism of receptor activation

A functional receptor complex for the TGF-β family is composed of type I and type II receptor

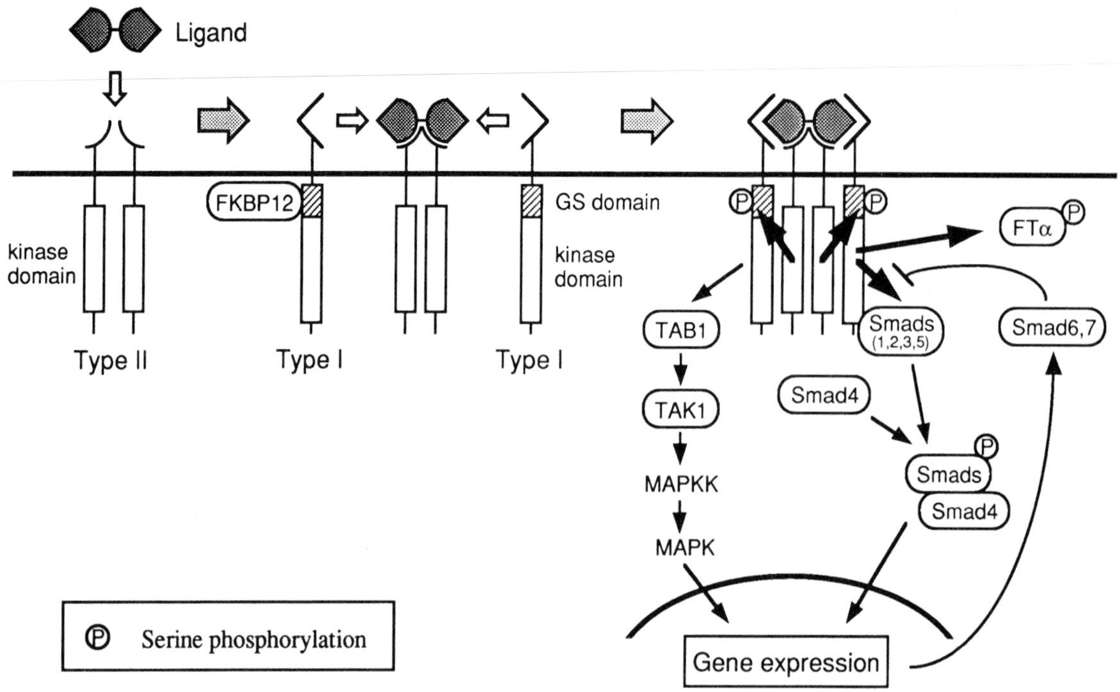

■ **Figure 5.** Model for the activation and signal transduction of the TGF-β receptor family.

subunits, both of which are membrane-bound serine/threonine kinases. The type II receptor functions as a primary binding subunit for the ligand, while the type I subunit functions as a signal transducer. Even in the absence of the ligand, the type II receptor is constitutively phosphorylated and activated; ligand binding does not alter its phosphorylation status. Functional TGF-β is a disulfidebonded dimer. After TGF-β family ligands have bound to the type II receptor, they form a highaffinity receptor complex with the type I receptor (Figure 5). This receptor complex is thought to be a heterotetramer: the type II receptor exists as a homodimer in the absence of the ligand, and the ligand–receptor complex contains two molecules of the type I receptor. The formation of this complex results in phosphorylation, by the type II receptor, of the type I receptor at multiple serine/threonine residues, mainly in the GS domain. The GS domain is a characteristic motif present in the membraneproximal region of the type I receptor cytoplasmic domain, and contains repetitive sequences of glycine and serine. Phosphorylation of the type I receptor in its GS domain results in an increase in its kinase activity, which leads to phosphorylation

of multiple cellular substrates. When a particular residue in the GS domain of the type I receptor is changed to an acidic residue, such as aspartic or glutamic acid, the type I receptor becomes constitutively activated and can transduce signals without the ligand or the type II receptor. Type I receptors with this kind of activating mutation can signal even in the presence of a dominant-negative type II receptor, suggesting that type I receptors function downstream of the type II receptors, and once activated can transmit signals independent of type II receptors.

5.2 Downstream signaling pathways

5.2.1 Smad family proteins

The TGF-β/TGF-β receptor system is highly conserved from higher eukaryotes to invertebrates. In the fruit-fly *Drosophila*, the *decapentaplegic* (*dpp*) gene encodes a bone morphogenetic protein (BMP)-like factor, which transduces signals through two type I receptors called Thick veins (Tkv) and Saxophone (Sax), and a type II receptor, Punt. Genetic approaches to identify genes related to *dpp*, *tkv* or *punt* resulted in the cloning of the

Mad (*Mothers against dpp*) gene. Similarly, the *sma* genes of the nematode *C. elegans*, which function downstream of a BMP type II receptor-like molecule, DAF4, were genetically identified and found to encode proteins with homology to Mad. These proteins, together with the subsequently identified vertebrate homologs, now constitute the 'Smad' family, whose name was derived from combining the names 'Sma' and 'Mad'.

At least nine Smads have been identified in vertebrates. Smads are cytoplasmic proteins, with no homology to other molecules. However, there are motifs in the N- and C-terminal regions that are conserved among the Smad family members. All the described null mutations in *Mad*, *sma2* and *sma3* are missense or nonsense mutations that fall within a highly conserved, short portion of the C-terminal domain. Mutations at this 'hot spot' have also been found in the human *DPC4* (*deleted in pancreatic carcinoma*) gene, which encodes Smad4. *DPC4* was characterized as a tumor-suppressor gene in human chromosome 18q21.1; it is mutated or deleted in half of all pancreatic malignancies. Tumor cells with a mutation in this gene do not respond to TGF-β, and introduction of normal Smad4 renders these cells responsive to growth suppression by TGF-β. Mutations in the gene encoding Smad2 have also been implicated in tumorigenesis. These Smads thus appear to play a critical role in growth suppression by members of the TGF-β family.

In *C. elegans*, three Smad-family genes, *sma-2*, *sma-3* and *sma-4*, act in the same cells and have non-redundant functions; their products may form a heterocomplex and function cooperatively. It is now evident that a some vertebrate Smads, namely Smad1, Smad2, Smad3, and probably Smad5, mediate specific signals from limited receptors, thereby functioning in a pathway-restricted manner. These pathway-restricted Smads form hetero-oligomeric complexes with another Smad molecule, Smad4 (see below). Experiments using *Xenopus* embryos have shown that *Xenopus* and/or mammalian Smad1 and Smad2 transduce signals from BMP and activin receptors, respectively. Thus, microinjection of the Smad1 mRNA into early-stage *Xenopus* embryos mimics the potential of BMP to induce ventral mesoderm formation, while microinjection of the Smad2 mRNA mimics the potential of activin to induce expression of dorsal mesoderm marker genes and formation of a secondary axis. Therefore, different TGF-β family members are likely to signal through different Smad members. After BMP2/4 treatment, Smad1 is phosphorylated by the BMP2/4 type I receptor, then translocates into the nucleus. Smad2 and Smad3 bind to the type I and type II TGF-β receptor complex, are phosphorylated by the type I receptor upon TGF-β stimulation, dissociate from the receptors, then translocate into the nucleus. Three serine residues in the C-terminal region identified as phosphorylation sites are conserved among the pathway-restricted Smads. Smad2 with mutations in these serine residues can no longer translocate into the nucleus and, when overexpressed, functions in a dominant-negative manner. Intriguingly, the C-terminal domains of Smads alone translocate into the nucleus more efficiently than the whole protein. The N-terminal domain is likely to have an inhibitory effect on nuclear translocation and/or the function of the C-terminal domain.

Among the target genes of TGF-β signaling is the plasminogen activator inhibitor-1 (PAI-1) gene. Using a reporter system including the PAI-1 gene promoter, the characteristic role of Smad4 in TGF-β signaling has been demonstrated. Transfection of Smad2 or Smad3 alone into cultured cells resulted in little induction of the PAI-1 gene, while co-expression of Smad3 and Smad4 strongly induced the PAI-1 gene, to a level comparable to that induced by TGF-β treatment. Furthermore, Smad4 associates with Smad2 following TGF-β stimulation. Therefore, Smad4 functions as a partner of Smad2 and 3 and cooperatively transduces the TGF-β and activin signals. Smad4 also cooperates with Smad1. Thus, Smad1 and Smad4 form a complex upon BMP2/4 stimulation. Overexpression of the inactive form of Smad4 blocks signals of TGF-β, activin and BMP2/4. In conclusion, Smad4 is involved in signals of both the TGF-β and activin receptors and the BMP2/4 receptors by functioning as a common partner for their specific, pathway-restricted mediators, Smad2/3 and Smad1, respectively.

In the nucleus, Smads are thought to function as transcriptional activators. The C-terminal domains of Smad1 and Smad4 have the potential to induce transcriptional activation when fused to a yeast GAL4 DNA-binding domain and bound to DNA. This activity is eliminated by a hot spot nonsense mutation found in various *Mad* and *DPC4* alleles. Interestingly, a full-length Smad1 fusion protein is inactive in this assay, but can be specifically activated in response to BMP2/4, in combination with BMP receptors. Thus, as in nuclear localization assays, activity of the C-terminal domains appears to be repressed by the N-terminal domain and this repression is eliminated by incoming BMP signals. Evidence indicates that Smads function *in vivo* as transcriptional regulators. Xenopus *Mix.2* is an immediate-early response gene induced by activin-like members of the TGF-β superfamily, and transcriptional activation of this gene depends on the presence of an activin-responsive element (ARE) in its promoter region. Molecular cloning of the activin-induced DNA-binding complex on ARE revealed that it is composed of a novel winged-helix transcription factor, FAST-1 (fork-head activin signal transducer 1) and Smad2. Smad2, which does not directly bind to DNA, forms a complex with FAST-1 in an activin-dependent fashion to generate an activated ARE-binding complex. Hence, Smads are latent cytoplasmic transcription factors, like STATs in the class I and class II cytokine receptor systems, which, after ligand stimulation, are phosphorylated and translocate into the nucleus where they bind DNA and activate transcription of target genes. Further studies are required to identify physiological targets for each Smad and their significance in TGF-β family functions.

In addition to the pathway-restricted Smads and common-partner Smad (Smad4), recent studies have identified a subset of Smads as inhibitory Smads; mammalian Smad6 and Smad7 and *Drosophila* Dad (encoded by the gene *Daughters against decapentaplegic*). Intriguingly, the expression of these inhibitory Smads is induced by TGF-β stimulation, suggesting that they constitute negative feedback loops that regulate signaling of TGF-β-family ligands. Smad6 and Smad7 differ from other Smads in their N-terminal regions and they form stable associations with type I receptors, but are not phosphorylated upon ligand stimulation. Thus, these inhibitory Smads interfere with the phosphorylation of Smad1 and Smad2 (and also Smad3, in the case of Smad7) by competing for binding to the type I receptor and blocking subsequent heterocomplex formation of these pathway-restricted Smads with Smad4, which results in inhibition of the signaling of TGF-β, activin and BMP.

Notably, Smad1, which mediates BMP signals, is also a target for mitogenic growth factor signaling through RTKs. Stimulation with EGF or HGF (hepatocyte growth factor) induces phosphorylation of Smad1 at specific serines within the linker region between the conserved domains in the N- and C-termini, and this phosphorylation is catalyzed by the ERK family of MAPKs. In contrast

to the BMP-stimulated phosphorylation of Smad1, which targets C-terminal serines and induces nuclear translocation of Smad1, ERK-mediated phosphorylation inhibits the nuclear accumulation and hence transcriptional function of Smad1. Thus, Smad1 receives two opposing regulatory signals through BMP receptors and RTKs, functioning as a merging point of the cross-talk between these two receptor systems.

5.2.2 TAB, TAK and MAPK cascade

A genetic screening based on a MAPK cascade in yeast led to identification of a novel mouse protein kinase, TAK1 (TGF-β-activated kinase 1), which was subsequently shown to function in the TGF-β signaling. TAK1 is a member of the MAP kinase kinase kinase family, with an N-terminal kinase domain. Stimulation with either TGF-β or BMP-2/4 results in increased kinase activity of TAK1, suggesting that some MAPK cascade(s) involving TAK1 exists downstream of the TGF-β family receptors. Deletion of the N-terminal region results in a constitutively activated form of TAK1, which when overexpressed can induce activation of the TGF-β-responsive PAI-1 gene promoter. Both *in vitro* and *in vivo* studies suggested that SEK1 (SAPK/ERK kinase 1)/XMEK1 (*Xenopus* MEK1), MKK3 and MKK6 can be target MAP kinase kinases (MAPKKs) of TAK1 phosphorylation. TGF-β stimulation activates these MAPKKs but not as potently as TAK1, suggesting the possible involvement of other MAPKK(s) downstream of TAK1 to transduce signals from TGF-β receptors.

Subsequently, TAB1 (TAK1-binding protein 1) was identified using the yeast two-hybrid system. TAB1 interacts with TAK1 in transfected mammalian cells. Overexpression of TAB1 increases the kinase activity of TAK1 and enhances induction of the PAI-1 gene promoter. In addition, a mutant TAB1 lacking the C-terminal TAK1-binding region cannot activate TAK1, and suppresses TGF-β-induced PAI-1 promoter activation. Therefore, TAB1 is likely to function as an activator of TAK1 and downstream MAPK cascade(s) in signaling from the TGF-β receptor (and probably from the BMP-2/4 receptor). Interaction of the receptor with either TAB1 or TAK1 has not been detected, and the mechanism of TAK1 activation by receptor signaling through TAB1 remains to be elucidated. Furthermore, since studies using dominant-negative molecules suggest that both the Smad and TAK1 pathways are necessary for TGF-β signaling, it has to be determined if there is cross-talk between these two pathways.

5.2.3 Others

Studies using a yeast two-hybrid system identified FKBP12 and farnesyltransferase-α as molecules which interact with the type I receptor. FKBP12, a protein well known to bind the immunosuppressant FK506 and rapamycin, inhibits TGF-β signaling; it binds to a Leu–Pro sequence located in the GS domain of the type I receptors and inhibits ligand-independent, spontaneous phosphorylation and activation of the type I receptors by the type II receptors. The farnesyltransferase-α and -β subunits together catalyse the farnesylation of low-molecular-weight G proteins such as Ras, a modification that is required for membrane localization of the small G proteins. (Only the FT-α/FT-β combination is catalytic, and the individual subunits do not each have activity.) The type I receptor phosphorylates farnesyltransferase-α, but does not alter its enzymatic activity; the significance of this molecule in TGF-β signaling remains unknown. It may be interesting to examine whether the small G proteins are involved in the activation of TAK1. A molecule that interacts with the type II TGF-β receptor, a WD domain-containing protein, TRIP-1, has been identified, whose functions also remain to be elucidated.

6. G-protein-coupled receptors: the IL-8 receptor

6.1 Involvement of pertussis toxin-sensitive and -insensitive G proteins in signaling

There are two distinct types of IL-8 receptors, designated α and β. Both of these receptors are members of the seven-transmembrane-domain receptor family, which are coupled to heterotrimeric G proteins. A heterotrimeric G protein consists of three subunits designated α, β and γ, the α-subunit is the guanine nucleotide-binding component and possesses intrinsic GTPase activity. Ligand binding to a G-protein-coupled receptor results in release of GDP complexed with the G protein α-subunit and subsequent loading of GTP; this causes dissociation of the heterotrimeric complex into α- and $\beta\gamma$-subunits. The released subunits, α and $\beta\gamma$, each bind to several effector molecules and modulate their functions. There are several subfamilies of G proteins, such as G_s, G_i, G_q, defined by functional differences due to the α-subunit. The G_q class of G protein α-subunits (comprising $G\alpha_q$, $G\alpha_{11}$, $G\alpha_{14}$, $G\alpha_{15}$ and $G\alpha_{16}$) can activate members of the β family of phospholipase C (PLC) isozymes, leading to phosphoinositide hydrolysis and release of inositides and diacylglycerol. Induction of PLC activity through the G_q family of G proteins is resistant to pertussis toxin inhibition, since $G\alpha_q$ subunits lack a target site of the toxin for modification which blocks ligand-induced exchange of GDP to GTP. Pertussis-toxin-sensitive pathways leading to induction of PLC activity also exist, and involve receptor-mediated release of $\beta\gamma$ subunits from members of the G_i class. The PLC-β2 isoform, but not the β1 isoform, can be activated by free $\beta\gamma$ subunits, and the toxin blocks activation of PLC-β2 by interfering with release of the $\beta\gamma$ subunits.

In neutrophils, IL-8 increases phosphoinositide hydrolysis by activating PLC-β, resulting in intracellular calcium mobilization and protein kinase C (PKC) activation. These IL-8-induced signaling events, as well as chemotaxis stimulation and the release of lysosomal enzymes, are sensitive to pertussis toxin treatment, suggesting involvement of G_i-mediated signaling. Studies using neutrophils and transfected 293 cells demonstrate that $G\alpha_{i2}$ physically associates with the IL-8 receptor and is likely to be at least one of the physiologically relevant G proteins coupled to the receptors. The IL-8 receptors, however, can functionally couple to several different $G\alpha$ proteins. In COS7 cells transfected with IL-8 receptors, members of the $G\alpha_i$ family, $G\alpha_{i2}$ and $G\alpha_{i3}$, and members of the $G\alpha_q$ family, $G\alpha_{14}$, $G\alpha_{15}$ and $G\alpha_{16}$, can be coupled to the receptors and the result is activation of PLC-β in a pertussis-toxin-sensitive and -resistant manner, respectively. Notably, expression of the $G\alpha$ subunits is regulated developmentally; for example, $G\alpha_{16}$ is apparently only expressed in hematopoietic lineage cells. Therefore, it is likely that IL-8 receptors interact with different members of the $G\alpha$ subunits and activate distinct signaling pathways, in a cell type-specific manner.

A new class of proteins, termed regulators of G protein signaling (RGSs), have been implicated in regulating signal transduction through heterotrimeric G-protein-coupled receptors. The RGS family proteins accelerate the intrinsic rate of GTP hydrolysis of $G\alpha$ subunits, thus functioning as GTPase-activating proteins (GAPs) toward $G\alpha$ and negatively modulating both the magnitude and duration of the receptor signaling. Several lines of evidence have indicated that RGS proteins stimulate the GTPase activity of $G\alpha_i$ and $G\alpha_q$, but not of other $G\alpha$ subunits. The introduction of RGS1, 2, 3 or 4 inhibits ERK1 activation in response to IL-8 stimulation in 293T cells permanently expressing the IL-8 receptor. Taken together with findings that the IL-8 receptors are coupled to G_i and G_q, RGS proteins may play a substantial role in negative regulation of IL-8 signaling.

6.2 Activation of MAPK cascades

G_i-coupled receptors can activate the guanine nucleotide exchange activity for Ras. IL-8 stimulation on human neutrophils activates Ras GTP loading, and the downstream Raf/ERK pathway. Both Raf-1 and B-Raf have been reported to be activated, yet it remains unclear if these two proteins play distinct roles in IL-8 signaling. Activation of ERK by IL-8 is inhibited by a specific inhibitor for MEK, PD098059, indicating that this activation depends on MEK activity.

Wortmannin, a PI3K-specific inhibitor, also inhibits the IL-8-induced activation of ERK, suggesting that PI3K activity is required for activation of the ERK pathway in response to IL-8 stimulation. Activation of Raf-1 and B-Raf, but not that of Ras, is also inhibited by wortmannin. Therefore, PI3K may be involved, either downstream of or in parallel to Ras, in regulating Raf-1 and B-Raf activation by IL-8 in neutrophils. The mechanism by which PI3K regulates Raf-1/B-Raf activation remains to be determined.

Various extracellular stimuli regulate parallel MAPK pathways which comprise, at present, ERKs, JNKs/SAPKs and p38/RK/Mpk2 in mammalian cells. In neutrophils, IL-8 activates $p38^{MAPK}$ also, but not JNK. The activation of $p38^{MAPK}$ was insensitive to PD098059 or to wortmannin. The pathway leading to $p38^{MAPK}$ activation by IL-8 stimulation is yet to be characterized.

The significance of MAPK activation in the biological functions of IL-8 remains unclear. IL-8 induces various cellular functions, such as shape change and migration, degranulation and respiratory burst. Neither wortmannin nor PD098059 affects IL-8-induced calcium mobilization and cell viability. However, wortmannin does inhibit the IL-8-induced granule release from human neutrophils. Neutrophil adherence was not inhibited by wortmannin to as great an extent as granule secretion was. Therefore, PI3K activity appears to regulate granule secretion and adherence, to different degrees. Activation of PI3K is also required for the induction of neutrophil migration, in response to IL-8, thereby indicating a critical role for PI3K in the multiple signaling pathways induced by IL-8. In contrast, the regulation of cell migration by IL-8 is independent of ERK and $p38^{MAPK}$ activity, since neither the MEK inhibitor PD098059 nor the $p38^{MAPK}$ inhibitor SK&F 86002 has any effects on IL-8-induced migration of human neutrophils

6.3 Others

Jaks and STATs are activated through the angiotensin receptor, a member of the G-protein-coupled receptor family. There is no available information as to whether they are also involved in IL-8 receptor signaling.

References

(This list includes selected review articles only.)

Aman, M. J. and Leonard, W. J. (1997). Cytokine signaling: cytokine-inducible signaling inhibitors. *Curr. Biol.* **7**, R784-8.

Darnell, J. E., Jr (1997). STATs and gene regulation. *Science* **277**, 1630–35.

Darnell, J. E., Jr, Karr, I. M., and Stark, G. R. (1994). Jak–STAT pathways and transcriptional activation in response to IFNs and other extracellular signaling proteins. *Science* **264**, 1415–21.

Heldin, C.-H., Miyazono, K., and ten Dijke, P. (1997). TGF-β signalling from cell membrane to nucleus through SMAD proteins. *Nature* **390**, 465–71.

Ihle, J. N., Witthuhn, B. A., Quelle, F. W., Yamamoto, K., and Silvennoinen, O. (1995). Signaling through the hematopoietic cytokine receptors. *Annu. Rev. Immunol.* **13**, 369–98.

Kehrl, J. H. (1998). Heterotrimeric G protein signaling: roles in immune function and fine-tuning by RGS proteins. *Immunity* **8**, 1–10.

Massague, J. (1996). TGFβ signaling: receptors, transducers, and Mad proteins. *Cell* **85**, 947–50.

Nagata, S. (1997). Apoptosis by death factor. *Cell* **88**, 355–65.

O'Shea, J. J. (1997). Jaks, STATs, cytokine signal transduction, and immunoregulation: are we there yet? *Immunity* **7**, 1–11.

Pawson, T. (1995). Protein modules and signalling networks. *Nature* **373**, 573–80.

Schlessinger, J. and Ullrich, A. (1992). Growth factor signaling by receptor tyrosine kinases. *Neuron* **9**, 383–91.

Stancovski, I. and Baltimore, D. (1997). NF-κB activation: the IκB kinase revealed? *Cell* **91**, 299–302.

Taniguchi, T. (1995). Cytokine signaling through nonreceptor protein tyrosine kinases. *Science* **268**, 251–5.

Ten Dijke, P., Miyazono, K., and Heldin, C.-H. (1996). Signaling via hetero-oligomeric complexes of type I and type II serine/threonine kinase receptors. *Curr. Opin. Cell Biol.* **8**, 139–45.

van der Geer, P., Hunter, T., and Lindberg, R. A. (1994). Receptor protein-tyrosine kinases and their signal transduction pathways. *Annu. Rev. Cell Biol.* **10**, 251–337.

Wallach, D. (1997). Cell death induction by TNF: a matter of self control. *Trends Biochem. Sci.* **22**, 107–9.

Watanabe, S. and Arai, K. (1996). Roles of the JAK-STAT system in signal transduction via cytokine receptors. *Curr. Opin. Genet. Dev.* **6**, 587–96.

Yuan, J. (1997). Transducing signals of life and death. *Curr. Opin. Cell Biol.* **9**, 247–51.

XIII Regulation of lymphokine gene expression

Edgar Serfling*, Andris Avots and Stefan Klein-Hessling

1. Introduction: T-cell activation and the inducible activation of lymphokine genes

The majority of peripheral CD4[+] T lymphocytes in blood and peripheral organs are in a resting state, i.e. they neither secrete lymphokines nor proliferate. The specific recognition of antigenic peptides presented in the context of major histocompatibility complex (MHC) molecules by the T-cell receptor complex (TCR, consisting of the TCR α- and β-chains and CD3 molecules) mediates activation of T cells, along with co-receptor signals. The TCR-mediated signals are transferred to downstream intracellular molecules, including a number of tyrosine protein kinases (PTKs). Activation of the PTKs p56[lck], p59[fyn] and ZAP70 leads to tyrosine phosphorylation of numerous targets, such as the zeta (ζ) chains of the CD3 complex and, in turn, to an increase in free, intracellular Ca^{2+} and the activation of small GTP/GDP-binding proteins, in particular of p21[ras].

The rapid release of Ca^{2+} from intracellular stores and the influx of extracellular Ca^{2+}, together with calmodulin, potently stimulate the cytoplasmic serine/threonine phosphatase calcineurin. The activity of calcineurin is crucial for the expression of lymphokines since inhibition of its activity by the immunosuppressants cyclosporin A (CsA) and FK506 blocks the induction of a large set of lymphokine genes, including those for interleukin 2 (IL-2) and IL-4. Both suppressants interact with low-molecular-weight cytosolic proteins, the immunophilins, and bind as complexes to the large subunit of calcineurin, thereby inactivating its phos-

phatase activity. Targets of calcineurin are the NF-AT factors which have to be dephosphorylated at multiple sites before their nuclear translocation (see Section 3.1). Additional downstream targets of calcineurin are the cytoplasmic inhibitors of NF-κB factors (IκBs) and, in primary T cells, the AP-1 factors. This might be due to the T-cell-specific activation of c-Jun N-terminal kinases (JNKs) by calcineurin and protein kinase C (PKC)-mediated pathways which could also contribute to the entry of resting, primary T cells into the cell cycle. For yeast cells, it has been shown convincingly that a loss of calcineurin blocks the transition from G_0 to the G_1 phase of the cell cycle.

p21[ras] and further members of the superfamily of small GTP/GDP binding proteins, e.g. Rac and Rho, are important transmitters of external signals leading to lymphokine activation. In their active (GTP-bound) state they initiate intracellular protein kinase cascades finally leading to the activation of MAP/ERK kinases, JNK/stress-activated kinases and p38 protein kinase. Members of these protein kinase cascades appear to be important for lymphokine production. Thus, inhibition of p38 protein kinase has a deleterious effect on Th1 responses, and mice transgenic for a constitutively-activated MAP kinase kinase 6, an upstream activator of p38 kinase, showed a distinct increase in the synthesis of interferon gamma (IFNγ), the marker lymphokine for Th1 cells. The p21[ras]-mediated activation of the serine/threonine protein kinase c-Raf, an upstream activator of the MAP/ERK cascade, was shown to be important for the induction of the IL-2 promoter.

This short review will focus on the induction of lymphokine promoters/enhancers in activated CD4[+] lymphocytes, especially on the induction of IL-2, IFNγ and IL-4 promoters. Apart from being one of the best-studied eukaryotic promoters, the IL-2 promoter is most active in naive CD4[+] T cells

*Corresponding author.

whereas the IFNγ and IL-4 promoters are most active in Th1 and Th2 cells, respectively.

2. Structure and activation of lymphokine promoters

To a large extent the expression of lymphokine genes in activated T lymphocytes is regulated at the transcriptional level. However, this does not imply that post-transcriptional control is without significance for the synthesis of lymphokines. Although still a matter of dispute, the CD28-mediated co-stimulation of IL-2 expression appears to increase (via JNK kinases) the stability of IL-2 mRNA. In addition, numerous lymphokine mRNAs, including the IL-2 and granulocyte–macrophage colony-stimulating factor (GM-CSF) RNAs, harbor several AU-rich sequences in their 3′ untranslated mRNA portions which have been shown to control the rapid turnover of these RNAs. It has been demonstrated for other lymphokines, such as IL-15 and IL-16, that RNA and protein processing mechanisms determine to a large extent their faithful expression in lymphocytes.

2.1 The IL-2 promoter

The expression of the IL-2 gene is restricted to activated T lymphocytes. A large number of studies from numerous laboratories have convincingly shown that the most important sequence control elements for the inducible and T-cell-specific transcription of murine and human IL-2 genes are located within their immediate upstream regions of approximately 300 bp. The promoter regions of the murine and human IL-2 genes exhibit extensive (80–90%) sequence homology and the multiple factor binding sites in the two promoters are almost identical in sequence. The 300 bp promoter works in an orientation- and distance-independent manner, corresponding to a transcriptional enhancer. In T cells a DNase I hypersensitivity site has been mapped to its proximal region.

A scheme of the topography of transcription factor binding sites for the IL-2 promoter is presented in Figure 1. Binding sites most important for the IL-2 promoter induction are those for NF-AT, NF-κB and Octamer/AP-1 factors. There are two high-affinity binding sites for NF-AT factors located around positions −145 and −285 (shown as black boxes in Figure 1). Mutations of these sites which abolish NF-AT binding suppress the promoter activity. In addition, three low-affinity NF-AT binding sites have been mapped near the transcriptional start site (labeled TATA-2 in Figure 1), immediately 5′ of the proximal Octamer/AP-1 (UPS or NFIL-2A) site and to the CD28-responsive element (CD28RE). Most, but apparently not all of these sites are bound by NF-AT together with AP-1 factors, thereby enhancing DNA binding.

The CD28RE is a low-affinity NF-κB binding site to which members of the Rel family of transcription factors bind upon CD28 co-stimulation. A second non-consensus NF-κB-binding site is located around position −205 (labeled TCE$_d$ or NFIL-2C in Figure 1). There are two Octamer-binding sites within the promoter, a high-affinity proximal (UPS/NFIL-2A/ARRE-1) site and a low-affinity distal (US2/3 or NFIL-2D) site. Octamer factors bind to the proximal site, in concert with AP-1 factors, to overlapping nucleotides creating novel functional properties. Similar to multiples of NF-AT sites, multiples of the Octamer/AP-1 site are strongly induced in T cells, and their induction can be inhibited by pharmacological doses of CsA. Mutations that suppress the binding of both Octamer and AP-1 factors to this site abolish IL-2 promoter induction. A further non-consensus AP-1 site is located around position −150. When this site was converted by one point mutation to a high-affinity AP-1 binding site, induction of IL-2 promoter activity was observed also in non-lymphoid cells. Similar results were obtained after converting the non-consensus NF-κB (at −205) and proximal Octamer site to consensus, high-affinity factor-binding sites indicating the importance of a fine-tuned control of transcription factor binding for the T-cell-restricted induction of IL-2 promoter activity.

Although the proximal IL-2 promoter harbors numerous control elements which guide IL-2 transcription, several lines of evidence indicate the existence of additional control sequences located outside of the proximal promoter:

(1) The 5′ regions of the murine and human IL-2 genes are highly homologous up to position −600, and *in vivo* footprinting studies have revealed that lymphoid-specific factors bind constitutively to the region between −300 and −600. A transcriptional enhancer controlled by GA-binding proteins (GABPs) has been identified within the sequences between −400 and −500. Additional DNA stretches of high homology are situated around the positions from −1300 to −1450 and −2000.

(2) Only one of 17 transgenic mouse founder lines carrying a *lacZ* reporter gene under the control

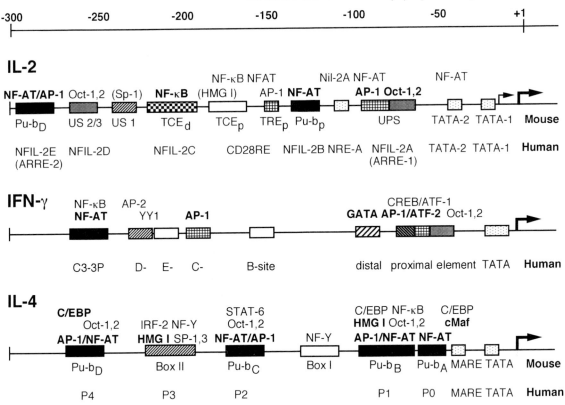

■ **Figure 1.** Organization of murine and human IL-2, IFNγ and IL-4 promoters. The figure shows the immediate upstream regions of genes which are known to contain numerous binding sites for transcription factors controlling the induction of the lymphokine genes in T cells. The transcription factors are indicated above the promoters. Those which have been determined to play a very important role in promoter induction are written in bold face letters. Below the promoters are given the names of the sequence elements, several of which are mentioned in the text. Factor binding sites are shown as boxes. Sites bound by the same factors are indicated by identical symbols; for example, high-affinity binding sites for NF-AT factors are indicated by filled boxes.

of a −600 bp promoter showed transgene expression indistinguishable from the endogenous IL-2 gene.

2.2 The IFNγ promoter

IFNγ is a prototypical Th1 lymphokine and is secreted by both CD4⁺ and CD8⁺ T cells and natural killer (NK) cells. Similar to the IL-2 gene, the immediate upstream region of the IFNγ gene, spanning approximately 500 bp of DNA, harbors a promoter which is strongly inducible upon T-cell stimulation. The IFNγ and IL-2 promoters are both induced in Jurkat T leukemic cells following activation, and share several sequence motifs, namely

the proximal Octamer/AP-1/ATF site next to the TATA box, and AP-1, NF-κB and NF-AT sites (Figure 1) which are also located within an intronic enhancer element. The activity of this enhancer is inducible in T cells and, like the IL-2 promoter/enhancer, harbors a DNase I hypersensitive site which can only be detected in lymphoid cells producing IFNγ. Surprisingly, the distal IFNγ promoter contains a sequence motif for the binding of GATA-3, a T-cell-specific GATA factor highly expressed in Th2 cells during differentiation and involved in the control of IL-5 expression (see Figure 1). Luciferase reporter genes controlled by multiples of fragments from the proximal and distal promoter portions (spanning nucleotides −47 to

−70 and −72 to −98, respectively) were selectively activated in memory CD4[+] T cells of transgenic mice, but not in naive CD4[+] or CD8[+] T cells. No marked differences in reporter gene expression were detected between CD4[+] T cells cultivated in the presence of IL-12 or IL-4 to induce differentiation to Th1 or Th2 cells.

As yet there is no indication that Th1-regulatory transcription factors, such as Stat4 or IRF-1, can bind to the IFNγ control regions, but methylation of a CpG dinucleotide within the Octamer/AP-1 site of the proximal IFNγ promoter has been observed in Th2 cells in which the IFNγ gene was inactive. CpG methylation renders promoters inactive for transcription, and recruitment of histone deacetylases appears to be one mechanism involved in methylation-dependent repression.

2.3 The IL-4 promoter and other promoters of Th2-type lymphokines

The genes encoding the lymphokines IL-4, IL-5 and IL-13 are highly expressed in Th2 cells and located on human chromosome 5q and mouse chromosome 11q where they are assembled in a linkage group spanning approximately 150 kb of DNA. While the IL-4 and IL-13 genes are only 13 kb apart, the IL-4 and IL-5 genes are more than 100 kb apart. Like gene families, such as the avian and mammalian α- and β-globin genes, which are organized in large linkage groups spanning several hundred kilobases of DNA, their expression appears to be controlled by locus control regions, i.e. remote DNA control sequences which are thought to allow, by binding of numerous transcription factors and histone acetylases, chromatin 'opening' and transcription. Although there is currently no direct experimental proof for this view, the common modulation of expression of Th2-type genes in NF-AT-deficient T cells (see Section 3.1 and Table 1) supports this conclusion.

A scheme of the proximal promoter of the murine and human IL-4 genes is shown in Figure 1. Like the proximal IL-2 promoter it spans approximately 270 bp and contains multiple NF-AT-binding sites. Although AP-1 binds to some of these sites, the NF-AT sites of the IL-4 promoter differ markedly from those of the IL-2 promoter. This is shown best for the two most important IL-4 NF-AT sites, namely the P1/Pu-b_B and P4/Pu-b_D sites (located around the positions −80 and −255 respectively) whose mutation strongly impairs IL-4

■ **Table 1.** Consequences of inactivation (or overexpression) of NF-AT genes in mice

Factor	Expression	Phenotype of NF-AT-deficient (or transgenic) mice
NF-ATp (NF-AT1/c2)	Constitutive synthesis in numerous cells; highest concentration in peripheral lymphocytes	Enhanced immune responses; incomplete termination of superantigen stimulation; increase in number of peripheral lymphocytes and size of spleen and lymph nodes; enhanced GC formation; no involution of the thymus. Modulation of Th2-lymphokine synthesis; no defect in IL-2 synthesis.
NF-ATc (NF-AT2/c1)	Inducible synthesis, particularly in T cells	Defects in generation of cardiac valves and septa: mice die before day 14.5 of development *in utero*. RAG[−/−] mice containing NF-ATc-deficient lymphocytes showed reduced number of thymocytes, impaired proliferation, impaired synthesis of Th2-lymphokines, no defect in IL-2 synthesis.
NF-AT3 (NF-ATc4)	Non-lymphoid cells, e.g. in myocardium	Cardiac hypertrophy after overexpression in cardiomyocytes.
NF-AT4 (NF-ATx/c3)	Constitutive synthesis; high expression in thymus	Defects in positive selection of thymocytes.

Mice deficient for NF-ATp were established in the laboratories of L. Glimcher, A. Rao and E. Serfling, mice deficient for NF-ATc in the laboratories of T. Mak, G. R. Crabtree and L. Glimcher, and mice deficient for NF-AT4 in the Glimcher laboratory. Overexpression of NF-AT3 in cardiomyocytes led to cardiac hypertrophy.

promoter activity. In addition to the binding of NF-AT, the −80 proximal site can be bound by Octamer, AP-1, CAAT element binding protein (C/EBP) and high mobility group (HMG) I(Y) proteins; this site in the human promoter can also bind NF-κB factors and differs from the site in the murine promoter in only one base pair. A stronger C/EBP binding site is located immediately adjacent to the −255 NF-AT site which, in DNase I footprinting assays, forms a large footprint spanning 35–40 nucleotides, allowing the binding of multiple factors. The NF-AT site at −150 (P2/Pu-b$_C$) overlaps with a binding site for STAT6, the IL-4-induced STAT factor, whereas the most proximal NF-AT site (labeled P0 in Figure 1), which is not bound by AP-1, is adjacent to a binding site for c-Maf, a leucine zipper factor that controls IL-4 trancription in Th2 cells. Finally, mast cells contain an intronic enhancer that controls IL-4 gene expession.

Several lines of experimental evidence indicate the existence of additional IL-4 control elements located outside of the promoter. A DNA element of approximately 40 bp located just 5′ to the promoter exerts a strong negative effect on promoter activity. This inhibition is released by DNA further upstream, indicating that negatively and positively acting factors bind to these 5′ regions. Binding sites for GATA-3 are located upstream and downstream from but not within the proximal promoter. A silencer/repressor element suppressing IL-4 transcription in Th1 cells has been mapped to a region downstream from the IL-4 gene. It contains a binding site for STAT6.

Although the synthesis of IL-5, a growth factor for eosinophils, is also typical of Th2 CD4$^+$ cells and strongly stimulated after T-cell activation, some stimuli, such as cAMP, glucocorticoids and, in particular, the drug OM-01, have selective positive or negative effects on the transcription of IL-5 but not IL-4 or other lymphokines. It remains to be shown how these specific mediators affect IL-5 transcription through its transcriptional control regions which have been mapped within less than 1.2 kb DNA of upstream DNA and shown to confer Th2-specific expression. Within the immediate upstream IL-5 promoter region spanning approximately 120 bp of DNA, the NF-AT/AP-1 binds between positions −95 and −115, GATA-3 binds between positions −74 and −55, and hitherto unidentified factors bind to a so-called CLE0 (conserved lymphokine element 0) element (between −54 and −38). It is very likely that, at least in part, the Th2-specific expression of IL-5 is mediated by the binding of GATA-3 to the proximal IL-5 promoter.

GATA-4, expressed in cardiomyocytes, has recently been shown to interact with NF-AT3, so perhaps a similar interaction between members of NF-AT family and the T-cell-specific GATA-3 factor contributes to IL-5 promoter regulation.

2.4 IL-3 and GM-CSF promoters/enhancers

The IL-3 and GM-CSF genes are only 10 kb apart on human chromosome 5 and murine chromosome 11; this linkage group is 500–600 kb from the cluster of Th2-type lymphokines. While the IL-3 gene is expressed almost exclusively in T cells, the GM-CSF gene is expressed by a number of cell types at sites of inflammation. A T-cell-specific enhancer has been identified approximately 14 kb upstream of the IL-3 gene's transcriptional start site. The T-cell-restricted activity of this enhancer is controlled by arrays of complex NF-AT/AP-1/Octamer binding sites which resemble (in both structure and activity) those of the IL-2 and IL-4 promoters. An enhancer has also been identified for the GM-CSF gene, located approximately 4 kb upstream from its transcriptional start site. Although the activity of this enhancer is also controlled by NF-AT factors in T cells, in contrast to the IL-3 enhancer the GM-CSF enhancer is active in endothelial cells which express GM-CSF. Interestingly, both the IL-3 and GM-CSF enhancer can cooperate well with the promoters of the two linked genes which contain binding sites for AP-1, C/EBP, Octamer and Ets-like transcription factors. By mapping of DNase I hypersensitive sites, further putative regulatory elements have been found just upstream and downstream of the IL-3 gene.

2.5 The TNF-α promoter

The genes encoding tumor necrosis factor alpha (TNF-α, or TNF), lymphotoxin alpha (LT-α, or TNF-β) and LT-β are located in a linkage group spanning not more than 12 kb DNA within the major histocompatibility complex (MHC) locus on mouse chromosome 17, approximately 70 kb centromeric to the H2D gene. The TNF-α promoter makes up much of the 2–3 kb of DNA between the polyadenylation site of LT-β and the transcriptional start site of the TNF-α gene. The main sources of TNF-α are macrophages/monocytes and activated T cells. The lipopolysaccharide (LPS)-mediated induction of human TNF-α promoter in monocytes has been studied in greatest detail, and it has been shown that multiple NF-κB-like binding sites located around positions −210, −510, −655 and −850, and C/EBP binding sites in the region from −101 to −189 appear to be of special importance. Tolerance of monocytes to LPS treatment is correlated with a change in the composition of NF-κB factors leading to the predominant binding of (transcriptionally inactive) NF-κB p50 proteins to the inactivated TNF-α promoter.

In T lymphocytes, the proximal TNF-α promoter (from −1 to −199) appears to be sufficient for maximal TNF-α induction. This region includes a complex CRE/NF-κB binding site (positions −88 to −106), called κ3, that has been studied in detail. It is bound by members of the NF-AT and AP-1/ATF families, particularly by NF-ATp and Jun/ATF-2 heterodimers, and confers CsA-sensitivity upon the TNF-α promoter in T cells. Additional NF-AT binding sites are important for promoter activity in T and B cells but it remains to be shown whether and how NF-AT factors contribute to gene activation in B cells.

2.6 The IL-6 promoter

Although IL-6 is synthesized by both B and T cells (e.g. Th2 cells), non-lymphoid cells are the main producers of IL-6, a multifunctional cytokine. The promoter regions of the murine and human IL-6 genes span approximately 350 bp of DNA, are highly conserved and contain a number of binding sites for inducible transcription factors: an NF-κB-binding site around position −70, an C/EBP-β (NF-IL6) binding site between positions −145 and −173, and multiple AP-1 and glucocorticoid receptor binding sites further upstream. The stimulatory effect of IL-1 and TNF-α on the IL-6 promoter is mediated by the proximal NF-κB and C/EBP sites. The activity of the C/EBP-binding site is also controlled by LPS and IL-6 itself. Thus, IL-6 can regulate its own expression by stimulating the activity of C/EBPβ.

3. Transcription factors

3.1 NF-AT factors

NF-AT ('nuclear factor of activated T cells') factors bind to and control the induction of numerous lymphokine promoters during T-cell activation. The four NF-AT factors cloned so far (Figure 2) share a highly conserved DNA binding domain of approximately 300 amino acids which, because of its structural similarity to the Rel DNA-binding domain of NF-κB factors, has been named Rel similarity domain (RSD). One further common property of NF-AT factors is their Ca^{2+}-dependent nuclear translocation. As mentioned in Section I, T-cell activation leads to very rapid mobilization of Ca^{2+} from intracellular stores, an increase of Ca^{2+} influx and activation of calcineurin. Numerous transfection studies showed that calcineurin activity has a strong stimulatory effect on lymphokine promoters by stimulating the nuclear transport and activity of NF-AT. It is very likely that calcineurin binds directly to the regulatory region in front of the RSD containing three conserved motifs of consensus sequence SPxxSPxxSPxxxxx(D/E) (D/E) and dephosphorylates the serine residues which are constitutively phosphorylated. Candidate NF-AT kinase(s) that are constitutively active in resting T cells include glycogen synthase kinase-3 (GSK-3) and casein kinase I (CKI).

Of the NF-AT factors, NF-ATp (also called NF-AT1 or NF-ATc2) and NF-ATc (NF-AT2, NF-ATc1) appear to play the most prominent role in gene induction during the activation of peripheral T lymphocytes. Both proteins are highly expressed in peripheral T lymphocytes. Their DNA-binding activities are almost indistinguishable, but inactivation of the NF-ATp and NF-ATc genes in mice has very different effects on cell differentiation and activation: whereas inactivation of the NF-ATc gene leads to severe defects in the morphogenesis of embryonic heart and the early death of embryos between days 13 and 17 of gestation, NF-ATp-deficient mice have no defects in embryonic heart development and are born healthy. Moreover, NF-ATc$^{-/-}$ thymocytes generated in RAG$^{-/-}$ mice after blastocyste injections of NF-ATc-deficient embryonic stem cells (ES cells) have severely defective intra-thymic development, leading to a reduction in the number of double-positive (CD4$^+$CD8$^+$) immature thymocytes and hypoplastic thymi. No defects in thymocyte differentiation were observed in NF-ATp$^{-/-}$ mice. Furthermore, the proliferation of peripheral T lymphocytes, their immune response and synthesis of Th2 lymphokines appear to be severely affected in NF-ATc-deficient mice whereas in NF-ATp$^{-/-}$ mice the same cells exhibited enhanced responses and, after an initial decrease, a pronounced increase in the synthesis of Th2 lymphokines and Th2-driven immune responses.

3.2 AP-1 and other members of the superfamily of basic leucine zipper factors

Binding sites for AP-1 factors have been identified in numerous lymphokine promoters (Figure 1).

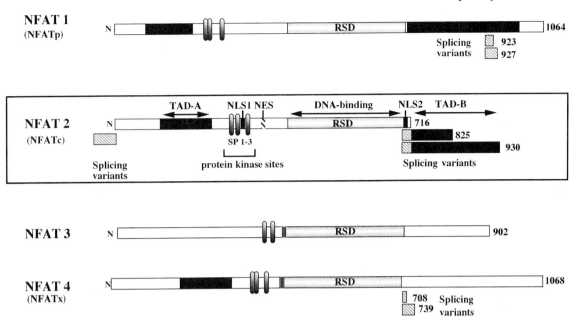

∎ **Figure 2.** Structure of NF-AT transcription factors. NF-AT1(p) and NF-AT2(c) are highly expressed in peripheral T cells, NF-AT4 in thymocytes and NF-AT3 in non-lymphoid cells, e.g. cardiomyocytes. The factors share a highly homologous DNA binding domain, the RSD (Rel similarity domain) which shows structural homology to those of Rel/NF-κB factors. All factors, except certain short splice versions of NF-ATc and NF-AT4, appear to contain two trans-activation domains, TAD-A and TAD-B, near their N- and C-termini, respectively. Between the RSD and TAD-A, there are regulatory sequence motifs, including numerous phosphorylation sites, mediating the binding of the Ca^{2+}/calmodulin-dependent phosphatase calcineurin and the nuclear localization and export of factors. NLS, nuclear localization signal; NES, nuclear export signal; SP 1–3, conserved phosphorylation sites.

AP-1 complexes consist of members of the Jun (c-Jun, JunB and JunD) and Fos (c-Fos, FosB, Fra1 and Fra2) protein families which, through coiled-coil interactions between their leucine zipper domains, generate homodimers or heterodimers for the binding to the TRE (consensus sequence TGA-G/C-TCA). Most of the TREs detected in lymphokine promoters and enhancers are modifications of this consensus site and, therefore, are either low-affinity binding sites or sites to which Jun binds in concert with other members of superfamily of basic leucine zipper factors, e.g. ATF-2. These sites are very often adjacent to binding sites for other factors, such as NF-AT and Octamer factors (Figure 1).

Several members of Jun/Fos families have been identified as part of NF-AT/AP-1 complexes generated after T-cell activation, varying depending on the cells and induction conditions used. This leads to the conclusion that the composition of Jun and Fos factors binding to the lymphokine promoters depends very much on the activation and differentiation status of cells. Thus, in Th2 cells an accumulation of JunB has been detected which appears to collaborate with c-Maf, a further member of the superfamily of basic leucine zipper factors, in the activation of the IL-4 promoter. Since JunB, in contrast to c-Jun, is not targeted by JNK/stress kinases, it is likely that JNKs play a minor role in IL-4 promoter induction.

Inactivation of the *c-jun* or *junB* gene in mice resulted in embryonic lethality. When ES cells deficient for c-Jun were injected into blastocytes from RAG-2-deficient mice, substantial numbers of normal peripheral T and B cells were observed in spite of a poor restoration of thymocytes. No obvious defects in IL-2 secretion and proliferation were seen when these c-Jun$^{-/-}$ T cells were stimulated. These results and the generation of 'normal' AP-1 complexes after stimulation of c-Jun-deficient T cells suggest that other Jun proteins can compensate for c-Jun deficiency. This seems to be true for c-Fos (and probably FosB) too, since c-Fos$^{-/-}$ T lymphocytes do not show obvious defects in lymphokine secretion and proliferation.

CREB (cAMP responsive element binding) protein and its relatives, CREM (modulator of CREB) and ATF-1, belong to a subclass of leucine zipper factors that bind as dimers to CRE motifs. Multiple CREs are part of transcriptional control elements of TCR genes, and the complete inactivation of CREB impairs fetal T-cell development and leads to perinatal lethality. While elevated cAMP levels suppress the induction of IL-2 and IFNγ promoters, they stimulate the induction of IL-5 and, to a lesser extent, the IL-4 promoter.

3.3 NF-κB factors

The family of Rel/NF-κB factors consists of five members, the transactivators p65/RelA, c-Rel and RelB, and the p50/NF-κB1 and p52/NF-κB2 proteins which are synthesized as larger precursor proteins, called p105 and p100, respectively. These proteins form homodimers and heterodimers that bind to palindromic sites of consensus sequence GGGATTTCCC which occur in slightly modified versions in the promoters of the IL-2, IFNγ, TNF-α and IL-6 genes. The DNA-binding of NF-κB factors is mediated by a highly conserved stretch of approximately 300 amino acids, the Rel homology domain, which also encompasses the factors' dimerization and nuclear localization motifs.

The activity of NF-κB factors is controlled by the interaction with cytoplasmic inhibitor proteins, IκBα, β and ε. These and the C-termini of p100 and p105 are composed of several ankyrin repeats, protein domains that are involved in protein–protein interactions. By masking nuclear translocation sequences the IκBs and the C-termini of p100 and p105 control the nuclear translocation of Rel proteins. Numerous stimuli that activate lymphocytes lead to the inducible degradation of IκBs, the release and subsequent nuclear translocation of Rel proteins. A prerequisite for the degradation of IκBs is the phosphorylation of two serine residues located at positions 32 and 36 near the N-terminus of IκBα. Mutations of these serine residues that suppress phosphorylation stablize IκBα and prevent the nuclear accumulation of all Rel proteins. Transgenic mice overexpressing such a dominant-negative version of IκBα in the thymus exhibit profound defects in T-cell development and a decreased number of peripheral T cells.

Mice defective for individual Rel factors show a large variety of immune defects. For example, p65- and p50-deficient mice show no defects in lymphokine transcription and secretion, whereas the lymphocytes of c-Rel-deficient mice show severe defects in the synthesis of IL-2, IL-3 and GM-CSF. In contrast, disruption of the C-terminus from p100, the precursor of NF-κB2/p52, which leads to the enhanced accumulation of nuclear p52, results in the hyperproliferation of lymphocytes and strongly enhances the synthesis of IL-2, IL-3, IL-4, GM-CSF, IL-10 and TNF-α. Disruption of the C-terminus of p105, on the other hand, leads to an increased transcription of several cytokines (G-CSF, GM-CSF, TNF-α and IL-2) in the thymus and the down-regulation of their secretion in peripheral T cells. A very strong decrease in RNA synthesis and secretion of GM-CSF, TNF-α and IL-6 was observed in macrophages indicating a cell-type specific contribution of p50/NF-κB1 to the regulation of these cytokine genes.

3.4 Octamer factors

Multiple versions of the Octamer consensus site, ATGCAAAT, have been identified in numerous lymphokine promoters as functionally important sequence motifs, in particular near the transcriptional start sites (Figure 1). The Octamer factors Oct-1 and Oct-2 are expressed in lymphocytes. While Oct-1 is constitutively expressed, Oct-2 expression is induced in pre-B cells and almost all types of T cells. But in some T-cell lines, such as El 4 cells, and in B cells, Oct-2 is also constitutively expressed in high concentrations. Activation of peripheral T cells results in a slow accumulation of Oct-2. Inactivation of Oct-2 in mice does not cause profound defects in T-cell differentiation and activation. The same was found in OBF-1-deficient mice. The expression of OBF-1 (also called Bob1 or Oca-B), a lymphocyte-specific co-factor of Octamer factors, is induced upon T-cell activation. In addition, T-cell stimulation leads to its phosphorylation-dependent activation, suggesting a specific role for this co-factor in T cells.

3.5 Ets factors

The family of Ets transcription factors is defined by a highly conserved 85 amimo acid region (the Ets domain) which directs DNA binding to the core sequence GGA. Ets motifs have been found in numerous lymphokine promoters. Within the IL-2 and IL-4 promoters, these motifs are bound by NF-AT (or NF-κB) factors, but for other promoters, such as the GM-CSF promoter, specific binding of Ets factors has been reported.

Ets-1 reduces IL-2 expression. In resting T-cells Ets-1 is highly expressed and undergoes a dramatic,

PKC- and Ca^{2+}-dependent decrease in concentration after antigen stimulation. In parallel, the DNA-binding activity of Ets-1 is inactivated by transient, Ca^{2+} dependent phosphorylation. Stable expression of Ets-1 antisense RNA in T-cell lines increases IL-2 production up to 20-fold and results in a five-fold increase in the activity of IL-2 promoter–reporter constructs. Studies using Ets-1-deficient mice demonstrated that Ets-1 is required to actively maintain lymphocytes in a resting (G_0) stage of the cell cycle, supporting the view that Ets-1 may regulate IL-2 gene expression indirectly.

The upstream region of the human IL-2 gene contains an enhancer element, located between positions −413 and −502, which is controlled by the binding of the ubiquitously expressed Ets factor GABPα, along with GABPβ containing a strong *trans*-acting domain. GABPα/β heterodimers bind with high affinity, and GABPα/α homodimers with lower affinity, to a palindromic sequence element consisting of two typical Ets-binding motifs, located in opposite orientation. A similar palindromic element, the so-called double symmetry element (DSE), has been detected within the proximal IL-16 promoter and shown to be crucially involved in its induction after the activation of Jurkat cells.

4. Conclusion and perspectives

The expression of eukaryotic genes is controlled by remote DNA sequences organized in locus control regions and enhancers, and proximal sequences organized in the promoter located immediately upstream from the transcriptional start site. There are numerous experimental data on the function of lymphokine promoters but little is known about the role of remote DNA sequences in lymphokine expression. Further analyses elucidating the chromatin domain structure of lymphokine genes, including the creation of transgenic mice containing long gene-control regions and conditional knockout mice models will help to elucidate the chromatin alterations that occur in lymphokine genes during their induction after T-cell activation *in vivo*. These and further data on the interaction of enhancer/promoter factors with non-DNA-binding (co-) factors — including the chromatin modifying enzymes and 'general' factors facilitating transcription — may enable us to interfere specifically with the expression of individual lymphokine genes, such as with IL-4 expression, which leads to high IgE levels in allergic patients.

References

(This list includes reviews and some recent important publications not covered by these reviews.)

Agarwal, S. and Rao, A. (1998) Long-range transcriptional regulation of cytokine gene expression. *Curr. Opin. Immunol.* **10**, 345–52.

Cantrell, D. (1996) T cell antigen receptor signal transduction pathways. *Annu. Rev. Immunol.* **14**, 259–74.

Chen, C.-Y., del Gatto-Konczak, F., Wu, Z., and Karin, M. (1998) Stabilization of interleukin-2 mRNA by the c-Jun NH_2-terminal kinase pathway. *Science* **280**, 1945–9.

de la Pompa, J. L., Timmerman, L. A., Takimoto, H., Yoshida, H., Elia, A. J., Samper, E., Potter, J., Wakeham, A., Marengere, L., Langille, B. L. *et al.* (1998) Role of the NF-ATc transcription factor in morphogenesis of cardiac valves and septum. *Nature* **392**, 182–6.

Ghosh, S., May, M. J., and Kopp, E. B. (1998) NF-κB and Rel proteins: evolutionarily conserved mediators of immune responses. *Annu. Rev. Immunol.* **16**, 225–60.

Ishikawa, H., Claudio, E., Dambach, D., Raventos-Suarez, C., Ryan, C., and Bravo, R. (1998) Chronic inflammation and susceptibility to bacterial infections in mice lacking the polypeptide (p)105 precursor (NF-κB1) but expressing p50. *J. Exp. Med.* **187**, 985–96.

Lee, H. J., O'Garra, A., Arai, K.-I., and Arai, N. (1998) Characterization of *cis*-regulatory elements and nuclear factors conferring Th2-specific expression of the IL-5 gene: a role for a GATA-binding protein. *J. Immunol.* **160**, 2343–52.

Molkentin, J. D., Lu, J.-R., Antos, C. L., Markham, B., Richardson, J., Robbins, J., Grant, S. R., and Olson, E. N. (1998) A calcineurin-dependent transcriptional pathway for cardiac hypertrophy. *Cell* **93**, 215–28.

Ranger, A. M., Hodge, M. R., Gravallese, E. M., Oukka, M., Davidson, L., Alt, F. W., de laBrousse, F. C., Hoey, T., Grusby, M., and Glimcher, L. H. (1998) Delayed lymphoid repopulation with defects in IL-4-driven responses produced by inactivation of NF-ATc. *Immunity* **8**, 125–34.

Ranger, A. M., Grusby, M. J., Hodge, M R., Gravallese, E. M., de la Brousse, F. C., Hoey, T., Mickanin, C., Baldwin, H. S., and Glimcher, L. H. (1998) The transcription factor NF-ATc is essential for cardiac valve formation. *Nature* **392**, 186–190.

Rao, A., Luo, C., and Hogan, P.G. (1997) Transcription factors of the NFAT family: Regulation and function. *Annu. Rev. Immunol.* **15**, 707–47.

Rincon, M. and Flavell, R.A. (1997) T-cell subsets: transcriptional control in Th1/Th2 decision. *Curr. Biol.* **7**, 729–32.

Schuh, K., Kneitz, B., Heyer, J., Bommhardt, U., Jankevics, E., Siebelt, F., Pfeffer, K., Müller-Hermelink, H.K., Schimpl, A., and Serfling, E. (1998) Retarded thymic involution and massive germinal center formation in NF-ATp-deficient mice. *Eur. J. Immunol.* **28**, 2456–66.

Serfling, E., Avots, A., and Neumann, M. (1995) The architecture of the interleukin-2 promoter: a reflection of T lymphocyte activation. *Biochem. Biophys. Acta* **1263**, 181–200.

Ward, S.B., Hernandez-Hoyos, G., Chen, F., Waterman, M., Reeves, R., and Rothenberg, E.V. (1998) Chromatin remodeling of the interleukin-2 gene: distinct alterations in the proximal versus distal enhancer regions. *Nucleic Acids Res.* **26**, 2923–34.

Yoshida, H., Nishina, H., Takimoto, H., Marengere, L. E. M., Wakeham, A. C., Bouchard, D., Kong, Y.-Y., Ohteki, T., Shahinian, A., Bachmann, M. *et al.* (1998) The transcription factor NF-ATc1 regulates lymphocyte proliferation and Th2 cytokine production. *Immunity* **8**, 115–24.

Part B

Cytokines and immune functions

XIV Cytokines in the development of lymphocytes

Chong-kil Lee, Kathrin Muegge and Scott K. Durum

1. Introduction

There are three principal lineages of lymphocytes: T cells, B cells and natural killer (NK) cells. All three lymphocyte lineages are produced throughout life from a common lymphoid precursor which is in turn continuously derived from a common hematopoietic precursor. Lymphoid development occurs in specialized microenvironments containing epithelial cells, cells of hematopoietic origin, and extracellular matrix. The signals these microenvironments provide to developing lymphocytes are not yet completely defined. This review will discuss those signals that are clearly defined, and that are considered to be 'cytokines'.

The types of cytokines involved in lymphopoiesis are generally not freely diffusing molecules, but are anchored, either to stromal cell membranes or extracellular matrix. These cytokines include interleukin 7 (IL-7) and stromal cell-derived factor 1 (SDF-1), which are probably bound to glycosaminoglycans in extracellular matrix, and stem-cell factor (SCF), Flt-3L, CD30 and Fas ligand (Fas-L), which are integral membrane proteins on stromal cells. Perhaps the advantage of anchoring these signals is that it confines lymphopoiesis to circumscribed anatomical regions with mechanisms for checking dangerous processes such as cell division, gene rearrangements and autoimmunity. The term 'cytokine' was once restricted to freely soluble molecules, but is now usually extended to the integral membrane proteins listed above. We will not discuss other types of membrane molecules traditionally termed 'co-stimulators' (like the B7s, which are involved in T-cell development) or 'adhesion substrates' (which are involved in B-cell development).

Historically, a much longer list of cytokines was once believed to participate in lymphopoiesis. This was particularly true of thymopoiesis, the origin of most T cells, because many different cytokines were capable of triggering thymocyte proliferation *in vitro*. However, a number of these cytokines are no longer believed to participate in thymopoiesis, based on the normal T-cell development pattern observed in mice lacking the respective cytokine or its receptor. The list of non-essential cytokines includes IL-1, -2, -3, -4, -5, -6, -8 and -12, tumor necrosis factor alpha (TNFα), lymphotoxin (LT), CD40-L, interferons, transforming growth factor beta (TGFβ), the colony-stimulating factors (CSFs) (granulocyte–macrophage (GM-), macrophage (M-) and granulocyte (G-), erythropoietin, thrombopoietin, and the family of cytokines sharing the gp130 receptor chain. We will restrict our discussion to those cytokines whose role in lymphopoiesis has been clearly established through mutation in mice—most of these strains were created through gene targeting, although there are mouse strains with natural mutations of SCF and its receptor, and of Fas-L and its receptor. Human lymphopoiesis generally resembles the murine pattern, with some exceptions that will be discussed.

2. Cytokines in T-cell development

T cells develop in the thymus from precursors originating in the fetal liver or in the bone marrow. Small numbers of precursors continually seed the thymus and initially can also produce B cells and NK cells. These precursors expand in number, become committed to the T-cell lineage, rearrange T-cell receptor (TCR) genes and further expand, and then are subject to a series of positive and negative selection events. If cells fail in the selection process, they die in the thymus; if they succeed they emigrate to the peripheral lymphoid organs.

The very early immature thymocytes are CD3⁻CD4⁻CD8⁻, and can be further distinguished on the basis of CD44 and CD25 expression. As seen in Figure 1, the sequence of maturation of these cells is as follows: $CD44^+CD25^- \rightarrow CD44^+CD25^+ \rightarrow CD44^-CD25^+ \rightarrow CD44^-CD25^-$. The two lineages of T cells, $\alpha\beta$ and $\gamma\delta$, diverge at the CD44⁻CD25⁺ stage. In $\alpha\beta$ cell development, cells that successfully rearrange the TCRβ gene in-frame then express the β-chain on the cell surface as part of a pre-TCR complex. Cells pass through a transient CD8⁺ or CD4⁺ immature stage, begin TCRα gene rearrangement, and then become CD3^{med}CD4⁺CD8⁺. The cells in this stage express intermediate levels of TCR, which partly mediates signals for positive and negative selection. After selection, cells express high levels of TCR and at the same time lose either CD4 or CD8, becoming single-positive mature T cells.

Differentiation of T-cell precursors to mature T cells is dependent upon both direct contact with thymic stromal cells and interaction with cytokines. Of the cell surface molecules involved in the direct contact with thymic stromal cells, the TCR plays a central role in that it determines positive selection or death. Cytokines, on the other hand, were long believed to provide proliferation signals for thymocytes. This was based on the observations that many cytokines induce proliferation of thymocytes *in vitro*. However, studies employing targeted gene knockout mice have shown that, of all the cytokines that have been proposed to control T-cell development, thus far only IL-7, thymic stroma derived lymphopoietin (TSLP), SCF, Fas-L and CD30-L play major roles in the development of T cells in the thymus.

2.1 IL-7 and TSLP

IL-7 was first identified and cloned based on its ability to induce proliferation of B-cell progenitors. Subsequently, it has also been shown to support survival and growth of early thymocytes. The specific requirement of IL-7 within the thymus for the differentiation of T-cell precursors has been elucidated from at least three different approaches. *In vivo* administration of neutralizing antibodies to IL-7 or IL-7 receptor (IL-7R) results in the inhibition of both T and B lymphopoiesis. IL-7- and IL-7R-deficient mice show a severe reduction in lymphoid development, confirming that IL-7 is indispensable for the normal development of T and B cells, as discussed below. Moreover, athymic mice (nude mice) are induced to develop T cells if given an IL-7 transgene.

IL-7⁻/⁻ mice display severe lymphopenia in the peripheral blood and lymphoid organs. Their thymic cellularity is reduced to 5% of normal, but the remaining subpopulations, however, are present in normal distribution as defined by CD3, CD4, and CD8, and proliferate in response to concanavalin A (ConA) stimulation. These results strongly suggest that survival and/or proliferation of early thymocytes is critically dependent upon IL-7, although there also remain possible roles of IL-7 in the generation of bone marrow lymphoid progenitors and

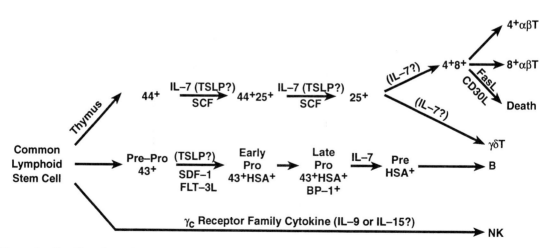

■ **Figure 1.** Cytokines in murine lymphoid development. Definition of terms: 4⁺, CD4⁺; 8⁺, CD8⁺; 25⁺, CD25⁺; 43⁺, CD43⁺; 44⁺, CD44⁺.

in the homing process to the thymic microenvironment. It has been shown that there is an extreme paucity of $\gamma\delta$ T cells in IL-7R$^{-/-}$ mice and that, for $\gamma\delta$ T cells, IL-7 (or IL-7R signaling) is critical for the initial step of γ gene rearrangement.

IL-7 signals through a heterodimeric receptor complex composed of IL-7Rα and the common γ-chain (γc) shared by receptors for IL-2, IL-4, IL-7, IL-9, and IL-15. IL-7R$\alpha^{-/-}$ mice are even more severely defective than IL-7$^{-/-}$ mice in the development of T cells. Individual mice of the IL-7R$\alpha^{-/-}$ strain show varying degrees of severity. Less severely affected mice are similar to the IL-7$^{-/-}$ mice. In the more severely affected phenotypes, thymic cellularity was less than 1 per cent of the number of cells of control thymi. In both of the IL-7R$^{-/-}$ phenotypes, the CD4$^-$CD8$^-$ cells arrested in the CD25$^{-/-}$ stage. Thus IL-7 may be particularly critical for the survival or expansion of this early stage (CD44$^+$CD25$^-$) in the thymus. Alternatively, IL-7 may induce maturation to the CD44$^+$CD25$^+$ stage, in which TCR β-chain genes begin to rearrange. In any case, the fact that IL-7R$\alpha^{-/-}$ mice are more severely defective in thymopoiesis than IL-7$^{-/-}$ mice suggested the existence of a second cytokine that utilizes the IL-7R. A possible candidate ligand is thymic stroma derived lymphopoietin (TSLP), identified and cloned from a thymic stromal cell line. TSLP was shown to bind to IL-7Rα and is presumed to share many activities in common with IL-7. Thus, IL-7R$\alpha^{-/-}$ mice may represent the functional inactivation of both IL-7 and TSLP signals.

Another deficiency in IL-7R$\alpha^{-/-}$ mice is their complete lack of $\gamma\delta$ T cells. No $\gamma\delta$ T cells were detected in fetal and adult thymus, spleen, skin, small intestine, and liver of IL-7R$^{-/-}$ mice. Analysis for the V–J recombination of TCR in these mice showed that TCR γ gene rearrangement was absent or severely reduced, while TCR α, β and δ genes showed a relatively normal pattern of rearrangements. Thus, it appears that for $\gamma\delta$ T cells, IL-7Rα signaling is critical for rearrangement of the γ gene, although the ligand of IL-7Rα required for γ gene rearrangement has not yet been clarified. It is possible that IL-7 may have primarily a survival and/or growth effect on $\gamma\delta$ T cells, whereas γ gene rearrangement may be induced by TSLP. Alternatively, TSLP and IL-7 may be fully redundant, and therefore exert additive effects on the rearrangement of γ genes.

Comparison of IL-7R$\alpha^{-/-}$, γc$^{-/-}$, and Janus kinase 3 (Jak3)$^{-/-}$ mice is useful for identifying the IL-7 signaling pathway that is crucial in T-cell development. Mutations of the γc chain have been shown to be associated with human X-linked SCID (X-SCID), and affect receptor-mediated activation of Jak3, which is the best characterized downstream signaling element of the γc. Mice lacking the γc chain have a hypoplastic thymus, and are deficient in NK cells, intestinal intraepithelial $\gamma\delta$ T cells, dendritic epidermal T cells, peripheral lymph nodes and gut-associated lymphoid tissue. Jak3$^{-/-}$ mice also show severe hypoplasia of the thymus, and deficient peripheral lymph nodes, NK cells, dendritic epidermal T cells and intestinal intraepithelial $\gamma\delta$ T cells. Overall, Jak3-deficient mice display defects in T-cell development which are similar to those seen in IL-7R$\alpha^{-/-}$ and γc$^{-/-}$ mice. These results, together with the findings that knockout of the IL-2 or IL-4 gene, or both genes, does not result in any defects in lymphopoiesis, suggest that the developmental abnormalities of T cells seen in X-SCID in humans is mainly due to defective IL-7Rα signaling.

Finally, although T lymphopoiesis is greatly reduced in IL-7$^{-/-}$, IL-7R$\alpha^{-/-}$, γc$^{-/-}$ and Jak3$^{-/-}$ mice, some $\alpha\beta$ T cells develop into mature phenotypes. T cells in peripheral organs of IL-7$^{-/-}$ mice have an apparently normal response to mitogenic stimuli. However, T cells from IL-7R$\alpha^{-/-}$ mice are hyporesponsive to alloantigens as well as receptor-independent stimuli such as phorbol myristate acetate (PMA) plus ionomycin. Thus, IL-7Rα signaling is also essential for functional maturation of T cells. In IL-7$^{-/-}$ mice, another cytokine, perhaps TSLP, must provide the functional maturation signal. Interestingly, T cells that develop in γc$^{-/-}$ mice are able to proliferate in response to PMA plus ionomycin, or anti-CD3 plus anti-CD28, even though the response is weaker than that in controls, and the cells have greatly increased apoptotic rates. However, T cells from Jak3$^{-/-}$ mice do not proliferate in response to PMA plus ionomycin or anti-CD3 plus anti-CD28. One explanation of these results would be that TSLP signals through IL-7Rα associated not with γc, but with an as yet unidentified signaling chain which also activates Jak3. In any case, it is apparent that the T cells generated in the absence of IL-7Rα signals are defective with respect to survival, proliferation and function.

2.2 SCF

Stem-cell factor (SCF), also known as c-Kit ligand, plays an important role in the generation of various hematopoietic stem cells. SCF interaction with its receptor, c-Kit, appears to be involved in the self-renewal of hematopoietic stem cells, and the expansion of early erythroid precursors and myeloid precursors. In the T-cell compartment, the cells expressing c-Kit include Thy1$^+$c-Kit^{low} pro-thymocytes in the blood, CD44$^+$CD25$^-$ and CD44$^+$CD25$^+$ early thymocytes, which are all stages in development preceding TCRβ chain rearrangements. In addition, precursor cells on the positive selection pathway, TCR^{low}CD4^{med}CD8^{med}, also appear to express high levels of c-Kit.

Since hematopoietic stem cells express c-Kit, injection of antagonistic antibodies to c-Kit may not distinguish whether the antibody inhibits development of thymocytes, or affects the generation and/or repopulation of thymic precursors homing to the thymus, or both. The direct evidence that SCF is important for expansion of very early thymocytes *in vivo* came from studies in mouse strains with natural (rather than targeted) mutations in the genes for c-Kit and for SCF. In c-*kit*-deficient (*W/W*) mice, the number of CD3$^-$CD4$^-$CD8$^-$ is reduced 40-fold compared with wild-type levels. When the fetal thymi of SCF-deficient (*sl/sl*) mice are grafted into wild-type recipient mice, the total number of double-negative thymocytes is 12-fold lower than that of wild-type grafts. Moreover, the proliferative rate of the immature thymocytes in *sl/sl* graft is halved. Thus, c-Kit–SCF interactions appear to be important for expansion of very early thymocytes. Since these mutant mice do not completely lack either ligand or receptor function, the role of SCF may be more essential in T-cell development than indicated by studies in these mice.

2.3 Fas-L

Fas, a member of the TNF receptor family, has been well characterized for its capacity to deliver apoptosis signals to mature T cells. Thus, it is reasonable to speculate that interactions between Fas and its ligand (Fas-L) might be involved in thymic selection processes, especially for negative selection. However, the role of Fas-L during thymic selection has been controversial. Several groups have reported that Fas–Fas-L interactions are not required during selection. This conclusion came from the observations that *lpr* mice, a mutant strain lacking functional Fas molecules, show normal thymic deletion of T cells specific for endogenous superantigens. However, in a recent study using a sensitive method to detect apoptotic cells, it was shown that blocking the Fas–Fas-L interaction reduced cell death following certain types of acute

stimuli that normally induce apoptosis in cortical thymocytes. These signals included cross-linking with anti-CD3, and injection of ovalbumin in a mouse transgenic for a TCR specific for ovalbumin. Thus, Fas–Fas-L interactions may be involved in negative selection of some autoreactive thymocytes, depending on the type of antigen.

2.4 CD30 and its ligand

CD30, another member of the TNF receptor family, was first identified as a marker on Reed–Sternberg cells in Hodgkin's disease and it has been found on many types of transformed lymphocytes. It is also been detected on some normal thymocytes, and on some mature lymphocytes. A ligand for CD30 has been identified whose expression has been reported on macrophages and activated T cells. Knockout of CD30 in mice suggests that CD30 promotes the apoptotic process of thymocytes undergoing negative selection. The dependency on CD30 for negative selection is, however, not absolute. Deletion of thymocytes expressing a transgenic $\alpha\beta$ TCR specific for H-Y or of a $\gamma\delta$ TCR specific for class I major histocompatibility complex (MHC) was reduced over ten-fold in CD30 knockout mice, whereas deletion of thymocytes specific for an endogenous viral superantigen was CD30-independent. Thus, as in the case of Fas, the role of CD30 in negative selection may depend on the type of antigen.

3. Cytokines in B-cell development

B cells differentiate in the bone marrow in four developmental stages: pre-pro-, early pro-, late pro- and pre-B (Figure 1). Pro-B cells express B220 and CD43, and can be further divided into at least three stages according to the expression of BP-1 and HSA (heat stable antigen): BP-1$^-$HSA$^-$, BP-1$^-$HSA$^+$, BP-1$^+$HSA$^+$. Analysis of immunoglobulin gene rearrangement shows that cells in BP-1$^-$HSA$^+$ and BP-1$^+$HSA$^+$ stages are rearranged only for the heavy chain D-J loci. Pre-B cells are B220$^+$CD43$^-$sIgM$^-$, and differentiate to immature B cells which are B220$^+$CD43$^-$sIgM$^+$. Mature B cells are B220^{++}CD43$^-$sIgM$^+$sIgD$^+$.

3.1 IL-7

IL-7, originally discovered based on its ability to stimulate the proliferation of B-cell progenitors, is the cytokine whose activity is best defined in the development of B cells. It stimulates proliferation of pre-B cells *in vitro*, and neutralizing antibodies to IL-7 added to Whitlock–Witte cultures blocks the generation of pre-B cells. The specific B-cell lineage populations that are critical to IL-7 were determined using IL-7$^{-/-}$ and IL-7R$\alpha^{-/-}$ mice. In IL-7$^{-/-}$ mice, B220$^+$IgM$^+$ mature B cells were almost completely absent in the bone marrow. However, more immature cells with the B220$^+$IgM$^-$ phenotype were detected in IL-7$^{-/-}$ mice. When these cells were further analyzed, it was shown that CD43$^-$HSA$^+$ pre-B cells were almost completely absent, whereas pro-B cells expressing CD43 and HSA were present. Thus, IL-7 appears to be essentially required for the transition of the late pro-B cells to pre-B cells.

IL-7R$\alpha^{-/-}$ mice also show a block during B lymphopoiesis. When the cells in the B220$^+$IgM$^-$ fraction of IL-7R$\alpha^{-/-}$ mice were analyzed in detail, it was found that CD43$^+$HSA$^+$ pro-B cells as well as CD43$^-$HSA^{++} pre-B cells were absent, while CD43$^+$ pre-pro-B cells were present in the bone marrow. Thus, B-cell development in IL-7R$\alpha^{-/-}$ mice is blocked before the pro-B cell stage, apparently a stage earlier than in the IL-7$^{-/-}$ mouse. Although the difference could be explained by different genetic background or environmental effects, it is also possible that there TSLP is required for development to the pro-B cell stage and IL-7 for development to the pre-B cell stage. Mice deficient in γc show deficiencies in B-cell development that are similar to those seen in IL-7 and IL-7Rα knockout mice.

Surprisingly, in humans that are deficient in γc (X-SCID patients), B cells develop almost normally. Furthermore, humans with mutations in Jak3, the downstream signaling element of the γc, also have normal or increased number of B cells. Thus evidence from genetic deficiencies suggests that human B-cell development is independent of IL-7, and this has been verified in cell culture. Anti-IL-7 antibodies failed to block CD34$^+$ hematopoietic stem cell development to surface-immunoglobulin-positive immature B cells when cultured on human fetal bone marrow stromal cells. Thus, IL-7 signaling is apparently not required for the development of human B cells, in contrast to the requirement in mice.

3.2 PBSF/SDF-1

Pre-B-cell stimulatory factor (PBSF)/stromal cell-derived factor 1 (SDF-1) is a member of the CXC (Cys–X–Cys) group of chemokines. The receptor for PBSF/SDF-1 is CXCR-4 (or 'fusin'), recently identified as a co-receptor for entry of human immunodeficiency virus. This chemokine has been shown to stimulate proliferation of B cells *in vitro*, and is constitutively expressed in bone marrow stromal cells. Studies of mutant mice lacking PBSF/SDF-1 confirmed the essential role of this cytokine in the development of B cells. B-cell progenitors (both pro-B and pre-B cells) were severely reduced in the fetal liver and bone marrow. Normal T-cell development was observed, but these mutant mice died perinatally, most probably due to abnormal cardiac development. Thus, compared with IL-7, which is essential for the transition of pro-B cells to pre-B cells, PBSF/SDF-1 appears to affect an earlier stage of B-cell precursors (perhaps comparable to IL-7Rα).

3.3 Flt3 ligand

The Flt3 ligand stimulates the proliferation of CD34$^+$ stem cells in human bone marrow. The receptor for Flt3 ligand, Flk2/Flk3, is expressed in all primitive hematopoietic cell populations and shows significant sequence and structural homology to the class III receptor tyrosine kinases including CSF-1 receptor, the SCF receptor (c-Kit) and the α and β platelet-derived growth factor receptors. Thus, Flt3 ligand is expected to exert activities on myeloid and lymphoid progenitors in combination with other cytokines. The role of Flt3 ligand in lymphoid progenitors has been clearly demonstrated by studies of the receptor, Flk2/Flk3, deficient mice. Flk2$^{-/-}$ mice developed normally and had a largely normal hematopoietic system. However, B-cell progenitors were significantly reduced in Flk2$^{-/-}$ mice. Specifically, the pro-B cells were reduced in Flk2$^{-/-}$ mice, whereas the levels of pre-B cells, immature B cells, and mature B cells were largely normal. Mice lacking both the *flk2* and the *c-kit* gene exhibit more severe defects in total numbers of B lymphoid progenitors. This finding suggests a functional redundancy between SCF and Flt3 ligand. In summary, Flt3 ligand appears to play a role in the commitment of more primitive progenitors to the B-cell lineage, as well as in the induction of proliferation of multipotent stem cells.

3.4 SCF

Interactions between SCF and its receptor, c-Kit, were generally believed to be involved in the development of B cells. *In vitro*, SCF synergizes with IL-7 to promote the proliferation of B-cell progenitors. Conversely, antibodies against c-Kit suppress the growth of B-cell precursors *in vitro*. However, studies of mice with natural mutations in c-*kit* suggested that SCF is not required for the normal development of B cells. The ratio of cell numbers at different stages of B-cell development appeared to be very similar between c-Kit-deficient mice and wild-type controls. Furthermore, transfer of c-Kit-deficient cells to RAG-2$^{-/-}$ c-Kit^{+} recipients gave rise to all stages of immature B cells in the bone marrow and subsequently mature B cells in the recipients. Thus, while SCF plays an important role in the proliferative expansion of very early thymocytes and multipotent stem cells, it may not be required for the development of B cells, although as noted above these natural mutations do not absolutely eliminate the protein.

4. Cytokines in the development of NK cells

Cytokines critical for the development of NK cells have not been elucidated thus far, nor have other aspects of NK development been closely detailed. However, studies utilizing γc-deficient mice provide clues to cytokines that are likely to be involved in the development of NK cells. Like X-SCID in humans, mice lacking functional γc expression do not develop NK cells. Because γc is the common component of the receptor for IL-2, -4, -7, -9 and -15, the clinical manifestation of γc mutation would be the concomitant inactivation of at least these five different cytokines. However, since NK cells have been found in IL-2$^{-/-}$, IL-4$^{-/-}$ and IL-7$^{-/-}$ mice, IL-9 or -15 may be important cytokines for NK cell development. Suggestively, IL-15 has been shown to have stimulatory activity for NK cells. We have observed that IL-9 induces NK cell development from the common thymic T/NK precursor, which is CD44^{+}CD25^{-}. It will be of interest to see whether IL-15- or IL-9-deficient mice produce NK cells.

5. Conclusions

Gene targeting has cast a bright light into the shadows of lymphoid development. It is now clear which cytokines are truly required in this process and their mechanisms will be the future directions of study. Major challenges lie ahead in determining

how each stimulus acts on lymphoid precursors. Lymphoid precursors have been difficult to obtain and work with. Multiple simultaneous stimuli are involved, making it difficult to study the important intracellular events triggered by cytokines such as IL-7, which normally works together with other stimuli such as SCF or unknown stromal factors. The IL-7 receptor incorporates γc and Jak3, but what is the unique contribution of IL-7Rα that triggers such unique responses as gene rearrangements and survival of precursors?

Lymphopoiesis probably depends on a cocktail of microenvironmental stimuli. The cytokines that are known to be involved are discussed in the preceding sections, but there are probably many more. In the thymus, there could be cytokines that initially attract precursors, control their slow passage through the organ and finally control their departure. There may be cytokines that induce cell growth, positive selection or death, cytokines that induce differentiation into different lineages, such as $\alpha\beta$ versus $\gamma\delta$, CD4 versus CD8, etc., and cytokines that induce production of the VDJ recombinase components (such as RAGs and TdT) and target the recombinase complex to certain genes. Extrathymic T-cell development may use different cytokines altogether. Human B-cell development appears to use stimuli that are quite different from those used in the mouse: they are not IL-7 and do not use the γc chain—what are they? The CD5 subset of B cells may use different developmental signals. How do NK cells normally develop? What signals impose a commitment upon the lymphoid precursor to become a T, B or NK cell? Solving these biological puzzles will undoubtedly advance our understanding of the basis of human diseases such as lymphoid cancers, autoimmunity and immunodeficiency.

Acknowledgements

We thank Drs J. Keller and J. J. Oppenheim for comments on the manuscript.

References

Amakawa, R., Hakem, A., Kundig, T. M., Matsuyama, T., Simard, J. J. L., Timms, E., Wakeham, A., Mittruecker, H.-W., Griesser, H., Takimoto, H. *et al.* (1996). Impaired negative selection of T cells in Hodgkin's disease antigen CD30-deficient mice. *Cell* **84**, 551–62.

Candeias, S., Peschon, J. J., Muegge, K., and Durum, S. K. (1997). Defective TCRγ gene rearrangement in interleukin-7 receptor knockout mice. *Immunol. Lett.* (in press).

Cao, K., Shores, E. W., Hu-Li, J., Anver, M. R., Kelsall, B. L., Russell, S. M., Drago, J., Noguchi, M., Grinberg, A. *et al.* (1995). Defective lymphoid development in mice lacking expression of the common cytokine receptor γ chain. *Immunity* **2**, 223–38.

Castro, J. E., Listman, J. A., Jacobson, B. A., Wang, Y., Lopez, P. A., Ju, S., Finn, P. W., and Perkins, D. (1996). Fas modulation of apoptosis during negative selection of thymocytes. *Immunity* **5**, 617–27.

DiSanto, J. P., Muller, W., Guy-Grand, D., Fischer, A., and Rajewsky, K. (1995). Lymphoid development in mice with a targeted deletion of the interleukin 2 receptor γ chain. *Proc. Natl. Acad. Sci.* **92**, 377–81.

Durum, S.K. and Muegge, K. (eds). (1997). *Cytokine Knockouts.* Humana Press, Totowa, NJ.

Friend, S. L., Hosier, S., Nelson, A., Foxworthe, D., Williams, D. E., and Farr, A. (1994). A thymic stromal cell line supports in vitro development of surface IgM+ B cells and produces a novel growth factor affecting B and T lineage cells. *Experimental Hematology* **22**, 321–8.

Grabstein, K. H., Waldschmit, T. J., Finkelman, F. D., Hess, B. W., Alpert, A. R., Boiani, N. E., Namen, A. E., Morrissey, P. J. (1993). Inhibition of murine B and T lymphopoiesis in vivo by an anti-interleukin 7 monoclonal antibody. *J. Exp. Med.* **178**, 257–64.

He, Y.-W. and Malek, T. (1996). Interleukin-7 receptor α is essential for the development of γδ⁺ T cells, but not natural killer cells. *J. Exp. Med.* **184**, 289–93.

Kisielow, P. and von Boehmer, H. (1991). Development and selection of T cells: facts and puzzles. *Adv. Immunol.* **58**, 87–209.

Kruisbeek, A. M. and Storb, U. (eds) (1996). Lymphocyte development. *Curr. Opin. Immunol.* **8**, 155–254.

Kuhn, R., Rajewsky, K., and Muller, W. (1991). Generation and analysis of interleukin-4 deficient mice. *Science* **254**, 707–10.

Mackarehtschian, K., Hardin, J. D., Moore, K. A., Boast, S., Goff, S. P., and Lemischka, I. R. (1995). Targeted disruption of the *flk2/flk3* gene leads to deficiencies in primitive hematopoietic progenitors. *Immunity* **3**, 147–61.

Nagasawa, T., Hirota, S., Tachibana, K., Takakura, N., Nishikawa, S.-I., Kitamura, Y., Yoshida, N., Kikutani, H., and Kishimoto, T. (1996). Defects of B-cell lymphopoiesis and bone-marrow myelopoiesis in mice lacking the CXC chemokine PBSF/SDF-1. *Nature* **382**, 635–8.

Nosaka, T., Van Deursen, J. M. A., Tripp, R. A., Thierfelder, W. E., Witthuhn, B. A., McMickle, A. P., Doherty, P. C., Grosveld, G. C., and Ihle, J. N. (1995). Defective lymphoid development in mice lacking Jak3. *Science* **270**, 800–2.

Park, S. Y., Saijo, K., Takahashi, T., Osawa, M., Arase, H., Hirayama, N., Miyake, K., Nakauchi, H., Shirasawa, T., and Saito, T. (1995). Developmental defects of lymphoid cells in Jak3 kinase-deficient mice. *Immunity* **3**, 771–82.

Peschon, J. J., Morrissey, P. J., Grabstein, K. H., Ramsdell, F. J., Maraskovsky, E., Gliniak, B. C., Park, L. S., Ziegler, S. F., Williams, D. E., Ware, C. B. *et al.* (1994). Early lymphocyte expansion is severely impaired in interleukin 7 receptor-deficient mice. *J. Exp. Med.* **180**, 1955–60.

Pribyl, J. A. R. and LeBien, T. W. (1966). Interleukin 7 independent development of human B cells. *Proc. Natl Acad. Sci. USA* **93**, 10348–53.

Rich, B. E. and Leder, P. (1995). Transgenic expression of interleukin 7 restores T cell population in nude mice. *J. Exp. Med.* **181**, 1223–8.

Rodewald, H.-R., Kretzschmar, K., Swat, W., and Takeda, S. (1995). Intrathymically expressed c-kit ligand (stem cell factor) is a major factor driving expansion of very immature thymocytes *in vivo. Immunity* **3**, 313–19.

Sadlack, B., Kuhn, R., Schorle, H., Rajewsky, K., Muller, W., and Horak, I. (1994). Development and proliferation of lymphocytes in mice deficient for both interleukin-2 and -4. *Eur. J. Immunol.* **24**, 281–4.

Schorle, H., Holtschke, T., Hunig, T., Schimpl, A., and Horak, I. (1991). Development and function of T cells in mice rendered interleukin-2 deficient by gene targeting. *Nature* **352**, 621–4.

Takeda, S., Shimizu, T., and Rodewald, H.-R. (1997). Interactions between c-kit and stem cell factor are not required for B-cell development *in vivo. Blood* **89**, 518–25.

Thomis, D. C., Gurniak, C. B., Tivol, E., Sharpe, A. H., and Berg, L. J. (1995). Defects in B lymphocyte maturation and T lymphocyte activation in mice lacking Jak3. *Science* **270**, 794–7.

von Freeden-Jeffry, U., Moore, T. A., Zlotnik, A., and Murray, R. (1998). IL-7 knockout mice and the generation of lymphocytes. In Durum, S. and Muegge, K. (eds), *Cytokine Knockouts*, pp. 21–36. Humana Press, Totowa, NJ.

XV Cytokines in the functions of dendritic cells, monocytes, macrophages and fibroblasts

Laurent P. Nicod and Jean-Michel Dayer

1. Definitions

Blood monocytes, tissue macrophages and dendritic cells are derived from the same CD34+ haematopoietic precursor cells. They gain access to various tissues via blood and lymph and, depending on their individual and organ-specific environment, e.g. the lung, they can differentiate and occasionally undergo some local self-renewal.

Monocytes circulate for 1–2 days before entering tissues at perivascular, interstitial or intraepithelial levels in lymphoid and non-lymphoid organs. They then mature into macrophages which have relatively long half-lives and retain considerable mRNA content and protein-synthesizing activity even as they become refractory to stimulation by growth factors. Resident macrophages display many membrane receptors that enable them to endocytose and phagocytose a wide range of soluble and particulate ligands. Their local secretory responses to cytokines and other mediators help to initiate early changes that amplify local inflammation. Thus macrophages can differentiate into 'elicited' or 'immunologically activated' macrophages, acquiring a distinct set of properties, including enhanced antigen-presenting or pathogen-destroying functions which are induced in macrophages by the interaction with foreign organisms or antigen-stimulated T lymphocytes.

Dendritic cells from different lymphoid and non-lymphoid organs have similar functions which are not readily observed in other cells:

(1) A very small number of dendritic cells elicit strong T-cell-dependent responses in standard tissue culture systems.

(2) They can stimulate quiescent cells, including naïve T cells.

(3) They appear to have a unique adjuvant action *in vivo*.

(4) *In vitro*, they appear to be the main cells that capture antigens for presentation to T cells.

Dendritic cells are large, non-adherent cells that extend motile, sheet-like processes, or 'veils', in several different directions. They express high levels of antigen-presenting major histocompatibility complex (MHC) products and also several accessory molecules that are involved in T-cell binding and co-stimulation such as CD40, leukocyte function-association antigen-3 (LFA-3/CD58 and B7-2/CD86). A small number of dendritic cell-restricted markers are known (DEC-205 in the mouse, OX62 in the rat, CD83 and p55 in the human system), but none of these antigens is specific to dendritic cells. Dendritic cells appear to be derived from CD34+ haematopoietic progenitors of marrow and blood before migrating to sites such as the skin and lung where they become non-proliferating dendritic cells capable of different degrees of activity.

Langerhans cells are potent accessory cells which belong to the dendritic cell lineage. They are thought to be derived from dendritic cells, and differ from the latter in their strong expression of CD1a antigens and in the presence of pentalaminar plate-like cytoplasmic organelles, Birbeck's granules, visible only by electron microscopy. The presence of Langerhans cells in the epidermis has been known since 1868. They are often found in normal mammalian epidermis, within the epithelium of human airways. The capacity of Langerhans cells to serve as accessory cells is highly dependent on their stage of maturation and differentiation. Langerhans cells are poorly phagocytic, although they can internalize and process antigens, but freshly isolated epidermal Langerhans cells, whilst capable of activating T-lymphocyte cell lines, lack the capacity to activate naïve T cells. Culture of Langerhans cells *in vitro*, particularly in the presence of interleukin 1 (IL-1) and granulocyte–macrophage colony-stimulating factor (GM-CSF), can dramatically improve the capacity of these cells to activate unprimed T cells. Based on these findings, it has been suggested that, in tissues, Langerhans cells serve as 'sentinels' which can pick up and process foreign antigens, but the initiation of an immune response to a previously unencountered antigen would require their migration to lymph nodes, during which time they 'mature' into potent accessory cells, acquiring features of dendritic cells.

2. Present concepts of differentiation/maturation: the role of the cytokines (Figure 1)

According to recent data, monocytes, macrophages and dendritic cells share CD34⁺ stem cells. Some of the cytokines/receptors required for early stages of

differentiation are: stem-cell factor (SCF), c-Kit ligand, and IL-1; IL-3 supports multiple lineages, including macrophages, whereas GM-CSF and macrophage colony-stimulating factor (M-CSF) act mainly at relatively late stages. Several cytokines, such as GM-CSF, IL-4 and tumor necrosis factor alpha (TNFα), have recently been shown to play a crucial role in the maturation of monocytes into dendritic cells.

CD34⁺ haematopoietic progenitors from human cord blood, cultured with GM-CSF and TNFα, generate a mixed cell population containing different types of dendritic cells. After a few days of culture, two subsets of dendritic cell precursors can be identified by the mutually exclusive expression of CD1a and CD14. After 1–2 weeks, CD1a⁺ precursors generate cells that express Birbeck granules, lymphocyte activation gene (Lag) antigen and E-cadherin, markers that are characteristic of epidermal Langerhans cells. In contrast, the CD14⁺ progenitors mature into CD14⁺/CD1a⁻ dendritic cells; they lack Birbeck granules, E-cadherin and LAG antigen but express CD2, CD9, CD68 and the coagulation factor XIIIA found in dermal dendritic cells. Interestingly, the CD14⁺ precursors, but not the CD1a precursors, are able to differentiate into macrophages in response to M-CSF.

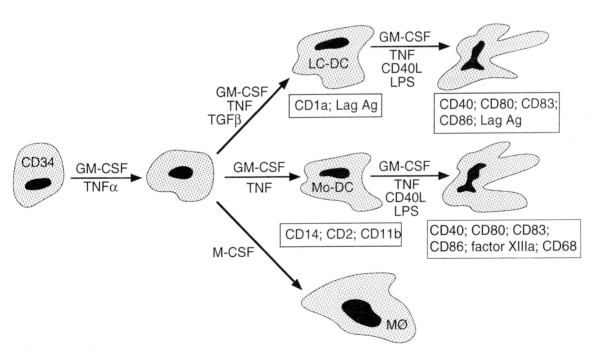

∎ **Figure 1.** Theoretical evolution of CD34 cells into macrophages (MØ) or into dendritic cells (DC) derived from monocytes (Mo) or Langerhans cells (LC).

Dendritic cells can also develop from CD14+ peripheral blood monocytes cultured with GM-CSF and IL-4. Under these conditions monocytes develop into a homogeneous population of dendritic cells without dividing, at an efficiency close to 100%. These cells have the characteristics of immature dendritic cells and can be further induced to mature by inflammatory stimuli such as TNFα, IL-1 or lipopolysaccharide (LPS), by monocyte-conditioned medium and also to some extent by IL-13. It is interesting that immature dendritic cells generated from monocytes should retain the M-CSF receptor, although they lose it after maturation has set in. Thus the concept has emerged that monocytes represent an abundant source of precursors that can develop into dendritic cells or macrophages, depending on the external stimuli. *In vitro*, this polarization can be driven by the addition of appropriate cytokines (GM-CSF plus IL-4 or M-CSF).

Transforming growth factor beta (TGFβ) has recently proved to be of particular importance in promoting dendritic cell development with high CD1a molecule expression from CD34 progenitors. Thus, while TGFβ is inhibitory on immature progenitor cells, it is also a critical growth and differentiation factor for dendritic cells. Macrophages express common as well as site-specific properties within different tissue micro-environments, including various marker antigens and receptors of soluble 'matrix' and cell-associated ligands. Mature macrophages display proficient phagocytic and endocytic activities and complex intracellular signalling pathways, linked by membrane traffic to their adoptive biosynthetic and secretory responses. The macrophages disperse in the body and are able to react to local injury, modifying the properties of all cells in their immediate vicinity. The modulation of macrophage functions is meant to meet specific needs. Their stage of activation differs depending on their stage of differentiation and micro-environmental factors. The acquisition of an activated phenotype is in many instances attributable to the regulation of gene expression through the activation of specific transcription factors such as NF-κB, STAT1α or interferon-responsive factor (IRF). Attention is focusing increasingly on reactive oxygen intermediates owing to their important contribution to the biology of macrophage activation.

A regulatory strategy that provides important checks and balances against the indiscriminate or excessive production of pro-inflammatory mediators by activated macrophages is the requirement that at least two stimuli be present in order to obtain the maximal expression of many different inducible genes. Thus interferons are provided by the host and other stimuli, such as LPS, come from the invading microorganisms. Prior exposure of macrophages to potential activating stimuli influences the profile of responsiveness upon subsequent challenge. It has been proposed that the more general term 'macrophage reprogramming' be employed rather than 'priming' or 'desensitization'.

3. Cell–cell interaction and the cytokine network

3.1 Antigen presentation

The interaction of antigen-presenting cells (APCs) with antigens is an essential step in immune induction because it enables lymphocytes to encounter and recognize antigenic molecules and to become activated. T cells recognize protein antigen molecules presented by APCs that express the class I and II MHC molecules. The activation of lymphocytes also depends on the type of accessory signals that are delivered by the APCs. In turn, the type of specific response depends closely on these co-stimulatory signals, delivered by soluble factors or surface molecules present on APCs, but also on other cells involved in the vicinity of an inflammatory reaction such as mesenchymal, epithelial or endothelial cells.

Antigen presentation requires internalization of most antigens by APCs, and the biochemical processing takes place intracellularly, in the vesicles that have a low pH. Protein antigens are taken up either by a receptor-mediated process or by fluid-phase internalization in the APCs. The carrier molecule, e.g. membrane immunoglobulin in B cells, or complement fragment receptor (FcR or lectin receptors in macrophages), may protect some of the protein antigen from normal catabolism. Processing consists of the unfolding of the protein antigen and/or its fragmentation to peptides. An immunogenic determinant is expressed on the APC surface 30–60 min after internalization. The exogenous antigens tend to be presented by class II MHC, and endogenous antigens by class I. In some cases, endogenously produced antigens can also be associated with class II MHC molecules. The antigen/MHC complex appears to be quite stable, thus leaving sufficient time for the accessory cell to encounter a T cell. A cell needs as few as 210–340 specific peptide/MHC class II complexes to activate T cells.

Two early signal transduction pathways are activated by stimulation of the T-cell receptors (TCRs) by MHC class I or II complexes: the inositol phospholipid second messenger pathway and a tyrosine kinase pathway. The CD4 and CD8 molecules appear to physically interact with Lck kinases. Cross-linking of CD4 results in increased Lck activity and phosphorylation of the TCR δ-chain.

During antigen presentation, APCs express important accessory molecules that regulate T-cell response. Thus, in addition to the signals transmitted through TCRs, membrane molecules or cytokines released by APCs regulate T-cell activation. Consequently, the type of APC and its degree of activity may be crucial to the local regulation of immunity.

Integrins are heterodimers consisting of non-covalently associated α- and β-subunits. They mediate both cell–substratum and cell–cell adhesion. A heterodimer composed of the α-chain (CD11a) and the β-chain (CD18) interacts with intracellular adhesion molecule 1 (ICAM-1). This interaction has been shown to improve adhesion of the interacting cells and, in association with TCR activation, to result in a co-stimulatory signal to lymphocytes. Other integrins may play a part in the adhesion or activation of T cells, including integrins of the β1 group, such as the very late activation antigen (VLA) 4 and VLA-5.

The natural ligand to the CD2 lymphocyte surface molecules ('sheep erythrocyte receptor'), present on virtually all T lymphocytes, is LFA-3. LFA-3 is a surface molecule present on many cell types, including macrophages, B lymphocytes and epithelial cells. The interaction between CD2 and LFA-3 produces transduction signals, provided that such interaction occurs in association with TCR triggering.

Recent results have shown that a major co-stimulatory activity required for the IL-2-induced proliferation of T cells is mediated by the interaction of the CD28 molecule on the T-cell surface with its ligands, members of the B7 family on the APC surface. CD28 is expressed on the surface of 95% of CD4$^+$ and 50% of CD8$^+$ human peripheral blood T cells. Both B7-1 (CD80) and B7-2 (CD86) are members of the immunoglobulin superfamily with extracellular regions containing two immunoglobulin-like domains and a short cytoplasmic domain. CD86 is apparently the most prevalent B7 form expressed on dendritic cells, although these also express CD80. CD86 is expressed at low levels on monocytes, and its expression is increased by interferon gamma (IFNγ) treatment; while this is said not to apply to alveolar macrophages, it is now apparent that in pathological conditions such as sarcoidosis, under circumstances poorly understood as yet, alveolar macrophages may express high levels of CD86, CD40 or CD30 ligand (CD30-L). CD40 and CD40-L interaction has also been shown to be a powerful inducer of CD86 expression on B cells and dendritic cells. These observations underline the importance not only of APC/T-cell interaction but also that of T cells in enhancing APC activity.

Cytokines such as IL-1α or -1β, IL-6 or TNFα are able to provide co-stimulatory signals to T cells. This network of soluble factors has recently been extended by important findings concerning IL-7, -10, -12, -13 and -15. IL-12 is a heterodimeric cytokine which enhances the proliferation and activation of the lytic activity of NK cells and it is one of the main factors to differentiate T-cell precursors into Th1 and favour Th1 cytokine response. IL-10 is in many respects an antagonist of IL-12 due to its negative effects on APCs and T-cell activation. The ability of the various APCs to produce these soluble factors may be crucial to controlling immunity.

3.2 Direct cell–cell contact

T lymphocytes are likely to play a pivotal role in the initial phase of the pathogenesis of chronic inflammatory diseases such as rheumatoid arthritis and multiple sclerosis. In rheumatoid arthritis, T lymphocytes that display a mature helper phenotype (i.e. CD3$^+$CD4$^+$) are the main infiltrating cells in the pannus, comprising between 16% of total cells in 'transitional areas' and 75% in 'lymphocyte-rich areas'. In rheumatoid arthritis, T lymphocyte extravasation occurs at the level of high endothelial venules. T lymphocytes are thought to be pathogenic in multiple sclerosis, since antigen-specific T lymphocytes induce experimental allergic encephalitis (EAE), a rodent model of multiple sclerosis. Indeed, activated, but not resting, T lymphocytes spontaneously cross the blood–brain barrier; when these activated T lymphocytes are specific for brain antigens such as myelin basic protein (MBP), they mediate EAE in rodents. Both multiple sclerosis and rheumatoid arthritis are thought to be Th1 cell-mediated diseases. The initiating event in autoimmune diseases, where exogenous and endogenous antigens are still unknown, may be a slowly developing differentiation of dendritic cells under the influence of cytokine release by target cells in a specific tissue in response to non-specific stimulation by minor trauma, infection, allergic reaction or immunological components. These dendritic cells eventually present autologous antigens to T cells. If this hypothesis is correct, T cells are not required during the initial stage of dendritic cell differentiation, nor is the presence of exogenous material for antigen presentation.

In multiple sclerosis and rheumatoid arthritis, as in many other diseases, tissue destruction takes place in areas where T lymphocytes are in close contact with monocytes and resident tissue cells, suggesting the importance of novel cell surface-mediated mechanisms potentiating inflammation.

3.2.1 T-cell signalling of monocytes/macrophages by direct cell–cell contact
(Figure 2)

The activation of effector cells mediated by T cells has been abundantly documented by the induction of B-cell proliferation and antibody secretion, which require both direct cell–cell contact and soluble signals. Indeed, B cells can be activated in the absence of antigen by direct contact with activated T cells; when the activation of both T and B cells was simultaneously examined a considerable amount of signalling 'cross-talk' was observed, that triggered a cascade of timed signals resulting in the activation of effector function in both cell types.

Considering that cells of the monocyte lineage have been considered 'classical' or 'professional' antigen-presenting cells (APCs or dendritic cells), it is surprising that, while a tremendous number of studies have addressed the question of APC signalling of T cells, very few have addressed the question of T-cell signalling of monocytes/macrophages in the context of direct cell–cell contact. Most studies dealing with the question of monocyte/macrophage activation have focused on the role of natural exogenous soluble factors such as LPS and other bacterial products or endogenous products such as IFNγ and colony-stimulating factors and, to some extent, TNFα and IL-1. Cytokines like TNFα and IL-1, which potently stimulate nearby connective tissue cells to release matrix metalloproteinases (MMPs) and prostaglandin E$_2$ (PGE$_2$), have very weak capacity to induce monocyte/macrophages to produce MMPs. It appears that most of the soluble cytokines mainly have an anti-inflammatory capacity on monocyte/macrophage functions.

In the murine system, Th1 cells produce mainly IL-2, IFNγ and lymphotoxin (LT), i.e. type 1 cytokines, whereas Th2 cells produce IL-4, -5, -6, -9, -10 and -13, i.e. type 2 cytokines. However, in the human system, Th1 and Th2 cells produce similar patterns, although cytokine synthesis is not as tightly restricted to a single subset as in the mouse. Cells other than CD4$^+$ T cells, such as CD8$^+$ cells, can be divided into similar subsets. In contrast to the monocyte/macrophage-activating functions of type 1 cytokines IFNγ, TNFα and IL-2, type 2 cytokines IL-4 and IL-10 have a strong inhibitory effect on almost every function of monocyte/macrophages. The expression of membrane-associated IL-1 (IL-1α) in mouse macrophages is mediated by both soluble factors and direct contact with T cells. T-cell–macrophage contact in IL-1 induction has been observed with both Th1 and Th2 cells in the absence of lymphokine release.

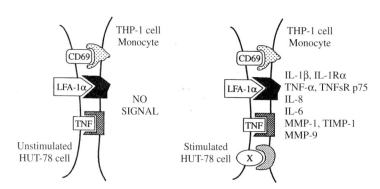

IL-1β, IL-1Rα
TNF-α, TNFsR p75
IL-8
IL-6
MMP-1, TIMP-1
MMP-9

■ **Figure 2.** Different T-cell surface molecules are involved in the signalling of various target cells. Although CD69, CD11a and TNFα are expressed by the human T-cell line HUT-78, they are not able to induce the activation of monocytic THP-1 cells.

We have also observed that direct contact with stimulated T cells was a potent stimulus of monocyte/macrophage activation. The production of IL-1β by human monocytes also requires direct contact with anti-CD3-stimulated T cells. Other observations have shown that the induction of macrophage effector functions mediated by T lymphocytes in living cell co-cultures involves signals delivered by cell–cell contact together with IFNγ. However, fixed, stimulated T cells induced TNF production in macrophages in the absence of IFNγ. Furthermore, by using isolated plasma membranes from different stimulated murine T-cell clones it has been demonstrated that stimulated Th1 and Th2 cells were both able to activate macrophages, establishing that direct contact with stimulated T cells was a potent mechanism in inducing monocyte/macrophage effector functions.

A critical question arises regarding the identity of the molecules on the T-cell surface that are involved in contact-dependent signalling of monocyte/macrophage activation as well as their counter-ligands. It has been postulated that T-cell membrane-associated TNF is involved in monocyte/macrophage activation. However, fixed, stimulated Th2 cells from a T-cell line that does not express membrane-associated TNF can induce both TNF and IL-1 production in monocyte/macrophages, ruling out the involvement of T-cell membrane-associated TNF in this monocyte/macrophage signalling. Similarly, although the LT-β receptor is expressed in macrophages, it is not likely that membrane-associated LT is involved in monocyte/macrophage signalling upon contact with stimulated T cells, since Th2 cells do not express LT protein or mRNA.

In addition to membrane-associated cytokines, other surface molecules have been assessed for their ability to activate monocyte/macrophages upon contact with stimulated T cells; these include LFA-1/ICAM-1, CD2/LFA-3, and CD40/CD40-L. Of these surface molecules, only CD40/CD40-L were clearly involved in the activation of murine macrophages by T cells. However, although CD40 ligation induces MMP expression in human monocytes, this pair of molecules cannot be involved in the signalling of monocyte/macrophages observed with T cells stimulated for 24–48 h, i.e. stimulated T cells in which CD40-L is strongly down-regulated. Other studies have shown that cytokine production was induced in monocytes by soluble CD23. In monocytes, the counter-ligands for CD23 are CD11b/CD18 and CD11c/CD18 rather than CD21. Studies by us and others have shown that LFA-1 (CD11a/CD18) and CD69 play a role in the activation of human monocytic cells by stimulated T cells. The latter data were recently confirmed in a study showing that IL-15 induced synovial T cells from rheumatoid arthritis patients to activate the production of TNFα by macrophages. This effect was inhibited by antibodies to CD69, LFA-1 and ICAM-1. However, neither anti-CD69 nor anti-CD11 antibody, nor both together, blocked more than 50% of the interaction between activated lymphocytes and monocytes/macrophages. Other molecules, referred to as surface-activating factors on stimulated T lymphocytes (SAFTs), which have not been wholly identified yet, are thought to be responsible for this stimulation by direct contact; one such factor, with an approximate M_r of 40 000 Da, triggers monocytic cell activation.

Recently we found that Th1 T-cell clones stimulated with OKT3, anti-CD28 or specific antigens preferentially induced IL-1β production in THP-1 cells, whereas Th2 clones mainly induced IL-1 receptor antagonist (IL-1Ra) production. Therefore, depending on T-cell type and T-cell stimulus, direct cell–cell contact with stimulated T cells can induce pro-inflammatory and tissue-destructive products without inducing their specific inhibitors, suggesting that this mechanism may account for any imbalance arising under pathological conditions. We have also demonstrated that, upon contact with stimulated T cells, the balance between IL-1β and IL-1Ra production in monocytes is regulated by serine/threonine phosphatase(s). Furthermore, cell–cell contact-activated THP-1 cells express membrane-associated protease(s) neutralizing TNFα activity both by degrading the latter cytokine and by cleaving its receptors at the cell surface. Thus the triggering of these mechanisms by direct contact with stimulated T lymphocytes may regulate the pro-inflammatory cytokines and their inhibitors; this balance dictates in part the outcome of the inflammatory process.

3.2.2 T-cell signalling of fibroblasts by direct cell–cell contact (Figure 3)

Signalling of stromal cells by contact with mononuclear cells has been studied very little. It has been shown that co-culture of living and resting mononuclear cells with autologous dermal fibroblasts results in a dose-dependent increase of collagen synthesis which is not observed with fibroblasts cultured alone. In addition, T cells cultured in the presence of mycobacterial antigen and IL-2 induce

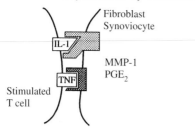

Figure 3. Fibroblasts and synoviocytes are activated by membrane-associated TNFα and IL-1.

outgrowth of synovial fibroblasts *in vitro*, and co-culture of T cells and fibroblasts induces the production of IL-6 and IL-8 by fibroblasts. These studies, however, did not distinguish the effects of direct cellular contact from those of soluble factors released by mononuclear cells. Contrasting results on the stimulatory or inhibitory effects of resting or mitogen-activated mononuclear cells are reported in the literature. Contact between fibroblasts and T lymphocytes also mediates *in vitro* survival of the latter cells.

Two pairs of cell adhesion molecules are involved in the adhesion of T lymphocytes to dermal fibroblasts: ICAM-1/LFA-1 and LFA-3/CD2. It has been shown that direct contact between fibroblasts and T cells induces the expression of adhesion molecules. Furthermore, steroids inhibit both ICAM-1 expression on fibroblasts and T-cell adhesion to fibroblasts. The capacity of binding to fibroblasts varies among T-cell subsets: $\gamma\delta$ T cells (Vδ1$^+$) adhere preferentially to fibroblasts, although CD8$^+$ T cells and T cells with memory phenotype are also able to adhere. It is therefore likely that part of T-cell signalling of fibroblasts occurs via direct contact, although there are no clear-cut data to verify this.

Contact with activated lymphocytes induces an imbalance between MMP-1 and TIMP-1 (tissue inhibitor of metalloproteinases 1) production by dermal fibroblasts. This imbalance is due to the length of time for which T lymphocytes are stimulated. Since direct cell–cell contact with stimulated T lymphocytes induces an imbalance between the production of MMPs and TIMP-1 by both monocytes and fibroblasts *in vitro*, it may, in analogy, also favour tissue destruction *in vivo*. In addition to MMP-1 and TIMP-1, direct cell–cell contact with stimulated T lymphocytes induces PGE$_2$ production on human dermal fibroblasts and fibroblast-like synoviocytes.

4. Monocytes/macrophages and tissue remodelling

4.1 Cytokines and cytokine antagonists

Cytokines are polypeptides or glycoproteins of low molecular weight. Their monomers generally have an M_r of $\leq$8–30 kDa, but some cytokines exert their biological activity in oligomeric forms of higher molecular weight. The production and secretion of most cytokines are of limited duration, and basal (constitutive) production is usually absent or occurs at low levels. Most cytokines show pleiotropy, exerting their influence on diverse cell types. Cytokines bind to specific receptors ($K_d = 10^{-9} - 10^{-12}$ M) which transduce signals into the target cell. Most cytokine receptors function as (i) receptor tyrosine kinases and tyrosine-kinase-associated receptors; (ii) receptor serine/threonine kinases; and (iii) G-protein-linked receptors. Contrasting with many other peptide hormones, cytokines are generally active over a short radius by binding to the cell of their origin (autocrine influence) or to a neighbouring target cell (paracrine influence).

The activity of pro-inflammatory cytokines may be counterbalanced by inhibitory molecules which can arbitrarily be classified into two main categories: (i) cytokine inhibitors that interfere with the binding of the cytokine to its specific receptor, or (ii) inhibitory (i.e. anti-inflammatory) cytokines which are biologically active in affecting the synthesis of pro-inflammatory cytokines and their inhibitors and/or the target cell activation. These are discussed in more detail below.

Cytokine inhibitors, or true cytokine antagonists, are generated either by the shedding of specific cytokine receptors or by competitive binding at the receptor binding site; thus, TNFα is counterbalanced by TNF soluble receptors (TNF-sR) p55 and p75, and IL-1 is inhibited by IL-1-sRII. In addition to soluble receptors, the pro-inflammatory effects of IL-1 are counterbalanced by IL-1Ra and IL-1-RII which functions as a decoy receptor at the cell surface.

Inhibitory cytokines such as IL-4, -10 and -13 are produced mainly by Th2 cells; they inhibit type 1 cytokine and MMP production and stimulate IL-1Ra production in monocytes/macrophages. Another example is IFNγ, a type 1 cytokine, which counteracts TGFβ-stimulation of collagen synthesis in fibroblasts. One of the main functions of IFNγ is to inhibit the proliferation of Th2 cells.

Cytokines are involved in the triggering of the immune response, induction of acute inflammatory events and transition to or persistence of chronic inflammation. There is now considerable evidence that cytokines such as TNF and IL-1 contribute to the pathogenesis of inflammatory autoimmune diseases like rheumatoid arthritis and multiple sclerosis.

Monocytes and macrophages produce many chemical signals, including cytokines with pro-inflammatory as well as anti-inflammatory activity. The principal pro-inflammatory cytokines produced by monocytes/macrophages are IL-1 and TNFα. A few others, such as IL-12 or IL-15, have a strong effect on T-cell immunity by enhancing Th1-like immunity, and still others, like IL-8 or MIPα and β, have chemotactic properties.

In the past, interest has focused on the balance between inflammatory mediators such as IL-1α, IL-1β or TNFα and their respective inhibitors, IL-1Ra and TNF-sR. The final biological result depends on the relative amount of pro-inflammatory cytokines and their respective inhibitors. IL-1Ra is a polypeptide of 17 kDa, with 26% identity with IL-1β and 19% identity with IL-1α. Apparently IL-1Ra has no agonist activity, but it inhibits the functions induced by both IL-1α and IL-1β. In lung macrophages the kinetics of production of IL-1Ra, IL-1α and IL-1β mRNA by alveolar macrophages reveal that the addition of phorbol myristate acetate (PMA) results first (after 15 min) in the expression of IL-1α or IL-1β, followed by the delayed expression (after 3 h) of IL-1Ra mRNA. Other investigators have found that to induce IL-1ra mRNA in peripheral blood monocytes, PMA alone is insufficient and a second signal, such as complexes of IgG or IL-4, is required. Although the anti-inflammatory effect of IL-1Ra has been clearly demonstrated, its role during antigen presentation or in allogeneic reaction cannot be demonstrated *in vitro*, even at high doses.

In contrast to blood monocytes, human macrophages are capable of enhanced TNFα production. The shedding of TNF-sR may be crucial in that it maintains the environment free of deleterious inflammatory processes. Several years ago we reported the presence of TNF inhibitor in the supernatant of alveolar macrophages. The link between this inhibitory activity and the shedding of TNF-sRs by alveolar macrophages has been demonstrated by a specific immunoassay measuring the two soluble fragments of TNF-R, TNF-sR75 and TNF-sR55. After 48 h of stimulation, the amount of TNF-sR75 released by monocytes was twice that released by alveolar macrophages. Thus, monocytes produce less TNFα and more TNF-sR than do alveolar macrophages. A five- to ten-fold excess of TNF-sR can decrease the inflammatory reactions induced by TNFα or lymphotoxin (TNFβ), but a thousand-fold higher concentration of these inhibitors is required to affect APC–T-cell interaction *in vitro*.

IL-10 is one of the most powerful anti-inflammatory molecules. It acts by inhibiting the synthesis of pro-inflammatory cytokines (i.e. TNFα, IL-1α, IL-1β, IL-5 or IL-8) by monocytes/macrophages, but also by polymorphonuclear leukocytes and eosinophils. IL-10 suppresses the release of free oxygen radicals and the activity of nitric oxide-dependent microbicidal activity of macrophages as well as their production of prostaglandins. Interestingly, *in vitro* IL-10 does not inhibit but enhances the production of IL-1 receptor antagonist by monocytes and polymorphonuclear leukocytes. However, IL-10 exerts an anti-inflammatory or anti-destructive action by stimulating TIMP production on macrophages.

4.2 Relationship between cytokines, metalloproteinases and their inhibitors

Inflammation involves the influx into the target tissue of migratory cells such as T and B lymphocytes, neutrophils and mononuclear phagocytes. As in the case of all processes involving cell migration and proliferation, the influx of inflammatory cells into the target tissue is associated with remodelling of the extracellular matrix. In normal biological processes, tissue remodelling involves the controlled breakdown and neosynthesis of extracellular matrix elements, requiring the action, limited in time and space, of extracellular proteases such as plasminogen activators and MMPs. The expression of these proteases and their inhibitors is controlled by soluble extracellular factors such as cytokines. In chronic inflammatory diseases, the production of cytokines by infiltrating and resident tissue cells escapes regulatory mechanisms, inducing tissue destruction either directly or indirectly through the activation of immune and inflammatory cells, e.g. by inducing them to produce inflammatory cytokines and proteases.

MMPs are a gene family of zinc metalloenzymes responsible for the degradation of extracellular matrix elements, e.g. collagens and proteoglycans. Three distinct groups of MMPs may be distinguished in terms of their substrate specificity:

(1) *collagenases*, degrading type I, II, III and X collagens, including interstitial collagenase (MMP-1) produced by fibroblasts and monocyte/macrophages, and neutrophil collagenase (MMP-8);

(2) *gelatinases*, degrading type IV, V and X collagens, denatured collagens (gelatins) and elastin; this groups includes 92 kDa gelatinase (MMP-9) from mononuclear phagocytes and 72 kDa gelatinase (MMP-2) from fibroblasts;

(3) *stromelysins* (MMP-3, MMP-7 and MMP-10), displaying a wide substrate specificity including proteoglycans, laminin, fibronectin and type IV and IX collagens.

MMP activity is counterbalanced by tissue inhibitors of metalloproteinases (TIMPs). To date, four TIMPs have been described, of which TIMP-1 displays a wide range of activities by physiologically complexing collagenases, stromelysins and gelatinases. The extent of matrix remodelling that occurs during inflammation and repair is believed to reflect not only the total amount of matrix-degrading enzymes present, but also the relative molar ratios of the proteinases and their specific inhibitors. Thus, any condition which up-regulates MMP expression but not TIMP-1 expression, increases matrix turnover. In rheumatoid arthritis, major producers of MMPs are monocytes/macrophages and stroma cells such as fibroblast-like cells. In multiple sclerosis, the best candidates are monocytes/macrophages, microglia and astrocytes.

Cytokines such as TNF and IL-1 are implicated in the pathogenesis of many destructive diseases such as rheumatoid arthritis and multiple sclerosis. Indeed, *in vitro* studies on synovial tissue from rheumatoid arthritis patients suggest that the effects of TNF are amplified due to its potential to induce other pro-inflammatory cytokines, such as IL-1 and GM-CSF. TNF and IL-1 are markedly produced by monocyte-macrophages in contact with Th1 lymphocytes which predominate in rheumatoid synovium and carry CCR5 receptors. These monokines participate in the induction of MMP secretion on fibroblast-like cells and mononuclear phagocytes. It is likely, therefore, that mononuclear phagocytes occupy a pivotal position in the control of extracellular matrix turnover, being capable of mediating joint destruction in rheumatoid arthritis directly by generating their own MMPs and also indirectly by releasing cytokines which then induce fibroblast MMP production. This activity is reflected at the systemic level, since stromelysin

(MMP-3) and TIMP-1 have been found in the serum and synovial fluid from rheumatoid arthritis patients where MMP-3 levels seem to correlate with disease activity.

In actively demyelinating lesions occurring in multiple sclerosis, macrophages are known to spread over myelinated fibres from the surface layers of which myelin lamellae are lysed and internalized by the cell until it becomes clogged with myelin debris. While numerous studies deal with the destruction of connective tissue by MMPs in rheumatoid arthritis, only a limited number of studies relate to the role of proteases in the pathogenesis of multiple sclerosis. MMP-9 has been found in the cerebrospinal fluid of multiple sclerosis patients, and the fact that it is able to release encephalitogens from human myelin basic protein (MBP) *in vitro* may indicate that it plays a part in the pathogenesis of multiple sclerosis.

4.3 Extracellular matrix synthesis, fibroblasts and cytokines

Although fibroblasts and fibroblast-like cells may participate in tissue destruction by producing MMPs and PGE_2 under the control of monocyte/macrophage cytokines, they are the principal producers of the extracellular matrix. Synthesis of collagens by fibroblasts is regulated by a number of cytokines and growth factors. Fibrogenic cytokines and growth factors include TGFβ, IL-4, IL-6, platelet-derived growth factor (PDGF) and fibroblast growth factor (FGF). In contrast, IFNγ, IFNα, IL-10 and relaxin suppress collagen synthesis. The effects of TNFα and IL-1 on collagen production by connective tissue cells are controversial and depend on the cell type. Furthermore, extracellular matrix components such as collagens and fibronectin act on fibroblasts by inhibiting their own synthesis when sufficient extracellular matrix has accumulated.

5. Conclusion

The purpose of this chapter is to give an integral view of the connection between APCs/dendritic cells and effector T lymphocytes acting by soluble products or by direct contact with monocytes/macrophages or fibroblasts (Figure 4). The following comments and hypotheses can be advanced.

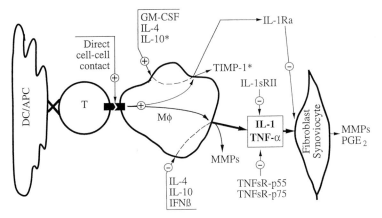

∎ **Figure 4.** Activation cascade from T lymphocytes (T) to macrophages (Mφ) and fibroblasts or synoviocytes. DC, dendritic cell; APC, antigen-presenting cell; see text for other abbreviations.

(1) The initial events are normally due to a well-defined exogenous or endogenous antigen presented by MHC molecule to specific receptors on T cells. Alternatively, the initiating event may be a slowly developing differentiation of dendritic cells under the influence of cytokines released by the specific target organ in response to non-specific stimulation. This scenario does not require T cells for the initial stage of dendritic cell differentiation, and dendritic cells can stimulate quiescent, naïve T cells.

(2) The co-stimulation signals between dendritic cells and T cells dictate which type of T cells (Th1 or Th2) proliferate. These signals may be controlled by the microenvironment (hormones, cytokines).

(3) The cells that are induced to proliferate may interact with monocytes and macrophages in an antigen-non-specific way by means of soluble factors or by direct contact. In general, direct contact with activated T cells generates a marked inflammatory response, while many soluble cytokines tend to exert an inhibitory control mechanism.

(4) Following lymphocyte interference, the monocytes/macrophages attempt to regulate the homeostasis between agonistic and antagonistic mediators of inflammation (e.g. cytokines/anti-cytokines or proteases/anti-proteases).

(5) Target connective tissue cells (e.g. fibroblasts) are part of the homeostasis of cytokines/ cytokine inhibitors and, depending on the balance of these mediators or their type, proceed to promote destruction or repair.

References

Arend, W. P. and Dayer, J. M. (1994). Naturally occurring inhibitors of cytokines. In Davies, M. E. and Dingle, J. T. (eds), *Immunopharmacology of Joints and Connective Tissue*, pp. 129–49. Academic Press, London.

Arend, W. P. and Dayer, J. M. (1995). Inhibition of the production and effects of interleukin-1 and tumor necrosis factor alpha in rheumatoid arthritis. *Arth. Rheumat.*, **38**, 51–160.

Birkedal-Hansen, H. (1995). Proteolytic remodeling of extracellular matrix. *Curr. Opin. Cell. Biol.* **7**, 728–35.

Burger, D. and Dayer, J.-M. (1998). Interaction between T cell plasma membranes and monocytes. In Miossec, P., van den Berg, W. and Firestein, G. S. (eds), *T Cells in Arthritis*, pp. 111–28, Birkhäuser, Basel.

Caux, C., Vandervliet, B., Massacrier, C., Dezutter-Dambuyant, C., De Saint Vis, B., Jacquet, C., Yoneda, K., Imanura, S., Schmitt, D. and Banchereau, J. (1996). CD34[+] hematopoietic progenitors from human cord blood differentiate along two independent dendritic cell pathways in response to GM-CSF + TNFα. *J. Exp. Med.* **184**, 695–706.

Cella, M., Sallusto F. and Lanzavecchia, A. (1997). Origin, maturation and antigen presenting function of dendritic cells. *Curr. Opin. Immunol.* **9**, 10–16.

Chizzolini, C., Chicheportiche, R., Burger, D. and Dayer, J. M. (1997). Human Th1 cells preferentially induce interleukin (IL)-1 beta while Th2 cells induce IL-1 receptor antagonist production upon cell/cell contact with monocytes. *Eur. J. Immunol.* **27**, 171–7.

Dayer, J.-M. (1998). Regulation of IL-1/TNF, their natural inhibitors, and other cytokines in chronic inflammation. *Immunologist* **5**, 192–6.

Gordon, S. and Hughes, D. (1997). Macrophages and their origins: heterogeneity in relation to tissue microenvironment. In

Lipscomb, M. F. and Russell, S. W. (eds), *Lung Macrophages and Dendritic Cells in Health and Disease*, pp. 3–31. Marcel Dekker, New York.

Harris, E. D. (1997). Enzymes responsible for joint destruction. In Harris, E. D. (ed.), *Rheumatoid Arthritis*, pp. 168–75. W. B. Saunders, Philadelphia.

Hohlfeld, R., Meinl, E., Weber, F., Zipp, F., Schmidt, S., Sotgiu, S., Goebels, N., Voltz, R., Spuler, S., Iglesias, A. and Wekerle, H. (1995). The role of autoimmune T lymphocytes in the pathogenesis of multiple sclerosis. *Neurology* **45**, S33–8.

Inaba, K. and Steinman, R. M. (1997). Dendritic cells. In Lipscomb, M. F. and Russell, S. W. (eds), *Lung Macrophages and Dendritic Cells in Health and Disease*, pp. 87–107. Marcel Dekker, New York.

Linsley, P. S. and Ledbetter, J. A. (1993). The role of CD28 receptor during T cell responses to antigen. *Annu. Rev. Immunol.* **11**, 191–212.

Loetscher, P., Uguccioni, M., Bordoli, L., Baggiolini, M. and Moser, B. (1998). CCR5 is characteristic of Th1 lymphocytes. *Nature* **391**, 344–5.

Miltenburg, A. M. M., Lacraz, S., Welgus, H. G. and Dayer, J. M. (1995). Immobilized anti-CD3 antibody activates T cell clones to induce the production of interstitial collagenase, but not tissue inhibitor of metalloproteinases, in monocytic THP-1 cells and dermal fibroblasts. *J. Immunol.* **154**, 2655–67.

Nicod, L. P. (1997). Function of human lung dendritic cells. In Lipscomb, M. F. and Russell, S. W. (eds), *Lung Macrophages and Dendritic Cells in Health and Disease*, pp. 311–34. Marcel Dekker, New York.

Nicod, L. P. and Isler, P. (1997). Alveolar macrophages in sarcoidosis coexpress high levels of CD86 (B7.2), CD40, and CD30L. *Am. J. Resp. Cell Mol. Biol.* **17**, 91–6.

Ries, C. and Petrides, P. E. (1995). Cytokine regulation of matrix metalloproteinase activity and its regulatory dysfunction in disease. *Biol. Chem.* **376**, 345–55.

Thomas, R. and Lipsky, P. E. (1996). Presentation of self peptides by dendritic cells: possible implication for the pathogenesis of rheumatoid arthritis. *Arth. Rheumat.* **39**, 183–190.

Unemori, E. N. and Amento E. P. (1996). Connective tissue metabolism in scleroderma, including cytokines. *Curr. Opin. Rheumatol.* **8**, 580–84.

Vey, E., Zhang, J. H., and Dayer, J.M. (1992). IFN-gamma and 1,25(OH)2D3 induce on THP-1 cells distinct patterns of cell surface antigen expression, cytokine production, and responsiveness to contact with activated T cells. *J. Immunol.* **149**, 2040–46.

Vey, E., Burger, D., and Dayer, J. M. (1996). Expression and cleavage of tumor necrosis factor-alpha and tumor necrosis factor receptors by human monocytic cell lines upon direct contact with stimulated T cells. *Eur. J. Immunol.* **26**, 2404–9.

Wesselius, L. J., Murphy, W. J., Morrison, D. C. and Russel, S. W. (1997). Macrophage activation: concepts and principles applicable to the lung. In Lipscomb, M. F. and Russell, S. W. (eds), *Lung Macrophages and Dendritic Cells in Health and Disease*, pp. 33–86. Marcel Dekker, New York.

Yang, Y. and Wilson, J.-M. (1996). CD40 ligand-dependent T cell activation: requirement of B7-CD28 signaling through CD40. *Science* **273**, 1862–4.

XVI Cytokines and T-cell responses

Hervé Groux* and Hans Yssel

1. Introduction

About three decades ago, Asherson and Stone showed that intravenous injection of alum-precipitated antigen into guinea pigs, prior to sub-cutaneous injection of the same antigen in complete Freund's adjuvant, which is effective in inducing both delayed-type hypersensitivity (DTH) and antibody responses, reduced the level of DTH without affecting the titres of antibody responses. The notion that resulted from these and other experiments was the existence of a dynamic balance between cell-mediated and humoral immune responses in which the dose, form of antigen and route of administration of antigen, play an important role. Moreover, this inverse relationship between antibody production and DTH reactions was observed in mice which showed striking, polarized pathological features, following infection with the parasite antigen *Leishmania major*. The mouse strain BALB/c, which succumbs to *L. major* infection, in general has elevated antibody titres, especially of the IgG1 and IgE isotype, but weak DTH responses. In contrast, strains of mice like D10 or C57B1/6, which are resistant to infection with *L. major*, show cell-mediated (DTH) immune responses, but low antibody titres. Importantly, the type of response observed in these different mouse strains, as well as the different clinical outcome of infection with *L. major*, are dependent on the generation of T cell populations with different functions and different cytokine production profiles. Using panels of T cell clones, Mosmann and co-workers demonstrated that CD4⁺ mouse T-helper cell clones could be divided into two mutually exclusive subpopulations, called T-helper type 1 (Th1) and Th2 cells, based on their cytokine production profile which is associated with their

functional properties. Mouse Th1 clones produce interleukin 2 (IL-2), interferon gamma (IFNγ), tumour necrosis factor beta (TNF-β), but not IL-4 and IL-5 and were capable of inducing inflammatory (DTH) reactions. In contrast, Th2 clones produce IL-4, IL-5, IL-6 and IL-13, but not IL-2 or IFN γ, and give optimal help for antibody production by B cells. In addition, the cytokine IL-10 was found to be produced exclusively by mouse Th2, but not Th1 cells. These findings have led to the hypothesis that Th1 cells induce cell-mediated responses, such as DTH reactions, whereas Th2 cells are more involved in humoral immunity, although this classification seems to be somewhat over-simplified.

For example, whereas the Th2 cytokines IL-4 and IL-13 are both switch factors for IgG1 and IgE production in the mouse (IgG4 and IgE in human), IFNγ, which is the hallmark Th1 cytokine, induces the switching of murine B lymphocytes to several IgG isotypes and thus must be involved in at least some humoral immune responses. It seems, therefore, that Th1 cells, by means of their production of IFNγ, are mainly involved in the protective immunity towards intracellular micro-organisms, stimulating the production of complement-fixing and opsonizing antibodies, and activating macrophages to enhance their microbicidal effector functions. As a consequence, Th1-mediated responses are generally characterized by inflammatory reactions which often lead to tissue damage and destruction.

In contrast, as a result of the production of IL-4 and IL-13, Th2 cells give help to B cells for the production of non-complement-fixing IgG1 (in the mouse), IgG4 (in human) antibodies, and IgE antibodies. In addition, IL-5 induces recruitment and differentiation of eosinophils. Th2 cells are therefore likely to play a role in the elimination of extracellular parasites, such as helminths, and are involved in allergic reactions. The Th2 cytokines

* Corresponding author.

IL-4, IL-13 and IL-10 have strong anti-inflammatory properties, counteracting the tissue-damaging effects of protective immune reactions, mediated by Th1 cells. Given the functional activities of the Th1 and Th2 cytokines, Th1-mediated responses are likely to dominate in early immune responses, whereas Th2-type reactions accumulate later during the immune response. This notion is supported by clinical observations in human disease, such as leprosy, in which the healing (tuberculoid) form, dominated by strong DTH responses, gradually makes way for the uncontrolled lepromatous form, characterized by weak DTH responses and high antibody levels.

Although initial studies of cytokine production by human CD4$^+$ T cell clones yielded mostly cells with a Th0 cytokine production profile, producing IL-4 and IFNγ upon activation, subsequent analysis of T cells clones obtained from patients with chronic infections or various immunological diseases have confirmed that in humans also there are CD4$^+$ T cell subsets with different cytokine production profiles. T cell clones obtained from atopic patients generally display a Th2-like phenotype, whereas those specific for intracellular pathogens, such as *Mycobacterium leprae* or *Borrelia burgdorferi*, or for purified protein derivative (PPD) derived from *Mycobacterium tuberculosis*, have a Th1-like profile of cytokine production. A number of differences in the cytokine production profile between mouse and human CD4$^+$ T cell clones have been observed, however. For example, the production of IL-10, which in the mouse is produced by Th2 clones, is not associated with a Th2 cytokine production profile in humans. Similarly, IL-13, a cytokine typically produced by mouse Th2 clones, is secreted by human Th1, Th0 and Th2 cell clones. However, both IL-10 and IL-13 share the anti-inflammatory function of IL-4, such as inhibiting the secretion of the pro-inflammatory cytokines IL-1α, IL-1β, IL-6 and INF-α by activated monocytes, whereas they induce the production of IL-1 receptor antagonist and in this respect both cytokines are typical Th2 cytokines. In addition, cytokine production profiles obtained by stimulation of T cell clones may be influenced by culture and stimulation conditions. For example, stimulation of most human allergen-specific Th2 clones with phorbol ester and calcium ionophore, via a pathway that does not involve the triggering of the TCR/CD3 complex, induces not only high levels of Th2 cytokines, but also the production of IFNγ. These observations indicate that different modes of activation, as well as relative, rather than absolute, cytokine secretion

levels are important in determining the cytokine production profiles of T cells.

The heterogeneity of cytokine production profiles observed among cloned human and mouse T cell lines has prompted some researchers to put into question the categorization of T-helper cell populations into discrete Th1 and Th2 subsets by arguing that the expression of cytokine genes is independently regulated in individual cells and that Th1 and Th2 cytokine production profiles represent only two extreme phenotypes among a continuous spectrum of possible combinations of cytokine production. However, transcripts representing Th1 and Th2 cytokine production profiles have been detected in whole tissue under experimental conditions which exclude the possible bias introduced by culture and cloning procedures, giving additional support for the notion that the degree of polarization of cytokine production may reflect the nature of the antigenic stimuli to which the cells have been exposed. It is of interest to note that IL-4 and IL-5 production is segregated in activated human peripheral blood T lymphocytes, whereas, human Th2 cell clones invariably produce both IL-4 and IL-5. The segregated production of IL-4 and IL-5, as well as that of IL-2 and IFNγ, was also observed in antigen-stimulated T cells generated from mice carrying an ovalbumin-specific TCR. These results suggest that the simultaneous production of Th1 and Th2 cytokines is a feature of T cell lines and clones that have been repeatedly activated during culture, and does not necessarily reflect those cytokine production profiles of individual T cells *in vivo*.

2. Development of Th1/Th2 subsets

Using naïve resting T-cells activated with polyclonal activators, such as phytohaemagglutinin (PHA) or anti-CD3 and anti-CD28 monoclonal antibodies (mAbs), or T cells from mice expressing transgenic antigen receptors of known specificities, it has become clear that, rather than being derived from separate committed lineage, Th subsets differentiate from a single precursor cell, which is a mature T cell capable of producing mainly IL-2 upon antigenic stimulation. However, whether polarized Th1 and Th2 cells differentiate via an intermediate population producing both Th1 and Th2 cytokines (Th0), or whether Th0 cells represent a separate, stably differentiated T cell popula-

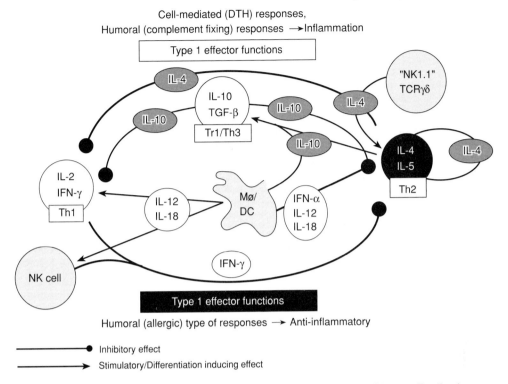

Cell-mediated (DTH) responses,
Humoral (complement fixing) responses →Inflammation

∎Figure 1. Cross-regulatory effects of cytokines in function and differentiation of human T-cell subsets.

tion remains unclear. Although many factors are involved in the differentiation of naïve T cells into mature polarized effector T cells, cytokines themselves have been recognized to be the most potent component of this process (Figure 1).

2.1 Cytokine-induced Th1 differentiation

Interleukin 12, which is produced by macrophages and dendritic cells, is the dominant cytokine, inducing the differentiation of IFNγ-producing Th1 cells from naive precursor T cells. The production of IL-12 is induced by adjuvants which activate macrophages, by microbial components such as bacterial endotoxin, by intracellular bacteria, like *Listeria* and mycobacteria, and by protozoa like *Toxoplasma*, resulting in the induction of Th1 immune responses which are dominated by the production of IFNγ. IL-12 can act directly on T cells, as shown by its capacity to drive Th1 differentiation in naïve human T cells as well as in T cells from transgenic mice. Interaction of IL-12 with its receptor results in the activation and phosphorylation of

three putative transcription factors, STAT 2, STAT 3 and STAT 4, of which only STAT 4 seems to be specific for IL-12-mediated signal transduction. The important role of IL-12 in Th1-differentiation is reinforced by the observations that mice deficient in the gene coding for STAT 4 or IL-12 show impaired Th1 responses.

Until recently, there has been some controversy as to the role of IFNγ in the process of Th1 differentiation. In many *in vitro* culture systems, IFNγ is unable to drive Th1 differentiation by itself when added to primary cultures. In addition, IFNγ$^{-/-}$ mice produce efficient antiviral cytolytic responses whereas IFNγR$^{-/-}$ mice are still able to generate Th1 cells, despite increased susceptibility to infection with *L. major*. However, in one study using an experimental transgenic mice model, IL-12-mediated Th1 differentiation was dependent on the presence of IFNγ, whereas in a similar model the Th1-inducing effects of IL-12 could not be blocked by the addition of a neutralizing anti-IFNγ mAb. Moreover, the involvement of IFNγ in Th1 differentiation seems to differ between mouse and human. Successful Th1 differentiation requires the expression of a functional IL-12R, consisting of two chains, designated IL-12R β_1 and IL-12R β_2. A functional IL-12R seems to be expressed only an recently activated, uncommitted, T cells and to Th1 cells; it is no longer expressed on fully differentiated Th2 cells and, as a consequence, polarized Th2 cells are no longer able to respond to IL-12. It has been shown that IFNγ, but not IFNα, can render naïve mouse T cells more responsive to the Th1 differentiation-inducing effects of IL-12, by maintaining the expression of the IL-12R β_2 chain on differentiating Th1 cells. Interestingly, whereas naïve human T cells, cultured in the presence of IL-4, quickly lose IL-12R β_2 chain mRNA, the addition of IFNα, but not of IFNγ, to these cultures induces transcripts for this component of the IL-12R, suggesting that an additional pathway is involved in the induction of a functional IL-12R on the cell surface of differentiating human Th cells, and that this pathway is not mediated by IL-12 or IFNγ. The latter results support the findings of earlier studies which showed that IFNγ has little or no effect on the differentiation of human CD4$^+$ cord blood T cells into populations of Th1 cells, although the mechanisms underlying the different effects of these two members of the interferon family on human and mouse T cells remain to be elucidated. Finally, an additional, albeit indirect, role for IFNγ has been demonstrated via its capacity to enhance the production of IL-12 by mouse macrophages in response to infection with *Listeria monocytogenes*, thus creating a positive feedback loop and acting in concert with IL-12 in the induction of Th1 responses, required for the elimination of these pathogens.

Other cytokines, notably IL-1α and IL-18, have also been reported to play a role in Th1 cell differentiation. IL-1α acts synergically with IL-12 in inducing Th1 differentiation of naïve cells obtained from BALB/c mice, but not from C57BL/6 mice. However, differentiated Th1 cells gradually seem to lose responsiveness to IL-1α and this cytokine does not bind to Th1 clones. IL-18, which is produced by Kupffer cells, activated macrophages and most likely by dendritic cells, was originally identified in an experimental *Propionobacterium acnes* model in mice, and later in humans, as IFNγ-inducing factor (IGIF). IL-18 strongly synergizes with IL-12 in the induction of IFNγ by T and natural killer (NK) cells and IL-18 together with IL-12 can even induce the production of high levels of IFNγ by B cells, thereby inhibiting their production of IgE. However, in spite of its strong IFNγ-inducing properties, IL-18 by itself

is not able to induce the differentiation of Th1 cells in mice or humans.

Structural analysis has suggested that IL-18 is highly homologous to IL-1, and the name IL-1γ has been proposed for this cytokine. Unlike IL-12, neither IL-1α nor IL-18 activates STAT-4, but interaction of both IL-1α and IL-18 with their respective receptors activates a receptor-associated kinase IRAK, resulting in the downstream activation of NF-κB. Th1 and Th2 cells differ not only in their response to the two members of the IL-1 family, but also in their surface expression of the receptor for these two cytokines. IL-1α, which does not bind to differentiated Th1 clones, activates NF-κB in Th2 clones and induces their proliferation. In contrast, the latter cells seem to lack a functional IL-18R, since they do not respond to IL-18. Signalling through NF-κB by these members of the IL-1 family is thus likely to affect Th1 and Th2 responses differentially.

2.2 Cytokine-induced Th2 differentiation

It has been conclusively demonstrated, in both *in vitro* and *in vivo* experimental mouse models, that the presence of IL-4 is critical for the generation of IL-4-producing Th2 cells. The effects of IL-4 in inducing Th2 development are dominant over Th1 differentiating cytokines so that as IL-4 levels reach a certain threshold at the beginning of an immune response, Th2 cells will progressively differentiate, leading to increasing levels of IL-4 production.

In humans, IL-4 has been shown to specifically and directly induce the differentiation of immunologically naïve cord blood or fetal thymocytes into IL-4-producing T-helper cells, following CD3-mediated stimulation. In view of the failure to detect IL-4 production by naïve T cells, an as yet unresolved problem is the origin of IL-4, which is required to be present at the priming of these cells for successful Th2 differentiation. It has been proposed that other cell types, notably mast cells and basophils, which produce IL-4 following triggering of their high-affinity FcϵR by IgE and antigen, might be involved in the priming of naïve cells for IL-4 production. In addition, there are several candidates for IL-4 production early in the immune response, which may be responsible for Th2 differentiation; these include major histocompatilitiy complex (MHC) class II-restricted CD4$^+$ T cells (memory, cross-reactive and possibly naïve), and the NK1.1$^+$ subset of CD4$^+$ and double-negative T cells.

Many researchers have suggested a critical role for NK1.1$^+$ CD4$^+$ T cells in Th2 development as a result of their ability to produce large amounts of IL-4 rapidly upon activation both *in vitro* and *in vivo*. These unusual T cells express some NK cell markers, have a restricted T cell receptor (TCR) Vα and Vβ usage, and are restricted by non-classical MHC class I molecules (CD1 or TL antigens). This specific interaction has been used to evaluate the role of NK T cells in the differentiation of Th2 cells *in vivo*. A number of recent studies using β_2-microglobulin (β2M)-deficient mice (which lack MHC class I, CD1 and TL molecules) have failed to support a role for these NK1.1$^+$ T cells in the development of Th2 responses to a variety of parasitic infections and protein antigens. These results were confirmed by infection and immunization of CD1$^{-/-}$ or TL$^{-/-}$ mice. These mice were found to lack the NK1.1$^+$ subset, but could nevertheless mount a typical Th2 response. Taken together, these studies confirm that CD1 is a selecting antigen for NK1.1$^+$ T cells and that these cells account for the early burst of IL-4 obtained after anti-CD3 activation. However, IL-4 production by NK1.1 cells is not required for the development of Th2 responses to specific antigens. This result is not unexpected since, even if immunogens, having associated antigens able to interact via CD1 with a fraction of NK1.1$^+$ T cells, were able to promote a primary pulse of IL-4, it is unlikely that all antigens leading to Th2 differentiation would necessary require prior activation of CD1-restricted T cells.

Another major candidate as the primary source of IL-4 are CD4$^+$ T cells themselves. Naïve CD4$^+$ T cells activated in the presence of fibroblasts, expressing co-stimulatory molecules, can develop into Th2 cells. Since this development is blocked by neutralization of IL-4, it has been suggested that Th2 differentiation can be driven by endogenous IL-4 produced by the naïve T cells. This hypothesis was confirmed in susceptible BALB/c mice infected by the parasite *L. major*. It has recently been demonstrated that the early burst of IL-4 peaking in lymph nodes after *L. major* infection occurs within CD4$^+$ T cells that express Vβ_4 Vα_8 TCRs. A previously identified antigen, *Leishmania* homologue receptor for activated C kinase (LACK), was found to be the focus of this initial response. The mechanism whereby LACK induces a Th2 response in BALB/c mice is still unclear. The early IL-4 secretion induced by LACK is so rapid that it is reminiscent of a memory response and may result from previous activations of T cells by cross-reactive antigens.

The strongest evidence of the pivotal role of IL-4 in the induction of Th2 differentiation comes from experiments with IL-4 knockout mice (IL-4$^{-/-}$) in which the gene for IL-4 has been disrupted by homologous recombination. Infection of these mice with the nematode *Nippostrongylus brasiliensis* results in strongly reduced levels of IL-4, IL-5 and IL-9, and in reduced eosinophilia, the presence of which is characteristic for Th2 responses. Similarly, mutant mice deficient for STAT-6, a tyrosine kinase that is specifically phosphorylated following the interaction of IL-4 with its receptor, show deficient Th2 responses, including impaired Th2 differentiation and immunoglobulin class switching to IgG1 and IgE, following nematode infection. It has to be noted, however, that Th2 responses in IL-4$^{-/-}$ mice are much weaker than those in wild-type mice, but not completely absent, indicating that other, as yet unidentified factors may be involved in Th2 differentiation.

2.3 IL-10 induces the differentiation of Tr1 cells

IL-10 knockout mice, in addition to developing intestinal inflammation, also exhibit enhanced and prolonged contact hypersensitivity, in some cases leading to severe tissue damage, indicating that host-derived IL-10 plays an important role in limiting the immune response and preventing induction of tissue damage. Precisely how IL-10 mediates its immune suppressive activities *in vivo* is not known. Studies *in vitro* have shown that it inhibits antigen-induced proliferation and cytokine secretion by mouse and human T cells. Inhibition is thought to be mediated primarily via effects on antigen-presenting cells (APCs), as IL-10 down-regulates monokine production and the expression of CD54 (ICAM-1, the ligand for LFA-1), and of CD80 and of CD86, which function as important co-stimulatory molecules for T cell activation. It has also been shown that IL-10, in the absence of professional APCs, has direct effects on CD4$^+$ T cells by suppressing IL-2 and TNF-α secretion.

We have recently described a hitherto unidentified property of IL-10 which may contribute to its immune suppressive activities *in vivo*. Repetitive antigenic stimulations *in vitro* of human and murine CD4$^+$ cells in the presence of IL-10 resulted in the generation of a T cell subset termed T regulatory-1 (Tr1) cells. Previous experiments had shown that, in addition to inhibiting antigen-specific responses, IL-10 could induce long-lasting antigen-specific anergy in human CD4$^+$ T cells which was characterized by a failure of these cells to proliferate or produce cytokines in response to antigenic stimulation. However, the addition of very high doses of cross-linked anti-CD3 mAb resulted in a partial restoration of the proliferative response. Interestingly, the cytokine production profile of the anergic human T cells following stimulation with high doses of cross-linked anti-CD3 mAb did not fit the Th0, Th1 or Th2 phenotype. The most striking feature of these cells was their low proliferative capacity and their unusually high level of IL-10 production. In addition, whereas they secreted detectable IL-4 and very low levels of IL-2 their levels of IL-5, IFNγ, TGF-β and TNF-α production were comparable to those of human Th0 clones. A similar type of clone was isolated from a population of ovalbumin-specific CD4$^+$ T cells obtained from ovalbumin-specific TCR transgenic mice that had been repeatedly stimulated with ovalbumin peptide in the presence of IL-10.

Both human and mouse Tr1 clones were found to be immune suppressive *in vitro*. Antigen-induced proliferation of naïve CD4$^+$ T cells was dramatically reduced following co-culture with activated Tr1 clones which were separated from the responding T cells by a trans-well insert. Suppression was reversed by addition of anti-TGF-β and IL-10 mAbs, implicating these cytokines in the mechanism of immune suppression. Suppression was characteristic of Tr1 clones: ovalbumin-specific Th1 or Th2 clones had no suppressive effects, but rather enhanced the ovalbumin-induced proliferation of naïve CD4$^+$ T cells.

3. Other factors involved in CD4$^+$ T-cell differentiation

Although, as discussed above, cytokines play a central role in governing the differentiation of CD4$^+$ T cells from naïve precursors, other factors, including the strength of the interaction mediated through the TCR and MHC–peptide complex, the concentration of antigen and the effect of co-stimulatory molecules, also play an important role in this process.

3.1 Effect of antigen concentration on T-helper cell differentiation

Several studies have demonstrated that the concentration or dose of antigen used to induce an immune response may have a strong effect on the cytokine production profile of the responding T cells. In general, low concentrations of antigens, or low-dose infections of experimental animals, seem to induce Th1 responses, whereas high antigen concentrations tend to result in the generation of Th2 cells. For example, in two different experimental models, naïve TCR transgenic T cells which were induced to differentiate in an antigen-specific manner, produced increasing amounts of IL-4 and reduced levels of IFNγ when either very high or very low doses of peptide antigen were used.

Similar findings were reported in a study in which human CD4$^+$ T cells from allergic donors were used and which showed that these cells produced high levels of IL-4 when stimulated with low concentrations of allergen, particularly when B cell-enriched populations presented the antigen. In contrast, the same responding CD4$^+$ T cell population produced little IL-4 when stimulated with high concentrations of allergen in the presence of monocytes as APCs. The mechanism underlying the effect of antigen dose on the differentiation of T cell is not clear yet. It is possible that, upon uptake of low doses of antigen, macrophages or dendritic cells with produce cytokines, (such as IL-12 or IL-18) that induce IFNγ production, whereas at high doses of antigen the production of such cytokines is low or absent. Alternatively, at high concentrations, antigens may be processed and presented by APCs, such as B cells, that do not produce IL-12 or IL-18. Other factors produced by APCs may complicate this picture; an example is prostaglandin E$_2$, which is produced by activated macrophages, but not by B cells, and which may favour the induction of Th2 responses, by means of their specific inhibition of IFNγ production by Th1 cells. One other published study on the effect of antigen dose on the cytokine production profile of human T cells has yielded somewhat different results. Carballido *et al.* reported that human T-cell clones, specific for the venom phospholipase A2 (PLA), derived from allergic and hyposensitized individuals, required higher critical amounts of antigen for IFNγ induction and produced increasing IL-4/IFNγ ratios following stimulation with increasing concentrations of PLA. Additional studies are required to substantiate this observation and to point out possible differences between the different experimental models.

3.2 Effect of different co-stimulatory molecules

In view of the important role of APCs in the generation of an immune response following antigenic challenge, it is likely that, in addition to cytokines produced by these cells, the expression of certain co-stimulatory molecules may promote the differentiation of T-helper cell subpopulations. The most studied co-stimulatory molecules to date include two structurally related members of the immunoglobulin superfamily, CD80 (B7.1, BB-1) and CD86 (B7.2). Both molecules are involved in T cell activation via binding to their specific receptor, CD28, which is constitutively expressed on many, but not all, T cells. Both CD80 and CD86 are expressed on activated B cells, dendritic cells and macrophages, as well as on activated T cells and NK cells. However, unlike CD80, CD86 is constitutively expressed on the surface of resting human monocytes. Optimal activation and differentiation of T cells requires interaction of CD28 with its ligands, as demonstrated by the observation that CD28⁻/⁻ mice have impaired proliferative responses to antigen. Moreover, it has been demonstrated in a number of mouse experimental models that the Th1 and Th2 differentiation pathways may be differentially activated by co-stimulation via CD80 or CD86, respectively, and that interaction of CD28 with either one of its ligands may result in a different clinical outcome in response to antigenic challenge. For example, in a model of experimental allergic encephalitis (EAE), a disease that is regulated by antigen-specific T cells, blockade of CD28–CD86 interactions increases disease severity. In contrast, administration of neutralizing anti-CD80 mAbs during immunization results in predominant generation of Th2 clones *in vitro*, and transfer of these into mice prevented both induction of EAE and abrogated established disease. However, evidence for the notion that co-stimulation via either CD80 or CD86 may directly affect initial cytokine secretion has not been confirmed in humans. Lanier *et al.* have reported that stimulation of human peripheral blood T cells with CD86-expressing L cell transfectants induces similar levels of Th1- and Th2-type cytokines. Furthermore, no differences in cytoline production profiles could be detected between human CD45RA⁺ or CD45RA⁻ T cells that were stimulated with transfectants, expressing comparable levels of CD80 or CD86. The latter results underscore the notion that there is as yet no proof that stimulation via CD80 or CD86 will result in different biochemical signals to T cells. In addition, there is little evidence that subsets of T-helper cells have different requirements for co-stimulation via CD28 during various stages of differentiation or activation.

T-helper cells that do not express the CD4 co-receptor are deficient in their capacity to differentiate into Th2 cells in response to antigen *in vitro*

and *in vivo*. This deficiency was evident using both CD4-negative cells from CD4⁻/⁻ mice and double-negative T cells that developed in MHC class II-restricted TCR transgenic mice. The defect in the ability to produce IL-4 has been observed in response to various challenges, including intestinal helminths and intracellular parasites. The lack of a Th2 response was also evident when CD4⁺ T cells were stimulated with APCs expressing mutated MHC class II molecules that were unable to bind to CD4. Thus, it seems that the interaction of the CD4 co-receptor with its ligand(s), perhaps by modulating the strength of the signal delivered to T cells, may influence T-helper subset development.

Another important molecule in T cells activation is CD40L (CD154), which is expressed on T cells upon activation. Interaction of CD154 with its ligand, CD40, constitutively expressed on B cells and various APCs was found to be required for CD4 T-cells to induce autoimmunity in an EAE model. In addition, it was reported that CD40-mediated activation of dentritic cells result in the production of IL-12 and in this way CD40–CD154 interactions could therefore indirectly play a role in the differentiation of Th1 cells.

A recent study has suggested that triggering of the ligand for OX40 (CD143), a molecule expressed on the surface of activated T cells, B cells and dendritic cells, and a member of the TNF/nerve growth factor superfamily, induced production of IL-4 and expression of the Burkitt's lymphoma receptor 1 (BLR-1) chemokine receptor. In addition, CD143/OX-40L interactions strongly reduce the production of IFNγ by IL-12-stimulated CD4⁺ T cells and CD8⁺ cells, suggesting that the interaction of CD134 with its ligand could play an indirect role in Th2 differentiation.

4. Stability of differentiated T-helper cell subpopulations

An important question, with far-reaching therapeutic implications, is whether the cytokine production profile of differentiated T-helper cells is stable or whether it can be reverted under opposing polarizing conditions. Studies using transgenic mice have indicated that, early in their differentiation, mouse Th1 and Th2 populations may be reverted to the opposite phenotype, but that during further differentiation the cells become irreversibly committed to a particular cytokine production profile. In particular, differentiated mouse Th2 cell clones seem to be resistant to the IFNγ-inducing effects of IL-12. A molecular basis for this loss of IL-12-responsiveness is that fully differentiated Th2 cells have lost the expression of a functional IL-12R, lacking the IL-12R β_2 chain and therefore are no longer able to signal via STAT. Further evidence for this notion has been presented using human T cells. However, a number of other studies have indicated that the phenotype of human Th2 populations is less stable. CD4$^+$ cord blood T cells that had been differentiated into a population of IL-4-producing Th2 cells, following repetitive stimulation in the presence of IL-4 and neutralizing anti-IFNγ mAbs, reverted to a Th0/Th1 phenotype after a single stimulation and subsequent culture for 4 days in the presence of IL-12. Similarly, when strongly polarized allergen-specific Th2 cells, obtained from skin biopsies of patients with atopic dermatitis, following the induction of late-phase cutaneous reactions, were stimulated and cultured with allergen in the presence of IL-4 and IL-12, about 50% of the cells produced IFNγ, as shown by an intracellular staining method. In contrast to supernatants of activated skin-derived T cells, generated in the presence of IL-4 only, those generated in the presence of IL-4 and IL-12 were no longer able to induce the synthesis of IgE by purified B cells *in vitro*, indicating that IL-12 was still able to functionally revert a population of human highly differentiated Th2 cells into a population containing Th1, Th0 and Th2 cells.

It has been argued that the presence of IFNγ during the initial priming may maintain the expression of the IL-12R β_2 chain on mouse T cells, thus creating a condition in which even differentiating Th2 cells can be reverted into IFNγ producers by the action of IL-12. However, the presence of IFNγ during priming of human naïve T cells has little effect on the transcription of the IL-12R β_2 chain, so fundamental differences, which remain to be elucidated, seem to exist between human and mouse populations. In addition, it is important to note that the above-mentioned studies were all carried out with populations and not with single cells, and therefore the possibility cannot be excluded that the observed changes in phenotype are the result of outgrowth of uncommitted T-helper cells, rather than a true reversal of differentiated cells. Indeed, it has been shown that IL-12, when given together with a standard drug to reduce the parasite load, could reverse established Th2 responses to chronic infection with *L. major* to healing Th1 responses, favouring the hypothesis of reversal by outgrowth and indicating that the polarized mouse Th2 responses still can be modulated under certain con-

ditions. Although the mechanisms underlying the differentiation of T-helper cells are not yet completely understood, the ability to manipulate established populations of T-helper cells with a polarized cytokine production profile, might be important for the treatment of infectious, autoimmune and allergic diseases.

5. Phenotypic differences between Th1 and Th2 populations

Although the operational definition of CD4$^+$ T-helper cell subpopulations, based on their different cytokine production profiles, has been useful in the analysis of the functional activity of these cells, the identification of a mutually exclusive and stable surface marker would be a valuable tool for use in the development and monitoring of clinical therapies, as well as for the isolation of polarized T-helper cell subsets for research purposes. Initial attempts to distinguish polarized mouse populations based on phenotype were unsuccessful: CD4$^+$ naïve precursors cells which are CD44^{high}, CD45RA^{high}, CD45RB^{high}, CD45RO^{low} and LECAM-1^{high} were all found to differentiate into effector cells with a CD44^{low}, CD45RA^{low}, CD45RB^{low}, CD45RO^{high} and LECAM-1^{low} cell surface phenotype, irrespective of their cytokine production profile. In a study using human T cells, it was reported Th1, Th2 and Th0 clones could be distinguished according to the amount of the Reed–Sternberg-specific molecule CD30 that they expressed. However this observation has not been

∎ **Table 1.** Phenotypic markers for Th1 and Th2 cells

Marker	Species	Th1 cell	Th2 cell
c-Maf	mouse	−	+
IFNγR β-chain	human/mouse	+	−
IL-12R β_2-chain	human/mouse	+	−
LAG-3	human	+	−
GATA-3	mouse	−	+
CCR3	human	−	+
CCR4	mouse	−	+
CXCR3	human	+	−
CCR5	human/mouse	+	−
CCR8	human/mouse	−	+
ST2	mouse	−	+

confirmed by others. Th1 clones, in contrast to Th2 clones, have been reported to have cytotoxic activity, so levels of expression of cells surface molecules involved in the lytic process could useful for distinguishing these two populations of cells. However, differential expression of cell surface molecules, like CD95 or its ligand (Fas-L), has not been reported yet.

5.1 Differential expression of cytokine receptors on subsets of T cells

In view of the (absence of) regulatory effects of certain cytokines, such as IFNγ and IL-12, on the cytokine production profile of polarized T-helper cells, much attention has been given to the differential expression of their corresponding cytokine receptors (Table 1). It was reported by Pernis *et al.* that murine Th1 clones, in contrast to Th2 clones, do not express transcripts for the β-chain of the IFNγR, which is involved primarily in signalling and not in binding of IFNγ. These results not only provided a molecular basis for the long-standing observation that IFNγ inhibits the proliferation of murine Th2 cells, but not that of Th1 cells, but also suggested a potential use of this receptor component for specific identification of T-helper cell subsets. IFNγ, produced during the differentiation of mouse Th1 cells can reportedly down-regulate the expression of the IFNγR β-chain on the latter cells, as well as on Th2 cells, suggesting that inactivation of IFNγ-mediated signalling in T cells is a dynamic process, depending on the cytokine environment and that the presence or absence of the IFNγR β chain does not discriminate between murine Th1 and Th2 cells *per se*.

Results from studies of human cells reveal a different picture: mRNA for IFNγR α- and β-chains are indeed expressed in polarized Th2, but not in Th1 clones. The expression of a non-functional IFN-γR on Th1 cells was confirmed by the inability of IFNγ to induce IFR-1 gene expression in Th1 clones. Moreover, human Th0 clones, producing IL-4 as well as IFNγ, were found to express a functional IFNγR, demonstrating that in human T cells, IFNγ does not seem to have a long-lasting down-regulatory effect on IFNγR β-chain expression. Human naïve CD4$^+$, CD45RA$^+$ cells also express the IFNγR β-chain; this ability was lost upon stimulation and culture of the cells in the presence of IL-12, but not in the presence of IL-4. It can be concluded that the loss of IFNγR β-chain expression on the surface of human T cells is intrinsic

to the Th1 differentiation process, confirming the original observation of Pernis *et al.*

As pointed out above, polarized Th2 cells, in contrast to Th1 cells, no longer respond to the IFNγ-inducing effects of IL-12. This commitment to the Th2 lineage is due to the extinction of signalling through the IL-12R, and is the result of loss of the β_2-chain of the IL-12R on the surface of mouse, as well as human T cells. This observation could potentially be useful to distinguish the two subsets. However, the above-mentioned studies were all carried out by analyzing the expression of these receptor components at the transcriptional level, and it is not yet known whether mAbs directed against these cell surface molecules will be useful for identifying and isolating polarized T-helper cell subsets. It is important to note that the number of IFNγ R β-chain and IL-12R β_2-chain molecules expressed at the cell surface of IL-4-producing CD4$^+$ (Th0/Th2) cells and IFNγ-producing Th1 cells, respectively, is likely to be too low to enable identification of these subsets using conventional flow-cytometric techniques. In addition, like the IFNγR β-chain, the IL-12R β_2-chain might also be expressed on populations of Th0 cells. Since it has been argued that populations of T-helper cells form a continuous spectrum in which the Th1 and Th2 phenotypes are only the two extremes, cytokine receptors are very likely to be expressed on 'intermediate' subsets as well, making the identification of polarized Th1 or Th2 cells very difficult, if not impossible.

5.2 Differential expression of chemokine receptors on subsets of T cells

Th1 and Th2 cells have distinct and opposite functions in the immune response, so it is important that their migration into sites of inflammation, such as inflamed lung wall epithelium or atopic skin in allergic disease, is tightly regulated. The molecular regulation of lymphocyte migration is complex and involves the interaction of cellular adhesion molecules, such as selectins and integrins, as well as a superfamily of chemokines and their receptors. Early studies have shown that subsets of T cells preferentially migrate to distinct anatomical sites, so a difference in expression of certain homing and chemokine receptors on the surface of these cells is to be expected. Indeed, T cells homing to the skin express the cutaneous lymphocyte-associated antigen (CLA), whereas gut-homing T lymphocytes preferentially express certain β7 integrins. Moreover, it has been shown that the homing of mouse Th1, but not of Th2, cells into sites of inflammation is mediated by the adhesion molecules E-selectin and P-selectin, probably as a result of differential expression of their ligand(s), such as CLA. However, this observation was not confirmed in a study in which human T cells were used and in which an independent regulation of the expression of CLA and cytokine synthesis was reported.

A number of recent studies have strongly suggested that expression of certain chemokine receptors, belonging to the CC (or β-chemokine) or the CXC (or α-chemokine) receptor family, may be restricted to subpopulations of T-helper cells with different cytokine production profiles (Table 1). According to Salusto *et al.*, CCR3 is preferentially expressed to Th2 lymphocytes, and its expression correlates positively with IL-4. Furthermore, polarized human Th1 cell were found to express transcripts for CXCR3 and CCR5, whereas, Th2 cells preferentially expressed CCR4 and, to a lesser extent, CCR3. The latter results were, in part, confirmed by a recent study in which T lymphocytes from rheumatoid joints, which acquire a Th1 phenotype *in vivo*, showed a preponderant expression of CCR5. In contrast to the results of the previous study, however, surface expression of CXCR3 was not limited to Th1 cells and could be detected on cultured Th2 cells as well. A new member of the CCR family, CCR8, has been identified recently; its cell surface expression is restricted to polarized Th2 cells. Although the T-helper subset-specific distribution of these various chemokine receptors has usually been analysed at the transcriptional level, these results were corroborated by the observation that Th1 and Th2 populations showed a selective migration in response to the chemokines corresponding to the type of receptor expressed at the surface and were consistent with results obtained with mouse Th1 and Th2 cells. It is to be expected that many other members of the rapidly expanding chemokine family with a restricted distribution on T cell subsets will be described.

It is important to note, however, that the expression of many chemokine receptors on the surface of T cells is not stable and shows significant variation, depending on the activation state of the cells, as well as on the presence of (endogenous) cytokines. For example, CXCR4 surface expression on T cells is specifically induced by IL-4, whereas IFNγ inhibits cell surface expression of CCR3 and up-regulates that of CXCR3 and CCR1. Polyclonal, but also TCR/CD3 complex-mediated,

T cell activation leads to down-regulation of a large number of chemokine receptors, including CXCR4, CCR1, CCR2, CCR5 and CCR3. Therefore, this transient and dynamic expression profile of many 'T-helper subset-specific' chemokine receptors may interfere with the usefulness of these surface molecules as phenotypic markers or Th1 and Th2 cells.

5.3 T1/ST2: a stable marker for Th2 cells?

Very recently, two groups independently reported the identification of a cell surface molecule that is stably expressed on mouse Th2, but not on mouse Th1 cells (Table 1). This molecule, designated ST2 or T1, had previously been cloned as an orphan receptor encoding a receptor-like molecule of the immunoglobulin superfamily. Although T1/ST2 has limited homology to the type I and type 2 IL-1 receptors, it does not bind IL-1α, IL-1β, IL-1 receptor antagonist or IL-18 (S.-i. Tominaga, personal communication). Injection of an anti-ST2 antibody into mice that had been challenged with *L. major* increased the production of IFNγ and decreased the production of IL-4 and IL-5 by T cells, thus enhancing resistance to infection with this intracellular parasite. In addition, in collagen-induced arthritis in mice, a disorder which is predominantly mediated by Th1 cells, the addition of an anti-ST2 mAb exacerbated the severity of the disease. Moreover, in a murine model of airway inflammation, induced by Th2 cells, the addition of an anti-T1/ST2 mAb or a T1/ST2 fusion protein reduced the eosinophilic inflammation of the airways. Together, these studies suggest that the T1/ST2 molecule is likely to play an important role in Th2 effector function and that the expression of this molecule can serve as a stable cell surface marker to distinguish mouse Th1 and Th2 cells. Although the human equivalent of ST2, which has 67% identity with mouse ST2 in a 327 amino acid overlap, has been cloned, no information is available at this time about the function and expression of human T1/ST2.

6. CD8⁺ T-cell subsets — Tc1 and Tc2

CD8⁺ T cells recognize antigen peptides presented on MHC class I molecules, and are able to kill the antigen-bearing APCs. As MHC class I molecules contain mainly peptides derived from the cytosol, this is an effective mechanism for killing cells infected with viruses or other intracellular pathogens. However, CD8⁺ T cells have also been associated with suppression of immune responses, and their ability to produce various cytokines suggests additional functions. The recent discovery of subsets of CD8⁺ T cells producing different patterns of cytokines suggests that, like CD4⁺ T cells, each subset of CD8⁺ T cells may mediate different functions.

The major cytokine pattern expressed by CD8⁺ effector T cells is similar to that produced by Th1 cells. The production of IFNγ, TNF and lymphotoxin by CD8⁺ T cells is consistent with their direct cytotoxic function, since these three cytokines activate cytotoxic function in macrophages and granulocytes. However, not all CD8⁺ T cells have a Th1-like pattern of cytokine production, as CD8⁺ T-cell clones can secrete both Th1 and Th2 cytokines. Some CD8⁺ T-cell clones derived from lepromatous leprosy patients secrete IL-4, but low amounts of IFNγ and CD8⁺ T-cell clones producing IL-4 can be obtained *in vitro* by mitogen or antigen stimulation in the presence of IL-4. Based on these findings, the names Tc1 and Tc2 were proposed for CD8⁺ T cells secreting IFNγ, but not IL-4 and IL-5, and for those secreting IL-4 and IL-5, but not IFNγ, respectively. It should be noted that the cytokine patterns of mouse CD8⁺ T-cell subsets are similar to those of human CD4 and CD8 T cells, i.e. IL-4 and IL-5 are specific for Th2/Tc2 cells, whereas IL-6 and IL-10 are preferentially (but not exclusively) produced by Th2/Tc2 clones.

6.1 Differentiation of CD8⁺ T-cell subsets

Unlike mouse CD4⁺ T cells, which can differentiate readily into either Th1 or Th2 cells, naïve CD8⁺ T cells show a strong preference for differentiating into Tc1 cells. As for the differentiation of CD4⁺ T cells into the Th1 phenotype, IFNγ and IL-12 promote differentiation of CD8⁺ Tc1 cells. Tc2 cells differentiation requires substantial amounts of IL-4 and the simultaneous inhibition of IFNγ-mediated effects with neutralizing anibodies. Similar to the differentiation of CD4⁺ T-cell subsets, preliminary studies have also pointed out the importance of APCs and the dose of the antigen, in the differentiation of CD8⁺ T cells.

6.2 Cytotoxicity

Although IL-4-secreting CD8⁺ T cells are generally less cytotoxic than Tc1 cells, they can also display a high cytotoxic ability, which probably depends on the conditions used for their growth and maintenance. Both Tc1 and Tc2 cells kill mainly by the perforin pathway, as shown by the abrogation of killing if calcium is chelated or if the cells are derived from perforin-deficient mice. Both subsets also kill Fas-expressing target cells.

6.3 B-cell help

Although it has been reported that Tc2 cells can give help to B cells, it is difficult to reconcile this function with the strong cytotoxic ability of Tc2 cells. Both Tc1 and Tc2 cells efficiently kill both resting and activated B cells upon recognition of specific alloantigens. Moreover, although allospecific Th2 cells provide strong help for IgM or IgG synthesis to small resting B cells, no help is provided by either Tc1 or Tc2 cells when they directly recognize specific alloantigens on the B cells.

In contrast to the outcome of direct B-cell recognition by T cells, some help could be provided if Tc2 cells were activated independently, by immobilized anti-CD3 antibodies. The inability of Tc1 and Tc2 cells to provide cognate help to B cells is consistent with the antigen-processing pathways used by MHC class I and class II presentation. Antigen captured by B-cell surface antibodies is processed and presented by MHC class II, whereas class I present peptides, which are derived from proteins synthesized within the B cell. Thus, by the regular antigen-processing pathways, foreign antigens would be presented on MHC class II if the B cells expressed specific antibodies, whereas the antigen would be expressed on class I if the B cells were infected. Interaction with CD4 and CD8 T cells, respectively, would ensure that antigen-specific B cells would receive help, whereas infected B cells would be killed. When Tc2 cells are activated following recognition and killing of infected cells, the Tc2 cells will express cytokines and surface molecules that may provide antigen non-specific help for neighbouring B cells, thereby augmenting the effects of antigen-specific Th2 cells.

7. T-cell subsets in disease

Based on results obtained from numerous animal models and human studies, there seems to be a direct association between the involvement of T-helper cell subsets and susceptibility to infectious, autoimmune and allergic disorders (Table 2). Th1 responses are thought to play an important role in the protection against intracellular microbes, induc-

▮ **Table 2** Involvement of Th1 and Th2 cells in diseases or experimental disease models.

	T helper type 1 responses	T helper type 2 responses
Type of immune response	cell-mediated-responses (DTH reactions); humoral responses (complement-fixing type); inflammation tissue destruction	humoral reponses (allergic type of responses); anti-infammatory
Protection against intracellular parasites	tuberculoid leprosy; tuberculosis (protective); Lyme's disease	lepromatous leprosy; tuberculosis (progression); progression to AIDS; atopic disorders; vernal conjunctivitis
Organ-specific autoimmunity	autoimmune encephalomyelitis; diabetus mellitus autoimmune thyroid diseases; acute graft-versus-host disease; Crohn's disease; acute Allograft Rejection	chronic graft-versus-host disease
Systemic autoimmunity	rheumatoid arthritis; multiple sclerosis	systemic lupus erythematosus; hypereosinophilic syndromes
Role in pregnancy	associated with recurrent spontaneous abortions	associated with successful pregnancy

ing cell-mediated inflammatory reactions, as well as the production of antibodies which bind to Fcγ receptors on APCs and which fix complement. In contrast, Th2 cells induce the production of IgE and the recruitment of eosinophils into locations, infiltrated by extracellular parasites. Importantly, Th2 cells, due to the nature of the cytokine that they produce, also limit the damage caused by Th1 cell-mediated inflammation. In addition to giving protection to various invading micro-organisms, Th1 and Th2 cells play a critical role in the induction and maintenance of other pathological conditions (Table 2).

7.1 The involvement of Th2 cells in allergy

Strong experimental evidence suggests that Th2 cells play an important role in the pathology of allergic diseases. T cells with a Th2 cytokine production profile accumulate at the site of cutaneous late-phase reactions in skin biopsies from patients with atopic dermatitis and mucosal bronchial biopsies or bronchial alveolar lavage fluid from patients with asthma. The importance of Th2 cells in allergic diseases is tightly associated with the functional activities of the cytokines that they produce. IL-4 and IL-13 are essential factors for the induction of ε-germline transcripts which, together with a signal mediated by CD40/CD40L interactions, following T–B cell contact, results in the subsequent switch-

ing of B cells to IgE-producing plasma cells. Interaction of allergen with specific IgE bound to high-affinity IgE receptors on mast cells and basophils results in receptor activation and in the release of soluble mediators, like histamine and leukotrienes which cause allergic reactions in the various target organs. IL-3, IL-4 and IL-10 have growth-promoting activities on basophils. whereas IL-5 induces the recruitment, differentiation and proliferation of eosinophils into sites of allergic reactions. The release of vasoactive mediators, chemotactic factors and cytokines promotes the cascade of events in allergic reactions resulting in allergic inflammation, mucosal tissue injury and vasodilation.

Not only T cells, but also mast cells, basophils and eosinophils produce IL-4, and probably also IL-13, raising the possibility that non-T-cell-derived IL-4 might contribute to IgE synthesis by B cells. Moreover, it has been demonstrated that eosinophils, lung mast cells and peripheral blood basophils express CD40L, and that the latter cells can induce IgE synthesis *in vitro*. However, results from a study on IL-4-deficient knockout mice that had been selectively reconstituted with either CD4+ or CD4− spleen cells from congenic wild-type animals, showed that no IgE production was detectable in mice in which only CD4−, IL-4-producing cells were present, suggesting that T-helper cells are the primary, if not the only, source for the initial production of IgE. Therefore, it can be concluded that the activation and recruitment of eosinophils, mast cells and basophils, as well as the production of IgE by B cells, which all participate in allergic reactions, are largely under the control of Th2 cells.

The mechanism(s) underlying the genetic predisposition for atopic disease have not yet been identified, but are likely to be the result of molecular changes in genes coding for the production of Th1 and Th2 cytokines, as well as their receptors. Asthma and other atopic disorders are multifactorial diseases, so multiple gene products seem to be involved in the pathology of these diseases. Much attention has been focused on human chromosome 5 (and its murine homologue, chromosome 11) in view of the cluster of Th2 cytokine genes found there, including the genes encoding IL-4, IL-5, IL-13, the IL-12 p40 chain and IFNγ response factor 1 (IRF-1). Although many genetic loci seem to predispose asthma, the genetic control of this disease, including the Th1/Th2 balance and the production of IgE, remains unclear. In addition to the genetic background of atopic individuals, another unresolved question is the contribution of the antigenic environment to the balance of Th1 and Th2 responses. In a recent epidemiological study, an inverse correlation was found between DTH responses against tuberculin and the presence of asthma and serum IgE concentrations, suggesting that infections such as those with *Mycobacterium tuberculosis* might have a protective effect on Th2 cell-mediated allergic disorders. Moreover, the results of this study suggest that vaccination to induce Th2 responses may be effective against atopic disorders.

7.2 CD4$^+$ T-cell subsets in autoimmunity

Evidence from animal models has strongly suggested that both Th1 and Th2 responses are implicated in the induction of organ-specific autoimmune diseases, such as insulin-dependent diabetes mellitus, autoimmune thyroid diseases and systemic autoimmune diseases, such as rheumatoid arthritis, EAE (all Th1-dominated) and systemic lupus erythematosus (SLE) (Th2-associated). Evidence from related diseases in human has shown an accumulation of a predominant Th1 response in the target organs of patients with autoimmune thyroid diseases and multiple sclerosis and in those suffering from granulomatous inflammation, such as arthritis and Crohn's disease.

7.2.1 Multiple sclerosis

Active lesions of multiple sclerosis (MS) patients are characterized by the infiltration of lymphocytes (mainly CD4$^+$ T cells) and macrophages in the brain, leading to an autoimmune reaction against myelin antigens, such as myelin basic protein (MBP), proteolipid protein (PLP) and possibly other proteins. The possible role of a Th1/Th2 imbalance has been investigated extensively in models of EAE. This condition can be induced in rodents by injection of MBP in complete Freund's adjuvant and has been shown to be dependent on the presence of CD4$^+$, but not CD8$^+$ T cells. There is evidence for a role of proinflammatory cytokines, secreted by Th1 cells in the pathogenesis of EAE. T-cell lines and clones producing Th1 cytokines (IFNγ, IL-2 and TNF-α) are able to transfer EAE in the rat and in the mouse. These cytokines are also found in the CNS of animals with active disease. The ability of IL-12 to enhance the pathology of EAE supports a role for Th1 cells in the pathology of this disease. However, the precise role of Th1 cytokines in the pathogenesis of EAE is still unclear, since mice with a disrupted IFNγ gene are still susceptible to the induction of EAE.

At the peak of disease severity, perivascular infiltrates in the brain are characterized by the presence of IL-2 and IFNγ, whereas recovery is associated with the appearance of IL-4 and IL-10. The presence of IL-4 in the brain of animals recovering from EAE led to the hypothesis that disease remission might be related to the presence of antigen-specific Th2 cells. Accordingly, MBP- or PLP-specific Th2 clones producing high amounts of IL-4 and IL-10 do not induce EAE, but suppress the induction of EAE when mixed during adaptive transfer with Th1 clones. However, other studies have shown that IL-4 administered at the time of disease induction or upon onset of disease exacerbated the disease in a model of EAE induced by PLP plus adjuvant. In addition, Lafaille *et al.* have recently shown that both Th1 and Th2 cells generated from MBP-specific TCR transgenic mice will transfer EAE in immunodeficient mice, whereas only Th1 cells will induce disease in normal mice. The Th2-induced EAE observed in immunodeficient mice manifested histopathological features of an allergic response.

The conflicting data described above suggest that the hypothesis that Th1 cells are pathogenic and Th2 cells are regulatory in the context of autoimmune disease may be too simplistic.

7.2.2 Insulin-dependent diabetes mellitus

Numerous studies, particularly on BB rats and *nod* (non-obese diabetic) mice, indicate that insulin-dependent diabetes mellitus is a T-cell-mediated disease, and observations that the mononuclear infiltrate of human pancreatic islets contains significant numbers of activated T cells suggest that this is probably true in human type-1 diabetes as well.

Several studies have correlated diabetes with a Th1 phenotype. For example, expression of genes encoding IFNγ and IFNα, under the control of the insulin promoter, resulted in inflammation in the islets and diabetes in transgenic mice. Furthermore, IFNγ production correlates with diabetes in *nod* mice and treatment with anti-IFNγ antibodies can prevent the development of diabetes induced in *nod* mice by cyclophosphamide or adoptive transfer of diabetogenic cells. In support of the hypothesis that Th1 cells promote diabetes, administration of IL-12 was found to accelerate insulin-dependent diabetes mellitus in genetically susceptible *nod* mice, but not in resistant BALB/c mice which correlated with increased Th1 cytokine production by islet-infiltrating cells.

Unlike Th1 cells, Th2 cells are relatively innocuous in the induction of diabetes, in that they invade the islets, but do not provoke disease. Although in these studies Th1-mediated diabetes was consistently induced, the results of other studies suggest that Th2 cells may in fact protect against the disease. In support of this, Rapoport *et al.* have shown that administration of IL-4 to prediabetic *nod* mice protects them from diabetes. Furthermore, transgenic *nod* mice which express IL-4 in their pancreatic β-cells, under the control of the human insulin promoter, were completely protected from insulitis and diabetes. However, more recent data have shown that Th2 cells are capable of inducing diabetes in immunocompromized *nod/scid* (severe combined immunodeficient) mice and insulitis in neonatal *nod* mice. This disease is inhibited by anti-IL-10 antibodies, but not by anti-IL-4 antibodies, suggesting that other regulatory cells, secreting IL-10, can protect against disease.

7.3 Regulatory CD4⁺ T-cell subsets

Multiple studies now suggest that alternative regulatory populations exist which are somehow associated with, but distinct from, Th2 cells. Regulatory CD4⁺ T-cell subsets have been described which can inhibit cell-mediated immune responses and/or inflammatory pathologies.

7.3.1 Regulatory cells in inflammatory bowel disease and the role of IL-10

Inflammatory bowel disease (IBD), including Crohn's disease and ulcerative colitis, is a chronic relapsing inflammatory disease of the gastrointestinal tract. A number of models of chronic intestinal inflammation in mice have recently been described which have provided excellent tools with which to study immune regulation in the gut. Thus, IBD with similarities to the human disease developed in mice with targeted disruptions of IL-2 or IL-10 genes, as well as in mice with alterations in T-cell subsets, including TCR α-chain deficient mice and *scid* mice restored with CD45RB^{high}CD4⁺ T cells. Colitis appears to be the result of the development of a dysregulated Th1 response against components of the enteric flora, as this disease could be inhibited by treatments known to inhibit Th1 responses and did not develop in animals raised in germ-free conditions. Taken together, the results from the various induced models of IBD have highlighted the important role that immune-mediated mechanisms play in normal intestinal homeostasis.

IL-10 is a key factor in the immunoregulatory mechanisms which control inflammatory responses in the intestine: IL-10-deficient mice develop severe colitis, but systemic administration of recombinant IL-10 can prevent the development of colitis in *scid* mice injected with CD45RB^{high}CD4⁺ cells. In addition, CD45RB^{high}CD4⁺ T cells, isolated from transgenic mice which expressed IL-10 under the control of the IL-2 promoter failed to induce disease in *scid* mice and were even able to inhibit development of colitis when mixed with pathogenic CD45RB^{high}CD4⁺ cells from wild-type mice. Direct evidence that immune responses in the intestine are actively regulated by a subpopulation of CD4⁺ T cells comes from the finding that colitis induced in *scid* mice by transfer CD45RB^{high}CD4⁺ T cells could be prevented by co-transfer of the reciprocal CD45RB^{low} population. The immune regulatory activities of the CD45RB^{low} population were shown to be dependent on TGF-β, but independent of IL-4, suggesting that these cells are functionally distinct from Th2 cells. IL-10 also plays an important role in the function of this population as CD45RB^{low}CD4⁺ cells isolated from IL-10 knockout mice failed to inhibit colitis. Indeed, it seems likely that IBD which develops spontaneously in IL-10 knockout mice is in part due to the absence of this population of regulatory T cells which normally act to inhibit the development of inflammatory responses in the gut. Whether IL-10 and

TGF-β are required for the differentiation of regulatory T cells or for their effector function, or both, is not known.

There is also evidence that immune responses to commensal enteric flora are actively regulated in humans. T cells isolated from the intestinal lamina propria have been found to be tolerant to an individual's own enteric bacteria, but not to bacteria isolated from other individuals. Recent data suggest that the state of non-responsiveness to the abundant presence of antigens of the intestinal flora is mediated by secretion of IL-10 and TGF-β after recognition of bacterial antigens by $CD4^+$ T cells.

7.3.2 Tr1 cells in IBD

Tr1 cells are immunosuppressive *in vivo*. Colitis induced in *scid* mice by transfer of $CD45RB^{high}CD4^+$ T cells can be prevented by co-transfer of murine Tr1 clones derived from $CD4^+$ T cells expressing a transgenic TCR specific for ovalbumin. The immunosuppression depends on antigen-induced activation of Tr1 cells *in vivo*, as these cells only inhibit colitis in animals that have received ovalbumin in their drinking water. One interpretation of this result, which we favour, is that Tr1 cells are activated as a result of oral administration of antigen and act via secretion of soluble mediators which suppress activation of pathogenic T cells (presumably responding to bacterial antigens) in the local environment by an antigen-driven bystander suppression mechanism. It is possible that Tr1 cells represent *in vitro*-generated counterparts of regulatory T cells which exist naturally within the $CD45RB^{low}CD4^+$ population and are dependent on IL-10 and TGF-β for their function. However, characterization of the immunosuppressive properties of Tr1 cells *in vivo* is at an early stage and further experiments are required to identify where Tr1 cells are activated *in vivo* and precisely how they mediate their immunosuppressive activities. These studies should further focus on whether these activities are attributable entirely to the production of soluble immunosuppressive molecules, such as IL-10 and TGF-β, or whether they involve cellular interactions between pathogenic and regulatory populations of T cells, either directly or via APCs.

7.3.3 Regulatory T cells in the control of the immune response

In addition to regulating inflammatory responses in the intestine, there is evidence that regulatory T cells also modulate systemic immune responses.

Indeed, the antigen-specific hyporesponsiveness induced by oral administration of antigen has been shown, under some circumstances, to involve the induction of regulatory T cells. In mice, given oral MBP prior to immunization with this protein, MBP-specific regulatory T cells have been cloned from the mesenteric lymph nodes. These cells, called Th3 cells, secreted high levels of TGF-β, but lower levels of IL-4 and IL-10, and were able to protect SJL mice from EAE. Protection, conferred by these MBP-specific clones applied both to MBP-induced disease and to PLP-induced EAE and could be inhibited by administration of anti-TGF-β mAb, indicating that Th3 cells could mediate TGF-β-dependent bystander suppression towards central nervous system (CNS) antigens. Similarly, T cell clones producing high levels of IFNγ and IL-10 have also been isolated from mice immunized with an altered peptide of PLP. These Tr1-type clones were able to protect mice from EAE, induced by the wild-type peptide. Similar types of T cell clones were isolated from lymphocytes infiltrating the pancreatic islets of *nod* mice. These cells, which responded to self-MHC class II determinants, had a significant immunosuppressive effect on proliferative responses of splenic effector T cells from *nod* mice through the secretion of soluble factors and prevented diabetes when injected into young *nod* mice. The observation that T-cell populations with similarities to Tr1 cells could inhibit pathogenic T-cell responses against CNS or islet cell antigens, suggests that regulatory T cells may also play a role in the maintenance of peripheral tolerance.

8. Concluding remarks

The recognition of the dual nature of the immune response, the classification of Th1 and Th2 $CD4^+$ subsets and their equivalent among $CD8^+$ T cells, as well as the discovery that other regulatory subsets with unique functions and cytokine production profiles exist have greatly contributed to our understanding of the importance of cytokines in the control of normal and pathological immune responses. Although many questions still have to be resolved, there is strong evidence, based on data obtained from genetically manipulated animal models, as well as human *in vitro* models, that modulation of the cytokine balance may alter the outcome of disease, providing the basis for future therapeutic intervention in human pathophysiological conditions.

References

Asherson, G. L. and Stone, S. H. (1965). Selective and specific inhibition of 24-hour skin reactions in the guinea pig. I. Immune deviation: description of the phenomenon and the effect of splenectomy. *Immunology* **9**, 205–17.

Bendelac, A., Rivera, M. N., Park, S.-H., and Roark, J. H. (1997). Mouse CD1-specific NK1 T cells: development, specificity and function. *Annu. Rev. Immunol.* **15**, 535–62.

Butcher, E. C. and Picker, L. J. (1996). Lymphocyte homing and homeostasis. *Science* **272**, 60–66.

Carballido, J. M., Carballido-Perrig, N., Terres, G., Heusser, C. H., and Blaser, K. (1992). Bee venom phospholipase A2-specific T cell clones from human allergic and non-allergic individuals: cytokine patterns change in response to the antigen concentration. *Eur. J. Immunol.* **22**, 1357–63.

Gauchat, J. F., Henchoz, S., Mazzei, G., Aubry, J. P., Brunner, T., Blasey, H., Life, P., Talabot, D., Flores-Romo, L., Thompson, J. *et al.* (1993). Induction of human IgE synthesis in B cells by mast cells and basophils. *Nature* **365**, 340–43.

Groux, H., Bigler, M., O'Garra, A., Rouleau, M., Antonenko, S., de Vries, J. E., and Roncarolo, M.-G. (1997). A CD4+ T-cell subset inhibits antigen-specific T cell responses and prevents colitis. *Nature* **389**, 737–42.

Hosken, N. A., Shibuya, K., Heath, A. W., Murphy, K. M., and O'Garra, A. (1995). The effect of antigen dose on CD4+ T cell phenotype development in an ab-TCR-transgenic mouse model. *J. Exp. Med.* **182**, 1579–84.

Julia, V., Rassoulzadegan, M., and Glaichenhaus, N. (1996). Resistance to *Leishamania major* induced by tolerance to a single antigen. *Science* **274**, 421–3.

Katz, J. D., Benoist, C., and Mathis, D. (1995). T helper cell subsets in insulin-dependent diabetes. *Science* **268**, 1185–8.

Lafaille, J. J., Van de Keere, F., Hsu, A. L., Baron, J. L., Haas, W., Raine, C. S., Tonegawa, S. (1997). Myelin basic protein-specific T helper 2 (Th2) cells cause experimental autoimmune encephalomyelitis in immunodeficient hosts rather than protect them from disease. *J. Exp. Med.* **186**, 307–12.

Lanier, L. L., O'Fallon, S., Somoza, C., Phillips, J. H., Linsley, P. S., Okumura, K., Ito, D., and Azuma, M. (1995). CD80 (B7) and CD86 (B70) provide similar costimulatory signals for T cell proliferation, cytokine production, and generation of CTL. *J. Immunol.* **154**, 97–105.

Liblau, R., Singer, S., and McDevitt, H. (1995). Th1 and Th2 CD4+ T cells in the pathogenesis of organ-specific autoimmune diseases. *Immunol. Today* **16**, 34–8.

Moore, K. W., O'Garra, A., de Waal Malefyt, R., Vieira, P., and Mosmann, T. R. (1993). Interleukin-10. *Annu. Rev. Immunol.* **11**, 165–90.

Mosmann, T. R. and Coffman, R. L. (1989). Th1 and Th2 cells: different patterns of lymphokine secretion lead to different functional properties. *Annu. Rev. Immunol.* **7**, 145–73.

Nabors, G. S., Afonso, L. C., Farrell, J. P., and Scott, P. (1995). Switch from a type 2 to a type 1 T helper cell response and cure of established *Leishmania major* infection in mice is induced by combined therapy with interleukin 12 and Pentostam. *Proc. Natl. Acad. Sci. USA* **92**, 3142–6.

Pernis, A., Gupta, S., Gollob, K. J., Garfein, E., Coffman, R. L., Schindler, C., and Rothman, P. (1995). Lack of interferon γ receptor β chain and the prevention of interferon γ signaling in TH1 cells. *Science* **269**, 245–7.

Powrie, F. (1995). T cells in inflammatory bowel disease: protective and pathologic roles. *Immunity* **3**, 171–4.

Sad, S., Marcotte, R., and Mosmann, T. R. (1995). Cytokine-induced differentiation of precursor mouse CD8+ T cells into cytotoxic CD8+ T cells secreting Th1 or Th2 cytokines. *Immunity* **2**, 271–9.

Sallusto, F., Lenig, D., Mackay, C. R., and Lanzavecchia, A. (1998). Flexible programs of chemokine receptor expression on human polarized T helper 1 and 2 lymphocytes. *J. Exp. Med.* **187**, 875–83.

Sornasse, T., Larenas, P., Davis, K. A., de Vries, J. E., and Yssel, H. (1996). Differentiation and stability of Th1 and Th2 cells derived from naive human neonatal CD4+ T cells, analyzed at the single-cell level. *J. Exp. Med.* **184**, 473–83.

Swain, S. L., Weinberg, A. D., English, M., and Huston, G. (1990). IL-4 directs the development of Th2-like helper effectors. *J. Immunol.* **145**, 3796–806.

Szabo, S., Dighe, A. S., Gubler, U., and Murphy, K. M. (1997). Regulation of the interleukin (IL)-12β2 subunit expression in developing T helper 1 (Th1) and Th2 cells. *J. Exp. Med.* **185**, 817–24.

Trinchieri, G. (1995). Interleukin-12: a proinflammatory cytokine with immunoregulatory functions that bridge innate resistance and antigen-specific adaptive immunity. *Annu. Rev. Immunol.* **13**, 251–76.

Ushio, S., Namba, M., Okura, T., Hattori, K., Nukada, Y., Akita, K., Tanabe, F., Konishi, K., Micalle, M., Fujii, M., Torigoe, K., Tanimoto, T., Fukuda, S., Ikeda, M., Okamura, H., and M., K. (1996). Cloning of the cDNA for human IFN-γ-inducing factor, expression in *Escherichia coli*, and studies on the biologic activities of the protein. *J. Immunol.* **156**, 4274–9.

Weiner, H. L. (1997). Oral tolerance: immune mechanisms and treatment of autoimmune diseases. *Immunol. Today* **18**, 335–43.

Windhagen, A., Scholz, C., Hollsberg, P., Fukaura, H., Sette, A., and Hafler, D. A. (1995). Modulation of cytokine patterns of human autoreactive T cell clones by a single amino acid substitution of their peptide ligand. *Immunity* **2**, 373–80.

Xu, D., Chan, W. L., Leung, B. P., Huang, F.-P., Wheeler, R., Piedrafita, D., Robinson, J. H., and Liew, F. Y. (1998). selective expression of a stable cell surface molecule on type 2 but not type 1 helper T cells. *J. Exp. Med.* **187**, 787–94.

Zingoni, A., Soto, H., Hedrick, J. A., Stoppacciaro, A., Storlazzi, C. T., Sinigaglia, F., D'Ambrosio, D., O'Garra, A., Robinson, D., Rocchi, M., Santoni, A., Zlotnik, A., and Napolitano, M. (1998). The chemokine receptor CCR8 is preferentially expressed in Th2 but not Th1 cells. *J. Immunol.* **161**, 547–51.

XVII Regulation of humoral immunity by cytokines

Clifford M. Snapper*,
Fred D. Finkelman and
Jacques Banchereau

1. Role of the humoral immune response in health and disease

1.1 Introduction

The humoral immune response is directly mediated through immunoglobulins of various isotypes, produced when B lymphocytes are stimulated by distinct combinations of activators and cytokines. The central topic of this chapter concerns the detailed regulatory pathways through which these stimuli act and interact on B cells either to induce, or to prevent, B cell clonal expansion, immunoglobulin secretion and immunoglobulin class switching. Unless otherwise indicated, the discussion will focus on the response of human B cells. Related murine studies will be briefly discussed only to highlight key similarities with and differences from the human system. The most important human B cell cytokines will be discussed in detail in separate sections.

1.2 Physiological role of immunoglobulins

Immunoglobulin production, when regulated appropriately, can provide a means by which the host prevents, controls or eliminates invading microbes and their pathogenic consequences. Thus, through antigen binding alone, immunoglobulins may neutralize microbial toxins, prevent a bacterium from specifically recognizing and hence penetrating a mucosal surface, or inhibit viral entry into cells by blocking the site on the virus that recognizes its specific cellular receptor. Further, depending upon its particular immunoglobulin isotype, the immunoglobulin molecules, through their Fc regions, may:

(1) mediate lysis of pathogens through complement fixation (classical or alternative pathways;

(2) target pathogens to phagocytes (opsonization), as well as activate phagocytes by binding to one or more of the various Fc or complement receptors that are expressed on the phagocyte cell membrane;

(3) aggregate pathogens for enhanced phagocytosis; or

(4) induce the production of cytokines through cross-linking various Fc receptors on cytokine-producing cells, and hence induce the activation of various immune cell types.

1.3 Immunoglobulin-mediated pathology

Aberrant regulation of the humoral immune response may lead to hypersensitivity or autoimmune responses that induce tissue injury and disease. Type I hypersensitivity involves the interleukin 4 (IL-4) and/or IL-13-dependent production of IgE which plays an important role in mediating allergic disease and perhaps the host response to certain multicellular parasites. Type II hypersensitivity involves the circulation of free antibody specific for a cellular target. Through binding its target the immunoglobulin can mediate injury by:

(1) blocking receptor binding of trophic signals (e.g. anti-acetylcholine receptor in myasthenia gravis);

(2) mimicking ligands for certain receptors, thus inducing constitutive and aberrant stimulation

* Corresponding author.

of the receptor-bearing cells (e.g. antibodies against thyroid-stimulating hormone receptor in Graves' disease (hyperthyroidism)); or

(3) destroying normal host cells through complement-mediated lysis and/or Fc-mediated phagocytosis (e.g. antibodies against red blood cells in various hemolytic anemias, or antibodies against platelets in idiopathic thrombocytopenic purpura).

Type III hypersensitivity occurs when immunoglobulin forms complexes with antigen that circulate in the blood and deposit in various tissues and organs. The tissue distribution and consequences of immune complex deposition will depend in large part on the physicochemical composition of the complex and on the particular characteristics of the organ and its blood supply, although not on the specificity of the immunoglobulin itself. Notably, immune complexes activate complement and, in so doing, induce tissue inflammation, including the attraction of neutrophils, and necrosis. Many diseases are mediated by type III hypersensitivity responses, including systemic lupus erythematosus (SLE), various glomerulonephritides, vasculitides and arthritides.

2. T cell-independent and T cell-dependent immunity

2.1 Introduction

Studies in mice have shown that humoral immune responses may show a relative or absolute dependence on T cells, or be completely T cell-independent, depending upon the nature of the immunizing antigen. Thus, soluble proteins are absolutely dependent upon T cell help (i.e. for eliciting antibody responses) and are referred to as T cell-dependent (TD) antigens. Immunoglobulin responses to non-mitogenic polysaccharide antigens may be positively or negatively influenced by T cells, but can occur in the absence of such cells (T cell-independent antigens, type 2 (TI-2 antigens)). Antigens possessing B cell mitogenic properties (e.g. lipopolysaccharide (LPS)) appear to be the most T cell-independent (TI-1 antigens). The importance of TI humoral immune responses in humans has not been clearly established.

2.2 T cell-dependent immunity

The nature of the T cell help in TD humoral immune responses comprises, in large part, the delivery of T cell-derived cytokines and provision of co-stimulatory T cell-derived membrane molecules to a specific B cell that is in physical contact with the T cell. This occurs when B cells specifically recognize a protein antigen, through membrane immunoglobulin binding, internalize and digest the antigen, and then present, to the T cell, the resultant antigenic peptides in combination with major histocompatibility complex (MHC) molecules. Specific recognition of this peptide–MHC complex by the T cell receptor in concert with T cell binding of various co-stimulating ligands on the B cell (e.g. B cell B7-1 or B7-2, and T cell CD28 or CTLA4) results in T cell activation. This includes release of B cell-activating cytokines, as will be discussed below, and induction on the T cell membrane of B cell-activating molecules, most notably CD40-ligand which binds CD40 on B cells. Of note, CD40 is one of a family of B cell-activating (or inhibiting) molecules (i.e. the tumor necrosis factor alpha (TNF-α) receptor family) for which a counter-ligand may be expressed by the T cell. These TNF-α receptor members include CD40, CD27, CD30, Ox40-ligand, TNF-α receptor, and Fas, all of which have been shown to regulate B cell function. In particular, the profound defect in TD induction of immunoglobulin class switching in patients with mutations in their CD40-ligand gene, and the profound disruption of TD humoral immunity in CD40 or CD40-ligand knockout mice or mice injected with anti-CD40-ligand antibody, underscores the central role of CD40 in TD B cell antibody responses. Indeed, as will be discussed below, CD40 activation of B cells can provide a critical co-stimulus for many cytokine-dependent B cell functions.

2.3 T cell-independent immunity

The cellular basis for TI humoral immunity in response to polysaccharide antigens (TI-2 antigens) is less well understood, although a model for its induction has recently been proposed. An essential difference between polysaccharide and protein antigens is that the former cannot associate with MHC molecules on antigen-presenting cells (APCs) and hence cannot effect cognate interactions with T cells, as described for TD responses to proteins. However, polysaccharides, unlike protein antigens, contain repeating, identical antigenic epitopes capable of cross-linking membrane immunoglobulin on B cells in a multivalent manner. This has been shown, in *in vitro* experimental systems in the mouse, to induce potent B cell signaling which can co-stimulate cytokine-dependent immunoglobulin secretion and class switching. Similar findings have begun to emerge using human B cells. Thus, potent signaling through membrane immunoglobulin may circumvent the need for CD40-ligand. Since polysaccharide antigens, as well as the bacteria that contain such antigens, can induce B cell-active cytokines from various non-T cells, the need for T cells as a source of cytokines is also obviated. When polysaccharide antigens are associated with a B cell mitogenic moiety, such as LPS (a TI-1 antigen), the mitogen may substitute for cytokines, and synergize with membrane immunoglobulin-mediated signals for induction of immunoglobulin secretion and limited immunoglobulin class switching.

In summary, induction of TD and TI humoral immunity will typically involve various combinations of cytokines, produced by different cell types, acting in concert with some type of B cell-activating signal provided by distinct membrane molecules present on immune cells and/or by the antigen itself.

3. Immunoglobulin class switching

3.1 Functional role of immunoglobulin isotypes

The immunoglobulin class switch is a process which confers biological diversity to a humoral immune response, since immunoglobulins of different classes mediate distinct effector functions. Thus, the particular Fc region of the immunoglobulin molecule, which defines its class (isotype), determines the ability of the immunoglobulin to: (i) fix complement (classical versus alternative), (ii) bind to distinct Fc receptors which are differentially expressed on multiple immune cell types, (iii) remain in the circulation, (iv) penetrate tissue and cross the placenta, (v) form pentamers (IgM) or dimers (IgA), (vi) resist proteolytic digestion and (vii) self-aggregate.

3.2 Cellular aspects of immunoglobulin class switching

At the cellular level, switching is seen to occur when B cells, which initially express IgM and/or IgD on their surface, switch upon activation to the expression of one of the non-IgM, non-IgD molecules as shown in Table 1.

∎ **Table 1.** Immunoglobulin classes and subclasses

Species	Immunoglobulins expressed after class switching
Human	IgG1, IgG2, IgG3, IgG4, IgE, IgA1 and IgA2
Mouse	IgG1, IgG2a, IgG2b, IgG3, IgE, IgA
Rat	IgG1, IgG2a, IgG2b, IgG2c, IgE, IgA

Under the influence of maturation (differentiation) factors the B cell can then secrete the immunoglobulin isotype to which it has switched. This new immunoglobulin class maintains the same antigen-specificity (VDJ) expressed by the original B cell but possesses a new biological effector function.

3.3 Molecular aspects of immunoglobulin class switching

At the molecular level, switching typically occurs by a process of looping out and deletion of all constant heavy (C_H) genes located 5' to the one that is to be expressed. This juxtaposes the recombined C_H gene with the VDJ that codes for antigen specificity and which is associated with a promoter and enhancer for efficient transcription of intact immunoglobulin molecules. The switch recombination event occurs between stretches of repetitive sequences, termed switch (S) regions, located 5' to each C_H gene except IgD. Although switching typically occurs between $S\mu$ and a downstream S region, sequential switching (e.g. IgM to IgG to IgE) is also possible under the appropriate stimulation conditions.

3.4 Specific functions of different immunoglobulin isotypes

The importance of immunoglobulin class switching is underscored by the distinct biological effector functions that individual immunoglobulin isotypes display, and their implications for mediating host defense against pathogens or inducing tissue damage and disease. The following briefly describes the main effector functions of the different *human* immunoglobulin isotypes:

3.4.1 IgM

IgM is the first immunoglobulin class elicited during a primary humoral immune response. It typically expresses germline-encoded variable regions that have not undergone somatic mutation. Thus

IgM binds antigens with low affinity. This low affinity, however, is compensated for by the presence of ten antigen binding sites per molecule (secreted IgM exists as a pentamer). This enables IgM to bind antigen with high avidity when the antigen possesses multiple representations of the same epitope (e.g. polysaccharides) or is present in multiple copies (e.g. immune complexes and in microbial walls). IgM antibodies can also destroy or opsonize targets through their ability to fix complement efficiently. These same characteristics, however, create the potential danger that even low-affinity antibodies (e.g. those specific to multivalent antigens contained within red blookd cells) can produce substantial damage. The pentameric structure of IgM also limits its ability to diffuse rapidly from sites of local production to distant sites. This, and its short *in vivo* half-life relative to IgG, probably make IgM antibodies less well adapted than IgG antibodies to protect against second infections by a pathogen.

3.4.2 IgD

Relatively little is known about this isotype, which represents 0.3% of serum immunoglobulins and can bind to both self and non-self antigens. Though IgD-secreting plasma cells have been identified, IgD has long been considered to be a cell-surface marker for naïve B cells which co-express membrane IgM bearing the same variable region. Recent studies revealed the existence of normal B cells expressing membrane IgD in the absence of membrane IgM. These cells indeed display highly mutated membrane IgD, have their $C\mu$ gene deleted and represent intermediates that will ultimately yield the plasma cells secreting highly mutated IgD. IgD may help for the development of humoral responses through various mechanisms including (i) their unique structural flexibility allowing for efficient antigen capture and (ii) their specific binding to $Fc\delta R$ on a subset of helper T cells which have been shown to enhance humoral immune responses.

3.4.3 Non-IgM, non-IgD isotypes

Induction of IgG, IgE, and IgA isotypes occurs later than that of IgM during a primary immune response, and the former are the predominant immunoglobulin isotypes produced during a memory response. Immunoglobulin isotype switching and affinity maturation tend to occur during a similar time-frame, so that IgG, IgE, and IgA typically possess a greater affinity for antigen than IgM.

IgG predominates in plasma and extravascular lymph, whereas IgA predominates in respiratory, digestive and urogenital secretions.

3.4.4 IgG subclasses

In contrast to IgM, IgG subclasses have a relatively long half-life, which facilitates the maintenance of high serum IgG levels. Complement is fixed most effectively by IgG1 and IgG3. IgG2 also has some ability to fix complement. $Fc\gamma RI/CD64$ (on monocytes, macrophages and neutrophils) binds IgG1 and IgG3 most avidly and IgG2 to a lesser degree. $Fc\gamma RII/CD32$ (on macrophages, monocytes, neutrophils and B cells) and $Fc\gamma RII/CD16$ (on monocytes, macrophages, natural killer (NK) cells, neutrophils and some T cells) selectively bind, with low affinity, IgG1 and IgG3. IgG isotypes also bind to placental Fc receptors, which facilitate transport of maternal IgG (but not other isotypes) into the fetal circulation.

Different antigenic stimuli induce the production of different IgG subclasses. Viruses induce mostly IgG1 and IgG3 responses. Carbohydrate antigens induce substantial IgG2. Nematode parasites and repeated immunizations are associated with the production of IgG4. To some extent, the functional properties of particular IgG subclasses appear to make them particularly well suited to the binding or destruction of particular types of antigen or parasite. However, functional properties that might be expected to enhance the ability of an immunoglobulin isotype to destroy a pathogen may also contribute to its potential to damage the host. The ability of IgG1 and IgG3 antibodies to bind to $Fc\gamma$ receptors makes them, rather than IgG2 and IgG4 antibodies, mediators of erythrocyte destruction in patients with hemolytic disease of the newborn.

The correlation between immunoglobulin isotype function and their expression in response to different antigens or pathogens does not necessarily imply that individual IgG subclasses are essential for protection against different pathogens. Humans with specific immunoglobulin isotype deficiencies suggest that *individual* isotypes are not always required for protection against infection. Although people with isolated IgG2 deficiency are more likely to have infections with encapsulated bacteria than are people with normal IgG2 serum levels, most people with isolated IgG2 deficiency are clinically normal. Furthermore, it is not clear whether the association between low IgG2 level and infection represents a specific lack of IgG2

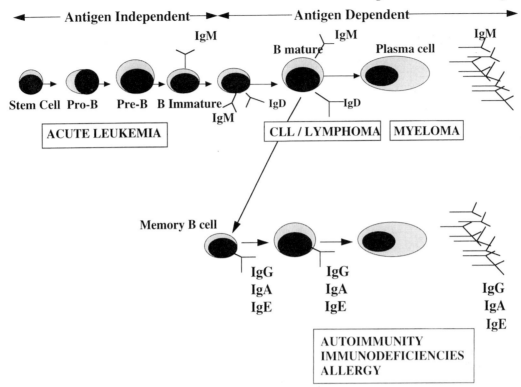

∎ **Figure 1.** Schematic representation of the pathways of antigen-dependent and antigen-independent B cell development and their associated diseases.

antibodies to carbohydrate epitopes on bacterial capsules or a lack of antibody specific for these epitopes, regardless of isotype. Similarly, it is unclear whether the association between low IgG1 and IgG3 levels and chronic lung disease in fact represents a specific requirement for these complement-fixing, Fcγ receptor-binding isotypes.

3.4.5 IgE

IgE binds with high affinity to IgE receptors (FcεRI) on mast cells and basophils, and induces degranulation and cytokine production by these cells when they are cross-linked by antigen. In addition to its central role in the pathogenesis of atopic disease, IgE may participate in the initiation of immune responses through induction of mast cell-derived mediators of inflammation. IgE may also play a role in the expulsion of gut and respiratory tract multicellular parasites, perhaps indirectly through mast cell degranulation-induced smooth muscle contraction and increased vascular permeability, and through promotion of eosinophil killing of parasites by IgE-mediated antibody-dependent

cellular cytotoxicity (ADCC). IgE may also facilitate presentation of soluble antigens through a mechanism that involves binding to FcεRII.

3.4.6 IgA

About 60 per cent of the total immunoglobulin produced in humans is IgA. Although IgA makes up only 10–15 per cent of serum immunoglobulin, it is the predominant immunoglobulin in mucosal secretions. There are two subclasses of IgA in humans: IgA1 and IgA2. Serum contains 80–90 per cent IgA1 whereas mucosal secretions contain up to 60 per cent IgA2.

The association of one or two IgA dimers with a molecule of secretory component facilitates transport of this isotype into the gut. The oligomeric nature of IgA enhances its ability to interact with high avidity with viruses and bacteria that are present in secretions. Although IgA lacks the ability to fix complement by the classical pathway, it is the most effective isotype at fixing complement by the alternative pathway. IgA antibodies bind well to Fc receptors on neutrophils. IgA is highly resistant to digestion by proteolytic enzymes, present in the digestive tract, allowing this antibody to function in this environment. An advantage of IgA2 over IgA1 is the lack of sensitivity to proteases secreted by several pathogenic microorganisms. IgA also has affinity for mucus and may help to reinforce the mucous barrier against pathogen penetration of the gut mucosa. Terminal mannose-containing oligosaccharides on the α-chain of IgA inhibit interactions between fimbriae on enteric bacteria and oligosaccharides on mucosal epithelial cell plasma membranes that would otherwise promote bacterial adherence to these cells. The frequency of isolated IgA deficiency in humans is approximately 0.1 per cent. Despite IgA's unique properties for protection at mucosal surfaces, most IgA-deficient individuals have no obvious clinical symptoms, although a minority of IgA-deficient individuals experience respiratory infections, malabsorption and autoimmune disorders. In most IgA-deficient individuals, IgM antibodies associate with secretory component and replace IgA antibodies as the predominant secretory isotype. In at least some of the IgA-deficient patients who have frequent infections, this substitution in secretions of IgM for IgA fails to occur, leading to a more global immune defect.

3.5 Cytokines and B cell activators direct immunoglobulin class switching

Immunoglobulin class switching is not a random process, but can be directed towards expression of particular immunoglobulin classes and subclasses by distinct cytokines and modes of B cell activation. Targeting of particular C_H genes for switch rearrangement is accomplished through the ability of various stimuli to selectively induce C_H gene transcription, prior to switching, thus making the C_H gene accessible to the recombination machinery. This transcriptional regulation by cytokines and B cell activators is mediated through promoters, termed I regions, that are located 5' to each S region. Thus, each I region has distinct sequences which lead to differential binding of regulatory factors. The capacity of different cytokines and B cell activators to selectively induce the transcription of particular C_H genes, and indeed suppress the transcription of others, is based upon the particular signaling pathways that are triggered by each cytokine or B cell activator, and hence the particular regulatory factors that are generated for binding to I regions. Cytokines and B cell activators may also regulate switch recombination, in a manner that is independent of alterations in germline

C_H RNA expression, but the mechanism(s) for this unknown. A more detailed discussion of immunoglobulin class switch regulation will follow under the headings of the individual cytokines which play a role in this process.

3.6 Development of humoral responses

3.6.1 Subpopulations of human B cells

Using anti-IgD and anti-CD38 antibodies, four tonsillar B cell populations can be isolated by flow cytometry. Single-positive IgD$^+$ cells correspond to naïve follicular mantle cells, while single-positive CD38$^+$ cells correspond to germinal center (GC) cells, as also observed by immunohistology. Double-negative (IgD$^-$ CD38$^-$) cells correspond to the memory B cell population while double-positive (IgD$^+$ CD38$^+$) cells represent a combination of cells at a transitional stage between follicular mantle cells and GC cells, as well as IgD$^+$ GC cells (Table 2).

The naïve IgD$^+$ CD38$^-$ B cells express the markers detected on tonsillar follicular mantles including IgM and Bcl2. Based on their CD23 expression, these cells can be further subdivided into Bm1 (CD23$^-$) and Bm2 (CD23$^+$) populations. The immunoglobulin variable region gene transcripts do not display somatic mutations. Germinal center founder cells (Bm2') express the IgM and IgD of naïve B cells as well as the CD38, CD10, CD71, CD77 and Fas of GC B cells. Some of the transcripts display a low number of mutations. The GC IgD$^-$ CD38$^+$ B cells are large cells which are

■ **Table 2.** Characteristics of human B lymphocyte subsets

	Bm$_1$	Bm$_2$	Bm$_2'$	Bm$_3$	Bm$_4$	Bm$_5$
IgD	+	+	+	– (+)	– (+)	–
CD38	–	–	+	+	+	–
IgM	+	+	+	– (+)	– (+)	– (+)
CD23	–	+	–	–	–	–
CD77	–	–	+	+	–	–
Ki67	–	–	+	+	–	–
Somatic Mutation	–	–	– (+)	+	+	+
Apoptosis	–	–	+	+	+	–

– (+) indicates expression by a minor population
Bm$_1$ + Bm$_2$: naive B cells
Bm$_2'$: germinal center founder cells
Bm$_3$: germinal center centroblasts
Bm$_4$: germinal center centrocytes
Bm$_5$: memory B cells

prone to undergo apoptosis. Based on their CD77 expression they can be divided into CD77$^+$ centroblasts (Bm3 cells forming the dark zone of the germinal center) and CD77$^-$ centrocytes (Bm4 cells forming the light zone of the germinal center). Bm3–Bm4 IgM transcripts accumulate an average of 5.7 point mutations whereas the IgG transcripts accumulate an average of 9.5 point mutations. While IgG-expressing B cells can be found within the Bm3 subset, cytoplasmic DNA circles produced during the switch recombination process are restricted to the Bm4 population, suggesting that somatic mutation precedes the initiation of isotype switching and that switched cells within the Bm3 compartment are the result of cell recirculation through the germinal center reaction.

A fraction of CD38$^+$ centroblasts and centrocytes also express surface IgD whose transcripts display a very high number of somatic mutations. These cells have the Cμ locus deleted and cannot produce any isotype other than IgD.

The memory IgD$^-$ CD38$^-$ B cells mostly express sIgG and some sIgA. They are resting cells accumulating within the tonsil mucosal epithelium. Their immunoglobulin transcripts have numerous mutations. Plasma cells express high levels of CD38 and do not express CD20. Even though they express Bcl2, they undergo very swift apoptosis *in vitro* unless cultured over bone marrow stroma.

3.6.2 Current view of antigen-induced B cell maturation

During T cell-dependent immune responses, naïve B cells that carry specific antigen receptors are activated in association with antigen-specific T cells and interdigitating cells within the extrafollicular areas. The activated B cell blasts undergo either terminal differentiation towards plasma cells or become GC founder cells (Bm2') that will migrate into the primary follicles or the dark zone of established GCs of secondary follicles.

There, GC founder cells undergo clonal expansion and differentiation into centroblasts which form and sustain the dark zones. Then, point mutations are introduced into the immunoglobulin variable (V) region genes in a stepwise fashion during the course of clonal proliferation. Three types of mutants can be generated: high-affinity, low-affinity, and autoreactive mutants. These mutants migrate into the basal light zone of the GC. The survival of these somatic mutants here seems to be dependent on their binding to the low levels of antigen–antibody immune complexes that are exposed on the surface of follicular dendritic cells. While low-affinity mutants that do not bind follicular dendritic cell (FDC)-bound antigen die by apoptosis, high-affinity mutants pick up antigen, process it, and present it to GC T cells which are mainly localized within the apical light zone. Autoreactive mutant clones may be deleted within the GC by two mechanisms: (i) a high concentration of soluble antigen within GCs may directly kill the cells, or (ii) Fas-ligand-expressing T cells, present within GCs, may kill autoreactive B cells or low-affinity B cells when their antigen receptors are not engaged by FDC-bound antigens.

High-affinity mutants retrieve antigen from FDCs, process it and present it to GC T cells which are mainly localized within the apical light zone. This zone contains strong CD23-expressing FDC networks. Here, the selected high-affinity centrocytes (Bm4) induce T cells to express CD40-ligand, which are key elements for survival, proliferation and isotype switching. The deletion of sterile transcripts within centrocytes (Bm4), and the co-localization of centrocytes (Bm4) with GC T cells within the apical light zone and outer zone, suggest that the cognate T cell/B cell interaction that results in the expansion and then isotype switch of high-affinity centrocytes occurs here. Finally, the high-affinity, isotype-switched centrocytes differentiate into memory B cells (Bm5) in the presence of prolonged CD40-ligand signaling, and into plasma cells when CD40-ligand signaling is removed. During secondary humoral immune responses, recirculating memory B cells can be activated in extrafollicular areas, giving rise to plasma cells and GC founder cells.

4. Interleukin-2

4.1 T cell-independent humoral immunity

IL-2 stimulates B cell activation, proliferation and differentiation to immunoglobulin synthesis. Resting B cells express low or undetectable levels of IL-2Rα and thus are unresponsive to low concentrations of IL-2. Stimulation with *Staphylococcus aureus* strain Cowan (SAC) induces the expression of IL-2Rα and makes the B cell IL-2-responsive. Thus, addition of IL-2 to SAC-activated, but not resting B cells, induces the secretion of immunoglobulin and augments SAC-mediated proliferation. Similar results have been obtained using anti-immunoglobulin antibodies or the TI antigen Trinitrophenol (TNP)-Ficoll, suggesting that SAC acts, at least in part, through its ability to cross-link the B cell antigen receptor. The continued expression of IL-2Rα by SAC-activated B cells, and hence maintenance of the IL-2-responsive state, is dependent upon continued IL-2-mediated signaling. IL-2 can augment the proliferation and induce the immunoglobulin secretion of SAC-activated B cells obtained from peripheral blood, spleen, tonsil and lymph node, although lymph node B cells are less responsive.

4.2 T cell-dependent humoral immunity

While IL-2 does not augment (or only marginally augments) the proliferation and immunoglobulin synthesis of B cells activated solely through their CD40 antigen, it does so when either IL-10 or dendritic cells is/are added to the cultures. In fact dendritic cells do not act through the release of IL-10, but rather through the release of IL-12. Finally, IL-2 stimulates immunoglobulin synthesis by B cells activated by influenza-specific T cells in the presence of influenza virus. In this latter system, T cells are induced to express CD40-ligand which then confers IL-2 responsiveness upon CD40-expressing B cells by inducing their expression of IL-2Rα.

4.3 Physiological importance of IL-2

Collectively, these studies suggest that IL-2 may play an inductive role in both T cell-independent and T cell-dependent humoral immune responses. The contexts in which IL-2 is critical for *in vivo* humoral immune responses in humans are not known. In this regard, a number of *in vitro* systems using SAC or activated, fixed T cells suggested a necessary role for IL-2 in induction of immuno-globulin synthesis. Nevertheless, B cells activated through CD40 in the presence of IL-4 or IL-10 can secrete immunoglobulin in the absence of IL-2, suggesting redundant pathways for B cell maturation. Indeed, as discussed below, studies on SAC-activated B cells suggest that IL-2 is perhaps more important in promoting B cell proliferation, whereas IL-10 is a more potent differentiation factor. Clinically, intravenous IL-2 has shown limited usefulness in promoting immunoglobulin secretion in patients with hypogammaglobu-linemia. Neoplastic B cells may also be regulated by IL-2. Thus B-CLL cell lines constitutively express IL-2R and can directly respond to IL-2 with enhanced proliferation and immunoglobulin secretion.

4.4 IL-2 interactions with other cytokines

As with other cytokines, IL-2 must be appreciated not in isolation, but as part of a complex cytokine network, with numerous functional interactions. Many of these cytokines appear to increase IL-2R expression and hence mediate their effects, at least in part, by augmenting IL-2 responsiveness. Inhibition of immunoglobulin secretion by anti-IL-R antibody underscores the critical role for IL-2 in many of these effects. Some of these key interactions are summarized in Table 3.

4.5 Immunoglobulin isotype production

IL-2 can promote the secretion of immunoglobulin of all isotypes in its capacity to act as a maturation factor for post-switch B cells; the immunoglobulin class switch itself being dependent upon the action of other cytokines and B cell activators. However, IL-2 may also play some role in class switching. Specifically, IL-2 induces germline C_H RNA specific for IgG and IgE in highly-purified SAC-activated adult B cells, thus targeting those genes for switch rearrangement. In another study, it was shown that IL-2 stimulates germline $C_H\gamma2$ RNA and up-regulates germline $C_H\gamma1$ RNA in IgM⁺ adult peripheral blood lymphocytes (PBLs) activated with SAC.

4.6 Murine studies

In the mouse, IL-2 by itself does not appear to directly induce immunoglobulin secretion by B cells activated through membrane immunoglobulin cross-linking, but does act as a powerful co-stimulant for immunoglobulin synthesis in the presence of other cytokines such as IL-3, granulo-cyte–macrophage colony-stimulating factor (GM-CSF) and interferon gamma (IFNγ). In contrast, IL-2 does not co-stimulate immunoglobulin secretion by murine B cells activated with CD40-ligand, even in the presence of other cytokines.

4.7 Interleukin 15, an IL-2-like cytokine

IL-15 is a recently discovered cytokine which mediates many of the same functional effects as IL-2. Their functional similarities are based on their both using the same β-chain and γ-chain for receptor signaling, but each having distinct α-chains for ligand binding. Whereas IL-2 is produced predominantly by T cells, IL-15 has a wide cellular distrib-

▌ **Table 3.** Interactions of IL-2 with other cytokines

Cytokine	Effect(s) when combined with IL-2
IL-1	Enhances proliferation and immunoglobulin secretion by SAC + IL-2-activated B cells through up-regulation of IL-2R.
IL-3	Increases immunoglobulin secretion by SAC + IL-2-activated B cells.
IL-4	Is antagonistic with IL-2 in many contexts. IL-4 suppresses proliferation and differentiation of B cells activated with (i) SAC + IL-2 $\pm$ IL-6, (ii) anti-immunoglobulin antibodies + IL-2, and (iii) antigen + IL-2. Additionally, IL-2 inhibits IL-4-mediated IgG4 and IgE production in vivo. These antagonistic effects may reflect the possibility of IL-2R/IL-4R interactions in the membrane through sharing of a common gamma chain (γc).
IL-5	Enhances immunoglobulin secretion by SAC + IL-2 or T-cell-activated B cells by making such cells more IL-2-responsive. This is accomplished through IL-5 up-regulation of IL-2R.
IL-6	Augments the synthesis of immunoglobulin by B cells stimulated with SAC + IL-2, through an apparent effect on differentiation.
IL-10	Augments the proliferation and immunoglobulin secretion of B cells activated through their antigen receptor or their CD40 receptor.
IL-12	Synergizes with IL-2 for proliferation and immunoglobulin secretion by SAC as well as CD40-activated B cells. This may be mediated in part through IL-12 up-regulation of IL-2Rβ on SAC or SAC + IL-2-activated B cells.
TNF-α	Enhances proliferation and immunoglobulin secretion by SAC-activated B cells through induction of increased IL-2 responsiveness.

ution, with the highest levels of IL-15 mRNA being detected in monocytes, epithelial cell lines, muscle and placenta. Similar to the effects of IL-2, IL-15 co-stimulates proliferation of B cells that have been activated with anti-IgM antibody or phorbol ester, but does not by itself stimulate DNA synthesis in resting B cells. IL-15, like IL-2, can also induce IgM, IgG1 and IgA production by CD40-ligand-activated B cells, but unlike IL-4 (see below) does not stimulate the secretion of IgG4 or IgE. IL-15 appeared to have activity in B cell proliferation and differentiation assays that was comparable to that of IL-2. Given the numerous non-T cell sources of IL-15, one may speculate on a potential role of IL-15 in T cell-independent humoral immune responses.

5. Interleukins 4 and 13

IL-4 and IL-13 exhibit closely overlapping functional effects on B cells. Although these cytokines bind to different receptors, these receptors have a common component important in cell signaling.

5.1 Phenotypic changes

CD23 (FcεRII) is an intermediate affinity receptor for IgE and exists in two forms, FcεRIIa and FcεRIIb. Resting B cells express only low levels of FcεRIIa which are enhanced by IL-4 or IL-13. IL-4 and IL-13 also induce the expression of FcεRIIb. Release of the soluble form of CD23 has been shown to play a role in enhancing IL-4-mediated IgE production. Specifically, CD23 is a ligand for CD21 which is expressed on B cells and serves as a receptor both for complement (complement receptor type 2) and for Epstein–Barr virus (EBV). Triggering of CD21 on B cells with recombinant CD23 up-regulates IL-4-induced IgE synthesis. CD23 may also serve to focus IgE–antigen complexes on to B cells for more efficient presentation to T cells. CD23 and CD21 also function as adhesion molecules for homotypic B cell aggregation and their interaction was found to be required for presentation of soluble protein antigen by lymphoblastoid B cell lines to specific CD4$^+$ T cell clones.

IL-4 and IL-13 also enhance the B cell expression of MHC class II molecules, CD71 (transferrin receptor) and CD72. Further, IL-4 augments B cell expression of CD40, a major activation molecule for co-stimulating proliferation, differentiation and immunoglobulin class. IL-4 and IL-13 may also play a role in enhancing T cell/B cell interactions through its ability to up-regulate membrane IgM and B7-1.

5.2 Proliferation

Both IL-4 and IL-13 stimulate proliferation by B cells activated either through membrane immunoglobulin cross-linking or through CD40 activation. However, IL-13 is significantly less effective than IL-4 at stimulating anti-IgM-induced proliferation. IL-4 has also been shown to augment DNA synthesis by antigen-activated B cells. Under certain conditions, IL-4 may also be suppressive for B cell proliferation. Thus, IL-4 inhibits IL-2-mediated B cell proliferation and can induce apoptosis in B-cell precursors.

5.3 Immunoglobulin isotype regulation

IL-4 induces IgG4 and IgE production by B cells derived from blood, tonsils and spleen. IL-4 also induces IgE by immature, cord blood B cells. Thus, the absence of IgE in cord blood and in newborns cannot be explained by an intrinsic inability of such cells to switch to IgE. In this regard, in contrast to adult peripheral blood mononuclear cells (PBMCs), cord blood mononuclear cells failed to produce IL-4 *in vitro* after appropriate activation possibly accounting for the absence of IgE in the newborn.

The ability of IL-4 to stimulate IgG4 and IgE production by naïve IgD^+ B cells in the presence of activated $CD4^+$ T cells indicates that IL-4 stimulates class switching to these immunoglobulin isotypes and not outgrowth of pre-switched cells. A similar observation has been made for IL-13. IL-4 and IL-13 act independently to induce IgE synthesis, since neutralizing anti-IL-4 antibody does not block this IL-13 action. However, IL-13 is two to five times less effective than IL-4 in inducing IgE synthesis *in vitro*. When used at optimal concentrations IL-4 and IL-13 were not found to act either synergically or additively. Mouse B cells, which also switch to IgE, as well as IgG1, in response to IL-4, fail to respond to any of the B cell-stimulatory actions of IL-13, perhaps due to their lack of IL-13 receptors.

IL-4 is sufficient for inducing switching to IgE by B cells activated with CD40-ligand and indeed, blocking CD40/CD40-ligand interactions between B cells and activated, IL-4-producing T cells, inhibits IgE synthesis. The switch to IgE, under certain conditions, may be preceded, within the same B cell or its progeny, by switching to IgG (i.e. sequential IgM to IgG to IgE switching). The ability of IL-4 to target $C_H\varepsilon$ rearrangement lies in its ability to induce transcriptional activation of the germline $C_H\varepsilon$ gene, prior to switch recombination,

which is manifest as an induction of germline (sterile) $C_H\varepsilon$ RNA. Thus, IL-4 can selectively confer a state of accessibility upon the germline $C_H\varepsilon$ gene for the switch recombination machinery.

The relative roles of IL-4 and IL-13 in atopic diseases remains to be determined. Nevertheless, PBMCs obtained from patients with allergic disease associated with high serum levels of IgE, produce greater amounts of IL-4, and less amounts of the IL-4 inhibitor, $IFN\gamma$ (see below) than cells from non-allergic donors. Further, local allergen challenge of patients with bronchial asthma, induces IL-4 by bronchoalveolar T cells, but not by cells from healthy donors. Allergen-induced late phase cutaneous reactions in atopic patients are also characterized by elicitation of IL-4 synthesis. Indeed, CD4+ T cell clones derived from atopic patients are typically of the IL-4- and IL-13-producing Th2 type. The importance of IL-4 in mediating IgE responses is further underscored by studies in the mouse, in which IgE production in response to parasite infection or to immunization with a TD antigen can be completely blocked with anti-IL-4 and anti-IL-4R antibodies or soluble IL-4 receptor.

A number of cytokines can *inhibit* IL-4-mediated induction of switching to IgE. These include $IFN\alpha$, $IFN\gamma$, TGF-β, IL-8 and IL-12, all of which have been shown to inhibit IL-4-induced IgE synthesis *in vitro* by human B cells. Of these cytokines, only TGF-β mediates sustained suppression of IL-4-mediated germline C_H RNA expression, indicating that it directly inhibits the switch to IgE. The other cytokines appear to act in an indirect fashion to inhibit IgE, by suppressing the function of the IL-4-producing $CD4^+$ T cell. IL-10 may also indirectly inhibit IL-4-induced IgE synthesis by freshly isolated PBMCs by inhibiting the accessory cell function of monocytes for T cells. In the mouse, $IFN\alpha$ or $IFN\gamma$ can inhibit *in vivo* IgE responses.

Other cytokines can *enhance* IL-4-mediated IgE synthesis; these include TNF-α, IL-5 and IL-6. The enhancing effect of TNF-α was associated with an increase in germline $C_H\varepsilon$ RNA, suggesting that it augmented switching to this immunoglobulin isotype. That, and its B cell growth-promoting properties could account for the increase in IgE synthesis. The enhancing effect of IL-5 was seen with suboptimal doses of IL-4, but the mechanism(s) of its action on IgE induction are not fully understood. IL-6-mediated enhancement of IgE synthesis appears to reflect its inductive effects on B cell growth and differentiation and is manifested even with optimal concentrations of IL-4.

5.4 Immature B cells

Several studies on the effect of IL-4 on human pre-B cell proliferation have been contradictory, with stimulation, suppression or no effect seen. The basis for these seemingly conflicting observations may lie with the particular *in vitro* systems employed, and suggest that additional stimuli can modulate the IL-4 effect on the immature B cell. As observed for mature B cells, IL-6 enhanced and IFNα, IFNγ, TGF-β and IL-12 inhibited IL-4 or IL-13, T cell-driven IgE synthesis by immature B cells. IgE responses to IL-13 were typically five- to 15-fold lower than that observed for IL-4, whether immature B cells were activated with anti-CD40 antibodies or CD4$^+$ T cells. However, IL-13-induced proliferation of immature B cells was only slightly lower than that observed for IL-4.

6. Interleukin-6

6.1 Differentiation

The major direct role of IL-6 in humoral immunity appears to be the maturation of B cells into immunoglobulin-secreting plasma cells. That IL-6 acts late as a maturation factor in the B cell activation program is reflected by the presence of IL-6R on activated, but not resting B cells and its ability to induce IgM, IgG and IgA synthesis without affecting proliferation. As discussed in the section on IL-4, IL-6 may also play a key role in stimulating the IL-4-dependent synthesis of IgE, by promoting B cell maturation, but not immunoglobulin class switching.

Data regarding the mechanism of IL-6 induction of immunoglobulin secretion are beginning to emerge. Thus, IL-6 stimulation of immunoglobulin synthesis by a human B cell line was secondary both to transcriptional up-regulation of the immunoglobulin gene, and to selection of the secretory versus membrane-specific polyadenylation sites by transcription/pausing termination. Transcriptional activation of immunoglobulin genes by IL-6 is associated with induction of the B cell-enriched transcription factor Oct-2 which binds an octamer motif (ATGCAAAT) present in the immunoglobulin promoter and enhancer. NF-IL-6, which binds to E-box motifs also present in the immunoglobulin promoter and enhancer, has also been implicated in IL-6 activation of the immunoglobulin gene.

6.2 B cell tumor growth

The early recognition that IL-6 was also a growth factor for plasmacytoma cells and the increased occurrence of plasma cell neoplasms in chronic inflammatory states, where there is sustained synthesis of IL-6, has suggested an important role for this cytokine in the generation of plasma cell tumors, such as multiple myeloma. Indeed, it has been demonstrated that:

(1) multiple myeloma cells secrete IL-6;

(2) this IL-6 can act in an autocrine fashion to stimulate their growth; and

(3) addition of neutralizing anti-IL-6 antibodies to cultures of myeloma cells inhibits their spontaneous growth.

IL-6 has also been implicated in the autocrine growth of non-Hodgkin lymphomas, chronic lymphocytic leukemias and acute myeloid leukemias.

6.3 Autoantibody production

In addition, the ability of IL-6 to stimulate immunoglobulin synthesis suggested a possible role for IL-6 in various pathological conditions associated with autoantibody production. Thus, cardiac myxoma (a benign intra-atrial neoplasm) is associated with hypergammaglobulinemia and autoantibody production. Removal of the tumor leads to resolution of the excess antibody synthesis. The observation that a cardiac myxoma made large amounts of IL-6 suggested a mechanism for this clinical association. Likewise the presence of elevated levels of IL-6 in the synovial fluid of patients with rheumatoid arthritis could help to account for the polyclonal plasmacytosis and autoantibodies seen in this inflammatory, autoimmune disease. In this regard, IL-6 transgenic mice exhibit hypergammaglobulinemia and polyclonal plasmacytosis. IL-6 may not however, be critical for stimulating immunoglobulin synthesis under many conditions, since IL-6-knockout mice, while exhibiting diminished percentages of IgA$^+$ plasma cells in the lamina propria of the gastrointestinal tract, can still mount immunoglobulin secretory responses to a number of immunogens. This underscores the existence of redundant pathways for stimulating humoral immune responses.

7. Interleukin-10

7.1 Survival

IL-10 promotes survival and enhances MHC class II expression in resting murine B cells, but similar effects of IL-10 on human B cells have not been demonstrated. Of interest, IL-10 was found to promote apoptosis of B cell chronic lymphocytic leukemia cells, but not non-Hodgkin lymphoma or hairy-cell leukemia cells. IL-10 has also been shown to promote apoptosis of normal human B cells activated with SAC.

7.2 Growth

IL-10 co-stimulates the growth of human B cells at various stages in their maturation. Thus, IL-10 can augment proliferation of membrane immuno-globulin-negative fetal bone marrow B cell precursors activated through CD40 in the presence of IL-3. IL-10 also augments proliferation of mature B cells activated through membrane immuno-globulin, either with anti-immunoglobulin antibodies or SAC, but this effect is not as great as that seen with IL-2 or IL-4. However, IL-10 is a potent growth factor for CD40-activated B cells, being as effective as IL-4. Indeed, the additive effects of IL-4 plus IL-10 for growth of CD40-activated B cells is the most effective cytokine-mediated mitogenic signal observed for human B cells *in vitro*. Although IL-2, by itself, is not an effective growth signal for CD40-activated B cells, it is co-stimulatory when present with IL-10. This may be due to the ability of IL-10 to up-regulate IL-2Rα on CD40-activated B cells, making them more responsive to the growth-promoting effects of IL-2. In contrast to IL-2 or IL-4, IL-10 can augment the growth of B cells exposed to EBV. Indeed, IL-10 acts as an autocrine growth factor for EBV-transformed B cells, in that anti-IL-10 antibodies added to culture, suppress their growth. In distinct contrast to what is observed for human B cells, IL-10 strongly suppresses DNA synthesis in murine B cells activated with anti-immunoglobulin antibodies or LPS.

7.3 Differentiation

Human B cells activated with SAC secrete immunoglobulins in response to IL-10. In comparative studies, IL-2 has been found to be more effective than IL-10 for promoting growth of SAC-activated B cells, whereas the reverse is true for induction of differentiation to immuno-globulin secretion. IL-10 is an even greater immunoglobulin-inducing factor for CD40-activated B cells, capable of promoting terminal maturation of B cells into plasmablasts. This explains the limited growth capacity of IL-10-containing cultures of CD40-activated B cells.

IL-10 stimulates immunoglobulin secretion by both IgM$^+$ and post switched B cells. Finally, IL-10 induces antigen-specific immunoglobulin secretory responses by antigen-primed memory B cells.

7.4 Immunoglobulin isotype production

Although cytokine-dependent induction of immunoglobulin isotype switching has been observed for murine B cells activated by a number of distinct means (e.g. by LPS, anti-CD40 or CD40-ligand, and dextran-conjugated anti-immunoglobulin antibodies), most studies of human immunoglobulin isotype switching have involved CD40-activated human B cells. Indeed, there is no evidence to date that membrane immunoglobulin crosslinking or LPS activation can induce switching in human B cells. The physiological relevance of CD40 activation for isotype switching in human B cells is underscored by the failure of patients with hyper-IgM syndrome to express significant amounts IgG, IgE and IgA. These patients have mutations in the gene encoding CD40-ligand.

Recent studies have indicated that IL-10 can promote switching to IgG1 and IgG3 in CD40-activated human B cells. Thus, addition of IL-10 to CD40-activated naïve IgD$^+$ B cells selectively induced the secretion of IgG1 and IgG3 *in vitro*, and this effect was further enhanced by addition of SAC. The switch-promoting activity of IL-10 was confirmed in molecular studies demonstrating the induction of switch recombination products by IL-10. IL-10 selectively increased germline Cγ1 and Cγ3 RNA expression by CD40-activated B cells, suggesting that it specifically targets these genes for switch recombination. IL-10 also regulates immunoglobulin class switching in the mouse, but in a distinct manner. Thus, IL-10 stimulates switching to IgG3 by LPS-activated murine B cells *in vitro*, while inhibiting switching to IgA by cells activated with LPS plus TGF-β in the presence of IL-4, IL-5 and anti-immunoglobulin–dextran antibodies.

7.5 Physiological role of IL-10

IL-10 is a cytokine produced by many cell types, including B cells. Although IL-10 is not detected in the serum of healthy individuals, it may be found in biological fluids from patients with certain pathological conditions. Thus, serum IL-10

∎ **Table 4.** Induction of isotype switching in the mouse

Cytokine	LPS (TI-1)	Anti-IgD-dextran (TI-2)	T cell	CD40 (TD)
IL-4	IgG1, IgE	IgG1	IgG1, IgE	IgE
IFN-γ	IgG2a	IgG2a, IgG3	IgG2a, IgG3	ND
TGF-β	IgA, IgG2b	IgA	IgA, IgG2b	ND

∎ **Table 5.** Induction of isotype switching in the human

Cytokine	Anti-CD40
IL-4	IgG4, IgE
IL-13	IgG4, IgE
TGF-β + IL-10	IgA1, IgA2
IL-10	IgG1, IgG3, IgA1
Activated T cell supernatant	IgG2

- SAC + IL-2;
- anti-CD40 antibody + PMA;
- CD40-ligand expressed by transfected fibroblasts;
- anti-CD20 + B cell growth factor; and
- EL-4 murine T thymoma cells + IL-1β.

Human B cell tumors may also produce, and be sensitive to, the growth-inhibitory properties of TGF-β.

TGF-β appears to act during the G_1 phase of the cell cycle to inhibit DNA synthesis and thus must be added early to culture for maximal anti-proliferative activity. It does not typically cause cell cycle arrest, but rather delays the progression of B cells through the cell cycle. However, under certain circumstances TGF-β may induce apoptosis; resting peripheral blood B cells cultured with TGF-β showed greater percentages of apoptotic cells than B cells cultured in medium alone. IL-4 can partially rescue these TGF-β-treated B cells from undergoing apoptosis. Further, TGF-β was able to induce cell cycle arrest and apoptosis in a number of murine B cell lymphomas. The TGF-β-induced effect on the cell cycle may entail its regulation of expression or activity of a number of signaling molecules involved in cell growth including the oncoprotein c-Myc, the anti-oncogene Rb (the retinoblastoma protein), the transcription factor E2F, and members of the family of cyclin-dependent kinases (CDKs).

has been detected in a substantial minority of patients with non-Hodgkin lymphoma and this is correlated with decreased survival. Lymphoma cells themselves have been shown to produce IL-10. Hence, IL-10 may be an autocrine factor for certain human B cell malignancies. Patients with early-, but not late-stage, multiple myeloma also have detectable serum IL-10, but in contrast to non-Hodgkin lymphoma, this is associated with a good prognosis.

IL-10 may also play a role in promoting autoantibody production. Thus, administration of anti-IL-10 antibodies into SCID (severe combined immune deficiency) mice, injected with mononuclear cells from SLE patients, inhibited the production of both autoantibodies and total human IgG *in vivo*. The spontaneous *in vitro* secretion of immunoglobulin by SLE-derived human mononuclear cells was also inhibited by anti-IL-10 antibodies.

8. Transforming growth factor beta

8.1 Proliferation

TGF-β is an anti-proliferative cytokine for both human and murine B cells, but does not reduce B cell viability. Thus, it inhibits the proliferation of human B cells activated with:

8.2 B cell development

TGF-β has been shown to inhibit the generation of membrane immunoglobulin-positive cells expressing λ, but not κ, chains in cultures of human bone marrow pre-B cells. Somewhat differing results were observed using murine bone marrow precursor B cells, in which TGF-β inhibited the development of membrane Igκ^+ B cells. Igλ^+ cells were not studied. TGF-β was also found to inhibit the generation of pre-B cells in this system.

8.3 B cell differentiation

B cells activated with SAC plus IL-2 showed reduced expression of membrane IgM, IgD, IgA, κ and λ chains in the presence of TGF-β. Further, TGF-β inhibited the secretion of IgM and IgG in this system. The inhibition of κ-chain expression was associated with a reduction of κ mRNA and κ gene transcription. In contrast, the TGF-β reduction of IgM secretion was associated with a marked loss of secretory μ mRNA, with only a modest reduction in the membrane form. In murine systems TGF-β can also inhibit the secretion of immunoglobulin of other isotypes as well. The inhibition of differentiation of B cells by TGF-β is independent of its anti-proliferative effects. A possible role for the transcription factors Oct-2 and AP-1 in TGF-β inhibition of differentiation has been suggested.

8.4 IgA class switching

TGF-β can promote switching to IgA in both murine and human B cells. Thus, tonsillar membrane IgD$^+$ human B cells cultured with anti-CD40, SAC and IL-10 can be induced to secrete increased IgA in response to TGF-β. These studies, using limiting dilution, indicated that TGF-β increased the precursor frequency of IgA-producing cells by over four-fold. Anti-IgM was able to substitute for SAC in this system. The requirement for IL-10 in this system probably reflected its ability to stimulate B cell maturation to immunoglobulin secretion. Under these conditions, naïve B cells are essentially induced to secrete IgA$_1$. The secretion of IgA$_2$ requires addition to B cell cultures of dendritic cells whose molecular contribution remains to be determined. Extensive data in the mouse also indicates a key role for TGF-β in inducing class switching to IgA. The mechanism of TGF-β induction of IgA appears to lie in its ability to induce germline $C_H\alpha$ RNA expression in both murine and human B cells. As yet, experimental evidence demonstrating a role for TGF-β for *in vivo* IgA class switching is lacking, but this role is strongly predicted from *in vitro* studies. In this regard, IgA deficiency is the most common primary immunodeficiency in humans. B cells from these mice lack germline $C_H\alpha$ RNA and evidence for $S\mu$–$S\alpha$ recombination relative to controls. However, the observation that PMA and TGF-β together can induce germline $C_H\alpha$ RNA in B cells from IgA-deficient patients, at levels comparable to that seen in controls, suggests that these patients may have a defect in producing active TGF-β.

9. Conclusion

Tables 4 and 5 show the various cytokines that control isotype switch in mouse and human.

References

Banchereau, J., de Paoli, P., Valle, A. *et al.* (1991). Long term human B cell lines dependent on interleukin 4 and anti-CD40. *Science* **251**, 70–72.

Clark, E. A. and Ledbetter, J. A. (1994). How B and T cells talk to each other. *Nature* **367**, 425–8.

Finkelman, F. D., Holmes, J., Katona, I. M. *et al.* (1990). Lymphokine control of *in vivo* immunoglobulin isotype selection. *Ann. Rev. Immunol.* **8**, 303–34.

Kelsoe, G. (1995). *In situ* studies of germinal center reaction. *Adv. Immunol.* **60**, 267–88.

Kipps, T. J. (1997). Human B cell biology. *Int. Rev. Immunol.* **15**, 243–64.

Kopf, M., Le Gros, G., Coyle, A. J., Kosco-Vilbois, M. and Brombacher, F. (1995). Immune responses of IL-4, IL-5, IL-6 deficient mice. *Immunol. Rev.* **148**, 45–69.

Liu, Y.-J. and Banchereau, J. (1996). The paths and molecular control of peripheral B cell development. *Immunologist* **4**, 55–66.

MacLennan, I. C. (1994). Germinal centers. *Annu. Rev. Immunol.* **12**, 117–39.

Mond, J. J., Lees, A., and Snapper, C. M. (1995). T cell-independent antigens type 2. *Annu. Rev. Immunol.* **13**, 655–92.

Parker, D.C. (1993). T cell-dependent B cell activation. *Annu. Rev. Immunol.* **11**, 331–60.

Rajewsky, K. (1996). Clonal selection and learning in the antibody system. *Nature* **381**, 751–8.

Snapper, C. M. (1997). Immunoglobulin class switching. In Paul, W. E. (ed.), *Fundamental Immunology*, 4th edn. Raven Press, New York (in press).

Snapper, C. M. and Mond, J. J. (1993). Towards a comprehensive view of immunoglobulin class switching. *Immunol. Today* **14**, 15–17.

Snapper, C. M. and Mond, J. J. (1996). A model for induction of T cell-independent humoral immunity in response to polysaccharide antigens. *J. Immunol.* **157**, 2229–33.

Snapper, C. M., Rosas, F., Moorman, M. A., Jin, L., Shanebeck, K., Klinman, D. M., Kehry, M. R., Mond, J. J. and Maliszewski, C. R. (1996). IFN-γ is a potent inducer of Ig secretion by sort-purified murine B cells activated through the mIg, but not the CD40, signaling pathway. *Int. Immunol.* **8**, 877–85.

Snapper, C. M., Marcu, K. B. and Zelazowski, P. (1997). The immunoglobulin class switch: Beyond 'accessibility'. *Immunity* **6**, 217–23.

Stavnezer, J. (1996). Antibody class switching. *Adv. Immunol.* **61**, 79–146.

Van Kooten, C. and Banchereau, J. (1996). CD40–CD40 ligand: a multifunctional receptor–ligand pair. *Adv. Immunol.* **61**, 1–77.

XVIII Cytokines and the cellular mechanism of inflammation

Jean-Marc Cavaillon and Gordon Duff

1. The inflammatory response

The inflammatory process is a physiological, rapid and temporary response to any encountered stress, including infection, foreign substances, chemical injury, radiation, trauma, burns or cold. Well recognized by the Egyptians in 1650 BC, inflammation was perfectly described by the Roman Cornelius Celsus during the first century: '*Notae vero inflammationis sunt quatuor: rubor et tumor cum calore et dolore*' ('In fact there are four signs of inflammation: redness and swelling with heat and pain.') It took twenty centuries to characterize the numerous mediators which contribute to the development and control of inflammation. The acute inflammatory response involves :

- vasodilation of capillary vessels (congestion)

- exudation of plasma proteins (oedema)

- adherence of circulatory leukocytes to endothelium

- chemoattraction of leukocytes and a local activation

- release of numerous mediators

- elimination of the foreign substance (phagocytosis)

- elimination of the recruited cells (apoptosis), and

- a healing process.

The inflammatory response results from a subtle balance between amplification and down-regulation (Figure 1). In the case of chronic inflammation, the inflammatory process is perpetuated, while the healing process may be exaggerated, leading to fibrosis. The local inflammation is accompanied by a systemic response as assessed by the presence in the bloodstream of acute-phase proteins released by hepatocytes upon activation by cytokines. After a lesion has occurred within connective tissue, the first cells to become involved include mast cells, fibroblasts and platelets from the injured vessels, while mediators — mainly histamine, serotonin and prostaglandins — are released. Other molecular processes are initiated, such as complement activation, which leads to the release of the C3a and C5a anaphylatoxins. Activation of the Hageman factor, in contrast, ends in the release of kinins. The combined effect of these factors is to favour vasodilation, increase the permeability of the endothelium, chemoattract circulating cells and induce smooth muscle contraction. These mediators also act on nerve-endings. leading to pain.

Later the endothelial cells become involved. Their increased permeability is responsible for the exudation of plasma proteins; their activation is associated with increased expression on their surface of adhesion molecules to which circulating cells bind. The chemoattractants released in the inflammatory foci recruit circulating neutrophils, eosinophils, basophils, monocytes, platelets and lymphocytes into the tissues. Red cells are also found within the injured tissues. Leukocytes are locally activated and release many more mediators, including cytokines, proteases, eicosanoids, platelet-activating factor (PAF), free radicals (O_2^- and NO) and heat-shock proteins, which can further induce a cascade phenomenon and an auto-amplification loop.

The initiation of the inflammatory process is very rapid. For example, following injection of interleukin 1 (IL-1) or anaphylatoxin into rabbit skin, the concentration of plasma protein in the tissues reaches a peak within 2h and that of neutrophils within 3h. Indeed, all steps of inflammation can be initiated by two major cytokines, namely IL-1 and tumour necrosis factor (TNF). Synergy between cytokines and amplification mechanisms is closely linked to the process.

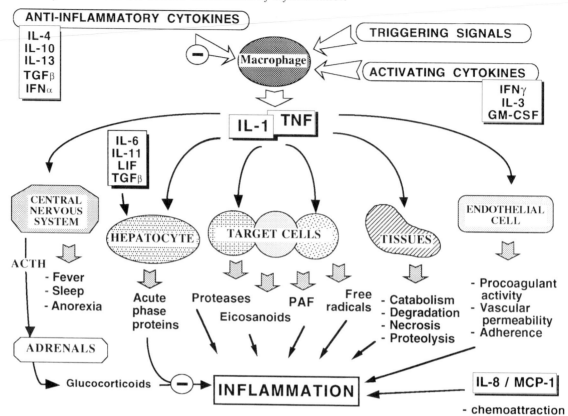

■ **Figure 1.** Inflammation is the consequence of a cascade of events initiated by IL-1 and/or TNF. The action of these cytokines on various target cells leads to the release of numerous inflammatory mediators. Anti-inflammatory cytokines, the activation of the neuroendocrine pathway and the effects of glucocorticoids as well as the enhanced production of acute-phase proteins control negatively the inflammatory process. (Adapted from Cavaillon, J.-M. (1996) *Les Cytokines.* Masson).

Gamma interferon (IFNγ) is one of the main amplifying cytokines, due to its capacity to activate the pro-inflammatory activities of macrophages (enhanced production of IL-1, TNF, IL-6, IL-8, and free radicals). Many other cytokines also contribute to the inflammatory response.

2. Induction of pro-inflammatory cytokines

When inflammation is initiated by an infectious process, the presence of microorganisms and their derived products (membrane compounds, released toxins, intracellular constituents following lysis) are potent activators of cytokine production. Macrophages are probably one of the major sources of cytokines. One substance derived from Gram-

negative bacteria, namely endotoxin or lipopolysaccharide (LPS), is a potent inducer of cytokines. During infection by Gram-positive bacteria, membrane compounds such as peptidoglycan or lipoteichoic acid are strong inducers. In addition, exotoxins behave as superantigens and trigger the release of T-cell-derived cytokines. Other cells can contribute to the release of cytokines. For example, the beneficial TNF observed in experimental peritonitis has been demonstrated to be released by mast cells (Echtenacher *et al.*). Thus, in addition to the well known contribution of mast cells to inflammation observed in acquired immunity via specific IgE antibodies, these cells play a central role during natural immunity by their capacity to release pro-inflammatory mediators in response to bacteria. Obviously, when the host is facing a stressful condition, many defence systems are activated, leading to the generation of newly synthesized or

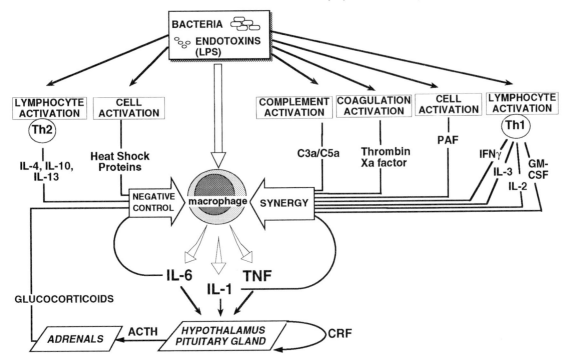

∎ Figure 2. The production of inflammatory cytokines by monocytes/macrophages upon activation by bacteria or microbially derived products involves many co-signals. These signalling events initiated during the infectious process can either amplify or inhibit the release of these cytokines. (Adapted from Cavaillon, J.-M. (1997). In Mège, J.L., Raoult, D. and Revillard, J.P. (eds), *Immunité et Infection*. Arnette.)

newly generated compounds as well as the release of preformed mediators. These factors have the capacity to modulate the generation of cytokines. As shown in Figure 2, the activation of the complement system and the activation of the coagulation cascade lead to the appearance of factors (anaphylatoxins, thrombin, factor Xa) which can further enhance the release of the pro-inflammatory cytokines by activated macrophages. Once generated, cytokines can perpetuate their own production. In addition, certain cytokines — such as those generated by T lymphocytes (e.g. IL-3 and GM-CSF) and particularly Th1 cells (i.e. IL-2 and IFNγ) — can amplify the release of the pro-inflammatory cytokines. A negative feedback involves the activation of the neuro-endocrine system which leads to the generation of corticotropin-releasing factor (CRF) by the hypothalamus, adrenocortinotrophic hormone (ACTH) by the pituitary gland and glucocorticoids by the adrenals. Other negative signals are due to heat-shock proteins. Finally, some of the Th2-derived cytokines can inhibit the release of pro-inflammatory cytokines.

3. Detection of cytokines in inflammatory diseases

3.1 Local production

Before techniques were available for identifying cytokines within tissues, the presence of cytokines was demonstrated within the fluids in close contact with inflammatory foci. In 1982, IL-1 bioactivity was demonstrated in crevicular fluids of patients with periodontal inflammation (Charon *et al.*) and in synovial fluids of patients with rheumatoid arthritis (Fontana *et al.*). Later, when direct measurement of cytokines became possible, the presence of IL-1 in these fluids was confirmed by enzyme-linked immunosorbent assay (ELISA) or radioimmuoassay (RIA). The presence of many other cytokines was been reported in synovial fluids and in other types of natural fluids associated with acute or chronic inflammation or infection (cerebrospinal fluid, pleural effusion, plasma, urine, sputum, stools). Cytokines can also be detected in fluids obtained artificially, e.g. by bronchoalveolar lavage (BAL) performed in patients with acute respiratory distress syndrome (ARDS) or asthma. The levels of detectable cytokines often correlate with the severity of the inflammatory disease. For example, it has been reported that high IL-8 levels in BAL correlated with poor outcome in ARDS patients (Miller *et al.*). In addition, a correlation between levels of IL-8, a well known chemokine for neutrophils (PMN), and the number of recovered PMN in BAL was observed. A correlation between high levels of TNF in plasma or cerebrospinal fluid and poor outcome was also reported for meningococcemia (Girardin *et al.* 1988). In sepsis syndrome, also known as systemic inflammatory response syndrome (SIRS), other cytokines such as IL-6, IL-8 or leukaemia inhibitory factor (LIF) correlate with the severity of the disease.

The detection of cytokines in inflammatory foci shows that their production by local cells increases. For example, higher than normal cytokine concentrations are found in BAL from patients with ARDS, asthma, pneumoconiosis or fibrosis. As assessed *ex vivo*, spontaneous release of pro-inflammatory cytokines by alveolar macrophages is greater than that by cells isolated from healthy volunteers. In contrast, in SIRS patients it has been reported that circulating monocytes and neutrophils have a reduced capacity to release cytokine upon *in vitro* activation (Muñoz *et al.*; Marie *et al.* 1988).

3.2 Systemic cytokines

Independently of the site of inflammation, a systemic response will be generated. An enhanced production of acute-phase proteins by hepatocytes and a reduced production of albumin and transferrin are initiated by many cytokines acting on hepatocytes. TNF, IL-1, TGFβ and the IL-6 super-family (i.e. LIF, ciliary neurotrophic factor (CNTF), IL-11, oncostatin M, cardiotrophin 1) are all capable of inducing the production of acute-phase proteins. In acute inflammation, circulating cytokines are detected very early. For example, the peak level of circulating TNF is observed 1.5 h after the injection of LPS, while the peaks of IL-1, IL-6 and IL-8 occur after 2 h. Similarly in trauma patients, during surgery, or following ischaemia–reperfusion, burn or haemorrhagic shock, circulating cytokines are detected very early. In most cases, IL-6 levels measured in plasma or in other biological fluids can be considered as a marker of severity and correlate with the degree of stress.

Acute-phase proteins are essentially protective and limit the inflammatory process. They have anti-protease activity (e.g. α-1 antitrypsin; α-1 antichymotrypsin) or scavenger activity (e.g. interaction of C-reactive protein (CRP) with membrane fragments, haptoglobulin with hemoglobin; ceruloplasmin with O_2^-, serum amyloid A with phospholipids . . .). Their protective properties have been demonstrated *in vivo*. For example, CRP can reduce the lung inflammation induced by instillation of C5a (Heuertz *et al.*), while α-1 acid glycoprotein can protect against a lethal injection of TNF. Interleukin 6 induces the largest panel of acute-phase proteins and IL-1 and TNFα can synergize with this cytokine. Compared with wild-type animals, IL-6 knockout animals produce reduced amounts of acute-phase proteins after local challenge with turpentine or systemic challenge with an injection of LPS (Kopf *et al.*). Certain acute-phase proteins can also be produced locally in various extrahepatic tissues (Meek *et al.*). Because of its capacity to induce mediators involved in the limitation of the inflammatory process, IL-6 can be considered as an anti-inflammatory cytokine. Furthermore, it has been reported that IL-6 can enhance the levels of circulating IL-1 receptor antagonist (IL-1ra) and soluble TNF receptors (sTNFR), two well known inhibitors of IL-1 and TNF, respectively. IL-6 can also inhibit the release of IL-1 and TNF (Schindler *et al.* 1990). IL-6 has been shown to be beneficial in endotoxic shock or during lethal streptococcal infections in animal models. IL-6's effects are not entirely beneficial, though: it is known to be responsible for osteolysis, anaemia and muscle atrophy.

4. Role of inflammatory cytokines

4.1 Activation of endothelial cells and leukocyte adherence to endothelium

As mentioned above, endothelial cells play an essential role in the inflammatory response. They constitute the interface between the injured tissues and the circulating leukocytes that need to be recruited. As well as being active producers, they are targets of the inflammatory mediators. In response to IL-1 and TNF, endothelial cells:

- synthesize phospholipase and cyclooxygenase, leading to the production of prostaglandins;

- express tissue factor, enhancing the coagulation process;

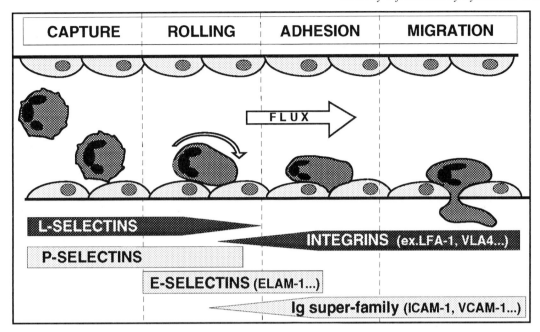

| CAPTURE | ROLLING | ADHESION | MIGRATION |

FLUX

L-SELECTINS

INTEGRINS (ex.LFA-1, VLA4...)

P-SELECTINS

E-SELECTINS (ELAM-1...)

Ig super-family (ICAM-1, VCAM-1...)

▮ **Figure 3.** Sequences of the events involved during the interaction of circulating leukocytes and endothelial cells. (Adapted from Cavaillon, J.-M. (1996). *Les Cytokines.* Masson.)

- release PAF, free radicals and a large panel of cytokines including IL-6 and IL-8.

In addition, these cells express on their surface enhanced levels of adhesion molecules, namely selectins and integrins. L- and P-selectins favour the capture of circulating leukocytes. The rolling of the cells on to the endothelium is mainly mediated by the E-selectins which, together with integrins, result in the firm adhesion of the leukocytes (Figure 3). Finally, integrins are involved in migration towards the tissues. The E-selectin ELAM-1 is expressed within 1 h following activation of endothelial cells by IL-1 or TNF. Monocytes and neutrophils, harbouring the counter-ligand Lewis X antigen, bind to the endothelium. VCAM-1 and ICAM-1 are expressed 4 h following IL-1 or TNF activation. *In vivo*, enhanced expression has been demonstrated in various inflammatory processes. For example, a high level of expression of ELAM-1 and ICAM-1 has been reported in skin biopsies from atopic patients locally challenged with the corresponding allergen (Kyan-Aung *et al.*). The counter-ligand for VCAM-1 is the VLA-4 molecule, expressed on the surface of monocytes and lymphocytes. The counter-ligands for ICAM-1 are the LFA-1 (CD11a/CD18) and CR3 (CD11b/CD18) molecules found on the surface of monocytes and neutrophils. The activation of circulating cells induces conformational changes in the LFA-1 and CR3 molecules, and this increases the strength of their interaction with their respective ligands.

Once leukocytes have adhered to endothelium, they will migrate in response to chemoattractant signals delivered by chemokines (Figure 4). Blocking the chemokines by specific antibodies is associated with reduction in the recruitment of circulating cells and reduced inflammation (Mulligan *et al.* 1993a). In infectious models, blocking chemokines is associated with a limited control of infection (Huffnagle *et al.*).

4.2 Cytokines and coagulation

As previously mentioned, IL-1 and TNF can induce pro-coagulant activity in endothelial cells; similar observations have been reported with monocytes. Pro-coagulant activity occurs following the membrane expression of tissue factor. Tissue factor combines with factor VII which, upon activation, initiates the coagulation cascade, activating factors X and IX. Injection of TNF in humans is associated with increases in the levels of factor-X-activated peptide and fragments 1 and 2 of thrombin (Van der Poll *et al.*). In fact, the initial response to TNF, as well as to IL-1, is activation of fibrinolysis, as assessed by increased levels of tissue plasminogen activator. Thrombin is released in response to

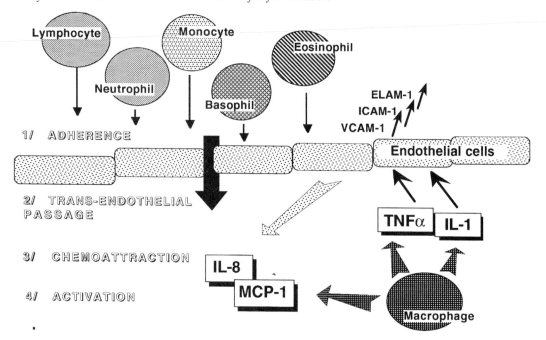

∎**Figure 4.** Leukocyte recruitment towards an inflammatory focus. (Adapted from Cavaillon, J.-M. (1996). *Les Cytokines*. Masson.)

this increased level of tissue plasminogen activator. TNF and IL-1 activate coagulation via the induction of tissue factor, whereas IL-6 can induce coagulation independently of the induction of tissue factor. In contrast to IL-1 and TNF, IL-6 cannot induce fibrinolysis.

4.3 Induction of lipid mediators and free radicals

Many cells, in response to activation by IL-1 or TNF, release mediators derived from the metabolism of arachidonic acid either via the cyclooxygenase pathway, leading to the release of prostaglandins and thromboxanes, or via the lipooxygenase pathway leading to the release of leukotriene. Arachidonic acid is itself generated from membrane phospholipids following the action of phospholipase A2. Neosynthesis of phospholipase A2 and inducible cyclooxygenase (COX2) occurs following the action of IL-1 or TNF. Prostaglandins are involved in smooth muscle contraction, mucosal oedema, increased vascular permeability, cellular infiltration and mucus secretion. Eicosanoids can also modulate the production of pro-inflammatory cytokines: For example, prostaglandin E_2 (PGE$_2$) inhibits TNF production,

but has no major effect on IL-1 and enhances IL-6 production, while thromboxane A2 favours TNFα and IL-1β production. Many of the activities generated by IL-1 and TNF can be suppressed by cyclooxygenase inhibitors (e.g. aspirin or ibuprofen), illustrating the contribution of PGE$_2$ to the activities initiated by IL-1 and/or TNF. The action of phospholipase A2 on phosphatidylcholine generates arachidonic acid and lyso-PAF. The latter is converted to PAF by acetyltransferase. This factor, which is known to activate and aggregate platelets, is also responsible for the release of vaso-active mediators, and increases permeability, vasoconstriction and bronchoconstriction.

IL-1 and/or TNF also induce endothelins. Four different endothelins have been described. They are short 21 amino acid peptides derived from a 39 amino acid precursor. They act (via a seven-transmembrane-domain receptor coupled to a G protein) on endothelial cells, leading to vasoconstriction and increased blood pressure.

Free radicals are also induced upon activation of the oxidative burst by IL-1 and TNF. Superoxide anion (O_2^-) generated upon the action of a membrane enzyme, NADPH oxidase, has antimicrobial activity but can also be toxic to nearby cells. It induces (i) the peroxidation of unsaturated fatty acids, which alters membrane fluidity and permeability, and (ii) the oxidation of amino acids, leading to alterations in proteins. Activation of the manganese superoxide dismutase by IL-1 and TNF generates hydrogen peroxide (H_2O_2). Many cell types, when activated by IL-1 and TNF (as well as by IFNγ and macrophage migration inhibitory factor (MIF)) produce inducible nitric oxide synthase (NOS 2), which produces nitric oxide (NO) from L-arginine. NO is toxic because of from its ability to inhibit glycolysis, the Krebs cycle, mitochondrial respiration, and DNA synthesis. It is very unstable, being rapidly transformed into nitrite (NO_2^-) in the presence of H_2O, or into nitrate (NO_3^-) by reaction with oxyhaemoglobin.

4.4 Cytokines and catabolism

IL-1 directly induces (i) degradation of cartilage and (ii) bone resorption, and this is reflected in one of its early names, catabolin and osteoclast activating factor. IL-6 is also involved in osteolysis, particularly during the post-menopausal period.

Muscle proteolysis is observed during inflammation. The release of amino acid from muscle increases the number of free amino acids available for the synthesis of inflammatory proteins. TNF and, to a lesser extent, IL-1, can induce this muscle proteolysis. The activation of neutrophils leads to the in degranulation and thus the release of proteases such as cathepsin and collagenase, which can degrade the extracellular matrix as well as other cellular constituents.

5. Negative control

5.1 Natural inhibitors and antagonists

IL-1 and TNF are the major directors of the inflammation process. As a consequence, natural specific inhibitors are directed against these two molecules. The main TNF inhibitors are the soluble forms of the two TNF receptors. They are naturally found in urine and plasma of healthy donors. Their circulating levels are increased by TNF infusion, following LPS injection and during any stressful conditions. Their levels often correlate with the severity of disease, as shown in plasma during severe meningococcaemia (Girardin *et al.* 1992) or in BAL during the development of ARDS (Evans *et al.*). Their beneficial effects have been demonstrated in various animal models either as recombinant sTNFRs or as chimeric molecules (Fcγ–sTNFR).

IL-1 is negatively controlled in a number of ways: by (i) a decoy receptor on the cell surface, (ii) two different soluble receptors and (iii) IL-1ra. The latter molecule is naturally present in plasma and its level increases during inflammation. It can combine with the IL-1 receptor without initiating any transducing signals. It competes with IL-1 for binding to the receptor and when IL-1ra is present at 100 times the concentration of IL-1 it can efficiently inhibit most IL-1-induced activities, including modification in haemodynamics, osteolysis, prostaglandin production, and can have beneficial effects in many inflammatory diseases such as rheumatoid arthritis, inflammatory bowel diseases, or sepsis. Many cells can produce IL-1ra, including those which synthesize IL-1. Many activators are shared, but it is worth noting that cytokines that can limit the production of IL-1 (IL-4, IL-10, IL-13, IFNα and TGFβ) simultaneously have the capacity to induce IL-1ra. Note that this statement is based on experiments performed with monocytes/macrophages, and that the nature of the producing cell and of the co-signals may influence the release of IL-1ra.

5.2 Anti-inflammatory cytokines

As mentioned above, certain cytokines repress IL-1 release. This is also true of TNFα and other pro-inflammatory cytokines such as the chemokines. In addition, the so-called anti-inflammatory cytokines can counteract some of the induction of the activities generated by IL-1 and TNF (Figure 5).

5.3 Interleukins 4 and 13

Interleukins 4 and 13 are produced by Th2 lymphocytes as well as CD8[+] cells, basophils and mast cells. These two cytokines share a common chain for their receptors and have many similar activities, except that IL-13 does not act on T cells, while IL-4 does. Both can reduce the release by activated monocytes/macrophages of pro-inflammatory cytokines (e.g. IL-1α, IL-1β, IL-6, IL-8, G-CSF, TNFα, MIP-1α), while they enhance the production of IL-1ra. However, *ex vivo* analysis of peripheral blood mononuclear cells from volunteers injected with IL-4 did not reveal a significant modification of the capacity of the cells to produce IL-1 and TNF (Wong *et al.*). In an *in vivo* model of alveolitis, IL-4 was shown to limit the inflammation induced by immune complex (Mulligan *et al.* 1993*b*). Intratracheal instillation of the causative antibodies together with IL-4 led to a reduced lung per-

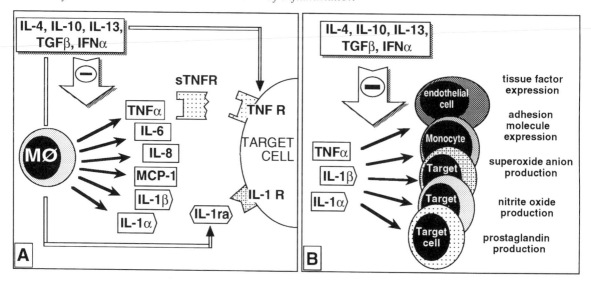

∎**Figure 5.** Anti-inflammatory cytokines can (A) block the production of pro-inflammatory cytokines, induce the production of the interleukin-1 receptor antagonist (IL-1ra), and favour the shedding of TNF receptors as well as (B) counteract some of the activities initiated by IL-1 and TNF acting on their target cells. (Adapted from Marie, C. and Cavaillon, J.-M. (1997). *Bull. Inst. Pasteur.* **95**, 41–54.)

meability index, a reduced haemorrhagic index, a reduced number of PMN and reduced level of TNF in BAL and a diminished expression of ICAM-1. However, the nature of the responding cells, the nature of the pretreatments of the cells and the nature of the activators may influence the modulating effects of IL-4 and IL-13. For example, IL-4 inhibits the production of IL-6 by monocytes/macrophages while it favours its release by B lymphocytes, fibroblasts and endothelial cells. Similarly, IL-4 reduces IL-8 production by monocytes/macrophages but enhances its release by LPS-activated endothelial cells. In addition, IL-4 and IL-13 can counteract IL-1- and TNF-induced functions. For example, IL-4 and IL-13 limit the capacities of IL-1 and TNF to induce osteolysis, prostaglandin release, pro-coagulant activity and tissue factor expression as well as expression of ICAM-1 and ELAM-1 on the surface of endothelial cells.

Some inflammatory diseases, such as Crohn's disease and ulcerative colitis, may be associated with a reduction in the number of IL-4-producing lymphocytes detectable in the lamina propria (West *et al.*).

5.4 Interleukin-10

IL-10, first called 'cytokine synthesis inhibitory factor' because of its capacity to block the produc-

tion of IL-2 and IFNγ by Th1 cells, in fact blocks a wide range of activities, limiting the release of most monocyte/macrophage-derived cytokines.

Besides limiting the inflammatory process induced by exogenous stress, IL-10 seems to play a role in homeostasis: IL-10 knockout mice develop chronic enterocolitis as assessed by intestinal mucosal hyperplasia, inflammatory cell infiltration and MHC class II antigen expression by colonic epithelium. When acting on monocytes/macrophages, IL-10 reduces the pro-inflammatory potential of TNF in three ways: (i) inhibiting TNF synthesis, (ii) reducing the expression of both forms of TNF receptors and (iii) enhancing the release of soluble TNF receptors (as a result not only of enhanced shedding of the receptors, but also of increased synthesis of TNFR mRNA). While TNF release by activated macrophages contributes to the production of IL-10, IL-10 in a feedback action represses the production of the various cytokines produced by activated monocytes/macrophages. This has been further demonstrated *in vivo* where treatment of mice with anti-IL-10 antibodies enhances the levels of circulating TNF and IFNγ. IL-10 inhibits the production of cytokines by other cell types, including mast cells and neutrophils. However, as for IL-4, the nature of the stimulus influences the efficacy of the inhibitory activity. For example, IL-8 production by LPS-activated PMN is reduced by IL-10 while this does not occur when the cells are activated by TNF (Marie *et al.* 1996). IL-10 can also counteract some of the activities induced by IL-1 or TNF. Once again, the inhibitory activity of this cytokine is highly dependent on the nature of the target cells. This is illustrated by the observations that, depending on the cell type and activator studied, IL-10 can inhibit NO production, or have no effect on it, or even enhance it.

IL-10 promotes apoptosis of neutrophils, favouring the clearance of these cells following their recruitment in inflammatory foci. In *in vivo* models, IL-10 inhibits antigen-induced cellular recruitment into the airways of sensitized mice, and like IL-4, limits the alveolitis induced by immune complexes. Furthermore, IL-10 has been demonstrated to protect mice in a lethal model of endotoxaemia or staphylococcal enterotoxin B induced shock. The presence of circulating IL-10 has been well documented in septic patients and the amount of circulating IL-10 correlates with the severity of the sepsis. IL-10 appears rapidly following the initiation of inflammation, as assessed by its early detection in patients undergoing cardiac surgery with cardio-pulmonary bypass who offer a good model of systemic inflammation (Dehoux *et al.*). Analysis of patients suffering from cystic fibrosis suggests that inflammatory diseases may be related to a limited capacity to produce IL-10: the levels of IL-10 in epithelial lining fluid and in BAL have been reported to be lower in these patients than in healthy individuals (Bonfield *et al.*).

5.5 Interferon alpha

Interferon alpha (IFNα) can inhibit the production of pro-inflammatory cytokines, although, as for the other anti-inflammatory cytokines, its activity can depend on the nature of the producing cell. For example, IFNα inhibits IL-8 production by LPS-activated PBMC but has no effect on LPS-activated neutrophils. *In vivo*, IFNα increases the levels of circulating soluble p55 TNFR. It can also prevent endotoxin-induced mortality in mice (Tzung *et al.*). Interestingly, in this model of septic shock, IFNα was beneficial even when used 6–10 h after the injection of a lethal dose of LPS.

5.6 Transforming growth factor beta

Transforming growth factor beta (TGFβ) knockout mice die within 24 days with multifocal inflammatory disease (Shull *et al.*). Inflammation (as assessed by cell infiltration and tissue necrosis) can be observed in their stomach, liver, pancreas,

myocardium, endocardium, striated muscles and serosa. PCR analysis has shown that IFNγ, TNFα and MIP-1α mRNAs are expressed in the spleen, liver and lung of these mice, but not in normal mice. Thus TGFβ seems to have anti-inflammatory properties.

The anti-inflammatory properties of TGFβ have also been demonstrated *in vivo* during inflammation-induced processes. For example, the injection of TGFβ in rats challenged with LPS leads to reduction in hypotension, reduction in inducible NOS mRNA expression in various tissues (heart, kidney, liver, lungs), and enhanced survival. Although TGFβ also contributes to the healing process, excess TGFβ production is associated with fibrosis and immunosuppression.

TGFβ, like other anti-inflammatory cytokines, cannot repress the production of the pro-inflammatory cytokines in all cell types. It has even been reported to *enhance* IL-1α and IL-8 mRNA expression in immortalized rat type II epithelial cells.

5.7 Corticosteroids

The two main pro-inflammatory cytokines, IL-1 and TNF, can initiate a neuroendocrine cascade leading to the release of glucocorticoids which can down-regulate their own production. IL-1 and TNF initiate the release by the hypothalamus of CRF which induce the production by pituitary gland of the ACTH. The latter, acting on adrenal glands, leads to the release of glucocorticoids. For example intaperitoneal injection of IL-1 leads to an increase of circulating corticosterone, which reaches a peak 2 h after injection and is four times higher than the normal level. Glucocorticoids can repress the production of numerous cytokines by various cell types, including lymphocytes, monocytes/macrophages, fibroblasts, endothelial cells and epithelial cells. The sensitivity to the inhibitory effects of glucocorticoids may decrease with ageing. It has also been reported that a few hours or days after the end of a treatment with cortisol, the production of TNF and IL-6 exceeded the level reached in the absence of any pre-treatment (Barber *et al.*).

The inhibitory activity of glucocorticoids on pro-inflammatory cytokine production is counteracted by the effect of MIF (Figure 6). MIF was first described in 1966, as a lymphokine associated with delayed-type hypersensitivity, then was rediscovered, as a hormone-like substance, in 1993 (Bernhagen *et al.*). It can be produced by macrophages in response to glucocorticoids and by

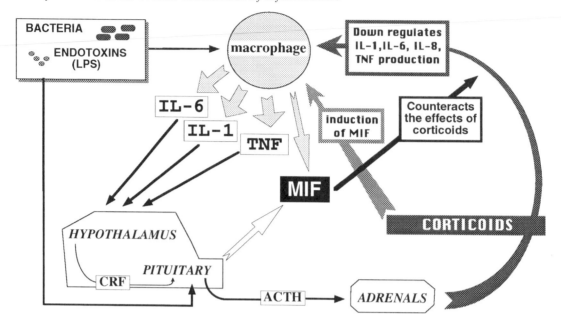

▪ **Figure 6.** Autoregulatory loops involving glucocorticoids. While glucocorticoids are induced by the inflammatory cytokines, they have the capacity to block their synthesis. Glucocorticoids acting on macrophages and microbial products acting via the pituitary glands can induce the production of macrophage migration inhibitory factor (MIF). MIF counteracts the inhibitory activity of glucocorticoids, particularly the inhibition of inflammatory cytokines. (Adapted from Cavaillon, J.-M. (1996). *Rev. Franç. Allerg.* **36**, 914–24.)

the pituitary gland in response to endotoxin. This illustrates the degree of complexity in the auto-regulatory loops involved in the inflammation process.

6. Individual heterogeneity

Heterogeneity between individuals occurs at two levels:

(1) Variation in the amount of cytokine produced by cells. This has been widely reported for both pro- and anti-inflammatory cytokines, and is under genetic control (Bailly *et al.*; Duff *et al.*; Pociot *et al.*). This type of variation may be responsible for the very varied range of symptoms seen in some diseases. For example, even though all cystic fibrosis patients have the same genetic defect, the outcome varies widely, probably because of variability in associated lung inflammation. The variation in suscepti-bility to many other diseases (e.g. cerebral malaria, diabetes mellitus, ulcerative colitis, SLE, juvenile rheumatoid arthritis) may have a similar explanation.

(2) Variation in how target cells respond to the signals delivered by the cytokines. This has been nicely demonstrated with genetically dis-tinct endothelial cells which express various levels of adhesion molecules in response to similar amounts of IL-1 or TNF (Bender *et al.*).

In conclusion, inflammatory response varies greatly from one individual to another. Envi-ronmental factors can also influence the resolution of the inflammation process. These include circa-dian rhythm, stress underlying disease and hor-mones. For example, both wound healing and the activity of TGFβ are highly influenced by oestrogens (Ashcroft *et al.*). Altogether, inflamma-tion is a sophisticated response which has to be highly regulated in order to eliminate the aggres-sive agent and to favour the return to homeostasis without damage.

References

Ashcroft, G.S., Dodsworth, J., van Boxtel, E., Tarnuzzer, R.W., Horan, M.A., Schultz, G.S. *et al.* (1997). Estrogen accelerates cyutaneous wound healing associated with an increase in TGFß1 levels. *Nature Med.* **3**, 1209–15.

Bailly, S., di Giovine, F.S., Blakemore, A.I., and Duff, G.W. (1993). Genetic polymorphism of human interleukin-1 alpha. *Eur. J. Immunol.* **23**, 1240–5.

Barber, A.E., Coyle, S.M., Marano, M.A., Fischer, E., Calvano, S.E., Fong, Y. *et al.* (1993). Glucocorticoid therapy alters hormonal and cytokine responses to endotoxin in man. *J. Immunol.* **150**, 1999–2006.

Bender, J.R., Sadeghi, M.M., Watson, C., Pfau, S., and Pardi, R. (1994). Heterogeneous activation thersholds to cytokines in genetically distinct endothelial cells: evidence for diverse transcriptional responses. *Proc. Natl Acad. Sci. USA* **91**, 3994–8.

Bernhagen, J., Calandra, T., Mitchell, R.A., Martin, S.B., Tracey, K.J., Voelter, W. *et al.* (1993). MIF is a pituitary-derived cytokine that potentiates lethal endotoxaemia. *Nature* **365**, 756–9.

Bonfield, T., Konstan, M., Burfeind, P., Panuska, J., Hilliard, J., and Berger, M. (1995). Normal bronchial epithelial cells constitutively produce the anti-inflammatory cytokine interleukin-10 which is downregulated in cystic fibrosis. *Am. J. Resp. Cell. Mol. Biol.* **13**, 257–61.

Charon, J.A., Luger, T.A., Mergenhagen, S.E., and Oppenheim, J.J. (1982). Increased thymocyte-activating factor in human gingival fluid during gingival inflammation. *Infect. Immun.* **38**, 1190–5.

Dehoux, M., Philip, I., Chollet-Martin, S., Boutten, A., Hvass, U., Desmonts, J.M. *et al.* (1995). Early production of interleukin-10 during normothermic cardiopulmonary bypass. *J. Thorac. Cardiovasc. Surg.* **110**, 286–7.

Duff, G.W. (1994). Interleukin-1 receptor antagonist and genetic susceptibility to inflammation. *J. Interferon Res.* **14**, 305.

Echtenacher, B., Männel, D., and Hültner, L. (1996). Critical protective role of mast cells in a model of acute septic peritonitis. *Nature* **381**, 75–7.

Evans, T.J., Moyes, D., Carpenter, A., Martin, R., Loetscher, H., Lesslauer, W. *et al.* (1994). Protective effect of 55 but not 75 kD soluble tumor necrosis factor receptor-immunoglobulin G fusion proteins in an animal model of Gram-negative sepsis. *J. Exp. Med.* **180**, 2173–9.

Fontana, A., Hengartner, H., Weber, E., Fehr, K., Grob, P.J., and Cohen, G. (1982). Interleukin 1 activity in the synovial fluid of patients with rheumatoid arthritis. *Rheumatol. Intl*, **2**, 49–53.

Girardin, E., Grau, G., Dayer, J., Roux-Lombard, P., and Lambert, P. (1988). Tumor necrosis factor and interleukin-1 in the serum of children with severe infectious purpura. *New Engl. J. Med.* **319**, 397–400.

Girardin, E., Roux-Lombard, P., Grau, G.E., Suter, P., Gallati, H., and Dayer, J.M. (1992). Imbalance between tumour necrosis factor-alpha and soluble TNF receptor concentrations in severe meningococcaemia. *Immunology* **76**, 20–3.

Heuertz, R.M., Piquette, C.A., and Webster, R.O. (1993). Rabbits with elevated serum C-reactive protein exhibit diminished neutrophil infiltration and vascular permeability in C5a-induced alveolitis. *Am. J. Pathol.* **142**, 319–28.

Huffnagle, G.B., Strieter, R.M., Standiford, T.J., McDonald, R.A., Burdick, M.D., Kunkel, S.L. *et al.* (1995). The role of monocyte chemotactic protein-1 in the recruitment of monocytes and CD4[+] T cells during a pulmonary *Cryptococcus neoformans* infection. *J. Immunol.* **155**, 4790–7.

Kopf, M., Baumann, H., Freer, G., Freudenberg, M., Lamers, M., Kishimoto, T. *et al.* (1994). Impaired immune and acute-phase responses in interleukin-6-deficient mice. *Nature* **368**, 339–42.

Kyan-Aung, U., Haskard, D.O., Poston, R.N., Thornhill, M.H., and Lee, T.H. (1991). ELAM-1 and ICAM-1 mediate the adhesion of eosinophils to endothelial cells *in vitro* and are expressed by endothelium in allergic cutaneous inflammation *in vivo*. *J. Immunol.* **146**, 521–8.

Libert, C., Brouckaert, P., and Fiers, W. (1994). Protection by alpha 1-acid glycoprotein against tumor necrosis factor-induced lethality. *J. Exp. Med.* **180**, 1571–5.

Marie, C., Pitton, C., Fitting, C., and Cavaillon, J.-M. (1996). Regulation by anti-inflammatory cytokines (IL-4, IL-10, IL-13, TGFβ) of interleukin-8 production by LPS- and /or TNFα-activated human polymorphonuclear cells. *Mediat. Inflamm.* **5**, 334–40.

Marie, C., Muret, J., Fitting, C., Losser, M.-R., Payen, D., and Cavaillon, J.-M. (1998). Reduced *ex vivo* interleukin-8 production by neutrophils in septic and non-septic systemic inflammatory response syndrome. *Blood* **91**, 3439–46.

Meek, R.L., and Benditt, E.P. (1986). Amyloid A gene family expression in different mouse tissues. *J. Exp. Med.* **164**, 2006–17.

Miller, E.J., Cohen, A.B., Nagao, S., Griffith, D., Maunder, R.J., Martin, T.R. *et al.* (1992). Elevated levels of NAP-1/Interleukin-8 are present in the airspaces of patients with the adult respiratory distress syndrome and are associated with increased mortality. *Am. J. Resp. Dis.* **148**, 427–32.

Mulligan, M.S., Jones, M.L., Bolanowski, M.A., Baganoff, M.P., Deppeler, C.L., Meyers, D.M. *et al.* (1993a). Inhibition of lung inflammatory reactions in rats by an anti-human IL-8 antibody. *J. Immunol.* **150**, 5585–95.

Mulligan, M.S., Jones, M.L., Vaporciyan, A.A., Howard, M.C., and Ward, P.A. (1993b). Protective effects of IL-4 and IL-10 against immune complex-induced lung injury. *J. Immunol.* **151**, 5666–74.

Muñoz, C., Carlet, J., Fitting, C., Misset, B., Bleriot, J.P., and Cavaillon, J.M. (1991). Dysregulation of *in vitro* cytokine production by monocytes during sepsis. *J. Clin. Investig.* **88**, 1747–54.

Pociot, F., Wilson, A.G., Nerup, J., and Duff, G.W. (1993). No independent association between a tumor necrosis factor-alpha promotor region polymorphism and insulin-dependent diabetes mellitus. *Eur. J. Immunol.* **23**, 3050–3.

Schindler, R., Mancilla, J., Endres, S., Ghorbani, R., Clark, S.C., and Dinarello, C.A. (1990). Correlations and interactions in the production of interleukin-6 (IL-6), IL-1, and tumor necrosis factor (TNF) in human blood mononuclear cells: IL-6 suppresses IL-1 and TNF. *Blood* **75**, 40–7.

Shull, M.M., Ormsby, I., Kier, A.B., Pawlowski, S., Diebold, R.J., Yin, M. *et al.* (1992). Targeted disruption of the mouse transforming growth factor-beta 1 gene results in multifocal inflammatory disease. *Nature* **359**, 693–9.

Suter, P.M., Suter, S., Girardin, E., Roux-Lombard, P., Grau, G.E., and Dayer, J.M. (1992). High bronchoalveolar levels of tumor necrosis factor and its inhibitors, interleukin-1, interferon, and elastase, in patients with adult respiratory distress syndrome after trauma, shock, or sepsis. *Am. Rev. Resp. Dis.* **145**, 1016–22.

Tzung, S.P., Mahl, T.C., Lance, P., Andersen, V., and Cohen, S.A. (1992). Interferon-alpha prevents endotoxin-induced mortality in mice. *Eur. J. Immunol.* **22**, 3097–101.

Van der Poll, T., Buller, H.R., ten Cate, H., Wortel, C.H., Bauer, K.A., van Deventer, S.J. *et al.* (1990). Activation of coagulation after administration of tumor necrosis factor to normal subjects. *New Engl. J. Med.* **322**, 1622–7.

West, G.A., Matsuura, T., Levine, A.D., Klein, J.S., and Fiocchi, C. (1996). Interleukin-4 in inflammatory bowel disease and mucosal immune reactivity. *Gastroenterology* **110**, 1683–95.

Wong, H.L., Lotze, M.T., Wahl, L.M., and Wahl, S.M. (1992). Administration of recombinant IL-4 to humans regulates gene expression, phenotype, and function in circulating monocytes. *J. Immunol.* **148**, 2118–25.

XIX Cytokines in the brain

Sophie Layé, Johan Lundkvist and Tamas Bartfai*

1. General introduction

Cytokines such as interleukins 1 and 6 (IL-1, IL-6) and tumour necrosis factor alpha (TNFα), although originally recognized for their signalling properties between immunocompetent cells, are also synthesized in the brain. Receptors for these exist on various cell types in the brain and occupancy of these receptors leads to profound changes in neuronal activity as assessed at the single neur one level. Cytokines also influence complex processes in the brain (neuroendocrine activity, sleep, and behaviour) and their involvement in neurodegeneration is now well documented. Thus cytokines can be regarded as a new class of neuronal signal substances. The mechanism of action and the route by which cytokines act on the brain have not been fully resolved and are the topic of intense investigation.

Cytokines such as IL-1 and IL-2 have been used to stimulate the immune system in patients receiving cancer therapy; although such treatment is only in the early phases of development, it is already clear that it is associated with a number of side effects, including symptoms typical of infection (fever, loss of appetite, etc.) and some behavioural and mental disturbances, and that these are caused by the cytokines.

There is now evidence that cytokines contribute to most, if not all, acute and chronic central nervous system (CNS) pathologies. Overexpression of cytokines has been observed in the brain of patients suffering from traumatic brain injury or from neurodegenerative diseases such as Alzheimer's disease, Parkinson's disease, multiple sclerosis and Down's syndrome, so cytokine imbalance may play an important role in the development of these pathologies. Changes in cerebrospinal fluid levels of several cytokines have also observed in affective disorders.

The major pro-inflammatory cytokine in the brain is IL-1; in this review the effects of IL-1 on various centrally mediated events of endocrine, behavioural and cognitive character will be used as examples of cytokine actions in the brain (Figure 1).

2. Expression of cytokines and their receptors in the nervous system

Cytokines in the periphery are produced by a wide range of cell types and have an extremely broad range of activities, making it difficult to identify their primary biological actions. The same difficulties occur with characterization of cytokine action in the nervous system.

The CNS is generally regarded as an 'immunologically privileged site' for two reasons. Firstly, it is devoid, for the most part, of a lymphatic system that captures potential antigens and produces T and B cell response. Secondly, it is separated by the blood–brain barrier from most of the circulating cells and proteins such as antibodies and cytokines of the immune system. Under certain circumstances the blood–brain barrier becomes selectively permeable to activated T cells and macrophages, and such cells may home in on targets in the brain (this phenomenon is now being investigated to determine whether it could be exploited to deliver genes to the CNS). During neurological trauma, stroke, encephalitis, etc., the blood–brain barrier may become leaky, in a non-specific manner, permitting influx of virtually any substance, including immunocompetent cells, immunoglobulins and cytokines from the periphery.

* Corresponding author.

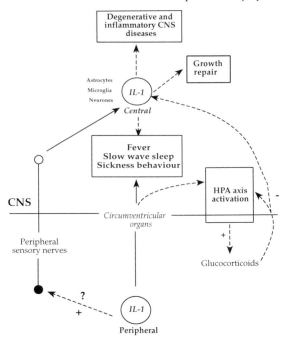

∎ **Figure 1.** Schematic diagram of the effects of interleukin 1 on the central nervous system (CNS) in physiological and pathophysiological conditions.

The normal brain also contains immunocompetent cells, namely the microglia and astroglia, which are capable of phagocytic function, can proliferate and migrate in response to injury, and are also capable of inducible expression of proinflammatory cytokines. Cytokine expression in the brain falls into two main categories: (i) constitutive expression of cytokines in the brain of healthy animals, and (ii) induction of cytokines during pathological conditions. The centrally expressed cytokines may be induced by peripheral inflammatory event(s) without any immediate change in the integrity of the blood–brain barrier, and/or by injury or pathological events within the CNS with an intact blood–brain barrier or subsequent to the breakdown of the blood–brain barrier and the access of immunocompetent cells to the brain.

Healthy, unchallenged mouse, rat, pig and human brains express low levels of the proinflammatory cytokine IL-1β as shown by immunohistochemistry of the gene product and/or by PCR and *in situ* hybridization for the encoding mRNA. In the hypothalamus of human subjects who died from non-neurological causes, IL-1 immunoreactiv-

ity has been detected in a network of axons rather than in cell bodies. Glial cells of white matter (oligodendrocytes) in different brain regions express both IL-1β and IL-6 mRNA but neurones are the main cell type expressing cytokines in healthy rat brain.

Measurement of the amounts of cytokine mRNA and protein during inflammatory events has shown that, when the peripheral immune system is stimulated, many cytokines and also some of their receptors are induced in the brain too. The most abundant source of cytokines in the rat brain, particularly after brain damage or subsequent to peripheral bacterial lipopolysaccharide (LPS) challenge, appears to be activated microglia, although neurones, astroglia, perivascular and endothelial cells can also produce cytokines within a few minutes to several hours after the challenge.

The neuroanatomical site of induced synthesis of cytokines within the brain is dependent on both the nature and magnitude of the peripheral immune stimulus: peripheral LPS treatment induces IL-1 immunoreactivity in microglial and perivascular cells in the ventricles, while a stab wound that penetrates the skull leads to the appearance of TNF and IL-1 immunoreactivity in neurones; this immunoreactivity spreads to adjacent regions of the brain, and expression levels are greatest in the hippocampus, independent of whether the mechanical injury involved the hippocampus. The immunoreactivity takes minutes rather than milliseconds to appear, and so may reflect a slow neurone-to-neurone induction of cytokines from one brain region to another, involving a paracrine type of signalling rather than a neuronal, synaptic type. Although there are probably multiple mechanisms of cytokine induction in the brain, it is likely that cytokine induction cascades in the brain involve more than one cell type (neurones and/or glial cells), and that cytokine production is enhanced by cytokine induction in an adjacent cell.

Cytokines have several pharmacological effects on the CNS. *In vitro* studies show that glial cells (astrocytes and microglial cells) express receptors for IL-1, IL-6 and TNFα. Radioligand binding studies, immunocytochemistry and *in situ* hybridization show that cytokine receptors are also localized on neurones in the hippocampus, hypothalamus and organum vasculosum laminae terminalis (OVLT), respectively. The electrophysiological effects of IL-1 on the firing rates of CNS neurones, and the effects of IL-1 and IL-6 on hypothalamic release of corticotropin-releasing factor (CRF) and on pituitary release of adrenocortinotrophic hormone, further support the data on neuronal localization of some of the IL-1 and IL-6 receptors.

These results strongly suggest that these cytokines act directly on neurones to trigger their effects, while glial cells are mainly producers rather than being major targets of cytokines in the brain. Glia, with its large cytokine synthesis capacity, may participate in the regulation of the cytokine network in the brain, in response to peripheral as well as local brain stimuli.

The discrete neuroanatomical and cellular localization of cytokines in neurones and glial cells indicates that cytokines may have vital physiological functions even under non-pathological conditions: transgenic animals lacking expression of these cytokines or showing over-expression may help to elucidate the role of cytokines in normal brain function.

3. Central actions of pro-inflammatory cytokines

3.1 Cytokines involvement in the febrile response

3.1.1 Fever

Fever is characterized by an alteration in homeostasis, in which the 'set-point' of body temperature regulation is higher than normal. The set-point is determined by central structures in the hypothalamus (medial preoptic area and amygdala). The higher body temperature achieved during fever stimulates proliferation of T cells and is unfavourable for the growth of many bacterial and viral pathogens. It is important to emphasize that, although fever probably evolved as an adaptive response to infection, fever is not necessarily beneficial in all cases.

Exogenous pyrogens (endotoxin, bacteria, etc.) are assumed not to act directly on the brain when they produce fever, but rather to induce the expression of endogenous pyrogens, such as pro-inflammatory cytokines (e.g. IL-1, IL-6 and TNFα), which are the true mediators of fever. Intracerebroventricular injection of cytokines has proved that the brain, specifically the preoptic area of the hypothalamus, is the major site of action of endogenous pyrogens:

(1) Injection of cytokines directly in the brain induces fever at concentrations that are 1000-fold lower than those required when these molecules are injected systematically (intravenously or intraperitoneally).

(2) Cytokines injected intravenously in human subjects produce brisk fever with short duration, but after a time lag which presumably is needed to activate the brain.

(3) Increased circulating concentrations of cytokines such as IL-1, IL-6 or TNFα have been found in patients with fever associated with a variety of different pathological conditions. Fever responses have been replicated by injection of a wide variety of inflammatory stimuli (exogenous pyrogens) in both humans and experimental animals.

IL-1 is regarded as the key or initator endogenous pyrogen because LPS-induced fever is partially or totally blocked by pretreatment with antibodies against IL-1 and/or with preteatment with IL-1 receptor antagonist (IL-1ra). Recently, however, it has been shown that IL-1β knockout mice can mount a fever response to LPS, suggesting that IL-1 is not the only endogenous pyrogen involved in LPS-mediated fever, but that its role can be circumvented by other cytokines, such as TNFα (Figure 2).

There is contradictory evidence in the literature concerning the effect of TNFα on the production of fever in experimental animals and humans. In humans, intravenous injection of TNFα is accompanied by fever and increased oxygen consumption. However, opposite effects of TNF on body temperature and thermogenesis have been reported in laboratory animals, suggesting that TNF can reduce the set-point instead of raising it and thus act as inhibitor of fever, or even as an agent that lowers the set-point (a cryogen) under certain circumstances.

The majority of evidence now points to IL-6 as a likely candidate for a circulating endogenous pyrogen. Injection (intraperitoneal or intracerebroventricular) of IL-6 into experimental animals induces fever. Moreover, IL-6 receptors are present in the thermoregulatory area of the hypothalamus. The IL-6-deficient mice could not mount a fever response after peripheral LPS or IL-1 treatment, while they still produce other pro-inflammatory cytokines (IL-1 and TNFα) suggesting that IL-6 acts downstream from both IL-1 and TNFα, which can both induce each other's biosynthesis and the biosynthesis of IL-6 (Figure 2). Peripheral administration of IL-6 does not elicit a fever response while a central injection of IL-6 induces a fever response, suggesting that it is the centrally

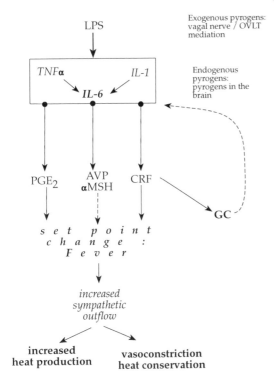

LPS

Exogenous pyrogens: vagal nerve / OVLT mediation

TNFα   IL-1

IL-6

Endogenous pyrogens: pyrogens in the brain

PGE$_2$ AVP αMSH   CRF

GC

s e t   p o i n t
c h a n g e :
F e v e r

increased sympathetic outflow

increased heat production   **vasoconstriction heat conservation**

∎ **Figure 2.** Generation of the fever response. Abbreviations: ACTH, adrenocorticitropic hormone; AVP, arginine vasopressin; CRF, corticotropin-releasing factor; GC, glucocorticoids; IL-1 and -6, interleukins 1 and 6; LPS, lipopolysaccharide; αMSH, melanocyte-stimulating hormone; PGE$_2$, prostaglandin E$_2$; TNFα, tumour necrosis factor α. ——→, stimulatory response; – – – →, inhibitory response.

expressed IL-6 which is required in the fever response.

Little is known about signal transduction mechanisms for any of the cytokines in the brain. Nevertheless, some of the mediators of cytokine action, such as prostaglandins, nitric oxide (NO) within the brain and CRF mediating the endocrine actions via the hypothalamus–pituitary–adrenal (HPA) axis, have been identified. Most cytokines cause release of prostaglandins in the brain, and also fail to elicit a fever response in presence of cyclooxygenase (COX) inhibitors. Prostaglandin E$_2$ (PGE$_2$) alters the firing rate of thermosensitive neurones and evokes fever, as do cytokines. Moreover, the preoptic area contains a high density of PGE$_2$ receptors. Many studies suggest that PGE$_2$, and also PGF$_2$, mediate the central febrile action of cytokines.

The biosynthetic, neuronal enzyme for NO production is induced by IL-1 and IL-6 and thus it is

likely that NO may mediate some paracrine actions of cytokines. CRF has also been suggested to be a mediator of some of the febrile effects of cytokines in the brain of rodents. Non-peptidic and peptidic CRF antagonists can block cytokine-induced fever, suggesting that CRF acts in parallel downstream from IL-1 in the fever response. Thus it has been proposed that CRF is involved as a mediator for IL-1 and IL-6 while prostaglandins are more involved in central action of TNFα.

3.1.2 Neuroendocrine effects of cytokines

Stress involves stimuli that disturb the physical state of the organism and result in a general and stereotyped somatic response known as the general adaptive syndrome. The HPA axis (see Figure 2) plays a critical role in the coordination of the nervous system, endocrine system and immune system.

The neuroendocrine cascade is triggered in vasopressin and/or CRF-expressing parvocellular neurones in the paraventricular nucleus PVN, in ACTH-expressing corticotropic cells of the anterior pituitary gland, and in the glucocorticoid-producing cells of the adrenal cortex. The resulting production of glucocorticoids by the adrenal gland down-regulates the activation of the HPA axis and affects the peripheral production of pro-inflammatory cytokines at the transcriptional and translational levels.

Cytokines act at several anatomical levels of the HPA axis. The HPA's response to endotoxin is mediated by the activation of hypothalamic CRF neurones. Likewise, IL-1, IL-6 and TNFα also stimulate ACTH secretion through stimulating hypothalamic CRF secretion. These observations suggest that IL-1 or IL-6 may be the mediators of the endotoxin-induced activation of the HPA axis, while TNF is less important.

Although the adrenals and pituitary have a role in the activation of the HPA axis by IL-1, both *in vivo* and *in vitro* studies have identified the hypothalamus as the major target of IL-1. Moreover, IL-1 has been identified as a centrally acting endogenous component of LPS-induced activation of HPA axis by showing that central, but not peripheral administration of IL-1ra, blocks the LPS-induced synthesis of CRF mRNA in the PVN (which is believed to reflect not only synthesis but also release of CRF).

IL-6 activates the HPA axis at the levels of the pituitary gland and adrenal gland, respectively, but is less potent than IL-1. In addition IL-6 is thought to act directly in the brain, as its injection into the ventricles induces the synthesis of ACTH and IL-6-induced neuroendocrine response is blocked by immunoneutralization of endogenous CRF. However, IL-6's effect does not seem to be exerted directly on the PVN since its injection into the PVN does not induce CRF mRNA synthesis, whereas IL-1 injection does have an effect. Prostaglandins synthesized elsewhere than in the hypothalamus, and acting in a paracrine manner on the hypothalamus, have been suggested to mediate central effects of IL-6 on CRF synthesis.

3.2 Behavioural modifications involving cytokines during the febrile response: sickness syndrome

The behavioural symptoms of sickness (depression, lethargy, anorexia, etc.) have been proposed to constitute — together with fever — a highly organized strategy of the organism to fight infection. The first evidence in favour of a possible role of cytokines in the non-specific symptoms of infection came from clinical trials with purified or recombinant cytokines in the treatment of viral diseases and cancer. Patients injected with these pro-inflammatory, immune-stimulating cytokine molecules develop not only 'flu'-like symptoms but also, on repetition of injection, acute psychotic episodes characterized by either depression or excitation. Toxicological studies carried out in laboratory animals confirmed the neurotropic activity of recombinant pro-inflammatory cytokines such as IL-1 or TNF. Animals injected acutely or chronically with these molecules usually appear lethargic and anorexic, and withdraw from their environment. The same behavioural effects are observed when animals are injected with a non-lethal dose of LPS.

The role of cytokines in the mental and behavioural symptoms occurring during pathological conditions has hardly been investigated. In some disease states, such as cancer or chronic infection, cachexia and anorexia are prominent. TNFα is elevated in the serum of patients suffering from these diseases and during malnutritional status (anorexia nervosa) and is thought to play an important role in the anorectic process. The pathophysiology of AIDS, anorexia and wasting syndrome includes impaired nutrient intake, decreased nutrient absorption and metabolic alterations, and could be linked to the overproduction of cytokines observed in the patients' plasma. Some experiments carried

out on volunteers clearly show that interferon alpha (IFNα) or LPS-injected humans displayed hyperthermia and fever and impaired performance in a task that measured reaction time. The chronic fatigue syndrome, which generally appears during viral infection, is thought to be linked to deregulation of the cytokine network. The lack of energy and loss of interest are also very frequent in depressed patients. The overproduction of cytokines by monocytes and lymphocytes during the depressive stage may play a major pathophysiological role in depression. The neuropsychiatric disorders observed in AIDS patients are likely to be due to the overexpression of cytokines in the brain of these patients. HIV can enter the brain and infect microglial cells which in turn express TNFα, IL-1, IL-6 and IL-2.

The influence of cytokine administration on animals' performance in classical behavioural tests has been studied. Locomotor activity, social activities and food intake are reduced in laboratory animals injected systematically with LPS, IL-1 or TNFα. These effects develop after 1–2h, and end 24h after injection. Central injection of cytokines is much more effective in inducing behavioural changes. In the rat, maximal behavioural changes are seen following intracerebroventricular injection of nanograms of IL-1, but several micrograms are required to achieve the same effect if administered intravenously. IL-1 also increases somnolence (measured in animals as slow-wave sleep), which is an important component of sickness behaviour.

LPS administration induces a decrease in both social and food-intake activities. Pretreatment with IL-1ra blocks LPS-induced modification in social activity but does not block LPS-induced decreased food intake. These results indicate that IL-1 is the main mediator of the social component of sickness behaviour but not of the effects on food intake. Other cytokines or other signal substances may mediate the diminished food intake in sick individuals. It has been proposed that CRF and prostaglandins have a role in those responses. It has been reported that the anorexic effects of IL-1 are attenuated by immunoneutralization of endogenous CRF or by indomethacin or other cyclooxygenase inhibitors. Somnogenic responses to IL-1, however, are not dependent on production of prostalandins and CRF.

Because of the pleiotropy and redundancy of the cytokine network, the relative importance of the central action of cytokines in the regulation of the different aspects of sickness behaviour has been studied. These studies have been made possible by the use of antagonists, specific antibodies and transgenic animal models.

4. Role of cytokines in communication between the immune and central nervous system: central versus peripheral cytokine action in the brain

Peripheral infection and inflammation both lead to activation of macrophages and endothelial cells with a resulting increase in circulating cytokines. Simultaneously, centrally governed symptoms — such as fever and sickness behaviour — also appear. There is thus little doubt that the brain somehow learns of changes in peripheral cytokine levels, but the mode of this communication is unclear presently. There has been much debate regarding how peripherally released cytokines influence brain functions. Since cytokines are large and hydrophilic proteins that cannot cross the blood–brain barrier, it has been suggested that: circulating cytokines act directly or indirectly on the brain at sites where the blood–brain barrier is incomplete, such as the OVLT, to trigger their effects, or that peripheral cytokines, via activation of afferent fibres projecting into the brain, induce the synthesis of central cytokines that act directly on the brain via their receptors.

It has been reported that chronically high circulating levels of IL-1 and TNF can disrupt the blood–brain barrier and enter the brain. Moreover, in endothelial cells of brain capillaries inflammatory cytokines may trigger the production of eicosanoids and NO, which cause vascular leakage and vasodilation. Released prostaglandins on the brain side of the blood–brain barrier can act as central mediators of peripheral cytokine effects. Other groups have proposed the existence of saturable transport systems for IL-1, IL-6, TNFα and, recently, soluble receptor of IL-1 in the blood–brain barrier. Circulating cytokines are also believed to enter the brain where the capillaries are fenestrated in circumventricular organs, but their access is limited. In any case the presence of the blood–brain barrier allows only a small amount of the circulating cytokines to penetrate into the brain. Thus it is unlikely that the the febrile response is triggered directly by circulating cytokines invading and acting on the brain. However, it has been suggested that, in the case of fever, the circulating cytokines entere in circumventricular organs and interact with target cells that transduce the immune signal into secondary signals in the form of prostaglandins, which freely diffuse to nearby target cells. However, this theory cannot explain all the multiple effects of cytokines in the brain.

There are several lines of evidence to support the view that centrally produced cytokines are responsible for the febrile response. For example, when cytokines are injected centrally, much lower doses are needed to achieve a given effect on the brain than if the cytokines are injected peripherally. Also, the actions of cytokines in the brain are inhibited by central injection of specific antagonists or antibodies neutralizing the cytokines or the receptors. Moreover, the above mentioned hypothesis does not account for the observations that cytokines are present with their receptors in the brain, and that their local expression is modulated by peripheral immune stimuli (see Section 2). It is important to note that the induced expression of cytokines in the brain is not limited to the circumventricular organs but is observed in other brain regions, such as hypothalamus and hippocampus. In addition, peripherally injected LPS induces cytokine expression in a transient fashion, parallel to its centrally mediated effects. To explain how the expression of cytokines in the brain is modulated by peripherally released cytokines, it has been proposed that the immune 'message' is transmitted to the brain via a neuronal route. In line with this finding, it has been shown that the sickness symptoms and the central IL-1 induced by a peripheral immune stimulus are abolished by cutting of the vagus nerve.

Taken together, these data suggest that peripherally-derived cytokines communicate with the brain via a number of mechanisms. Whether these mechanisms are triggered individually or together is a question that remains to be answered. During chronic infections or inflammatory disease it is clear, however, that there is permanent damage to the blood–brain barrier.

5. Regulation of expression of cytokines in the brain and impact on the febrile response

A variety of specific inhibitory mechanisms have been identified for cytokines, including circulating binding proteins, soluble receptors, natural antagonists, anti-inflammatory cytokines (IL-4, IL-10 and IL-13) and glucocorticoids. The importance of these anti-inflammatory systems in the brain is unknown, although some preliminary studies show that some of them are present in the brain. Since overproduction of cytokines in the brain leads to tissue damage, it would not be surprising if the mechanisms that are known to inhibit cytokines in the periphery are also active in the brain. One of the most interesting of these, IL-1ra, is present in the rat brain and inhibits many of the central actions of IL-1 when applied as a pharmacological agent. IL-10 is produced by cells in the brain and inhibits LPS-induced IL-1 and TNF in the brain. However, this inhibitory effect of IL-10 on brain cytokine production does not reduce the HPA axis induction by IL-1. Nevertheless, this anti-inflammatory cytokine inhibits experimental allergic encephalomyelitis (EAE), suggesting a role in the suppression of overproduced pro-inflammatory cytokines in the brain.

There are other regulatory mechanisms involved in the down-regulation of central cytokine synthesis and action. Some of them have been identified on the basis of their ability to oppose the pyrogenic effects of cytokines. The most notable antipyretic peptides in the brain are arginine vasopressin (AVP) and melanocyte-stimulating hormone type alpha (α-MSH). AVP is produced by hypothalamic neurones and released in the general circulation to act as an antidiuretic hormone and in response to fever. Endogenous AVP may play a physiological modulatory role in the behavioural effects of IL-1.

Some of the most potent and effective antipyretic and anti-inflammatory molecules are the glucocorticoids. These molecules suppress fever via several independent molecular mechanisms. The antipyretic effect of glucocorticoids is probably due to suppression of prostaglandin synthesis. Glucocorticoid effects may also be due to inhibition of CRF synthesis, since glucocorticoids also abolish $PGF_2\alpha$-induced fever, which is not affected by cyclooxygenase 2 inhibitors. Glucocorticoid-mediated regulation of the phospholipid-binding protein, lipocortin 1, which inhibits inflammation, fever, pituitary–adrenal activation and neurodegeneration, probably via inhibition of CRF release, is also of importance. Antipyretic effects of lipocortin 1 may reflect inhibition of prostaglandin synthesis as well as modification of CRF synthesis/release. Lipocortin 1 inhibits febrile responses evoked by IL-1, IL-6 and $PGF_2\alpha$, but not by TNFα, although TNFα fever is assumed to

involve prostaglandins. Finally, the mechanism underlying the antipyretic effects of glucocorticoids on the febrile response is in part due to the down-regulation of the synthesis and release of cytokines both in the periphery and CNS.

6. Cytokine involvement in CNS pathologies

There is a wealth of data indicating that many pro-inflammatory cytokines can exert neurotrophic, neuroprotective and neurotoxic actions. IL-6 promotes neuronal survival but transgenic mice overexpressing IL-6 in astrocytes show marked neurodegeneration. Similarly, IL-1 promotes repair from neuronal damage and is neuroprotective in low concentration where it can induce expression of nerve growth factor (NGF). NGF has also been shown to induce IL-1α expression, reflecting the close interregulation between cytokines and neurotrophic factors. The CNTF, an important neurotrophic factor, is a potent pyrogen acting centrally on nerve cells. The involvement of IL-1 in stroke is indicated by a reduction of infarct size upon administration of IL-1ra. TNFα is not toxic for astrocytes but causes demyelination and death of oligodendrocytes. The neurotoxic effect of cytokines is more likely to be indirect. NO produced by astrocytes is one of the potential mediators of the toxicity indirectly induced by IL-1, which up-regulates neuronal NO synthetase.

In Alzheimer's disease, aberrant splicing of β-amyloid precursor protein (β-APP) leads to the production and deposition of β-amyloid. Amyloidosis seems to be a self-propelling process as IL-1 and TGFβ induce the synthesis of β-APP, while β-amyloid itself stimulates the release of IL-1 from astroglia. A single injection of systemic LPS modifies the ratio of APP isoforms in the brain, demonstrating that inflammatory cytokines are crucial in inducing a shift from the physiological pathway of APP processing to the pathological accumulation of β-amyloid. It has also been suggested that head injury is an important risk factor for Alzheimer's disease. As IL-1, TNFα and APP are strongly induced in injury, a surge in IL-1 could be a causative factor in Alzheimer's disease.

IL-1β-converting enzyme (ICE), a mammalian homologue of the *Caenorahbditis elegans* protein Ced3, is involved in neuronal apoptosis which is prevalent in the brain-injured, and inhibitors of ICE have been successfully used to block the neurodegenerative process induced by ischaemia and excitotoxic amino acid release. Thus ICE inhibitors may be part of therapeutic cocktails to be used in stroke.

These few examples barely show the full importance of cytokine regulation in the brain, since these molecules contribute to most if not all acute and chronic CNS pathologies. Modulation of pro-inflammatory cytokine activity by antagonists, monoclonal antibodies or soluble receptors, or treatment with anti-inflammatory cytokines, may well come to be used in neuroprotective therapy. The specific inhibition of cytokine production by anti-inflammatory cytokines and/or by inhibitors of ICE is being studied, for example. However, the extreme complexity of the cytokine network in the brain means that future therapeutic strategies will need to be used with caution.

References

Bartfai, T. and Ottoson, D. (eds) (1992) *Neuroimmunology of Fever*. Pergamon Press, Oxford.

Bartfai, T. and Schultzberg, M. (1993) Cytokines in neuronal cell types. *Neurochem. Int.* **22**, 435–44.

Benvenise, E. N. (1992) Inflammatory cytokines within the central nervous system: sources, function and mechanism of action. *Cell. Physiol.* **32**, 1–16.

Besedovsky, H. O. and Delrey, A. (1992) Immune-neuroendocrine circuits — integrative role of cytokines. *Frontiers Neuroendocrinol.* **13**, 61–94.

Dantzer, R., Bluthé, R. M., Kent, S. and Kelley, K. W. (1991) Behavioural effects of cytokines. In Rothwell, N. and Dantzer, R. (eds), *Interleukin-1 in the Brain*, pp. 135–50. Pergamon Press.

Dantzer, R. (1994) How do cytokines say hello to the brain? Neural versus humoral mediation. *Eur. Cytokine Netw.* **5**, 271–3.

Dinarello, C. A. (1996) Biologic basis for interleukin-1 in disease. *Blood* **87**, 2095–147.

Elmquist, J. K., Scammel, T. E. and Sapper, C. B. (1997) Mechanisms of CNS response to systemic immune challenge: the febrile response. *Trends Neurosci.* **20**, 565–70.

Hopkins, S. J. and Rothwell, N. J. (1995) Cytokines and the nervous system. I: Expression and recognition. *Trends Neurosci.* **18**, 83–8.

Kent, S., Bluthe, R.-M., Kelley, K. W. and Dantzer, R. (1992) Sickness behaviour as a new target for drug development. *Trends Neurosci.* **13**, 24–8.

Kluger, M. J. (1991) Fever: role of pyrogens and cryogens. *Physiol. Rev.* **71**, 93–127.

Morganti-Kossmann, M. C., Kossmann, T. and Wahl, S. M. (1992) Cytokines and neuropathology. *Trends Pharmacol.* **13**, 286–91.

Pennisi, E. (1997) Tracing molecules that make the brain–body connection. *Science* **275**, 930–31.

Ransohoff, J. and Benveniste E. (eds) (1996) *Cytokines and the CNS*. CRC Press, Boca Raton.

Rivier, C. (1993) Effect of peripheral and central cytokines on the HPA axis of the rat. In Tache, Y. and Rivier, C. (eds), *Corticotropin-Releasing Factor and Cytokines: Role in the Stress Response*, pp. 97–105. New York Academy of Sciences, New York.

Rothwell, N. J. and Hopkins, S. J. (1995) Cytokines and the nervous system. II: Actions and mechanisms of action. *Trends Neurosci.* **18**, 130–86.

Schöbitz, B., De Kloet, E. R. and Holsboer, F. (1994) Gene expression and function of interleukin-1, interleukin-6 and tumor necrosis factor in the brain. *Prog. Neurobiol.* **44**, 397–432.

Tilders, F. J., De Rijk, R. H., Van Dam, A.-M., Vincent, V., Schotanus, K. and Persoons, J. H. (1994) Activation of the hypothalamus-pituitary-adrenal axis by bacterial endotoxins: routes and intermediate signals. *Psychoneuroendocrinology* **19**, 209–32.

Deregulated immune responses in cytokine and cytokine receptor transgenic and knockout mice

Anneliese Schimpl* and
Thomas Hünig

1. Introduction

The increasing use of transgenic and gene targeting technologies during the last few years has made it possible to study the consequences of overexpression or deletion of individual cytokines *in vivo* and *in vitro* and thus to assess their relative contributions and importance in responses of the innate and adaptive immune system. This is a vastly expanding field, in which additional insights are currently being gained from mutant mice in which the downstream effectors of cytokine receptor signals are ablated or overexpressed. Interbreeding of the various mutants has revealed hitherto unsuspected mechanisms controlling lymphoid development and function. In a number of instances it has been possible to establish links between the newly established mouse models and human diseases. The availability of mouse mutants with genetic defects analogous to those found in humans will facilitate the evaluation of therapeutic approaches and disease management.

Several reviews have recently dealt with cytokine knockout mice, so we will not attempt to give a comprehensive overview of all mutant mice now available, but will focus on a few principles that have emerged from these studies. So far they suggest the following conclusions:

(1) Overexpression of an isolated cytokine, particularly in non-lymphoid organs and immunologically privileged sites, is usually quite detrimental.

(2) Of the cytokines previously suggested to be important in the development of T and B lym-

phocytes, only interleukin 7 (IL-7) seems to serve a non-redundant function.

(3) Cytokines known to regulate peripheral immune responses seem to be quite redundant: in the absense of one cytokine there always seems to be a back-up system which allows activation of B and T cells in the periphery. Back-up systems may involve the use of alternative cytokines or even cytokine-independent induction of cell cycle progression. However, several cytokines have non-redundant functions in stimulating a particular defence system.

(4) In several instances, the ablation of cytokines leads to a failure to dampen or terminate immune responses which in the long run is as deleterious as the failure to activate a particular type of response.

2. Lymphopenia caused by cytokine deficiency

In the first part of this section we shall deal briefly with central failures to generate lymphocytes in cytokine-defective animals; this subject is discussed in depth in Chapter XIV.

Cytokines have been implicated in lymphoid development because their receptors are expressed at various stages of thymocyte and B-cell development. In most cases, however, cytokine gene inactivation has no effect on the generation of the major lymphoid subsets. Exceptions are mice with disrupted genes coding for the CXC (Cys–X–Cys) chemokine PBSF/SDF-1 and for IL-7: in PBSF/

* Corresponding author.

■ **Table 1.** Defective organogenesis in cytokine-deficient and cytokine-receptor-deficient mice

Cytokine/receptor deficiency	Effects	References
LT-α	Lack lymph nodes and Peyer's patches	Banks *et al.* (1995). *J. Immunol.*, **155**, 1685
LTα, TNF-α double deficient	Defective segregation of B and T cells in spleen	Eugster *et al.* (1996). *Int. Immunol.*, **8**, 23; De Togni *et al.* (1994). *Science*, **264**, 703; Matsumoto *et al.* (1996). *Science*, **271**, 1289
TNFR-1	Defective generation of Peyer's patches and germinal centres; normal lymph nodes	Neumann *et al.* (1996). *J. Exp. Med.*, **184**, 259

SDF-1-deficient mice, lymphopoiesis of B cells (but not that of T cells) and myelopoiesis are impaired, whereas the absence of IL-7 results in a drastic decrease in both T and B cells. IL-7 signals through interaction with the IL-7Rα chain and the common gamma (γc) chain of the IL-2R. In keeping, mice deficient for either IL-7Rα or γc have few B and T cells. Although the sharing of the γc by the receptors for IL-2, IL-4, IL-9 and IL-15 would also make defective signalling by these cytokines a plausible explanation for defective lymphoid development in γc, this has been ruled out at least for IL-2 and IL-4, since even IL-2/IL-4 double-deficient mice generate lymphocytes in normal numbers.

In view of previous *in vitro* studies, it was suspected that IL-7 acted as a growth factor of early T and B progenitor cells. Several recent reports indicate, however, that part of the function, at least in T cells, may be to induce *bcl-2* and thereby allow survival at an early stage of thymopoiesis preceding expression of the pre-TCR. This was documented by the fact that crossing *bcl-2* transgenic mice to IL-7Rα- or γc-deficient animals rescued $\alpha\beta$ T-cell development provided TCR rearrangement was possible. Peripheral T cell (but not B cell) numbers normalized and, in case of *bcl-2* transgenic IL-7R α-deficient mice, the cells were functional in *in vitro* proliferation assays, responding almost as well to concanavalin A (ConA), anti-CD3 plus anti-CD28 and alloantigens as wild-type mice.

2.1 Role of lymphotoxin and tumour necrosis factor in organogenesis

A most surprising finding emerging from studies of cytokine-deficient mice is the importance of the tumour necrosis factor (TNF) family for organogenesis and architecture of peripheral lymphoid organs (for references see Table 1). Thus, lymphotoxin-α (LT-α) deficient mice have no morphologically detectable lymph nodes or Peyer's patches and fail to properly segregate B and T cells in the white pulp of the spleen. Mice deficient for tumour necrosis factor receptor type I (TNFR-1) have defects in the generation of Peyer's patches and of germinal centres but develop lymph nodes, while TNFR-II-deficient mice are overtly normal with respect to the architecture of lymphoid organs. Expectedly, ablation of both TNF-α and LT-α also led to animals which were completely devoid of lymph nodes and Peyer's patches and showed structural alterations of the spleen with defective germinal centre formation. The thymus was intact in all types of mutants. Adoptive transfer experiments indicated that LT is required in the development of lymph nodes but not for homing of lymphocytes to existing nodes while the ability to form germinal centres in the spleen was transferable with normal bone marrow to LT-$\alpha^{-/-}$ mice.

3. Lymphocyte activation

3.1 *In vitro* activation of T cells from cytokine-deficient mice

The major cytokines known to induce *in vitro* proliferation of peripheral T cells are those that signal through the γc (i.e. IL-2, IL-4, IL-7, IL-9 and IL-15) and, in addition, IL-12. Knockout mice affecting the IL-2/IL-2R system, IL-4, IL-7 (in the context of the *bcl-2* transgene, since otherwise T cells do not seed the periphery) and the IL-12 p40 chain have shown that there is a high degree of redundancy with respect to T-cell proliferation. Following *in vitro* stimulation with T-cell mitogens, ablation of IL-2 and its receptor has the most serious effect, particularly in the induction of proliferation and cytotoxic activity of CD8⁺ T cells, both in primary reactions to mitogenic or allogenic stimuli and in the secondary *in vitro* propagation of virus-specific cytotoxic T lymphocytes (CTLs). The importance of IL-2 for the *in vivo* activation of CD8⁺ cells will be further discussed below.

3.2 Non-redundant roles of cytokines in the modulation of immune responses *in vivo*

Gratifyingly, the roles of cytokines in the regulation of specific lymphocyte functions have been mostly borne out by studies on gene-deficient mice. Thus, IL-4⁻/⁻ mice show a greatly reduced IgG1 response and IgE is virtually absent. In keeping with the important role of IL-12 in inducing Th1 cytokines, p40 IL-12-defective mice have grossly reduced interferon gamma (IFNγ) production but intact IL-2 and CTL induction, indicating that IL-2 and IFNγ production in Th1 cells are independently regulated. A clear defect in most cases of cytokine knockout animals thus becomes visible when specific functions (generation of Th1 or Th2 cells, production of particular isotypes, CD8-mediated CTL responses, natural killer (NK) activity) become decisive for an effective immune response. Consequently, cytokine-deficient animals have provided excellent tools for verifying the role of particular defence strategies in combating pathogenic microorganisms which had previously been inferred from studies in mice following depletion of various lymphoid subpopulations or lymphokines themselves with monoclonal antibodies. There are several excellent reviews on the antiviral and antibacterial responses of mice lacking IFNγ or the receptors for IFNγ and IFNα/β or other cytokines (see reference list).

3.3 Is IL-2 needed for CTL responses?

A somewhat controversial issue has been the generation of CD8⁺ T cells in IL-2-deficient mice. Most of the reported discrepancies between whether or not IL-2 is indispensable seem, however, to be due to the different antigens and strategies employed to induce CD8⁺ cytotoxic effector T cells; we shall briefly discuss these (for summary and references see Table 2).

In vitro studies had suggested that IL-2 was of particular importance for the proliferation and effector function of CD8⁺ cytotoxic T cells. It came as a surprise, therefore, that IL-2-deficient mice seemed to mount near-normal *in vivo* CTL responses to lymphocytic choriomeningitis virus (LCMV) and vaccinia virus, and exhibited early delayed-type hypersensitivity (DTH) responses previously shown in normal mice to be dependent on CD8⁺ cells. However, a closer inspection of the anti-LCMV response provided evidence that its magnitude in IL-2⁻/⁻ mice was profoundly altered. There was very little sustained proliferation of CD8⁺ cells in the LCMV-infected animals, a reduction in the numbers of IFNγ-producing cells and impaired viral clearance. However, IL-2 deficiency did not lead to a complete loss of CD8 cell activation and proliferation, since inhibitors of DNA synthesis reduced the remaining CTL activity and still allowed the induction of activation markers.

In TCR-transgenic animals expressing a class I restricted TCR specific for an influenza nucleoprotein-derived peptide, CD8⁺ T cells proliferated following injection of the relevant peptide but did not convert to CTL effector cells. In this system, peptide-induced proliferation of transgenic T cells *in vivo* was not radically different between wild-type and IL-2⁻/⁻ mice. However, the extent of proliferation in a system in which so many cells participate (due to the homogeneous receptor expression) is probably restricted to a few cell cycles, while proliferation will be extensive following virus infection in an animal with a full TCR repertoire and consequently few cells which will recognize the antigen initially.

Defects in CD8 responses have also been observed in IL-2-deficient mice with allogeneic transplants or immunized with allogeneic tumour cells. Again, as in the viral systems, the defects were not absolute, but sufficient to delay transplant rejection from 7 to 27 days and to reduce CTL activity analysed *ex vivo* by a factor of 10.

In all *in vitro* studies, CD8⁺ cells from IL-2-deficient mice were even more severely compromised than after *in vivo* treatment. Polyclonal activation by anti-CD3 stimulation with allogenic stimulator cells, peptide-loaded antigen-presenting cells (APCs) co-incubated with T cells from TCR-transgenic mice and IL-2-deficient T cells, and secondary stimulation *in vitro* of T cells from virus-infected mice, all failed to generate CTL.

It is likely that the major difference between the moderate-to-good induction of CTL activity *in vivo* following virus infection and the greatly reduced

∎ **Table 2.** Effects of IL-2/IL-2R deficiency on T cell responses

Deficiency	Antigen	*In vivo* response	*In vitro* response	References
IL-2$^{-/-}$	LCMV, vaccinia virus	CTL generated	No secondary response	Kündig *et al.* (1993). *Science*, **262**, 1059
	LCMV	Proliferation and frequency of CD8$^+$ cells reduced; defective viral clearance		Cousens *et al.* (1995). *J. Immunol.*, **155**, 5690
	Peptide from influenza virus nucleoprotein in TCR-transgenic mice	Proliferation but no CTL generation		Krämer *et al.* (1994). *Eur. J. Immunol.* **24**, 2317
	Islet allografts P815 (allogeneic tumour cells)	Rejection possible but delayed CTL activity reduced		Steiger *et al.* (1995) *J. Immunol.*, **155**, 489
IL-2$^{-/-}$	Staphylococcal enterotoxin A and B (SEA, SEB)	Normal expansion but incomplete deletion	Reduced primary response, enhanced secondary response, impaired deletion	Kneitz *et al.* (1995). *Eur. J. Immunol.*, **25**, 2572
IL-2R$\alpha^{-/-}$	SEB	Incomplete deletion		Willerford *et al.* (1995). *Immunity* **3**, 521
IL-2R$\beta^{-/-}$	SEB	Normal expansion and deletion		Suzuki *et al.* (1997). *Int. Immunol.*, **9**, 136

activity after allogeneic or peptide stimulation *in vivo* and *in vitro* lies in the recruitment of alternative cytokines induced in the massive inflammatory response following the confrontation of the host with viruses. The two most obvious candidates for alternative cytokines in the induction of the effector phase of CD8$^+$ T cells *in vivo* are IL-12 and IL-15. As discussed above, IL-15 uses the same signal transduction machinery as IL-2. *In vitro*, IL-15 is partially able to replace IL-2 in the induction of proliferation of CD8$^+$ cells and their conversion to CTL. The response is further augmented by IL-12, reaching levels as high as those observed after addition of IL-2. It is likely that viruses (and other pathogens) will efficiently induce both cytokines at the same time and will do so more efficiently than peptides injected into TCR transgenic mice or allogeneic tumour cells. Mice living in different environments and being adjusted to different bacteria may vary in their ability to produce IL-15 and IL-12 following virus injection or confrontation with other antigens activating CD8 T cells. This may lead to a whole range of different responses in the absence of IL-2, depending on a number of parameters, such as the virus used and the route of infection.

4. The IL-2/IL-2R system controls lymphocyte homeostasis

The number of B and T cells allowed to survive in the periphery following activation and clonal expansion is strictly controlled by homeostatic mechanisms. The crucial importance of apoptosis for this clonal trimming is illustrated by the severe systemic autoimmune syndromes exhibited by mice with defective CD95 (Fas, Apo-1) or CD95-ligand genes, i.e. the well-studied *lpr* and *gld* mice, respectively.

Studies on cytokine-deficient and cytokine-receptor-deficient mice have revealed a role for cytokines in the control of clonal contraction following activation. This has been most clearly documented for mice lacking transforming growth factor β1 (TGF-β1) or IL-2, which fail to dampen immune responses and develop autoantibodies and a lymphoproliferative syndrome.

Mice deficient for IL-2, the IL-2Rα and the IL-2Rβ chain (see references in Table 2) normally develop all major lymphocyte subsets but suffer from a so-called 'IL-2 deficiency syndrome' characterized by the accumulation of T cells with an activated phenotype in the periphery, multiorgan infiltration, inflammatory bowel disease (IBD), a transient increase in activated B cells and production of autoantibodies of various specificities. At later stages of the disease, B cells are lost in IL-2$^{-/-}$ mice. Hyperactivation of B cells and their subsequent loss is due to an indirect effect exerted by activated T cells and not due to an essential role of IL-2 in maintaining B cell numbers. This was shown by the ameliorating effects of treatment with antibodies against CD4 and the CD40-L and the survival of B cells in IL-2$^{-/-}$ *nu/nu* mice. In general, the IL-2 deficiency syndrome is lethal within a few weeks to months, depending on the genetic background.

One possible explanation for the accumulation of activated T cells in the periphery of IL-2/IL-2R-deficient mice was a possible in negative selection of the T-cell repertoire. This was ruled out in IL-2-deficient animals for CD8$^+$ (see reference in Table 2) and CD4$^+$ cells. In the latter case, T cells expressing Vβs reactive with endogenous superantigens were efficiently deleted in the appropriate mouse strains. On the other hand, mice kept in a specific-pathogen-free (SPF) environment with a reduced antigenic load do not develop IBD, but this can be induced following injection of normally innocuous antigens in adjuvants. Futhermore, reducing the ability to respond to antigens by restricting the T-cell repertoire in TCR-transgenic IL-2$^{-/-}$ mice also greatly delays disease induction and progression.

Thus, rather than being due to incomplete tolerance induction in the thymus, IBD is initiated by an overreaction to antigens contained in the normal flora of the gut, food, etc., and may then spread to autoantigens as well. The recent observation that induction of disease in germ-free animals by a massive antigenic stimulus results in thymic atrophy and the appearance of mature T-cells in the thymus which are able to transfer IBD is, in our view, not indicative of defective thymopoiesis but rather is a result of the dysregulated response of mature T-cells with a normal repertoire.

A link between the inappropriate response to normal antigens and defective homeostasis was directly established by the analysis of superantigen-induced T-cell activation in IL-2$^{-/-}$ and IL-2R$\alpha^{-/-}$ but not IL-2R$\beta^{-/-}$ mice (see Table 2 for references). In the two former types of mutants, superantigen-reactive CD4$^+$ T cells expand normally but the subsequent deletion is incomplete. For IL-2-deficient

mice a failure of activation-induced cell death was directly shown: CD4$^+$ cells from these animals express CD95 but die less readily following its ligation by monoclonal antibodies. The exact molecular defects in the responses to the ligation of the death receptors remain to be determined.

This role of IL-2 in homeostasis of the peripheral T-cell compartment seems to be unique and is not restricted to T cells growing in the somewhat artificial environment of IL-2/IL-2R deprivation. Thus *in vitro* studies on mouse T cells have shown that IL-2 is required for setting up the apoptotic machinery and that it cannot be replaced by other cytokines signalling through the γc chain.

5. Inflammatory disorders without lymphoproliferation

Another cytokine whose inactivation by gene targeting leads to chronic inflammatory bowel disease (IBD) is IL-10. The pathology seen in IL-10 knockout mice is quite similar to that of human IBD. So far no evidence for defective T-cell homeostasis in IL-10 deficient animals has been presented and the defect is thus mechanistically quite distinct from IBD in IL-2$^{-/-}$ mice. Similarities do, however, exist in that mice doubly deficient in IL-10 and B cells also develop disease, as do B-cell-deficient IL-2$^{-/-}$ mice. Furthermore, transfer experiments showed that TCR$\alpha\beta^+$ Th1 cells mediated IBD in IL-10-deficient mice and that the disease-promoting lamina propria T cells could be either CD4$^+$ or CD4$^+$CD8$^+$. Th1-like cells have also been implicated in the IBD of IL-2$^{-/-}$ mice.

6. Future directions

The 'first-generation' gene targeted and transgenic mice discussed in this review, which lack or overexpress cytokines or their receptors throughout life, have confirmed some previously held concepts but also yielded some surprising results such as the role of IL-2R-mediated signals in clonal contraction of T cells. While offering the advantage of an *in vivo* situation, these models may, however, also have given misleading answers about the cytokine requirements of a normal mature immune response because of possible adaptations of the developing immune system to the engineered abnormalities and the pre-existence of severe systemic disorders like the IL-2 deficiency syndrome at the time experiments are initiated. Thus, it is well accepted that lymphopoiesis involves activation events as part of 'positive selection' of lymphocytes with a functional signalling machinery. Since in addition to the antigen receptors themselves, a multitude of cell interaction molecules and cytokines contribute to signalling during lymphopoiesis, developing lymphocytes are likely to adapt to defects in stimuli given by the microenvironment. The obvious answer to this caveat would be the development of 'second-generation' knockout and transgenic technology which would allow the selective ablation or induction of the genes of choice during *adult* life. This approach will provide mice with an immune system which has developed in a normal cytokine evironment. Moreover, it will prevent spontaneous disease in cytokine knockout mice which would normally develop inflammatory disorders, paving the way for a more direct analysis of the roles of IL-2, TGFβ, and IL-10 in the mature immune response.

References

Cosman, D., Kumaki, S., Ahdieh, M., Eisenman, J., Grabstein, K.H., Paxton, R. *et al.* (1995). Interleukin 15 and its receptor. *Ciba Found. Symp.* **195**, 221–9, 229–33.

Dang, H., Geiser, A.G., Letterio, J.J., Nakabayashi, T., Kong, L., Fernandes, G. *et al.* (1995). SLE-like autoantibodies and Sjogren's syndrome-like lymphoproliferation in TGF-beta knockout mice. *J. Immunol.* **155**, 3205–12.

Di Santo, J.P., Kühn, R. and Müller, W. (1995). Common cytokine receptor γ chain (γ_c)-dependent cytokines: understanding *in vitro* function by gene targeting. *Immunol. Rev.* **148**, 19–34.

Horak, I., Löhler, J., Ma, A. and Smith, K.A. (1995). Interleukin-2 deficient mice: a new model to study autoimmunity and self-tolerance. *Immunol. Rev.* **148**, 35–44.

Hünig, T. and Schimpl, A. (1998). The IL-2 deficiency syndrome: a lethal disease caused by abnormal lymphocyte survival. In Durum, S.C. and Muegge, K. (eds), *Cytokine Knockouts*, pp. 1–19. Humana Press, Totowa, NJ.

Kopf, M., Le Gros, G., Coyle, A.J., Kosco-Vilbois, M. and Brombacher, F. (1995). Immune responses of IL-4, IL-5, IL-6 deficient mice. *Immunol. Rev.* **148**, 45–69.

Krämer, S., Schimpl, A. and Hünig, T. (1995). Immunopathology of interleukin (IL) 2-deficient mice: thymus dependence and suppression by thymus-dependent cells with an intact IL-2 gene. *J. Exp. Med.* **182**, 1769–76.

Kühn, R., Löhler, J., Rennick, D., Rajewsky, K. and Müller, W. (1993). Interleukin-10-deficient mice develop chronic enterocolitis. *Cell* **75**, 263–74.

Kühn, R., Schwenk, F., Aguet, M. and Rajewsky, K. (1995). Inducible gene targeting in mice. *Science* **269**, 1427–9.

Leonard, W., Shores, E.W. and Love, P.E. (1995). Role of the common cytokine receptor γ chain in cytokine signaling and lymphoid development. *Immunol. Rev.* **148**, 97–114.

Ludviksson, B.R., Gray, B., Strober, W. and Ehrhardt, R.O. (1997). Dysregulated intrathymic development in the IL-2-deficient mouse leads to colitis-inducing thymocytes. *J. Immunol.* **158**, 104–11.

Magram, J., Sfarra, J., Connaughton, S., Faherty, D., Warrier, R., Carvajal, D. *et al.* (1996). IL-12-deficient mice are defective but not devoid of type 1 cytokine responses. *Ann. N.Y. Acad. Sci.* **795**, 60–70.

Maraskovsky, E., O'Reilly, L.A., Teepe, M., Corcoran, L.M., Peschon, J.J. and Strasser, A. (1997). Bcl-2 can rescue T lymphocyte development in interleukin-7 receptor-deficient mice but not in mutant rag-1-/- mice. *Cell* **89**, 1011–9.

Nagasawa, T., Hirota, S., Tachibana, K., Takakura, N., Nishikawa, S., Kitamura, Y. *et al.* (1996). Defects of B-cell lymphopoiesis and bone-marrow myelopoiesis in mice lacking the CXC chemokine PBSF/SDF-1. *Nature* **382**, 635–8.

Rolink, A. and Melchers, F. (1991). Molecular and cellular origins of B lymphocyte diversity. *Cell* **66**, 1081–94.

Sadlack, B., Merz, H., Schorle, H., Schimpl, A., Feller, A.C. and Horak, I. (1993). Ulcerative colitis-like disease in mice with a disrupted interleukin-2 gene. *Cell* **75**, 253–61.

Sadlack, B., Löhler, J., Schorle, H., Klebb, G., Haber, H., Sickel, E. *et al.* (1995). Generalized autoimmune disease in interleukin-2 deficient mice is triggered by an uncontrolled activation and proliferation of CD4$^+$ T cells. *Eur. J. Immunol.* **25**, 3053–9.

Schimpl, A., Hünig, T., Elbe, A., Berberich, I., Krämer, S., Merz, H. *et al.* (1994). Development and function of the immune system in mice with targeted disruption of the interleukin 2 gene. In Bluethmann, B. and Ohashi, P. (eds), *Transgenesis and Targeted Mutagenesis in Immunology*, pp. 191–201. Academic Press, San Diego.

von Freeden Jeffry, U., Vieira, P., Lucian, L.A., McNeill, T., Burdach, S.E. and Murray, R. (1995). Lymphopenia in interleukin (IL)-7 gene deleted mice identifies IL-7 as a nonredundant cytokine. *J. Exp. Med.* **181**, 1519–26.

Wang, R., Rogers, A.M., Rush, B.J. and Russell, J.H. (1996). Induction of sensitivity to activation-induced death in primary CD4$^+$ cells: a role for interleukin-2 in the negative regulation of responses by mature CD4$^+$ T cells. *Eur. J. Immunol.* **26**, 2263–70.

Zlotnik, A. and Moore, T.A. (1995). Cytokine production and requirements during T-cell development. *Curr. Opin. Immunol.* **7**, 206–13.

Part C

Cytokines in pathology

XXI Cytokines and genetic immunodeficiencies

Alain Fischer, James Disanto, Geneviève De Saint Basile, Jean-Pierre De Villartay, Françoise Le Deist and Jean-Laurent Casanova

It is estimated that there are over 80 different primary immunodeficiencies. By studying these natural models we can gain important insight into the development and function of all arms of the immune system. Advances in the understanding of the molecular mechanisms underlying these diseases will be important for developing effective treatment. The molecular basis of more than 20 of these primary immunodeficiencies is now known, including some involving cytokines and/or their receptors; these findings are reviewed in this chapter.

1. γc deficiency/X-linked severe combined immunodeficiency

Severe combined immunodeficiency (SCID) X1 is a rare, X-linked form of immunodeficiency characterized by faulty differentiation of T cells and natural killer (NK) cells. Patients typically lack mature T and NK cells but have higher than normal numbers of mature B cells in the periphery. The affected locus has been mapped to Xq1.3, and it is now recognized that all patients with SCID-X1 have mutations of the gene encoding the interleukin 2 receptor (IL-2R) γ-chain (γc). That mutation of this gene causes SCID-X1 has been demonstrated in two ways: (i) the SCID-X1 locus and the γc gene co-localize on the X chromosome, and (ii) γc gene transfer restores high-affinity IL-2 receptors on patients' B cells, and NK cell differentiation from marrow progenitor cells. γc is a member of the cytokine receptor family, expressing four conserved cysteine residues in the extracellular domain and a repeated WS motif. It is a component not only of the IL-2 receptor but also of the IL-2, IL-4, IL-7, IL-9 and IL-15 receptors. It provides high-affinity binding to the respective cytokine, and signal transduction through activation of the Jak3 kinase (see Chapter XII).

A number of mutations of the γc gene have been described in SCID-XI patients. They mostly affect the extracellular domain of the molecule, preventing either (i) association with cytokine receptor subunits and membrane γc expression or (ii) appropriate cytokine binding. A few mutations have been described that result in an inability of γc to recruit Jak3.

What roles do the five impaired cytokine pathways play in creating the disease phenotype of SCID-X1? Defects in IL-2 and IL-4 signalling cannot play a major role, since T-cell differentiation is normal in IL-2⁻ mice, in IL-2⁻, IL-4⁻ mice and in immunodeficient patients with impaired IL-2 production. In contrast, mice in which IL-7⁻, IL-7Rα⁻ or γc⁻ has been inactivated all show the same block in T-cell differentiation. This abnormality differs slightly from the human SCID-X1 phenotype, but consists of a partial block at the double-negative (CD4,CD8) thymocyte stage, before T-cell receptor (TCR) β gene rearrangement and expression.

Since the IL-7 receptor is expressed early on during haematopoietic progenitor differentiation on the lymphocytic lineage, these results strongly suggest that T-cell lymphopenia in SCID-X1 results from the impairment of a prothymocyte proliferation wave mediated by IL-7. The NK cell differentiation block mainly results from faulty IL-15/IL-15 receptor interaction. Indeed, several recent reports have indicated that NK cells can develop *in vitro* from haematopoietic progenitor cells in the presence of IL-15 (and Kit-ligand). We have shown that, following γc gene transfer to the bone marrow cells of γc⁻ patients (using a defective retrovirus), one can restore NK cell diffentiation in the presence of IL-15 (and Kit-ligand). These results do not, however, exclude a role for other cytokines requiring the γc cytokine receptor subunit for binding and signal transduction in the NK cell differentiation pathway. In contrast to γc⁻ mice, SCID-X1 patients exhibit an increased number of mature B cells in the periphery, showing that, at least in humans, IL-7 is dispensable for B-cell differentiation. There is still a debate regarding how γc⁻ B cells function. As expected, B-cell proliferation induced by γc-dependent cytokines is impaired. Nevertheless, in the presence of IL-4 or IL-13, γc⁻ B cells can switch to produce IgE. This demonstrates that a second IL-4 receptor (IL-4Rα–IL-13R) can transduce a signal to B cells. In rare cases, incomplete γc deficiency leads to oligoclonal T-cell differentiation, indicating that a limited number of T-cell clones can be rescued when suboptimal γc-dependent signals are provided.

It has been known for some time that a phenotype identical to SCID-X1 could be inherited as an autosomal recessive disease. It has recently been found that the T/NK cell differentiation block observed in this group of patients is the consequence of impaired Jak3 expression due to *jak3* gene mutations. This finding is important since it shows that γc-induced signal transduction is fully Jak3-dependent and that Jak3 has no other major function, since these patients have no abormalities other than those seen in SCID-X1 patients. Jak3⁻ and γc⁻ mice also exhibit an identical phenotype. Interestingly, there is a minor subset of T⁻ B⁺ human SCID in which γc and *jak3* genes are normal. Thus studies of SCID-X1 and of Jak3-deficiency have been instrumental in advancing our understanding of the role of γc-dependent cytokines in lymphocytic differentiation pathways.

There are some rare cases of immunodeficiencies in which there is poor TCR-triggered T-cell proliferation but a normal T-cell phenotype. In some of these instances, T-cell proliferation can be restored by adding exogenous IL-2. In one case it was shown that there was defective IL-2, IL-4 and GM-CSF production by the patient's T cells. Interestingly, an abnormal migration pattern of the NF-AT transcription complex from nuclear lysates of one patient's cells has been described. Thus it is possible that an as yet unrecognized genetic defect in a component of the NF-AT complex (or in a NF-AT-regulatory pathway) is responsible for this immunodeficiency.

2. CD40-ligand deficiency

Male patients with the condition called X-linked hyper-IgM syndrome (HIGM) are prone to infection and have low serum concentrations of IgG, IgA and IgE but normal or increased serum levels of serum IgM and IgD. When an ill-defined T-cell lymphoma supernatant was added to HIGM B cells *in vitro*, it induced these cells to produce IgG and IgE, leading to the suggestion that HIGM cells have a T-cell defect. Similar results were reported recently using a combination of agonist anti-CD40 antibody and cytokines (IL-4 or IL-10).

Together with the finding that the disease locus lies on Xq 2.6, these results led several groups to demonstrate that X-linked HIGM patients had CD40 gene mutations impairing CD40 expression or function. A variety of mutations have now been described mostly affecting the gene encoding the extracellular domain of CD40. CD40-L is a type II membrane protein belonging to the tumour necrosis factor (TNF) family. It functions as a trimer and can be also secreted. Soluble forms of trimeric CD40-L can induce HIGM B cells to switch to IgG, IgA or IgE production in the presence of relevant cytokines. These observations were instrumental in demonstrating the role of CD40-L/CD40 T-cell/B-cell interaction in inducing immunoglobulin class switching. HIGM patients lack germinal centres, like CD40-L/CD40 knockout mice, showing that CD40-L/CD40 interaction is required to prevent antigen-activated germinal centre B cells from dying by apoptosis.

X-linked HIGM patients not only suffer from infections favoured by defective IgG and IgA production but also exhibit frequent neutropenia and opportunistic infections caused by mucosal intracellular pathogens, i.e. *Cryptosporidium* and *Pneumocystis carinii*. This finding has made it clear that CD40-L/CD40 interaction must be involved in many more cell-to-cell signalling pathways, such as T-cell/haematopoietic progenitor cell and T-cell–macrophage and/or epithelial cell interactions. It is now known that activated T cells can trigger macrophages to produce IL-12 through CD40-L/CD40 interaction. In mice, IL-12 is a major component of immunity to *Pneumocystis carinii*: it triggers Th1 cell differentiation resulting in interferon γ (IFNγ) production by Th1 cells.

An autosomal, recessively inherited form of HIGM has been described. B cells from these patients are unable to secrete IgG, IgA and IgA even (*in vitro*) in the presence of anti-CD40 antibodies and cytokines. The mechanism(s) underlying this B-cell defect is presently unknown. CD40 expression is normal but, in some patients there is a defect in a CD40-induced signal, i.e. phosphatidylinositol 3'-kinase activation and/or NF-κB activation.

3. Lymphoproliferative syndrome and autoimmunity (Fas/CD95 deficiency)

Lpr/lpr mice are characterized by the progressive occurrence of a non-malignant T- and B-cell proliferative syndrome associated with autoimmune manifestations. These mice, as well as the similar *lpr*^cg^ mice, have a defect in lymphocyte apoptosis as a consequence of *fas* gene mutations. Fas is a member of the TNFR family sharing with TNFR-1 the so-called intracytoplasmic death domain. A similar syndrome has now been described in a number of patients. In most of them, a heterozygous *fas* gene mutation, often localized in the part of the gene encoding the death domain, is associated with partially defective Fas-mediated activated lymphocyte apoptosis. These patients exhibit marked lymphoproliferation leading to lymph-node enlargement, splenomegaly and an elevated number of circulating CD4$^-$,CD8$^-$,TCR $\alpha\beta^+$ T cells. In many patients, autoimmunity mostly directed against blood cells has been observed. Existence of a phenotype in patients with single allele mutations is suggestive of a negative transdominant effect. This was shown in some cases. However, approximately half of the heterozygotes in affected families do not develop lymphoproliferation, suggesting that the disease results from mutations in two separate genes; perhaps other gene products interacting with Fas, or close to and involved in the apoptosis signal (e.g. FADD/MORT-1 or MACH-1/FLICE), contribute. In one family, a patient with a homozygous *fas* deficiency was found to exhibit severe lymphoproliferation of prenatal onset. In this patient, Fas expression could not be detected on activated lymphocytes, while anti-Fas antibody-triggered lymphocyte apoptosis was undetectable. Early onset of T- and B-cell proliferation is suggestive of uncontrolled autoimmune activation of T- and B-cell clones in the periphery.

A Fas ligand (Fas-L) deficiency has been shown to account for a similar lymphoproliferative syndrome in *gld* mice. It is possible that Fas-L defects could cause genetic predisposition to lymphoproliferation and autoimmune manifestations, as recently shown in one patient with systemic lupus erythematosus.

4. Inherited interferon gamma receptor deficiency

Interferon gamma receptor (IFNγ-R1) deficiency is an autosomal recessively inherited disorder which has recently been identified in three kindreds. Although IFNγ-R1 is ubiquitously expressed in healthy individuals, children with inherited IFNγ-R1 deficiency have no overt developmental defects. The sole clinical manifestations observed thus far are related to impaired immunity. This results in opportunistic infections which constitute the hallmark of inherited IFNγ-R1 deficiency. All of the affected children identified to date show severe and apparently selective susceptibility to weakly pathogenic mycobacteria, either bacillus Calmette–Guérin (BCG) or non-tuberculous mycobacteria. This condition has revealed the importance of IFNγ in the control of mycobacteria in humans.

The susceptibility to mycobacteria associated with IFNγ-R1 deficiency seems to be severe, since most of the children in these kindreds died as a result of mycobacterial infection. In addition, it is profound, since even one of the least virulent non-tuberculous mycobacteria, *Mycobacterium smegmatis*, caused disseminated disease in one child. Finally, it seems to be selective, since no other opportunistic infections have yet been documented in any child. Remarkably, the pathogenic organisms identified to date are limited to poorly pathogenic mycobacteria species, either BCG or non-tuberculous mycobacteria (Table 1). Nevertheless, other potential pathogens may be recognized when additional cases of IFNγ-R1-deficient children are diagnosed, especially intracellular microorganisms, such as *Salmonella*. The prognosis for IFNγ-R1-deficient children is poor: of the nine affected children in the three affected kindreds, only one survived.

A characteristic feature of IFNγ-R deficiency is the failure to form mature granulomas in response to mycobacteria. Biopsies of tissues infected by

■ **Table 1.** Clinical features of children with inherited IFNγR1 deficiency[a]

Kindred	Child	Onset	Outcome	Mycobacterium
Malta	1	1 year	death 3.5 year	*M. chelonei*
	2	3 years	death 8 years	*M. fortuitum*
	3	1.25 years	alive 6 years	*M. avium*
	4	2.75 years	death 6 years	*M. avium*
Tunisia	1	2.5 months	death 10 months	BCG
Italy	1	nk	death 11 years	nk
	2	nk	death 3 years	nk
	3	nk	death 6 years	nk
	4	3 y	death 8 years	*M. smegmatis*

a. nk = not known. In the Maltese kindred, none were vaccinated with BCG. In the Tunisian kindred, the affected girl was vaccinated with BCG at one month of age. In the Italian kindred, four out of eight siblings (1 boy, 3 girls) were affected. The age of onset, the BCG status, the pathogenic mycobacterium and the genetic defect were not available for cases 1 to 3 who died earlier of diffuse granulomatous disease most similar to that of their sister (case 4).

non-tuberculous mycobacteria may show only non-specific inflammatory lesions, with neutrophils, macrophages and foamy vacuolated cells. Furthermore, in many cases non-tuberculous mycobacteria are not visible by microscopy on initial biopsy. Biopsies of BCG-infected tissues reveal an ill-circumscribed lepromatous-like granuloma, with poorly differentiated macrophages loaded with acid-fast rods.

There is no detectable developmental defect of the immune system in children with IFNγ-R1 deficiency. Immune cell subsets from affected children appear to be qualitatively and/or quantitatively normal when currently available immunological investigations are performed.

The four mutations of the IFNγ-R1 gene identified thus far all preclude expression of the receptor at the cell surface. There is an apparent clustering of mutations in the upstream region of the gene, and there is a relatively high allelic diversity, including a nonsense mutation, a frameshift deletion, a frameshift insertion and a splicing mutation (Table 2).

Although the genotype and phenotype of IFNγ-R1-deficient patients are now clearly delineated, the cell types that are directly responsible for the susceptibility to poorly virulent mycobacteria remain to be more precisely determined (Table 3). A strongly positive delayed-type hypersensitivity to tuberculin has been noted in several children with IFNγ-R1 deficiency. These results probably suggest a normal or moderately altered Th1 response. The macrophage may have a causal role in the pathogenesis of mycobacterial infections associated with IFNγ-R1 deficiency, but definitive experimental evidence for this is still lacking.

The diagnosis of IFNγ-R1 deficiency should be considered in any child with idiopathic disseminated infection due to mycobacteria of low pathogenicity, or attenuated (BCG). One should also consider the diagnosis of IFNγ-R1 deficiency despite the lack of histological or bacteriological evidence, and it is possible that additional cases of IFNγR1 deficiency may be identified in the absence of mycobacteria.

In a kindred where IFNγ-R1 deficiency has not been established, the screening can be approached by genotyping the family for a highly polymorphic haplotype encompassing the IFNγ-R1 gene, provided that the family is sufficiently informative. Analysis of cell surface IFNγ-R1 expression on peripheral blood cells using monoclonal antibodies is possible in all kindreds. Definitive diagnosis requires protein analysis and/or mutation detection. In a kindred where inherited IFNγ-R1 deficiency has been established, a secure diagnosis of IFNγ-R1 deficiency in any family member may be achieved by comparing the haplotype encompassing the IFNγ-R1 gene with the haplotype of the proband. This may enable either prenatal diagnosis early during pregnancy or diagnosis early in infancy.

▮ **Table 2.** Genetics of inherited IFNγR1 deficiency[a]

Kindred	Origin	Status	IFNγR1 mutation	IFNγR1 expression
1	Malta	H	S116X	mAb
2	Tunisia	H	131delC	mAb, ^{125}I-IFNg
3	Italy	Ch	107ins4	mAb, HLA-II
			200 + 1G → A	

a. Affected children were either homozygous (H) or compound heterozygous (Ch) for the IFNγR1 gene mutations. All are null mutations that preclude expression of the receptor at the cell surface. The absence of cell surface expression of the IFNγR1 was documented in either of three ways: 1. staining of blood mononuclear cells or transformed B-cell lines with a panel of specific monoclonal antibodies (designated as mAb); 2. binding of iodinated IFNγ to the same cells (designated as ^{125}I-IFNg); 3. testing the induction of HLA class II molecules at the cell surface of a fibroblastic cell line by exogenous addition of IFNγ (designated as HLA-II).
S116X = mutation introduces a stop codon at position 116.
131delC = delesion of a cytodine at position 131.
107ins4 = insertion of 4 nucleofides as position 107.
200 + 1g → A = mutation in position 1 of intron following position 200.

▮ **Table 3.** Main characteristics of the four genetic immunodeficiencies discussed in this chapter

Diseases	phenotype	gene
SCID X1	lack of mature T and NK lymphocytes	γc, common cytokine receptor subunit to IL-2, -4, -7, -9 and -15
Hyper Ig M syndrome (X.L.)	lack of Ig switch defective control or intracellular mucosal parasites	CD40 ligand
lymphoproliferative syndrome with autoimmunity	lymphoprolyferation (T and B cells) accumulation of CD4(–) CD8(–) T cells Autoimmunity	fas
Susceptibility to mycobacterial infections	severe mycobacterial infections (BCG, atypical mycobacterias, salmonella infections)	interferon γ receptor

Therapy primarily relies on antimycobacterial drugs, directed at each mycobacterial species identified. The only potentially curative therapeutic option available for children with IFNγ-R1 deficiency may be bone marrow transplantation. This has not yet been tried but the disease is lethal and affected cells responsible for the illness are bone marrow-derived. In the future gene therapy may become an alternative treatment, provided blood monocytes, bone marrow precursor cells or bone marrow stem cells can be stably transfected with the wild-type IFNγ-R1 gene.

In conclusion, inherited IFNγ-R1 deficiency demonstrates that IFNγ is obligatory in humans for both an appropriate granuloma structure and an efficient macrophage antimycobacterial activity. Whereas a number of the effects of IFNγ may be compensated for by other cytokines, IFNγ is essential for the control of mycobacteria in humans.

References

Casanova, J.L., Blanche, S., Emile, J.F., Jouanguy, E., Lamhamedi, S., Altare, F., Stéphan, J.L., Bernaudin, F., Bordigoni, P., Turck,

D. *et al.* (1996). Idiopathic disseminated bacillus Calmette–Guérin infection: a French national retrospective study. *Pediatrics* **98**, 774–8.

Castigli, E., Pahwa, R., Good, R.A., Geha, R.S. and Chatila, T.A. (1993). Molecular basis of a multiple lymphokine deficiency in a patient with severe combined immunodeficiency. *Proc. Natl Acad. Sci. USA* **90**, 4728–33.

Conley, M.E., Larche, M., Bonagura, V.R, Lawton, A.R., III, Buckley, R.H., Fu, S.M., Coustant-Smith, E., Herrod, H.G. and Campana, D. (1994). Hyper-IgM syndrome associated with defective CD40-mediated B cell activation. *J. Clin. Invest.*, **94**, 1404–12.

Disanto, J.P., Muller, W., Guy-Grand, D., Fischer, A. and Rajewski, K. (1995). Lymphoid development in mice with a targeted deletion of the interleukin-2 receptor gamma chain. *Proc. Natl Acad. Sci. USA* **92**, 377–81.

Drapa, J., Vaisunaw, A.K., Sullivan, K.E., Chu, J.L. and Elkon, K.B. (1996). Fas gene mutations in the Canale–Smith syndrome, an inherited lymphoproliferative disorder associated with autoimmunity. *New Engl. J. Med.* **335**, 1643–9.

Durandy, A., Hivroz, C., Mazerolles, F., Chiff., C., Bernard, F., Jouanguy, E., Revy, P., Di Santo, J.P., Gauchat, J.F., Bonnefoy, D.Y. *et al.* (1997). Abnormal CD40 -mediated activation pathway in B lymphocytes from patients with hyper-IgM syndrome and normal CD40 ligand expressin. *J. Immunol.* **158**, 2576–84.

Emile, J.F., Patey, N., Altare, F., Lamhamedi, S., Jouanguy, E., Boman, F., Quillard, J., Lecomte-Houcke, M., Verola, O., Mousnier, J.F. *et al.* (1997). Correlation of granuloma structure with clinical outcome defines two types of idiopathic disseminated BCG infection. *J. Pathol.* **181**, 25–30.

Fischer, A., Cavazzana-Calvo, M., De Saint Basile, G., de Villartay, J. P., Disanto, J. P., Hivroz, C., Rieux-Laucat, F. and Le Deist, F. (1997). Naturally occurring primary immunodeficiencies of the immune system. *Annu. Rev. Immunol.* **15**, 93–124.

Fisher, G.H., Rosenberg, F.J., Straus, S.E., Dale, J.K., Middleton, L.A., Lin, A.Y., Strober, W., Lenardo, M.J. and Puck, J.M. (1995). Dominant interfering Fas gene mutations impair apoptosis in a human autoimmune lymphoproliferative syndrome. *Cell* **81**, 935–46.

Fraser, A. and Evan, G. (1996). A licence to kill. *Cell* **85**, 781–4.

Grewal, J.S., Xu, J. and Flavell, R.A. (1995). Impairment of antigen-specific T-cell primary in mice lacking CD40 ligand. *Nature* **378**, 617–20.

Jouanguy, E., Altare, F., Lamhamedi, S., Revy, P., Emile, J.F., Newport, M., Levin, M., Blanche, S., Seboun, E., Fischer, A. and Casanova, J.L. (1996). Interferon gamma receptor deficiency in an infant with lethal BCG infection. *New Engl. J. Med.* **335**, 1956–61.

Kamanaka, M., Yu, P., Yatsui, T., Yosha, K., Kawabe, J., Horii, T., Kishimoto, T. and Kikutani, H. (1996). Protective role of CD40 in leishmania major infection at two distinct phases of cell immunity. *Immunity* **4**, 275–81.

Levin, M., Newport, M., D'Souza, S., Kalabalikis, P., Brown, I.N., Lenicker, H.M., Agius, P.V., Davies, E.G., Thrasher, A., Klein, N. *et al.* (1995). Familial disseminated atypical mycobacterial infection in childhood: a human mycobacterial susceptibility gene. *Lancet* **345**, 79–83.

Leonard, W.J., Noguchi, M., Russel, S.M. and McBride, O.W. (1994). The molecular basis of SCID-X1 : the role of the IL2R γ chain as a common γ chain (γc). *Immunol. Rev.* **138**, 61–86.

Macchi, P., Villa, A., Giliani, S., Sacco, M.G., Frattini, A., Porta, F., Ugazio, A.G., Johnston, J.A., Candotti, F., O'Shea, J.J., Vezzoni, P. and Notarangelo, L.D. (1995). Mutations of JAK-3 gene in patients with autosomal severe combined immunodefiency. *Nature* **377**, 65–8.

Nagata, S. and Goldstein, P. (1995). The Fas death factor. *Science* **267**, 1449–65.

Newport, M., Huxley, C.M., Huston, S., Hawrylowicz, C.M., Oostra, B.A., Williamson, R. and Levin, M. (1996). Mutation in the interferon-gamma receptor gene and susceptibility to mycobacterial infection in man. *New Engl. J. Med.* **335**, 1941–9.

Noguchi, M., Yi, H., Rosenblatt, H.M., Filipovitch, A.W., Adelstein, S., Nodi, W.S., McBride, O.W. and Leonard, W.J. (1993). Interleukin-2 receptor γ chain mutation results in X-linked severe combined immunodeficiency in humans. *Cell* **73**, 147–56.

Notarangelo, L.D., Peitsch, M.C., Abrahamsen, T.G., Bachelot, C., Bordigoni, P., Cant, A.J., Chapel, H., Clement, M., Deacock, S., De Saint Basile, G. (1996). CD40L base: a database of CD4OL mutations causing X-linked hyper-IgM syndrome. *Immunol. Today* **17**, 511–16.

Puck, J.M., De Saint Basile, G., Schwarz, K., Fugmann, S., and Fischer, R.E. (1996). IL2RGbase: a database of γc-chain defects causing human X-SCID. *Immunol. Today* **17**, 507–11.

Rieux-Laucat, F., Le Deist, F., Hivroz, C., Roberts I., Debatin, K., Fischer A. and De Villartay, J. (1995). Mutations in fas associated with human lymphoproliferative syndrome and autoimmunity. *Science* **268**, 1347–9.

XXII Cytokines in infectious diseases

Robert L. Coffman

1. Introduction

The primary role of the immune system is to protect the host from attack by infectious microorganisms. Although this is a simple and obvious statement, it is easy to lose sight of it in discussions of experimental immunology in which single proteins, peptides or haptens are used as the test antigens. From this perspective, virtually all cytokines produced as part of an innate or specific immune response are, directly or indirectly, involved in the response to infection. However, a smaller number of cytokines appear to have major roles in the response to pathogens that are direct and easily demonstrated in experimental models. Accordingly, this chapter will concentrate on the principal actions of these major cytokines as revealed by clinical and experimental studies of infectious disease responses. Emphasis will be given to a relatively small selection of diseases which have been well studied and which illustrate general principles that apply broadly to infectious disease responses.

Several subdivisions of immune responses are central to understanding different responses to infection. These are outlined in the following sections.

1.1 Innate versus antigen-specific responses

Innate immune responses are mediated by many of the same cell types that serve as ultimate effector cells for specific T cell- and antibody-mediated functions, including, macrophages, granulocytes, mast cells and basophils. A cell type unique to innate responses is the natural killer (NK) cell. NK cells can function much like T cells, by both cytokine production and direct cytotoxicity, but have a different mode of antigen recognition and activation. The innate immune system can be triggered rapidly either by certain classes of microbial macromolecules (e.g. endotoxins or various cell wall components) or by recognition of specific changes in infected host cells, especially the loss of class I major histocompatibility complex (MHC) expression that is recognized by NK cells. Although innate immune responses act as a first line of defense against invading microorganisms, they are transient and have a rather limited repertoire of recognition. Specific immunity mediated by immunoglobulins and T cells, in contrast, requires at least 3–4 days to begin to be effective against a primary infection. Specific responses, however, persist and are usually more intense and rapid upon a second encounter with the same pathogen.

1.2 T-cell-dependent and T-cell-independent responses

Successful responses to some types of extracellular bacteria require specific antibacterial antibodies, but have little or no requirement for specific T-cell responses, either for B-cell help or for other effector functions. In general, these responses are initiated either by bacterial molecules that are direct mitogens, such as lipopolysaccharides, or by repetitive polymeric antigens such as capsular polysaccharides. Although T-independent B-cell responses do still require cytokines such as interleukin 2 (IL-2), IL-4, IL-10 and interferon gamma (IFNγ) for optimum growth and differentiation, there appear to be sufficient sources of these in T-cell-deficient animals to allow for protective antibody responses to many bacteria.

1.3 CD4$^+$ and CD8$^+$ T-cell responses

This fundamental dichotomy among T cells is based primarily on differences in antigen recognition: the nature of the antigen binding MHC molecule (class I MHC for CD8$^+$ T cells versus class II MHC for

CD4$^+$ T cells), the constraints on the peptide epitope and the subcellular sites of antigen processing. However, CD4$^+$ and CD8$^+$ T cells have substantially different effector functions as well. CD4$^+$ T cells are otherwise known as helper T cells for their ability to stimulate B-cell growth and differentiation, but the majority of their many other effector functions are mediated by cytokines that stimulate growth and function of a variety of hematopoeitic cells. In contrast, CD8$^+$ T cells are generally regarded as functioning primarily as cytotoxic cells, using cytotoxic mechanisms that require cell–cell contact. This functional distinction is not as clear as it once seemed, as some CD4$^+$ T cells (especially human) are able to kill other cell types *in vitro*, and CD8$^+$ T cells secrete large amounts cytokines, especially IFNγ and tumor necrosis factor (TNF) -α and -β. Nonetheless, it remains useful to contrast the functions of CD4$^+$ and CD8$^+$ T cells as cytokine-mediated and cell-mediated, respectively.

1.4 Th1 and Th2 responses

This subdivision of CD4$^+$ T cells is based primarily on the pattern of cytokines secreted by T cells upon stimulation with antigen. The principal cytokines that distinguish the two subsets are IFNγ, TNF-β and IL-2 for Th1 cells, and IL-4, IL-5 and IL-10 for Th2 cells — all cytokines with prominent roles in CD4$^+$ T cell responses to infection.

2. Th1 and Th2 responses in infectious diseases

CD4$^+$ T cells play central roles in the responses to most pathogenic microorganisms. As nearly all of the regulatory and direct effector functions of CD4$^+$ cells are mediated by cytokines, it is not surprising that Th1 and Th2 cells have quite distinct functions and quite different implications for the success of an anti-pathogen response. Th2 cells stimulate the production of the three hallmark features of allergic diseases: mast cells and eosinophils and the IgE antibodies which mediate the degranulation of the two cell types. The cytokines responsible for these activities are IL-4 for IgE production , IL-5 for eosinophilia, and the combination of IL-3, IL-4 and IL-10 for mast-cell production. These Th2-specific responses are prominent in infections with metazoan parasites, although it is controversial whether these responses are always beneficial to the host.

The cytokines produced by Th1 cells mediate a very different set of immune responses. Central to these responses is the activation of macrophages by IFNγ, and, to some extent, by TNF and granulocyte–macrophage colony-stimulating factor (GM-CSF), for enhanced antigen presentation, phagocytosis, Fc receptor expression and nitric oxide (NO) and superoxide production. These changes greatly enhance the ability of macrophages to kill many types of intracellular and extracellular pathogens. Th1, but not Th2 cells, also mediate the complex cellular inflammatory response known as delayed-type hypersensitivity (DTH) and, by the secretion of IFNγ and TNF, are directly cytotoxic to some cell types. Thus, each Th subset induces and regulates a coherent set of effector functions targeted at specific types of antigens and pathogens.

A third Th subset, Th0, which can produce both Th1- and Th2-specific cytokines, has been reported and may dominate in the earliest stages of some immune responses. In addition, there is increasing evidence that stable CD4$^+$ T-cell populations and clones with other discrete cytokine profiles may exist. In humans, there is now considerable evidence for CD4$^+$ T cells with cytokine patterns and functions that are comparable to murine Th0, Th1 and Th2 cells, although the expression of a few cytokines, such as IL-2 and IL-10, may be less restricted. In both humans and the mouse, most CD8$^+$ T-cell populations and clones have a Th1-like cytokine pattern, but there is recent evidence that CD8$^+$ T cells can develop *in vitro* into cells with a Th2-like cytokine profile.

In addition to stimulating a specific subset of effector responses, Th1 and Th2 cells also inhibit the functions mediated by the opposite subset. This property, which has been termed cross-regulation, provides a series of molecular checks and balances that are essential for the proper functioning of an anti-pathogen response. Cross-regulation occurs at two distinct levels: reciprocal inhibition of Th cells themselves and reciprocal inhibition of subset-specific effector functions (Figure 1). IFNγ, a major product of Th1 cells, is a potent inhibitor of Th2, but not Th1, proliferation and the basis for this is the inability of Th1 cells to respond to IFNγ, because of a loss of the β-chain of the IFNγ receptor. Th2 cells, in return, produce IL-4 and IL-10, both potent and preferential inhibitors of Th1 cytokine production and, to a lesser extent, proliferation. This activity represents the sum of a number of specific activities, especially of IL-10, on both Th1 cells and antigen-presenting cells. Furthermore, these same three cytokines are inhibitors of many of the effec-

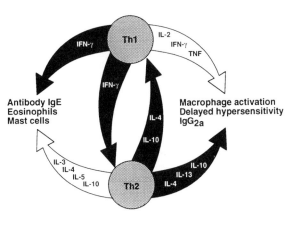

▮ Figure 1. Regulatory interactions between Th1 and Th2 CD4+ T cells. {blkarr}, inhibitory interactions; {whtarr}, stimulatory interactions.

tor functions mediated by the opposite subset. Thus, IFNγ is a potent inhibitor of IL-4-mediated IgE production, IL-5-mediated eosinophilia and Th2 granuloma formation, whereas IL-4 and IL-10 inhibit Th1-mediated DTH responses and IFNγ-stimulated macrophage activation and pathogen killing. Cross-regulation between Th subsets is central to understanding the consequences of making an inappropriate T-cell response to a pathogen. As is illustrated below in the example of infection with *Leishmania*, an inappropriate Th2 response is not only ineffective, but actively inhibits development of a protective Th1 response.

3. Th1 and Th2 responses in leishmaniasis

The consequences of producing anti-pathogen responses dominated by either the Th1 or Th2 subset are nowhere more clearly illustrated than in the responses of humans and mice to infection with protozoan parasites of the genus *Leishmania*. These organisms live and replicate intracellularly and do so primarily in cells of the monocyte–macrophage lineage. The most common disease form in humans is cutaneous leishmaniasis, a skin lesion with low numbers of parasites that are confined largely to the local lesion. A second form of disease is visceral leishmaniasis, with very high numbers of parasites distributed throughout the body, especially visceral organs, such as spleen and liver. This progressive disease is usually fatal if left untreated. The domi-

nant T-cell response in cutaneous leishmaniasis is a typical Th1 response, whereas the response in visceral leishmaniasis has many of the features of a Th2 response, although many patients do not have elevations in IL-4 or IgE responses.

A similar dichotomy of disease states can be observed by infection of different inbred mouse strains with *Leishmania major*. In this well-studied model, most strains respond with a strong Th1 response, which leads to containment of the parasite to the site of injection followed by resolution of the lesion and long-lived resistance to reinfection. These mice exhibit the hallmarks of a Th1 response *in vivo* with strong DTH reactivity, low antibody production and no IgE response. CD4+ T cells from these mice respond *in vitro* to *L. major* antigens with high levels of IFNγ, but no IL-4 or other Th2-specific cytokines. A few strains, the prototype being BALB/c, respond to infection with a mixed response, but one that is dominated by the Th2 subset. BALB/c mice exert little effective control of parasite replication and the disease spreads to the visceral organs, frequently leading to death of the animal. These mice make high levels of antibody to the parasite, including IgE, but have little or no DTH response. *In vitro*, the T cells make both IL-4, IL-10 and other Th2 cytokines, but also produce some IFNγ. Both *in vivo* and *in vitro*, however, the potential for IFNγ production is underestimated, as the Th2 cytokines, IL-4 and IL-10, actively inhibit cytokine production by *L. major*-specific Th1 cells.

The consequences of the different T-cell responses to *Leishmania* infection dramatically illustrate the differences in the effector functions mediated by Th1 and Th2 cells. The principal effector mechanism for control of this parasite is IFNγ-mediated macrophage activation, leading to the production of microbicidal molecules, especially NO. Th2 responses are not simply ineffective because of the absence of IFNγ production, but strongly inhibit Th1-mediated parasite killing. This is due both to the inhibition of Th1 cells and to the inhibition of macrophage activation and NO production by IL-4 and IL-10 (Figure 2). The protective effects of Th1 cytokines, principally IFNγ, and the counterprotective effects of the Th2 cytokines IL-4, IL-10 and IL-13 are general features of the host responses to a wide range of nonviral intracellular infections, including bacterial, protozoan and fungal infections (Table 1). In certain of the examples in this table, a substantial fraction of the protective IFNγ derives from CD8+ in addition to CD4+ T cells and the most relevant microbicidal

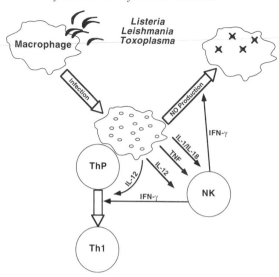

■ Figure 2. The innate immune response to intracellular nonviral pathogens and the link of the innate response to the development of a specific Th1 response.

■ Table 1. Examples of pathogens controlled by IFN-γ dependent mechanisms

Protozoa
Toxoplasma gondii
Leishmania donovani
Leishmania major
Leishmania mexicana
Trypanosoma cruzi
Cryptosporidium parvum

Fungi
Histoplasma capsulatum
Candida albicans
Cryptococcus neoformans
Pneumocystis carinii

Bacteria
Listeria monocytogenes
Legionella pneumophila
Mycobacterium tuberculosis
Salmonella typhimurium
Francisella tularensis
Brucella abortus
Yersinia enterocolitica

activity is not necessarily NO production, but the consequences for the host of an inappropriate Th2 response remain the same. In most cases of nonviral infections that require CD8+ T cells for full protection, it is the cytokine production rather than the cytolytic activity of these cells that appears most important. However, many of these disease models are being re-evaluated with knockout mice deficient in one or more key components of cell-mediated cytolysis.

4. Complex pathogens can induce more complex T-cell and cytokine responses

The response to *L. major* in humans and mice is useful as an example because it is a relatively simple one. Resistant animals make a response that is dominated by highly polarized and very stable Th1, CD4+ T cells that protect by a single, dominant effector mechanism. Both appropriate and inappropriate host responses to many other pathogens are not as simple. As mentioned earlier, different intracellular pathogens (including viruses) can induce responses with different balances of CD4+

and CD8+ T cells and similarly can induce mixed Th1 and Th2 responses. Many pathogens, especially among the protozoan, metazoan and fungal parasites, have complex life cycles, each with distinct antigens and anatomical locations, and it is not uncommon for different T cell and cytokine responses to be made to these different developmental stages.

Malaria, both in human disease and mouse models, is a well-characterized example in which both Th1 and Th2 responses can contribute to control and elimination of the pathogen, but at different stages in its life cycle. Th1 responses are most effective at controlling the initial acute infection by killing of blood-stage parasites. Subsequent complete clearance of the organism, however, is dependent upon both Th2 cells and B cells and is mediated, at least in part, by antibody. The natural pattern of response to malaria infection in resistant animals is, thus, an initial Th1 response followed by a transition to a dominant Th2 response and increased antibody production. In some situations, such as visceral leishmaniasis, successful antimicrobial drug therapy can result in acquisition of a significant Th1 response, which effectively protects the patient from disease long after the end of drug therapy.

5. Th2 responses to helminth parasites

The responses of many mammals to infection with parasitic helminths display all of the hallmark features of a Th2-mediated immune response. Prominent features are large IgE responses, elevated blood eosinophil levels and eosinophil-rich granulomas, and increased mast cell numbers, especially in cases of intestinal parasites. It has been far more difficult to show experimentally or epidemio-logically that these responses afford significant protection to the parasitized host. Depleting one or more of these Th2-mediated mechanisms, either by depleting the relevant cell type or cytokine (IL-4 for IgE, IL-5 for eosinophils) with antibodies or gene knockouts, can be shown to lead to higher parasite numbers in only a few mouse infection models, including *Trichuris muris* and *Heligmosomoides polygyrus* (Table 2). In many other experimental models, however, such depletion of a prominent arm of the Th2-mediated response has little or no effect on parasite numbers

▮ **Table 2.** Examples of mouse and human pathogens controlled by different types of Th responses

Organism	Protection can be mediated by:		In Human/Mouse
	Th1	Th2	
Viruses			
Vaccinia	+		MO
Bacteria			
Listeria monocytogenes	+		MO
Mycobacterium tuberculosis	+		HU/MO
Mycobacterium intracellulare	+		MO
Mycobacterium leprae	+		HU/MO
Salmonella typhimurium	+		MO
Francisella tularensis	+		MO
Brucella abortus	+		MO
Yersinia enterocolitica	+		MO
Protozoa			
Leishmania donovani	+		HU/MO
Leishmania major	+		HU/MO
Leishmania mexicana	+		HU/MO
Trypanosoma cruzi	+		HU/MO
Cryptosporidium parvum	+		MO
Plasmodium falciparium	+	+	HU/MO
Plasmodium chabaudi	+	+	MO
Fungi			
Histoplasma capsulatum	+		MO
Candida albicans	+		MO
Cryptococcus neoformans	+		MO
Coccidiodes immitis	+		HU/MO
Paracoccidiodes brasiliensis	+		MO
Pneumocystis carinii	+		HU/MO
Helminth parasites			
Schistosoma mansoni	+		HU/MO
Trichuris muris		+	MO
Heligmosomoides polygyrus		+	MO
Nippostrongylus brasiliensis		+	MO

or parasite-induced pathology. These observations contrast sharply with analogous experiments used to demonstrate critical roles for IFNγ, IL-12, NO and other Th1 components in control of nonviral intracellular infections. This does not necessarily mean that Th2-mediated responses are not relevant in other helminth infections. Indeed, there are very few examples, one being murine schistosomiasis, of helminth parasites for which Th1 responses can be shown to confer any resistance. Clearly, much remains to be learned about the functional importance of Th2 responses and Th2-specific cytokines in helminth infections.

6. Antiviral responses

Among viral diseases, control and clearance of the initial infection and resistance to re-infection can show a dependence on CD4$^+$ Th1 cells, CD8$^+$ T cells, antibody or varying combinations of all three. A useful distinction in understanding antiviral responses is between cytopathic and noncytopathic viruses. In general, control of cytopathic viruses, such as influenza and vaccinia, are highly dependent on Th1-like cytokines, especially IFNγ and TNF-α. The IFNγ and TNF-α-producing cells may be a mixture of CD4$^+$ and CD8$^+$ T cells or predominantly CD8$^+$ T cells, but an absence of these cytokines in antibody-treated or gene knockout mice substantially impairs control of these viruses. The effects of IFNγ and TNF-α may be partly due to the direct cytotoxic effects of this combination on many cell types, but control of vaccinia infection in mice has been shown to require NO production, implying a role for activated macrophages in some instances. Elimination of either the perforin or Fas-mediated pathway of cell killing, however, has little effect on the control of such viruses. In the case of many such viruses, notably influenza and vesicular stomatitis virus, antibodies play significant roles in viral clearance and resistance to re-infection. Viral infections rarely induce predominantly Th2 responses; however, Th2 responses have been described for measles and respiratory syncitial virus infections. In both cases, Th2 responses do not confer protection, but instead cause significant immunopathology.

Control of many noncytopathic viruses, in contrast, is dependent upon CD8$^+$ T cells and direct cytotoxic killing. Cytotoxic T lymphocytes (CTLs) can kill *in vitro* using several mechanisms; the most prominent are direct perforin-mediated cell lysis

and induction of apoptosis in Fas-positive target cells by Fas-ligand-expressing CTL. In experiments with a prototypical noncytopathic virus, lymphocytic choriomeningitis virus (LCMV), however, inactivation of the perforin gene renders mice highly susceptible to the virus, whereas Fas-deficient or Fas-ligand-deficient mice remain as resistant as wild-type mice. Similarly, inactivation of the IFNγ or TNF-α genes does not lead to a significant increase in susceptibility to LCMV.

7. Pathology versus protection

A key to understanding the immunology of infectious diseases is the realization that, for many diseases, the immune response is responsible for more of the pathology than is the pathogen itself. This is the case with infections with most helminth, fungal and protozoan pathogens and with the great majority of noncytopathic viruses and bacteria. The paradox is that, in most instances, the pathology is mediated by some of the same effector functions that control and clear the infectious agent. For example, the granulomas that form around foci of infection with many intracellular pathogens are effective at containing and resolving the infection, but they can also lead to destruction of uninfected tissue and extensive fibrosis. Similarly, protective Th1-like responses to neurotropic viruses can control and eventually clear the virus, but can simultaneously cause demyelination and encephalomyelitis. In some infections, however, the most severe form of disease can result from an inappropriate, nonprotective immune response. Examples include the Th2-like responses in visceral leishmaniasis and lepromatous leprosy.

Just as most beneficial effector functions are mediated and regulated by cytokines, so most pathological responses are cytokine-dependent. Table 3 lists the most frequently encountered immune pathologies associated with infections and, in cases where clear assignments can be made, the cytokines directly responsible for pathology. This is certainly not a comprehensive list, but serves to illustrate the types of mechanisms that can be involved.

8. Innate responses

A number of cell types participate in a set of rapid responses to infection that do not involve recogni-

∎ **Table 3.** Mechanisms of immune-mediated pathology in infectious diseases

Disease manifestation	Organisms	Important cytokine(s)[1]
Dominant Th1 responses		
Granuloma/fibrosis	*Mycobacterium tuberculosis*	TNF-α, IFN γ
Arthritis (Lyme disease)	*Borrelia burgdorferi*	
Demyelination, encephalitis	Theiler's virus, Sindibis virus	
Hepatitis	Hepatitis B virus	
Myocarditis	Coxsackie virus, *Trypanosoma cruzi*	
Cerebral malaria	*Plasmodium falciparum*	TNF-α
Cell death/apoptosis	HIV, *Listeria*	
Immunosuppression	*T.cruzi, Salmonella typhimurium*	IFN γ
Cachexia	African trypansomaisis	TNF-α
Fever	Malaria, gram-negative bacteria	TNF-α, IL-6, IL-1
Dominant Th2 responses		
Granuloma/fibrosis	*Schistosoma mansoni*	TNF-α
Tropical pulmonary eosinophilia	*Filaria*	IL-5
Acute bronchopulmonary aspergillosis	*Aspergillus fumigatus*	IL-5
Immune suppression	Schistosomaisis, filariasis	IL-4, IL-10

[1] Important cytokines, where defined experimentally.

tion by highly diverse and specific immunoglobulin and T-cell receptors. These are collectively referred to as 'innate immune responses' and play two very important roles in infectious disease control: the initial control of pathogen replication and the induction of the subset of specific T cells most appropriate for long-term control and resolution of infection. Three cell types are prominent, although in varying degrees, in most innate responses — macrophages, NK cells and neutrophilic granulocytes.

The response to the intracellular bacterium *Listeria monocytogenes* in mice is a particularly well-studied model that illustrates the functional roles and regulatory interactions between these three cell types. The effectiveness of the innate response can be shown in mice with the *scid* (*severe combined immunodeficiency*) mutation, lacking both B and T cells. Such mice, nevertheless, are able to control infection with moderate doses of *Listeria* and this control is eliminated by depletion of NK cells or neutralization of IFNγ. The initial response to the infection is not, however, by NK cells, but by macrophages. Macrophage uptake of *Listeria* results in the production of a number of cytokines, principally IL-12, TNF and IL-1. IL-12 is a potent stimulator of IFNγ production by NK cells and this

can be further enhanced by TNF and IL-1 (Figure 2). Recently, the IL-1-like molecule IL-18 has been shown to have an activity equivalent to that of IL-1β on NK cells. This macrophage activation does not require live, infectious *Listeria* and can be approximated with heat-killed organisms. The combination of these macrophage-produced cytokines activates NK cells to produce cytokines, principally IFNγ. This IFNγ, in turn, can act on macrophages to induce potent microbiocidal activities, such as NO and oxygen radicals, as well as higher levels of IL-12 and TNF. This entire circuit of events can be shown to take place within 48h, both in *scid* mice and in tissue culture. The strong positive feedback loop demonstrated in this model could spiral dangerously out of control, were it not for the later induction in macrophages of several potent macrophage 'deactivating' or 'anti-inflammatory' cytokines, principally IL-10, TGF-β and IL-1 receptor antagonist.

The macrophage–NK pathway of innate immunity can be stimulated by a wide variety of intracellular pathogens, including bacteria, protozoa, fungi and a few viruses. The majority of viruses, however, trigger a distinct pathway that also involves NK cells. Viral infection of many cell types, both hematopoeitic and non-hematopoeitic, stimulates production of the type-1 interferons, IFNα and IFNβ. These molecules act in two ways: direct antiviral effects and the stimulation of NK-cell cytotoxicity. The relative contribution of these two actions of IFNα and IFNβ have not been clarified in most viral infection models, but mice rendered unresponsive to both types of interferon by deletion of the gene for the IFNα receptor succumb quite rapidly to most viral infections.

Neutrophilic granulocytes represent the third major arm of the innate immune system and play an important role in the initial response to *Listeria*. Mice depleted of neutrophils die rapidly after exposure to low numbers of *Listeria* or *Candida albicans*, but do not have increased susceptibility to some other intracellular pathogens, such as *Leishmania*. Neutrophils act largely independently of macrophages and NK cells; however, a number of macrophage and NK products, especially IFNγ and several chemokines, can enhance both the activation state and homing pattern of neutrophils. Although most of the antimicrobial activities of neutrophils appear not to be cytokine mediated, neutrophils are capable of making a number of inflammatory cytokines themselves.

It has recently become apparent that the cytokines produced during an innate immune response play an important role in the differentiation of the subsequent specific T-cell response. This has been shown most clearly in studies in the mouse with some of the intracellular pathogens discussed above: *Listeria*, *Leishmania* and *Toxoplasma*. Pathogens that stimulate the macrophage/NK pathway described above most commonly induce strong Th1 responses and, in some cases, significant CD8+ CTL responses. These type 1 responses mediate largely the same antimicrobial effector functions as the innate response, but with greater antigen specificity and the ability to generate long-lived memory responses to specific microbial antigens. Two of the prominent cytokine components of this innate response, IFNγ and IL-12, are the cytokines responsible for inducing T cells responding to microbial antigens to differentiate into Th1 cells or CTLs. Thus, the cells that mediate innate immunity can also act to 'sense' the nature of the pathogen and guide the subsequent specific response down the pathway leading to the most effective response. It has also been suggested that neutrophils, by the production of IL-12, can also help induce Th1 responses to some microorganisms. IFNα and IFNβ, the cytokines more commonly made rapidly in response to virus infection, may have a similar ability to promote the induction of type 1 responses, especially in humans.

9. Subversion of immune responses by virus-encoded cytokines and cytokine receptors

Pathogenic microorganisms have collectively developed an impressive range of strategies to evade the host immune response. Of particular relevance to this chapter is the recent discovery of

■ **Table 4.** Cytokines and cytokine receptors encoded by DNA viruses

Virus	Gene	Homolog to	Virulence factor
Poxviruses			
Shope fibroma &	T2	TNF R	Yes
rabbit myxoma	T7	IFN γR	
Vaccinia & cowpox	A53R	TNF R	
	B8R	IFN γR	
	B15R	IL-1 RII	Modulates virulence, inhibits fever
	B18R	IL-1 R/IL-6R	not known, but binds to IFN α/β, not IL-1 or IL-6
Herpesviruses			
Epstein-Barr virus	BCRF-1	IL-10	Yes, *in vitro*
Equine Herpesvirus	BCRF-1 homolog	IL-10	
Cytomegalovirus	US28	Chemokine receptor RANTES, MIP1α, IL-8	

cytokine and cytokine receptor homologs produced primarily by large DNA viruses, principally poxviruses and herpesviruses (Table 4). A number of these genes have biological activities that would be expected to inhibit one or more antiviral effector mechanisms and at least one has been shown to be an important virulence factor *in vivo*. Deletion mutants of the myxoma T2 gene, a TNF receptor homolog, are much less virulent in rabbits than is wild-type virus. The best characterized of these genes, listed in Table 4, have significant sequence and structural homologies with their mammalian counterparts, indicating that the genes were 'captured' from the host genome at some point in evolution. In addition, the host range of the biological activities can reflect the narrow or broad host range of the virus itself. Thus, the soluble IFNγ receptor homolog of rabbit myxoma virus is specific for rabbit IFNγ, whereas the IFNγ receptor in vaccinia, a virus with a broad host range, will bind human, rat, rabbit and bovine IFNγ. It is not known whether DNA viruses are the only group of pathogens to manipulate cytokine-mediated regulatory pathways in this direct a fashion. Most of these viral cytokine and receptor homologs were first identified by sequence homology rather than by biological activity, and the absence of examples in prokaryotic and eukaryotic pathogens may reflect the fact that the complete genomes of relatively few such pathogens have been sequenced. A possible example of direct immune regulation by a eukaryotic pathogen is the recent report that *Trypanosoma cruzi*, the parasite responsible for Chagas' disease, can bind to and signal through a cellular receptor for TGF-*β*, a potent immunosuppressive cytokine.

References

Abbas, A.K., Murphy, K.M., and Sher, A. (1996) Functional diversity of helper T lymphocytes. *Nature* **383**, 787–93.

Aichele, P., Bachmann, M.F., Hengartner, H., and Zinkernagel, R.M. (1996) Immunopathology or organ-specific autoimmunity as a consequence of virus infection. *Immunol. Rev.* **152**, 21–45.

Alcami, A. and Smith, G.L. (1995) Cytokine receptors encoded by poxviruses: a lesson in cytokine biology. *Immuno. Today* **16**, 474–8.

Billiau, A. (1996) Interferon-gamma: biology and role in pathogenesis. *Adv. Immunol.* **62**, 61–130.

Biron, C.A. (1997) Activation and function of natural killer cell responses during viral infections. *Curr. Opin. Immunol.* **9**, 24–34.

Fearon, D.T. and Locksley, R.M. (1996) The instructive role of innate immunity in the acquired immune response. *Science* **272**, 50–53.

Guidotti, L.G. and Chisari, F.V. (1996) To kill or to cure: options in host defense against viral infection. *Curr. Opin. Immunol.* **8**, 478–83.

Kagi, D. and Hengartner, H. (1996) Different roles for cytotoxic T cells in the control of infections with cytopathic versus noncytopathic viruses. *Curr. Opin. Immunol.* **8**, 472–7.

Kagi, D., Ledermann, B., Burki, K., Zinkernagel, R.M., and Hengartner, H. (1996) Molecular mechanisms of lymphocyte-mediated cytotoxicity and their role in immunological protection and pathogenesis *in vivo*. *Annu. Rev. Immunol.* **14**, 207–32.

MacMicking, J., Xie, Q.W., and Nathan, C. (1997) Nitric oxide and macrophage function. *Annu. Rev. Immunol.* **15**, 323–50.

McFadden, G., Graham, K., Ellison, K. *et al.* (1995) Interruption of cytokine networks by poxviruses: lessons from myxoma virus. *J. Leukocyte Biol.* **57**, 731–8.

Mosmann, T.R. and Coffman, R.L. (1989) Heterogeneity of cytokine secretion patterns and functions of helper T cells. *Adv. Immunol.* **46**, 111–47.

Mosmann, T.R. and Moore, K.W. (1991) The role of IL-10 in crossregulation of TH1 and TH2 responses. *Immunol. Today* **12**, A49–53.

Mosmann, T.R. and Sad, S. (1996) The expanding universe of T-cell subsets: Th1, Th2 and more. *Immunol. Today* **17**, 138–46.

Murray, H.W. (1994) Interferon-gamma and host antimicrobial defense: current and future clinical applications. *Am. J. Med.* **97**, 459–67.

Reiner, S.L. and Locksley, R.M. (1995) The regulation of immunity to *Leishmania major*. *Annu. Rev. Immunol.* **13**, 151–77.

Romagnani, S. (1997) The Th1/Th2 paradigm. *Immunol. Today* **18**, 263–6.

Scharton-Kersten, T.M. and Sher, A. (1997) Role of natural killer cells in innate resistance to protozoan infections. *Curr. Opin. Immunol.* **9**, 44–51.

Sher, A. and Coffman, R.L. (1992) Regulation of immunity to parasites by T cells and T cell-derived cytokines. *Annu. Rev. Immunol.* **10**, 385–409.

Unanue, E.R. (1997) Inter-relationship among macrophages, natural killer cells and neutrophils in early stages of *Listeria* resistance. *Curr. Opin. Immunol.* **9**, 35–43.

XXIII Cytokines and chemokines in acquired immunodeficiencies

Dominique Emilie* and Pierre Galanaud

1. Introduction

Human immunodeficiency virus (HIV) infection is characterized by chronic hyperactivity of the immune system contrasting with deficiency of several lymphocyte functions. This reflects the simultaneous polyclonal activation of several immune compartments and an abnormal immune response against specific antigens. This contrast is also found when analyzing the pattern of cytokine production in HIV-infected patients, in which there is increased spontaneous production of several cytokines but a defect in cytokine production during antigen-specific responses. These abnormalities in cytokine production contribute directly to dysfunction of the immune system. Cytokines can also affect the infectivity of cells. In this review, we will consider the production and role of cytokines and chemokines in HIV infection, except during the complications of HIV infection such as malignancies, opportunistic infections and wasting syndrome.

2. Imbalanced cytokine production during HIV infection

2.1 Monocyte-derived cytokines (monokines)

The chronic hyperactivation of the monocyte/macrophage compartment in HIV-infected patients induces spontaneous hyperproduction of interleukin 1 (IL-1), IL-6 and tumor necrosis factor alpha (TNFα). Although there is a consequent increase in the serum concentration of these cytokines, the hyperproduction is not of the same magnitude as that observed in lymphoid organs. Hyperproduction has been demonstrated by direct visualization of monokine-producing cells in lymphoid organs of HIV-infected patients, either by *in situ* hybridization or by reverse transcription–polymerase chain reaction (RT-PCR).

During HIV infection, as in other conditions, the hyperproduction of monokines originates not only from monocyte/macrophages, but also from other cell types such as endothelial cells and fibroblasts. Monokine-producing cells are probably induced to hyperproduce not only by virally derived proteins (such as envelope proteins and the Tat transactivating protein) but also by other cells of the immune system. It has also been shown that Tat amplifies the effects of TNF by synergizing with the intracellular messengers induced by the binding of this cytokine to its receptor.

The hyperproduction of pro-inflammatory monokines (IL-1, IL-6 and TNFα) is well known. Recently, though, it has been suggested that this hyperproduction is counteracted, at least in part, by additional abnormalities, namely the simultaneous hyperproduction of anti-inflammatory monokines, and defects in the production of other monocyte-derived cytokines.

Two anti-inflammatory monokines, IL-1 receptor antagonist (IL-1ra) and IL-10, are hyperproduced during HIV infection. IL-1ra binds the IL-1 receptor but does not transduce an activating signal, so it behaves as a competitive antagonist. Increased serum concentrations of IL-1ra have been detected in HIV-infected patients. IL-10 is a cytokine mainly produced by monocyte/macrophages, and it inhibits the production of pro-inflammatory monokines and of IL-12. HIV stimulates the production of IL-10 by macrophages and T cells *in vitro*, and gp120 or Tat mediate this

* Corresponding author.

stimulation. *In vivo*, IL-10 production has been found to be increased in some HIV-infected patients. Increased production of soluble TNF receptors (sTNFRs) induced by HIV has also been reported both *in vitro* and *in vivo*. These sTNFRs counteract the effects of TNF by preventing its binding on cell membrane receptors. Interestingly, the serum level of sTNFRs, especially sTNFRII, tightly correlates with viral load, and a high level appears to be associated with a poor prognosis. Thus, during HIV infection, there is a simultaneous increase of pro-inflammatory cytokines and of their natural antagonists.

While there is increased production of several monokines, the production of IL-12 is decreased in HIV infection. When monocytes from HIV-infected patients are stimulated by *Staphylococcus* extracts, they produce five to ten times less IL-12 than monocytes from HIV-uninfected people. This is due to decreased expression of the gene encoding the p35 chain of IL-12 (which is independent of decreased CD4+ cell counts), and (ii) decreased production of the p40 chain, which is linked to the level of immune deficiency. IL-12 (or, rather, the IL-10/IL-12 equilibrium) has a triggering role in the initiation of cell-mediated immune reactions, so decreased IL-12 production is very significant. The cell-mediated immune reactions are the first to be affected during the disease, and their impairment accounts for most of the clinical manifestations of the immune deficiency.

2.2 T-cell-derived cytokines

One of the first abnormalities to have been discovered during HIV infection is decreased production of IL-2, a cytokine produced mainly by CD4+ T cells. This defect has several causes, including reduction in the number of CD4+ T cells, functional abnormalities in these CD4+ T cells and a deficiency in antigen-presenting cells. The decreased IL-2 production presumably plays a role in the failure of the immune system, as IL-2 stimulates many immune functions. This explains the current interest in IL-2-based therapeutic trials in HIV-infected people.

It has been suggested there is imbalanced production of T-cell-derived cytokines in HIV infection. One hypothesis proposed that the synthesis of IL-2 and of interferon gamma (IFNγ), produced by the Th1 subpopulation of helper T cells, was decreased during HIV infection, whereas the production of IL-4 and of IL-10 (defining the Th2 subpopulation) was increased, and that this imbalance became more pronounced as the disease progressed. Observations that were taken as support of this 'Th1–Th2 switch' hypothesis were that it seemed to explain:

- the deficiency of cell-mediated immune responses (which are largely dependent on the presence of IL-2 and of IFNγ);

- B-cell hyperactivity (which can be stimulated by IL-4 and IL-10); and

- the increase of serum IgE levels seen in a few patients, because IL-4 is the main cytokine stimulating the production of this immunoglobulin isotype.

Indeed, the first results obtained from a subgroup of (highly selected) patients showed increased production of IL-4 and decreased IL-2 production, and seemed to support this hypothesis. However, this hypothesis did not take into account several previous findings.

(1) We have shown, and others have subsequently confirmed, that IFNγ production is increased rather than decreased in lymphoid organs of HIV-infected patients.

(2) CD4+ T cells are not the exclusive producers of the cytokines defining the Th1 and Th2 subgroups: these cytokines are often produced in larger amounts by other cells. For example, CD8+ T cells are the main producers of IFNγ during HIV infection. Also, monocytes/macrophages and, to a lesser extent, CD8+ T cells, are the main sources of IL-10 in this condition.

(3) Most immune-cell populations (monocyte/macrophages, B cells and CD8+ T cells) except CD4+ T cells are hyperactivated in this disease, so it seems surprising that this latter cell population has been proposed to account for the increased cytokine production.

Another criticism of the hypothesis is that, with respect to the abnormalities described in the initial study, the hyperproduction of IL-4 was not confirmed by subsequent studies at the blood cell levels or in lymphoid organs. The hyperproduction of IL-10 was also inconsistent.

Was the initial hypothesis wrong? Perhaps not, but it is presumably more subtle than initially stated. Defective production of the cytokines involved in cell-mediated immunity and a relative excess of cytokines that inhibiting the former presumably does play a role in the immune dysfunction of the disease. Analysis at the single-cell level of cytokines by cells activated with phorbol esters plus calcium ionophore has shown that IFNγ production is lower in HIV-infected patients than in healthy controls, while IL-4 production is preserved. It is also possible that the profile of IL-4 production dramatically differs between the early and the chronic phases of the infection. We indeed have recently shown that in macaques infected with simian immunodeficiency virus (SIV), the IL-4 gene expression in lymph nodes is significant during primary infection but drops between weeks 8 and 13 after infection. This contrasts with the expression of IL-13, a cytokine sharing many functional properties with IL-4. IL-13 gene expression in lymph nodes from SIV-infected macaques is also increased during primary infection, but this increase persists during the chronic phase of the infection. IL-13 gene expression is thus dissociated from that of IL-4 in SIV-infected macaques. This also applies in HIV-infected patients, in whom the amount of IL-13 mRNA molecules is 10 times and 100 times higher than that of IL-4 mRNA molecules in the blood and in lymph nodes, respectively. This raises the possibility that IL-13 rather than IL-4 may be responsible for some of the clinical manifestations previously attributed to the latter. Supporting this hypothesis, we recently showed that IL-13 administration to SIV-infected macaques induces skin rash, increased sCD23 serum levels and digestive epithelium injury, all manifestations frequently observed in HIV-infected patients.

3. Role of unbalanced cytokine production in the immune deficiency associated with AIDS

3.1 Cytokines and the failure of cell-mediated immunity

Although it is difficult to summarize the complexity of immune abnormalities of HIV infection in a few words, they are mainly characterized by a deficiency of cell-mediated immunity and by polyclonal activation of humoral immunity and CD8$^+$ T cells. The main consequence of the cell-mediated immunity failure is to affect the delayed-type hypersensitivity (DTH) reactions, which contribute mainly to the control of intracellular microorganisms. The two main partners in DTH reactions are CD4$^+$ T cells and macrophages. In addition, cell-mediated immunity is crucial in the control of viral infections and malignancies, through the action of both natural killer (NK) cells and cytotoxic T lyphocytes (CTLs). In this context, the recent observation of the key role of the IL-10/IL-12 equilibrium in the induction of an efficient cell-mediated immunity is critical. IL-10, produced mainly by monocyte/macrophages *in vivo* in humans, directly inhibits the IL-12 production by macrophages. Moreover, IL-10 also inhibits the effects of IL-12. In contrast, IL-12 stimulates the production of IFNγ by T cells and NK cells, and IFNγ inhibits IL-10 production (Figure 1). The equilibrium between IL-10 and IL-12 during the initiation of an immune response (presumably within the first hours following the introduction of the antigen) determines the nature of the immune response that will subsequently develop. Indeed, once the immune response has embarked in a given direction (for example a DTH reaction, or a humoral response), self-perpetuating mechanisms maintain this orientation.

Therefore, recent demonstrations of defective IL-12 production during HIV infection, and a correction of many immunological abnormalities of the disease by this cytokine, have stimulated interest. In the same line, neutralization of IL-10 in cultures of cells from HIV-infected patients partially restores immune cell responses. Notably, the cellular response against *Mycobacterium tuberculosis* antigens is decreased in HIV-infected patients, and is restored by the addition of an anti-IL-10 monoclonal antibody. Heterogeneity of IL-10 and IL-12 production among patients at an early stage of the disease may explain why, independently of CD4$^+$ cell counts, only a few patients at this stage display negative skin tests and defective T-cell proliferation and cytokine production *in vitro*. Such abnormalities are markers of poor prognosis in HIV-infected patients, consistent with a critical role of the IL-10/IL-12 imbalance in the progression of immune deficiency. Importantly, this imbalance, if it contributes to the functional abnormalities of the T-cell compartment, originates from the dysfunction of the monocyte/macrophage compartment, underlining the role of the latter in AIDS. Only HIV strains with a tropism for monocytes/macrophages inhibit the production of IL-12. Also, several animal models have suggested that macrophage infection have a key role in the destruction of CD4$^+$ T cells.

Decreased IL-2 production also contributes to the failure of cell-mediated immunity in AIDS, as the administration of even low doses of IL-2 often restores skin reactivity to various antigens for

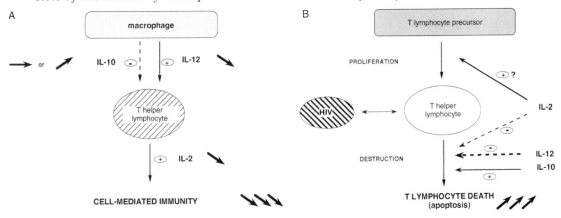

▮**Figure 1**. Cytokine dysregulation in HIV infection. (A) In HIV-infected patients, IL-10 production by antigen-presenting cells is unchanged or increased, whereas IL-12 production decreases. These two cytokines have opposing effects on the activation of T lymphocytes. The IL-12/IL-10 imbalance results in impairment of T-helper cell function, including decreased production of IL-2 and decreased proliferation. (B) In HIV-infected patients, there is n increased rate of destruction of CD4+ T cells, which presumably results from an interaction between CD4 and the gp120 envelope protein of HIV. The increased destruction of CD4+ T cells does not, however, result in a dramatic decrease in the peripheral CD4+ T cell count, as it is counterbalanced by increased renewal from T cell precursors. IL-12 and IL-2 reduce this increased rate of apoptosis of CD4+ T cells, whereas IL-10 enhances it. Some cytokines, such as IL-2 or IL-7, may modify the turnover of CD4+ T cells by increasing the rate at which new cells are generated from precursors. Arrows indicate the modification of cytokine production cell-mediated immunity and apoplosis observed in HIV-infected patients.

several weeks or months. It is not clear whether this results from an increase in the number of CD4+ T cells or from an improvement in their function.

There are two main reasons for the failure of cell-mediated immunity in AIDS: (i) a functional defect in CD4+ T cells and of antigen-presenting cells and (ii) a reduction in the number of CD4+ T cells. At an early stage of infection, only the first mechanism applies, whereas both occur later in infection. We have already discussed the pivotal role of the IL-10/IL-12 imbalance and decreased IL-2 production in the functional defect of immune cells. We will now address the role of deregulated cytokine production in the progressive decline in CD4+ T-cell numbers.

The number of circulating CD4+ T cells results from an equilibrium between two dynamic processes: cell production and cell destruction. The flux is continuous, with several billion CD4+ T cells being produced each day during HIV infection. Cytokines may theoretically affect both processes. There is as yet no direct proof that cytokines affect the production of CD4+ T cells. However, IL-2 is the main factor stimulating the proliferation of T cells, and this may contribute to the increase in the number of circulating CD4+ T cells seen in AIDS

patients treated with IL-2. Deregulated cytokine production can, though, contribute to the increased *destruction* of CD4+ T cells: two studies have shown that IL-2 and especially IL-12 prevent apoptosis of CD4+ T cells in HIV-infected patients, whereas IL-10 increases apoptosis. Deregulated production of IL-2, IL-12 and IL-10 is probably not sufficient to lead to CD4+ T-cell depletion; comparable deregulations have been reported for other disorders where CD4+ T cells are less decreased. However, this deregulated cytokine production should accelerate the destructive process induced by the virus. If so, potential markers of cytokine deregulation, such as skin tests, could be important prognostic factors early in the disease.

In some other infectious process, such as leishmaniasis and leprosy, individuals vary in how they respond to the infectious agent, and variation in the pattern of cytokine production has been proposed to explain this heterogeneity. Therapy aimed at re-establishing a cytokine equilibrium beneficial for the patient has been reported for the abovementioned infections. The positive effect of IL-2 in HIV-infected patients may be the first demonstration that this phenomenon also applies during HIV infection.

3.2 Role of CD8⁺ T cells in the dysfunction of the immune system during HIV infection

In addition to abnormalities in the function of monocytes/macrophages during HIV infection, there is also a dysfunction of the CD8⁺ T-cell compartment. This compartment is in part represented by CTLs which kill cells recognized as non-self (virus-infected cells, allogeneic cells). A more recent discovery defined a new subpopulation of CD8⁺ T cells which produce IL-4 and IL-10.

During HIV infection, there may be a progressive shift of CD8⁺ T cells from CTLs to suppressive T cells. The cytotoxic functions, especially those directed toward HIV, are initially strong and efficient, but they progressively decline despite a considerable expansion of the CD8⁺ T-cell compartment. In contrast, there is evidence of IL-4 and IL-10 production by CD8⁺ T cells in HIV-infected patients, observed at the level of either cloned T cells or freshly isolated cells.

3.3 Cytokines and B-cell hyperactivity in HIV infection

Abnormalities in cytokine production not only disturb cell-mediated immunity, but also contribute directly to B-cell hyperactivity. The main cytokine responsible for this is IL-6. We have shown that IL-6 neutralization *in vivo* in patients partially prevents the increased production of IgG and of IgA. Other cytokines, such as IL-10, may also contribute to the chronic and polyclonal activation of B cells.

The imbalanced cytokine production seen during HIV infection, culminating at the AIDS stage, reflects the abnormal behavior of several compartments of the immune system, including CD8⁺ T cells, monocytes/macrophages and other antigen-presenting cells. Although CD4⁺ T cells are the main victims of such an imbalance, they contribute only partially to the abnormal cytokine production observed during the disease.

4. Cytokines and regulation of HIV replication

A number of studies have reported the effects of cytokines on the *in vitro* replication of HIV, sometimes with contradictory findings (Table 1). Some findings are well established, such as the stimula-

∎ **Table 1.** Cytokine effects on the in vitro replication of HIV

Cytokine	effect on HIV replication replication
IL-1	↑
IL-2	→↑
IL-4	↑
IL-6	↑
IL-10	↓
IL-12	↑
IL-13	↓
IL-15	?
IL-16	↓
TNFα	↑
IFNγ	↓↑
IFNα	↓
G-CSF	→
GM-CSF	↑
	↓ (in the presence of AZT)
M-CSF	↑
TGFβ	↓ ↑

tory effect of TNFα on HIV replication, or the inhibitory effect of IFNα. Rather than describing these studies in detail, we emphasize two points here. First, the experimental models used to test the effects of cytokines on HIV replication are limited and biased, either because they use chronically infected cell lines or because they study replication of laboratory HIV strains, which differ from wild-type strains. Second, the clear findings often reported from *in vitro* studies with cytokines or cytokine antagonists often fail to be reproduced *in vivo*. For example, TNF and IL-6 stimulate HIV replication in *in vitro* studies but neither they nor their antagonists affect viral load when tested *in vivo* in patients. Similarly, IFNα, which down-regulates HIV replication *in vitro*, does not have a clear effect when tested *in vivo* in HIV-infected patients.

What, then, is the role of cytokines on the rate of HIV replication *in vivo*? Generally speaking, the effect of a cytokine is highly dependent on the microenvironment, including the simultaneous production of other cytokines. Obviously, *in vitro* models lack the complexity of the situation *in vivo*. Another important parameter not taken into account by *in vitro* studies is the modulation of the antiviral immune response by cytokines. This is best exemplified by IL-2 treatment of HIV-infected patients. Several *in vitro* studies have shown that IL-2 stimulates HIV replication, but results are less clear *in vivo*. Following IL-2 administration, a transient peak of HIV levels is observed. This contrasts with the sustained improvement seen in CD4⁺ cell counts. We already discussed the possible role of IL-2 on the turnover of CD4⁺ T cells. Another explanation for its favorable effect is that IL-2 stimulates CD8⁺ anti-HIV CTLs, and that this effect persists several weeks after IL-2 treatment. IL-2 is indeed a potent activator of CTLs, which play a critical role in the control of HIV replication since (i) they are cytotoxic for HIV-infected cells and (ii) they produce soluble mediators which interfere with the replication of the virus. These mediators include IFNγ, and CD8⁺ cell antiviral factor (CAF). Although the gene encoding this molecule has not been cloned yet and the effect of this factor is rapidly lost when the supernatant of CD8⁺ T cells is diluted, this activity has been reported by several investigators, either in humans or in SIV-infected monkeys. Interestingly, the non-cytotoxic antiviral activity of CD8⁺ T cells has been reported to be increased by IL-2.

IL-16 has been reported to inhibit HIV and SIV replication *in vitro*. It is a chemoattractant specific for cells expressing its receptor, CD4. The inhibitory effect of IL-16 on HIV replication may rely not on competition between IL-16 and the virus for binding to CD4, but rather on inhibition of the HIV LTR activity. However, high concentrations of IL-16 are required for such an effect, and it is unlikely that this antiviral effect operates *in vivo*.

5. Chemokines and HIV infection

It was demonstrated several years ago that CD4 expression by a cell was not sufficient to allow its infection by HIV. Indeed, murine cells expressing the human CD4 molecule, although able to bind the virus, could not be infected, showing that additional membrane proteins of human origin are required for HIV entry into the cell. Recently it has been shown that the proteins required for HIV entry are chemokine receptors. Chemokines are cytokines with chemoattractant activity. A number of cells (particularly cells of the immune system) express chemokine receptors of one type or another. Chemokines and their receptors thus play an important role in the development of all immune and inflammatory responses, through the recruitment of circulating immune cells.

Two groups of chemokine receptor are involved in HIV entry. One is CXCR4, initially named fusin or LESTR (for leucocyte-expressed seven-transmembrane-domain receptor), which is the receptor for the Cys–X–Cys (CXC) chemokine called stromal-derived factor-1 (SDF-1). Murine cells expressing CD4 and CXCR4, both of human origin, can be infected with HIV, thus identifying CXCR4 as one of the co-receptors of the virus. CXCR4 is also involved in the fusion of infected cells with uninfected cells, hence its original name 'fusin'. Importantly, it is selective for viral strains with a tropism for T-cell lines (T-tropic strains), which are able to form syncytia and are predominant in the advanced stages of infection. Addition of high concentrations of SDF-1 prevents infection of cells by T-tropic strains, but not by M-tropic strains (which infect not only CD4⁺ T cells, but also cells of the monocyte/macrophage lineage). This can be explained because CXCR4 does not allow entry of M-tropic strains, which predominate early in infection. Such M-tropic strains, which are unable to form syncytia, are also responsible for most contamination of patients by HIV.

Some chemokines (RANTES, MIP-1α and MIP-1β), when present at high concentrations, were found to inhibit cell infection by M-tropic strains of HIV. These three chemokines share a receptor, the Cys–Cys (CC) chemokine receptor CCR5, and it is now known that this is the main receptor for M-tropic strains of HIV. Two other chemokine receptors, CCR3 and CCR2b, also allow infection by HIV, although their involvement is restricted to few strains by a broader cellular tropism. T-tropic strains using the CXCR4 receptor can also use CCR5.

The mechanism of action of chemokine receptors in HIV infection is poorly understood. It may rely on a cascade of events involving binding of the viral envelope protein gp120 on CD4, resulting in a twist of gp120 exposing another part of the molecule, the V3 loop, which then interacts with the chemokine receptor. Such an interaction is consistent with the well described role of the V3 loop polymorphism and of chemokine receptors in determining the cellular tropism of HIV. Evolution of the V3 loop throughout infection may thus select for the chemokine receptor(s) used by the virus, and thus for its cellular targets. This phenomenon may be a critical factor governing the rate of progression of the disease. Binding of the V3 loop of gp120 to a chemokine receptor would in turn induce a conformational change of the other envelope protein, gp41, which contains a fusion domain.

Interaction of this with an as yet uncharacterized membrane component, by fusing virus and cellular membranes, would then lead to delivery of the viral genome and proteins into the cell. For some viruses, such as HIV-2, expression of CD4 may be dispensable for envelope proteins binding to chemokine receptors.

Several mechanisms could account for the initial demonstration that chemokines inhibit HIV replication in CD4$^+$ T cells:

- competition for receptor binding;
- down-modulation of receptor expression; or
- desensitization of the cell to receptor triggering.

The relationship between chemokines and HIV is, however, not simple. Firstly, it is unlikely that production of chemokines by CD8$^+$ T cells accounts for the non-cytotoxic antiviral activity of such cells, because (i) there is no correlation between antiviral activity and chemokine concentration in the supernatant of CD8+ T cells, and (ii) a number of cells other than CD8$^+$ T cells produce substantial amounts of chemokines although they display no antiviral activity. Secondly, HIV strongly induces production of the chemokines RANTES, MIP-1α and MIP-1β by monocytes/macrophages *in vitro*, and its simian counterpart, SIV, has the same properties *in vivo*. These chemokines have no effect on HIV replication in macrophages, contrasting with their effect on CD4$^+$ T cells. It is thus possible that the total effect of chemokines on viral replication *in vivo* is not as beneficial as initially expected. This is not to say that chemokine receptors do not have a key role in HIV entry. In fact, it is possible that in macrophages, the binding of HIV envelope or of the chemokine itself to the chemokine receptor triggers a stimulating signal favoring subsequent steps of the viral infectious cycle. Truncated chemokines have been designed which can still bind their receptor but do not generate an activation signal. These chemokine antagonists inhibit HIV infection *in vitro*, either by competing with the virus for the receptor or by preventing induction of a receptor-mediated activation signal.

Together, these recent discoveries restore hope that it may be possible to develop new therapies based on molecules that interfere with the interaction between the virus and the cellular membrane. This was strengthened by the demonstration that, *in vivo* in humans, chemokine receptors play a critical role in HIV infectivity. It has long been known that a number of individuals, although repeatedly exposed to HIV-infected people, remain free of infection, and even developed an anti-HIV cellular immunity. This resistance to HIV infection largely relies on an inability of HIV to infect CD4$^+$ cells from such individuals. It has been recently shown that a fraction of such exposed uninfected individuals display a 32 bp deletion in the CCR5 chemokine receptor. Approximately 1% of the Caucasian population is homozygous for this deletion. However, only very few homozygotic individuals have been identified among HIV-infected patients, suggesting that expression of a functional CCR5 is absolutely required for HIV infection. In addition to showing that previous *in vitro* findings on the role of CCR5 in HIV infectivity can be extended to the *in vivo* condition, these results also emphasize the role of M-tropic strains in transmission of the virus. This apparently applies for all modes of transmission, as homozygotic CCR5 deletion prevents HIV infection regardless of the mode of contamination. However, it should be added that such a homozygous deletion only accounts for a fraction (around one in 10) exposed uninfected individuals in the Caucasian population, and that it does not apply for non-Caucasian exposed uninfected people: this deletion is apparently not found in African or, Asiatic populations. There may be other factors that protect against infection, some of which may rely on as yet unidentified mutations in chemokine receptors.

People heterozygous for the CCR5 gene deletion are not protected from infection, but the rate of progression of infection in such people seems to be than lower than that in people with two functional alleles, based on the fraction of patients surviving as well as of patients without AIDS ten years after contamination.

Altogether, these important findings indicate that, by interfering with the binding of the virus with chemokine receptors, for example with chemokine antagonists, it may be possible to prevent infection in recently exposed individuals, and to slow down the rate of progression of the disease in patients who are already infected.

6. Conclusions

This review illustrates the complexity of the pathophysiology of the HIV infection. Cytokine network abnormalities are simultaneously a consequence of the immune activation and the immune destruction and also a potential mechanism contributing to this destruction. However, most of our knowledge in this field has been obtained from *in vitro* studies, and their results often seem not to be replicated *in vivo*.

An increasing number of cytokines and cytokine antagonists will be available to physicians in the next few years in order to enhance the positive aspects of the cytokine network and minimize the negative ones. A restoration of cytokine equilibrium may allow the immune system to function better, and to stimulate antiviral and anti-infectious defenses. Results obtained so far with IL-2 treatment are quite encouraging, and new research focuses on chemokines for their potential role in immunotherapy.

References

Arenzana-Seisdedos, F., Virelizier, J.-L., Rousset, D., Clark-Lewis, I., Loetscher, P., Moser, B. and Baggiolini, M. (1996). HIV blocked by chemokine antagonist. *Nature* **383**, 400.

Baier, M., Werner, A., Bannert, N., Metzner, K. and Kurth, R. (1995). HIV suppression by interleukin-6. *Nature* **378**, 563–4.

Chehimi, J., Starr, S.E., Frank, I., D'Andrea, A., Ma, X., MacGregor, R.R., Sennelier, J. and Trinchieri, G. (1994). Impaired interleukin 12 production in human immunodeficiency virus-infected patients. *J. Exp. Med.* **179**, 1361–6.

Clerici, M., Hakim, F.T., Venzon, D.J., Blatt, S., Hendrix, C.W., Wynn, T.A. and Shearer, G.M. (1993). Changes in interleukin-2 and interleukin-4 production in asymptomatic, human immunodeficiency virus-seropositive individuals. *J. Clin. Invest.* **91**, 759–65.

Clerici, M., Lucey, D., Berzofsky, J., Pinto, L.A., Wynn, T.A., Blatt, S.P., Dolan, M.J., Hendrix, C.W., Wolf, S.F. and Shearer, G.M. (1993). Restoration of HIV-specific cell-mediated immune responses by interleukin 12 *in vitro*. *Science* **262**, 1721–4.

Clerici, M., Wynn, T.A., Berzofsky, J.A., Blatt, S.P., Hendrix, C.W., Sher, A., Coffman, R.L. and Shearer, G.M. (1994). Role of interleukin-10 in T helper cell dysfunction in asymptomatic individuals infected with HIV. *J. Clin. Invest.* **93**, 768–77.

Dean, M., Carrington, M., Winkler, C., Huttley, G.A., Smith, M.W., Allikmets, R., Goedert, J.J., Buchbinder, S.P., Vittinghoff, E., Gomperts, E. *et al.* (1996). Genetic restriction of HIV-1 infection and progression to AIDS by a deletion allele of the CKR5 structural gene. *Science* **273**, 1856–62.

Emilie, D., Peuchmaur, M., Maillot, M.-C., Crevon, M.-C., Brousse, N., Delfraissy, J.-F., Dormont, J. and Galanaud, P. (1990). Production of interleukins in human immunodeficiency virus-1-replicating lymph nodes. *J. Clin. Invest.* **86**, 148–59.

Graziosi, C., Pantaleo, G., Gantt, K.R., Fortin, J.P., Demarest, J.F., Cohen, O.J., Sékali, R.P. and Fauci, A.S. (1994). Lack of evidence for the dichotomy of Th1 and Th2 predominance in HIV-infected patients. *Science* **265**, 248–52.

Kovacs, J.A., Vogel, S., Albert, J.M., Falloon, J., Davey, R.T., Walker, R.E., Polis, M.A., Spooner, K., Metcalf, J.A., Baseler, M. *et al.* (1996). Controlled trial of interleukin-2 infusions in patients infected with the human immunodeficiency virus. *New Engl. J. Med.* **335**, 1350–56.

Lane, H.C. (1994). Interferons in HIV and related diseases. *AIDS* **8**, S19–23.

Lane, H.C., Depper, J.M., Greene, W.C., Whalen, G., Waldmann, T.A. and Fauci, A.S. (1985). Qualitative analysis of immune function in patients with the acquired immunodeficiency syndrome. Evidence for a selective defect in soluble antigen recognition. *New Engl. J. Med.* **313**, 79–84.

Levy, J.A. (1996). Infection by human immunodeficiency virus CD4 is not enough. *New Engl. J. Med.* **335**, 1528–30.

Liu, R., Paxton, W.A., Choe, S., Ceradini, D., Martin, S.R., Horuk, R., Mac Donald, M.E., Stuhlmann, H., Koup, R.A. and Landau, N.R. (1996). Homozygous defect in HIV-1 coreceptor accounts for resistance of some multiply-exposed individuals to HIV-1 infection. *Cell* **86**, 367–77.

Mackewicz, C.E., Blackbourn, D.J. and Levy, J.A. (1995). CD8⁺ T cells suppress human immunodeficiency virus replication by inhibiting viral transcription. *Proc. Natl Acad. Sci. USA* **92**, 2308–12.

Mackewitz, C.E., Levy, J.A., Cruikshank, W.W., Kornfeld, H. and Center, D.M. (1996). Role of IL-16 in HIV replication. *Nature* **383**, 488–9.

Romagnani, S., Del Prete, G., Manetti, R., Ravina, A., Annunziato, F., De Carli, M., Mazzetti, M., Piccinni, M.P., D'Elios, M.M., Parronchi, P. *et al.* (1994). Role of Th1/Th2 cytokines in HIV infection. *Immunol. Rev.* **140**, 73–92.

Samson, M., Libert, F., Doranz, B.J., Rucker, J., Leisnard, C., Farber, C.M., Saragosti, S., Lapouméroulie, C., Cognaux, J., Forceille, C. *et al.* (1996). Resistance to HIV-1 infection in caucasian individuals bearing mutant alleles of the CCR-5 chemokine receptor gene. *Nature* **382**, 722–5.

Zou, W., Dulioust, A., Fior, R., Durand-Gasselin, I., Boué, F., Galanaud, P. and Emilie, D. (1997). Increased Th2-type cytokine production in chronic HIV infection is due to IL-13 rather than IL-4. *AIDS* **11**, 533–4.

Zou, W., Lackner, A.A., Simon, M., Durand-Gasselin, I., Galanaud, P., Desrosiers, R.C. and Emilie, D. (1997). Early cytokine and chemokine gene expression in lymph nodes of macaques infected with simian immunodeficiency virus is predictive of disease outcome and vaccine efficacy. *J. Virol.*, **71**, 1227–36.

XXIV Allergic responses and cytokines

Enrico Maggi*, Francesca Brugnolo, Salvatore Sampognaro and Paola Parronchi

1. Introduction

Atopy is a genetically determined group of disorders characterized by an increased ability of B lymphocytes to synthesize IgE antibodies against ubiquitous antigens (allergens) able to activate the immune system after inhalation or ingestion, and perhaps after penetration through the skin. IgE antibodies can bind high-affinity Fcε receptors (FcεRI) present on the surface of mast cells/basophils, and allergen-induced FcεRI cross-linking triggers the release of vasoactive mediators, chemotactic factors and cytokines, which are responsible for the allergic events. Eosinophils also appear to be involved in the pathogenesis of allergic reactions, since they usually accumulate at the site of allergic inflammation and their toxic products significantly contribute to the induction of tissue damage.

The mechanisms by which IgE-producing B cells, mast cells/basophils and eosinophils contribute to the pathogenesis of allergic reactions were unclear until distinct subsets of CD4+ T-helper (Th) cells, based on their profile of cytokine secretion, were discovered. At least three different subsets of T-helper cells have been described in both mouse and humans:

(1) Th1 cells, which produce interleukin 2 (IL-2), interferon gamma (IFNγ) and tumour necrosis factor beta (TNF-β);

(2) Th2 cells, which produce IL-4, IL-5 and IL-10; and

(3) Th0 cells, which produce both Th1- and Th2-type cytokines.

IL-3, granulocyte–macrophage colony-stimulating factor (GM-CSF) and TNF-α are variably produced by all subsets of T-helper cells. IL-3, IL-4 and IL-10 are growth factors for mast cells, while IL-5 is a selective activating and differentiating factor for eosinophils. It has been clearly shown that IgE synthesis results from the collaboration between Th2 cells and B cells. Recently, a less restrictive definition has been introduced for Th1/Th2 cell subsets: Th2 cells are defined as being able to produce IL-4 actively, whereas Th1 cells have lost this function. Knowledge of T-helper cell properties has allowed clarification of the mechanisms linking IgE-producing B cells, mast cells/basophils and eosinophils, leading to the proposal of the 'Th2 hypothesis' for the pathogenesis of allergic reactions. In this chapter we examine the activity of cytokines in allergic response, the molecules and signals operating in the regulation of IgE synthesis, the role of allergen-reactive Th2 cells in the pathogenesis of human allergic disorders and the possible mechanisms involved in the regulation of Th2-cell development. Finally, we will review novel strategies for therapy of atopic diseases.

2. Regulation of IgE synthesis

The investigation of the molecular events underlying IgE synthesis has provided an interesting model for identifying and characterizing the signals involved in isotype-specific regulation of immunoglobulin synthesis in humans.

2.1 Signals required for IgE producton

C_H genes encode the constant region of the heavy chains of immunoglobulins. The C_H genes for the

*Corresponding author.

various immunoglobulin isotypes are found in a cluster, all in the same transcriptional orientation, and downstream of the V_HDJ_H gene complex, which encodes specificity for antigen. Human C_H genes are located in a 300 kb region of chromosome 14, in the following order: μ, δ, $\gamma3$, $\gamma1$, $\varepsilon2$, $\alpha1$, $\gamma2$, $\gamma4$, $\varepsilon1$ and $\alpha2$ (where μ denotes the heavy chain of IgM, δ the heavy chain of IgD, etc.). Upstream of each C_H gene, there are 2–10 kb sequences called switch regions (or S regions) which are composed of short tandem repeats of GAGCT or GGGCT. During an immune response, the same B lymphocyte can express different C_H region genes in association with the same V(D)J region; this phenomenon is known as heavy-chain class switching, or isotype switching.

Two processes are required for the induction of immunoglobulin isotype switching: (i) expression of germline RNA for the C_H gene to which switching will occur, and (ii) DNA synthesis (i.e. B-cell proliferation). Germline C_H RNA differs from mRNA encoded by a rearranged C_H gene in that it lacks the exon that encodes V_HDJ_H, and it is 'sterile', i.e. not translated. All germline transcripts initiate 5' of the switch region, suggesting that transcription through a targeted region may make it accessible to a recombinase. This 'accessibility model' predicts that an increase in steady-state levels of germline C_H RNA results from an increase in the rate of C_H gene transcription. IL-4, IFNγ and transforming growth factor beta (TGF-β) increase the steady-state levels of germline murine $\gamma1$ and ε, $\gamma2a$ and $\gamma3$, and $\gamma2b$ and α, respectively.

In 1986 Coffman and Carty showed that IL-4 was able to induce production of IgE and IgG1 by murine B cells stimulated *in vitro* with lipopolysaccharide (LPS) and that its activity was strongly inhibited by IFNγ. This was later confirmed *in vivo* by injection of a monoclonal anti-IL-4 antibody that abolishes IgE production and, more importantly, in IL-4 knockout mice which exhibited no IgE synthesis. The role of IL-4 in the induction of human IgE synthesis was shown by the use of T-cell clones. Subsequent observations confirmed that human recombinant IL-4 induces IgE synthesis by peripheral blood lymphocytes, and that this effect is inhibited by recombinant IFNγ.

In humans IL-4 induces transcription through the Sε region, resulting in the synthesis of germline ε transcript by B cells. It consists of an approximately 134 bp Iε exon directly spliced to Cε by removal of the intervening sequences from the primary transcript. The germline transcript contains all four Cε exons, as well as the 3' untranslated

region. The human Iε does not contain an initiation codon in-frame with the Cε region, and stop codons are present in all three reading frames, suggesting that it is a sterile transcript. Germline ε transcripts can be directly induced by IL-4 in resting B cells. Recently IL-13, a cytokine showing poor homology with IL-4 (30%) but having similar IgE-switching activity, was discovered (see Chapter I). Human IL-13 has IL-4-like effects, inducing both B-cell proliferation and switching to IgE and IgG4 production. IL-13 also induces germline ε mRNA expression in highly purified B cells, suggesting that it can direct switching to the IgE isotype. It has recently been shown that receptors for IL-4 and IL-13 are distinct but share a common subunit (γ-chain). However, their kinetics of production differ markedly: IL-13 mRNA expression peaks 2 h after activation and fairly high levels of IL-13 mRNA are still observed after 72 h, whereas IL-4 mRNA peaks after 4–6 h and is undetectable after 24 h. Based on this finding, it was suggested that IL-13 may play an important role in the enhanced IgE synthesis seen in allergic patients.

Even though both IL-4 and IL-13 are sufficient to begin germline transcription through the ε locus, it is thought that physical interaction between T and B cells is needed for the expression of mature ε mRNA transcripts and for the production of IgE protein; this is called the 'two-signal model' for the induction of human IgE synthesis. The molecules involved in the contact-mediated non-cognate signalling required for IgE production have recently been clarified. The most critical signalling is provided by the interaction of CD40 present on the B cell with its ligand (CD40L) expressed on the activated T cell. CD40 is a 50 kDa cell-surface glycoprotein expressed on all human B cells (but not on T cells or monocytes), whereas CD40L encodes a 39 kDa type II glycoprotein on activated T cells. Transfectants expressing CD40L induce B cells to proliferate and differentiate into IgE-secreting cells in the presence of IL-4, indicating that all signals required for IgE switching can be delivered by CD40L and IL-4. Mutations of the CD40L gene (mapped to the X-chromosomal location q26.3-q27.1) have recently been identified in patients with X-linked hyper-IgM syndrome (see Chapter IV), resulting in defective expression of CD40L and impaired isotype switching *in vivo*. CD40 triggering does not *per se* induce either germline transcript or mature Cε transcripts, but synergizes with IL-4 to enhance ε germline transcript accumulation. A mature mRNA is observed only in B cells stimulated with both IL-4 and anti-CD40 monoclonal antibody (mAb) for 10 days. The 26 kDa membrane form of TNF-α, expressed on activated CD4[+] cells, is also associated with productive T-cell/B-cell interactions. Recently, another member of the TNF superfamily, the CD30 ligand (CD30L) was found to be very active in inducing CD40L-independent IgE secretion. Besides the above-mentioned model of IgE synthesis, other T-cell-independent systems of IgE synthesis have been described, such as IL-4 plus Epstein–Barr virus infection and IL-4 plus hydrocortisone.

Mast and basophilic cell lines (as well as normal purified lung mast cells and blood basophils) have been reported to increase IgG and to induce IgE synthesis in purified B cells stimulated with IL-4. This IgE synthesis is inhibited by CD40 fusion protein, suggesting that mast or basophilic cell lines can express CD40L for the induction of IgE production. Basophils can induce IgE synthesis even in the absence of exogenous IL-4, since they are able to release preformed IL-4, whereas the ability of normal mast cells to release IL-4 is more controversial. However, the *in vivo* relevance of this *in vitro* model is doubtful, since no antigen-specific receptor molecules have been identified on mast cells, and we do not know how the specificity of the mast cell-induced IgE synthesis could be controlled. In any case it is a very efficient amplifying system to locally increase IgE.

2.2 Modulation of IgE synthesis by soluble factors

T-cell/IL-4-dependent IgE synthesis can be modulated by cytokines other than IL-4: IL-2, IL-5, IL-6, TNF-α and IL-9 enhance IL-4-induced IgE synthesis. IL-6 is known to act at a late stage in B-cell differentiation with no isotype preference. The mechanisms responsible for the enhancing effect of IL-2 and TNF-α on IgE synthesis are not completely clear. Both cytokines indeed stimulate proliferation of both T and B cells, as well as the differentiation of B cells into antibody-producing cells. Interestingly, IL-2 has opposite effects on IgE helper activity by naïve and memory T cells: it inhibits the IgE helper activity of naïve T cells but enhances the IgE helper activity of memory T cells. Factors such as IFNα , IFNγ, TGF-β, IL-8, IL-10, IL-12, platelet activating factor (PAF-acether) and prostaglandin E_2 have been shown to down-regulate IgE synthesis. IFNγ profoundly suppresses the expression of ε germline transcripts in murine B cells stimulated with LPS plus IL-4, whereas no inhibition is observed when IFNγ is added to highly purified human B cells stimulated with IL-4, sug-

gesting that activation of ε germline transcription and switch recombination may not be associated. The ability of IL-8 to block spontaneous IgE synthesis appears to be mediated by its ability to decrease the production of IL-6 and TNF-α. On the other hand, IL-10 blocks IgE synthesis by peripheral mononuclear cells by inhibiting some monocyte functions, but, conversely, it also directly stimulates B cells cultured in the presence of IL-4 and anti-CD40 mAbs cross-linked to CD32 (FcγRII) on murine L cells. The mechanism responsible for the inhibitory activity of IL-12 on IgE synthesis is probably mediated by IFNγ, since IL-12 is a powerful inducer of IFNγ by T cells and natural killer (NK) cells.

3. The 'Th2 hypothesis' in atopy

As mentioned above, Th2 cells produce IL-4 and IL-13 (which stimulate IgE and IgG1 antibody production), IL-5 (which recruits and differentiates eosinophils) and IL-10 (which, together with IL-4 and IL-13, inhibits several macrophage functions). Therefore, the Th2 cell can induce a phagocyte-independent host defense response, e.g. against certain nematodes, as well as being an excellent candidate to explain why the mast cell/eosinophil/IgE-producing B-cell triad is involved in the pathogenesis of allergy.

3.1 Allergens preferentially expand T-helper cells showing a Th2 profile of cytokine secretion

The study of the cytokine profile of T-cell clones specific for different antigens has clearly shown that, in contrast to clones specific for bacterial antigens showing a prevalent Th1/Th0 phenotype, the great majority of allergen-specific T-cell clones generated from peripheral blood lymphocytes of atopic donors express a Th0/Th2 profile, with high production of IL-4 and IL-5 and no or low production of IFNγ.

3.2 Th2 cells accumulate in target organs of allergic patients

Several pieces of evidence have been obtained, using either cloning techniques or *in situ* hybridization, to show that Th2-like cells accumulate in target organs in various allergic disorders. For example, the majority of T-cell clones generated from the conjunctival infiltrates of patients with vernal conjunctivitis develop into Th2 clones. Using *in situ* hybridization, cells showing mRNA for Th2, but not Th1, cytokines have been detected at the site of late-phase skin reactions in skin biopsies from atopic patients, in mucosal bronchial biopsies or bronchoalveolar lavage (BAL) from patients with asthma and after local allergen challenge in nasal mucosa of patients with allergen-induced rhinitis. Likewise, increased levels of IL-4 and IL-5 have been measured in the BAL of allergic asthmatics, whereas in non-allergic asthmatics IL-2 and IL-5 predominates.

3.3 Allergen-challenge results in the activation and recruitment of allergen-reactive Th2 cells in target organs

Inhaled allergens induce activation and recruitment of allergen-specific Th2 cells in the airway mucosa of patients with respiratory allergy. Biopsy specimens were obtained from the bronchial or nasal mucosa of patients with grass pollen-induced asthma or rhinitis 48 h after positive bronchial or nasal provocation test with allergen. High proportions of T-cell clones derived from stimulated airway mucosa of these patients were specific for grass allergens and exhibited a Th2 profile. Similarly, high proportions of *Dermatophagoides pteronyssinus* (DP)-specific Th2-like CD4+ T-cell clones were generated from the skin of patients with atopic dermatitis taken after contact challenge with DP, suggesting that transcutaneous sensitization to aeroallergens may be essential in the induction of skin lesions in patients with atopic dermatitis.

Very recently it has been proposed that CCR3, the ligand for the β-chemokine eotaxin, is expressed not only on eosinophils, basophils and mast cells, but also on Th2-type cells; the attraction of Th2 cells by eotaxin, produced by endothelial and phagocytic cells, could represent a key mechanism in allergic reaction, since it favours the locally allergen-driven production of IL-4 and IL-5 essential to activate basophils and eosinophils.

3.4 Grass-specific Th2 cells expressing membrane CD30 are present in the peripheral blood of allergic patients during seasonal exposure to pollens

Recently, we have shown that CD30, a member of the TNFR superfamily, is preferentially expressed by T-cell clones able to produce Th2-type cytokines.

No CD4+CD30+ cells were detected in any of the non-atopic or atopic donors examined before the grass pollination season, whereas the majority of grass-sensitive donors assessed during the season showed small proportions of circulating CD4+CD30+ cells (from 0.08% to 0.3%). Only CD30+ cells proliferated in response to Lol p 1 and produced IL-4 and IL-5, whereas CD30- cell fractions produced mainly Th1-cytokines. These findings demonstrate that grass allergen-reactive CD4+CD30+ Th2 cells can circulate in the peripheral blood of grass-sensitive patients during the *in vivo* natural exposure to grass pollen allergens.

3.5 Specific immunotherapy can modify the cytokine profile of allergen-specific T-helper cells

Further evidence for the role of Th2-like cells in allergic disorders has been provided by studies on the effect of allergen-specific immunotherapy on the cytokine profile of T cells. Pollen immunotherapy did not affect the expression of Th2-type cytokine pattern in response to allergen exposure at the level of allergen-induced late-phase cutaneous reactions, but mRNA expression of Th1-type cytokines was enhanced. In another study, successful immunotherapy was found to reduce IL-4 production by allergen-specific CD4+ T cells, whereas production of IFNγ was not affected. Finally, both decreased production of IL-4 and increased production of IFNγ was observed in bee venom-sensitized patients treated with specific immunotherapy. Although partially discordant, these reports support the concept that the cytokine profiles of allergen-specific CD4+ T cells are not fixed and can be manipulated by *in vivo* therapies. This possibility is also supported by studies in mice injected with chemically modified allergen or infected with *Leishmania major* and treated with IL-12 and Pentostam.

4. Mechanisms involved in the regulation of Th2 development

The mechanisms responsible for the preferential development of allergen-reactive Th2 cells in atopic subjects have not yet been completely clarified. Attention has been focused on the possible role of antigen-presenting cells (APCs), the T-cell repertoire and soluble factors present in the microenvironment at the time of allergen presentation.

4.1 APCs and the T-cell repertoire

It is well known that Langherans cells in the skin, as well as dendritic cells in the respiratory mucosa, are the primary point of contact between the immune system and allergens coming through the skin or the respiratory airways, respectively. Langherans cells and dendritic cells are probably involved in allergen transport to regional lymph nodes where allergens are presented to allergen-specific CD4$^+$ T cells. It has been suggest that atopic patients with asthma have higher numbers of intraepithelial dendritic cells than non-asthmatic subjects and that these cells (in the presence of allergen) can trigger T cells to release IL-4 and IL-5. However, the actual role of APCs in driving the development of allergen-reactive Th2-like cells remains obscure.

Recently, co-stimulary signals, namely the interactions between CD28/CTLA and their ligands expressed on APCs (CD80 and CD86), have been found to have a clear-cut effect on modulating Th1 and/or Th2 development. Alteration of CD30L expression and/or activity (another co-stimulatory signal) seems to have some role in the development of allergen-specific Th2 responses in atopic individuals. Blocking of CD30L on APCs may shift the *in vitro* differentiation of allergen-specific T cells from the Th0/Th2 profile to the Th0/Th1 profile.

More recently, it has been reported that varying either the density or the affinity of peptide for MHC class II molecules could determine the type of T-cell response. A high density of MHC class II peptides on the APC surface favoured Th1-like responses, while low densities favoured Th2-like responses. Moreover, by using a panel of ligands with varying affinity for MHC class II binding, it was shown that stimulation with the highest affinity ligand resulted in IFNγ production while stimulation with a lower affinity ligand induced IL-4 secretion. These findings suggest that the MHC-binding affinity of antigenic peptides leads to differential interactions at the T cell–APC interface which is crucial for the differential development of cytokine patterns in T cells.

The role of T-cell repertoire in determining the development of Th1 or Th2-type responses is still controversial. In mice infected with *Leishmania major*, Th1 and Th2 cells displaying the same repertoire and recognizing the same peptide have been demonstrated, suggesting that cells with identical TCRs can differentiate into either the Th1 or the Th2 phenotype. However, there is evidence that specific Vβ-expressing T-cell subsets have a pivotal role in the stimulation of IgE production and increased airway-responsiveness induced by ragweed allergen. Thus, the possibility cannot be excluded that allergen recognition by the TCR provides a signal driving T cells towards the production of either IL-4 or IFNγ.

4.2 Microenvironmental factors

4.2.1 Hormones

It has also been suggested that hormones have a role in promoting the differentiation of T-helper cells or in favouring the shifting of differentiated T-helper cells from one cytokine profile to another. Glucocorticoids enhance Th2 activity, and synergize with IL-4, whereas dehydroepiandroster one sulfate enhances Th1 activity. Another major prohormone, 25-hydroxycholecalciferol (25-OH-vitamin D$_3$) may have opposite effects on the Th1/Th2 balance. The intense conversion of 25-OH-vitamin D$_3$ to 1,25-(OH)$_2$-vitamin D$_3$ (calcitriol) decreases secretion of IL-2 and IFNγ and increases the Th2 pattern of response. Calcitriol analogues can also antagonize cyclosporin A in their ability to prolong survival of skin grafts by inhibiting Th1 activity. Lastly, progesterone favours the *in vitro* development of human T cells producing Th2-type cytokines and promotes both IL-4/IL-5 production and membrane CD30 expression in established human Th1 clones. This may be one of the mechanisms involved in the Th1/Th2 switch occurring at the maternal–fetal interface in order to promote successful pregnancy.

4.2.2 Th1-skewing factors

IFNα, IL-12 and TGF-β produced by macrophages and B cells play an important role in the induction of Th1 expansion. IFNγ produced by T and NK cells promotes the differentiation of Th1 cells. Likewise, IL-12, which is a powerful IFNγ inducer, appears to be the most important natural initiator of Th1 responses by acting either directly or indirectly via the induction of IFNγ production. Recent *in vivo* data from IL-12 knockout mice show that (i) Th1 responses were impaired, but not completely absent, (ii) the magnitude of DTH was substantially decreased and (iii) secretion of IL-4 was enhanced.

Addition of IFNγ to cultures containing optimal amounts of IL-4 failed to inhibit the priming of CD4$^+$ T cells from TCR-transgenic mice to develop into IL-4-producing T cells. When sub-optimal concentrations of IL-4 were used, IFNγ caused a significant decrease in the amount of IL-4 produced after restimulation. This finding suggests that few IL-4-producing T-cell clones emerge from a culture containing IFNγ. Indeed, IFNγ can regulate the development of Th1 cells independently of IL-12 and, at relatively low concentrations, it can inhibit the proliferation of murine Th2 clones. Although not absolute, the inhibitory effect of IFNγ on proliferation of murine Th2 cells is significant and seems to be sufficient to limit the clonal expansion of such cells. The proliferation of human Th2 clones is also inhibited by IFNγ.

IFNα was found to decrease the level of splenic IL-4 mRNA induced either by treatment with anti-IgD mAb or by infection with *Nippostrongylus brasiliensis*, whereas in both conditions the levels of IFNγ mRNA were increased. In the absence of IFNγ, IFNα augments IL-12 effects on inhibition of subsequent IL-4 production rather than enhancing IFNγ-driven by IL-12 production. Recently, it has been shown that, like IL-12, IFNα can induce tyrosine phosphorylation and DNA-binding of STAT4. In humans, IFNα inhibits the development of allergen-specific T cells into Th2-like cells, and up-regulates the expression of the IL-12 receptor β-chain in naïve human T cells.

4.2.3 Th2-skewing factors

Cytokines produced by macrophages and/or B cells seem to have a less critical effect on the development of Th2 cells. IL-10 favours the development of Th2 cells both in mouse and in humans. IL-1 is a selective co-factor for the growth of some murine Th2 clones and can favour the development of human Th2-like clones *in vitro*. Recently IL-6 produced by APCs has been found to be one of the Th2-skewing factors in the primary response. In both murine and human systems, IL-4 appears to be the dominant factor determining the likelihood of Th2 polarization in cultured cells. Accordingly, IL-4-gene-targeted mice fail to generate mature Th2 cells *in vivo* and fail to produce IgE antibodies, suggesting that early IL-4 production by other cell types is involved.

The source of IL-4 in the primary response able to modulate Th2 differentiation is not known. Possible candidates include mast cells and basophils, the CD4+ NK1.1+ T-cell subset or T-helper cells themselves.

Human mature mast cells and basophils produce IL-4 in response to several secretagogues, and activated human eosinophils can also release high IL-4 concentrations. Thus, at least potentially, FcεR+ non-T-cells might be an amplification system for Th2 cells *in vivo* during allergic reactions and parasitic infestations. However, it is unlikely that parasites or allergens would be able to cross-link FcγR and FcεR before parasite-specific IgG and IgE had been produced. This has been confirmed by two observations: mast cell-deficient mice develop normal Th2 responses, and IL-4-producing T cells (but not IL-4-producing non-T-cells) are able to reconstitute the antigen-specific IgE response in IL-4-deficient mice. Thus, IL-4 production by mast cells triggered by IgE-containing immune complexes may amplify secondary responses to parasites, but cannot induce the Th2 development in primary responses.

The role of CD4+ NK1.1+ T cells in the development of Th2 responses has been suggested on the basis of various observations:

(1) They are selected by the non-polymorphic MHC class I molecule, CD1.

(2) They express large amounts of IL-4 mRNA after intravenous injection of anti-CD3 antibody, that is at the onset of an immune response.

(3) β_2-microglobulin-deficient mice and SJL mice (which are deficient in CD4+NK1.1+ T cells) cannot produce IL-4 quickly in response to anti-CD3 antibody.

However, other studies suggest that these cells are not involved in Th2 responses; indeed, it is unlikely that all antigens that promote the differentiation of naïve T-helper cells into the Th2 pathway are capable of activating CD4+ NK1.1+ T cells. It has been suggested that another subset of NK1.1+ cells (the CD4− CD8− γδ+ subset) provides high levels of IL-4 during priming. Recently, in our laboratory, NK1.1+ cells producing IL-4 were found in the thymus of 9–12 week human fetuses; some of them were CD4+, suggesting the existence of human equivalents of murine IL-4-producing CD4+ NK1.1+ T cells.

A more likely possibility is that the maturation of naïve T cells into the Th2 pathway depends mainly on the level of IL-4 production by naïve T cells themselves at priming. That is supported by several findings:

(1) Low-intensity signalling of TCR mediated by low peptide doses or mutant peptides leads to secretion of low levels of IL-4 by murine naïve T cells.

(2) Naïve T cells require two or more stimulatory events to produce IL-4 and this is blocked by

anti-IL-4 mAb, suggesting a role for endogenous IL-4 produced by the naïve T cells themselves.

(3) Human adult and neonatal CD45RA⁺ T cells develop into IL-4-producing cells in the absence of any source of IL-4 or anti-IL-4 antibodies.

(4) High proportions of Th2 clones can be generated from single CD4⁺ $\alpha\beta^+$ T cells isolated from the thymus of young children.

A significant fraction of uncommitted T cells may be primed for a Th2 phenotype (independent of antigen and IL-4) if they are exposed to IL-2 and interact with accessory cells bearing the natural CD28 ligands B7-1 and B7-2. When stimulated by specific antigen, such primed Th2 precursor cells provide a source of IL-4 to promote a Th2-type response.

Thus, there is strong evidence that the maturation of naïve T cells into the Th2 pathway mainly depends upon the levels and the kinetics of autocrine IL-4 production at priming.

5. Mechanisms possibly responsible for allergen-specific Th2 responses in atopic subjects

New insights into the pathogenesis of allergic disorders have led to the proposals that the preferential development of allergen-specific Th2-like responses in atopic subjects is determined by the nature and the intensity of TCR signalling by the allergen peptide ligand on both CD4 and CD8⁺ T cells, and by altered regulation of IL-4 production by T-helper cells, which is probably controlled at genetic level. There are several pieces of evidence that in atopic subjects there is genetic dysregulation at the level of Th-cell-derived IL-4:

(1) CD4⁺ T-cell clones from atopic individuals can produce IL-4 and IL-5 in response to bacterial antigens that usually evoke Th1-like responses in non-atopic individuals.

(2) Atopic donors have a higher frequency of IL-4-producing T cells than normal subjects.

(3) T-cell clones generated from cord blood lymphocytes of newborns with atopic parents produce higher IL-4 concentrations than these from newborns with non-atopic parents

(4) Peptide-specific T-cell clones generated from donors with high IgE serum levels produce high amounts of IL-4 and low IFNγ whereas T-cell clones generated from 'low IgE' donors show a Th1-like profile.

The role of CD8⁺ T cells in allergic sensitization is still unclear and probably complex. Some studies suggest a suppressive function for CD8⁺ T cells, associating IFNγ-producing CD8⁺ T cells with the inhibition of IgE production, and consequent down-regulation of allergic sensitization. Recent data support the possible role of allergen-specific CD8⁺ T cells in controlling the T-helper cell response to allergens in humans. Firstly, it has been shown that a nonapeptide expands higher numbers of CD8⁺ T-cell clones in 'low' than IgE producers than in 'high' IgE producers. Secondly, lactalbumin expands more CD8⁺ T-cell clones in non-atopic milk-sensitive donors than in atopic milk-sensitive donors, suggesting that allergen-specific CD8⁺ T cells may play an important role (via IFNγ production?) in preventing the differentiation of allergen-specific Th2 cells in non-atopic people.

On the whole these data suggest that allergen peptide ligand can influence the cytokine profile of T-helper cells. However, mechanisms underlying non-cognate regulation of IgE responsiveness are overwhelming. These mechanisms may be either stimulatory (up-regulation of factors that promote IL-4 gene expression in human T cells, favouring the development of Th2-like cells), or inhibitory (down-regulation of factors that dampen IL-4 production and/or Th2-cell development), or both types of regulation may occur.

6. Genetic alterations favouring allergen-specific Th2-responses in atopic patients

Genetic mechanisms underlying heightened IgE responsiveness seen in atopic diseases may be divided into two types, antigen-specific and non-antigen-specific. In the past the possible role of MHC class II molecules has been extensively investigated by both genetic and molecular approaches. Significant associations between some HLA haplotypes and specific immune responses to different allergens (such as Amb a 5 and 6, or Lol p 1, 2 and 3) have been observed. However, weaker associations for other allergens have been reported, suggesting that class II molecules on APCs can play a permissive role in the binding of allergen peptides, but other non-MHC-associated genes are more important in the overall IgE immune responsiveness to allergens (prevalent Th2-type response). Recent evidence has been provided for a linkage of overall IgE to markers in chromosome 5q31.1, expecially to the IL-4 gene.

The entire IL-4 gene has been scanned for possible atopy-associated polymorphisms based on the hypothesis that they may reside in transcriptional regulatory elements. The IL-4 promoter contains functionally important binding sites for several transcription factors (including NF-AT, the CCAAT box binding protein NF-Y, Oct1, HMGI(Y), AP-1 members, NF-κB, and an as yet unpurified factor termed PCC) and silencer elements (which bind factors termed NRE). It has been shown that nuclear extracts from individuals with atopic dermatitis have higher affinity for a consensus P-element, suggesting that polymorphism in NF-AT family members (either in their amino acid sequence or in post-translational modification) may be involved. More recently, it has been shown that the product of the proto-oncogene c-*maf*, a transcription factor, controls tissue-specific expression of IL-4. c-Maf is expressed in Th2 clones but not in Th1 clones and is induced during normal precursor cell differentiation along a Th2 lineage but not a Th1 lineage. These data indicate that c-Maf is responsible for dictating Th2-selective IL-4 gene transcription and make c-*maf* an obvious candidate for an atopy gene. Another good candidate for such an atopy gene is the one that encodes STAT6, a protein that binds to DNA sequences found in the promoters of IL-4-responsive genes. Experiments on knockout mice have shown that this protein is required for the development of Th2 cells.

Down-regulation of mechanisms dampening IL-4 production may also be involved in the development of prevalent Th2 responses in atopy. IL-12 induces tyrosine phosphorylation and DNA-binding of STAT3 and STAT4. STAT-deficient mice have impaired Th1 development and enhanced Th2 development. A locus that controls the maintenance of IL-12 responsiveness, and therefore favours the preferential development of Th1 cells, has recently been described in B10.D2 mice. This locus maps to a region of chromosome 11 which is syntenic with the locus on human chromosome 5q31.1, shown to be associated with elevated serum IgE levels. Several other genes map within 5q31.1, including possible candidates which might influence IgE production, namely *irf1* (whose gene

product up-regulates IFNα, which down-regulates IgE production and inhibits Th2-cell development) and *il12β* (which encodes the β-chain of IL-12, another down-regulator of Th2 cells).

Direct alterations in the regulation of IL-4 gene expression, or deficient regulation of cytokines responsible for inhibiting Th2-cell development, or both may account for the preferential Th2-type response in atopic people, and for the production by Th2 cells of the cytokines involved in the allergic inflammation, and hence explain the persistent histological, pathophysiological and clinical aspects of allergic disorders (Figure 1).

7. Novel strategies for immunotherapy of allergic disorders

These new insights, as well as biotechnological advances, provide novel opportunities for the treatment of allergic diseases. Immunotherapy may be either allergen-specific (targeting allergen-specific T cells) or non-allergen-specific (targeting effector molecules). Allergen-specific immunotherapy might be aimed at induction of T-cell anergy with allergen-derived peptides, or at induction of Th1 switch. By contrast, non-allergen-specific immunotherapy is based on antagonizing Th2 cytokines, or on anti-IgE therapy.

7.1 Allergen-specific immunomodulation

Anergy in allergen-specific Th2 cells *in vitro* has been achieved by incubating DP group 1 (Der p 1)-specific Th2 clones with high doses of the relevant Der p 1-derived peptides in the absence of APCs. Der p1-specific T-cell clones have lost the ability to respond to subsequent stimulation with the intact antigen and APCs, and the ability to provide B-cell help for IgE production even in the presence of exogenous IL-4 and IL-13. The activation of allergen-specific T cells and immunoglobulin synthesis in mice can be inhibited *in vivo* by intranasal, oral or subcutaneous administration of allergens or their peptides. This and other studies suggest that it may be possible to use peptide-induced T-cell anergy to treat human allergic diseases. The major advantage of this type of specific immunotherapy over traditional types is its potential safety. So far, a trial with bee venom phospholipase A2-derived

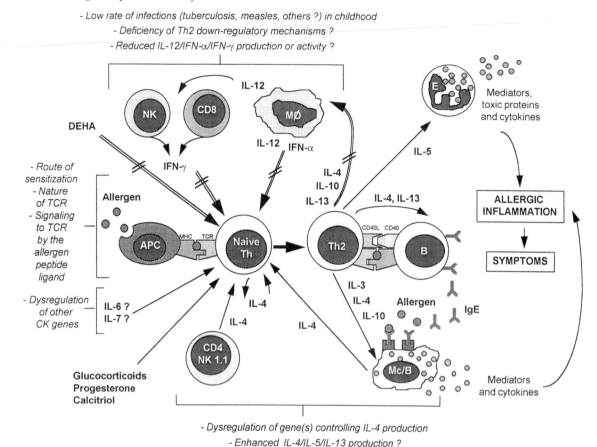

▮ **Figure 1.** Mechanisms responsible for the development of allergen-specific T cells into Th2 cells in atopic individuals. Allergens preferentially stimulate the development of Th2-like T cells in genetically determined subjects in whom IL-4 family genes are overexpressed and/or there is deficient regulation of the activity of cells (macrophages (MØ), NK cells and CD8[+] cells) producing Th2-type inhibitory cytokines. The latter scenario may result from genetic dysregulation of mechanisms regulating Th2 responses or to a low rate of childhood infections which down-regulates potential Th1 immunity. Other micro-environmental cells (CD4[+] NK1.1[+], mast cells, etc.) and hormones can be locally involved in Th2 development. Allergen-specific Th2 cells produce IL-4/IL-13 (which promote IgE switching, mast cell recruitment and survival), IL-5 (which promotes eosinophil recruitment and survival) and IL-10 (which, together with IL-4 and IL-13, inhibits macrophage functions). Mediators, toxic proteins and cytokines released by both eosinophils and mast cells/basophils initiate and locally sustain the allergic inflammation, leading to symptoms.

peptides has shown good clinical efficacy with no adverse reactions. Another clinical trial with immunodominant peptides derived from the major cat allergen Fel d 1 showed that the treatment is safe and may result in clinical improvement.

Another strategy to neutralize allergen-specific Th2 responses would be to change the cytokine profile of allergen-specific CD4[+] T cells. It might be possible to act directly on allergen-specific Th2 cells, but fully polarized murine Th2 cells seem to be resistant to a reversal in their cytokine produc-

tion patterns. It is reasonable that reversibility of polarized Th2 cell populations is lost after long-term stimulation. However, memory CD4[+] T cells can be induced to change to the opposite phenotype. *In vivo* administration of IL-12 to mice reverses detrimental *Leishmania major*-specific Th2 responses when it is given with a drug that reduces the parasite load. This finding confirms our studies suggesting that memory cells previously committed to the Th2 pattern of cytokine secretion can be modulated to produce cytokines of the opposite phenotype. Moreover, naïve human T cells primed to develop in the Th2 cell phenotype by IL-4 are not stable and rapidly revert to a population containing Th1 cells after restimulation with IL-12. Recently, successful specific immunotherapy *in vivo* associated with changes in the cytokine profile of allergen-reactive Th2 cells has been reported. Thus, a potential approach might be to prime both naïve and memory allergen-specific T-helper cells to select for prevalent Th1 phenotype. Predominant secretion of Th1-type cytokines in response to allergen would result in different effector responses (such as IgG production, macrophage activation), including the inhibition of the established Th2 responses as a result of a cross-regulatory circuit. Up-regulation of allergen-specific Th1 responses has been achieved *in vitro* by using cytokines, such as IFNγ, IFNα and IL-12. Several regulatory effects of Th1-inducing cytokines have also been demonstrated *in vivo*:

(1) Knockout mice lacking the IFNγ receptor have impaired ability to resolve a lung eosinophilic inflammatory response associated with infiltration of Th2 cells.

(2) Nebulized IFNγ decreases IgE production and normalizes airway function in a model of allergen sensitization.

(3) IL-12 can block antigen-induced airway-hyperresponsiveness and pulmonary inflammation *in vivo* by suppressing Th2 cytokine expression.

These data provide evidence that the injection of selected allergen peptides plus Th1-inducing cytokines may represent the basis for a novel immunotherapeutic strategy.

The effects of varying the density of antigenic peptide or its affinity for MHC class II molecules have been studied in various experimental models. A high density of MHC class II peptide favours Th1-like responses, while a low density favours Th2-like responses. p28–40 analogues (with alanine residues at positions 34 and 36 of the dominant T-cell epitope of the group 2 mite allergen) can alter the ratio of IFNγ/IL-4 produced by human Th0 cells by selectively enhancing IFNγ secretion. Also replacement of arginine 21 by lysine in the peptide fragment 18–31 from *D. farinae* group I allergen result in a significant increase in IFNγ production by a specific human Th0 clone.

Another novel approach is based on vaccination with plasmid vectors. These vaccines, called naked DNA vaccines or polynucleotide vaccines, consist of a desired gene inserted into a plasmid which can enter cells near the injection site, where the gene is transcribed and translated, causing expression of the gene product. The advantages of this treatment are that:

(1) It could bypass the numerous problems associated with other vectors, mainly the immune responses against the delivery vectors.

(2) It could be used to express antigens resembling native epitopes.

(3) Genes for several different allergens could be included on the same plasmid, thus potentially decreasing the number of vaccinations required.

(4) The protein encoded by the injected gene could enter the cells' MHC class I pathway, thus resulting in the stimulation of CD8⁺ cytotoxic T cells alone.

Novel gene constructs and methods of delivery are likely to be active areas of research in the future. We also need to develop a deeper understanding of the immune system and how it is regulated. For example, vectors with a short palindromic DNA sequence containing CpG motifs could be potentially more effective, since these unmethylated oligodeoxynucleotides act as adjuvants that switch on Th1 immune response.

7.2 Non-allergen-specific immunomodulation

Additional potentially therapeutic strategies are based on targeting Th2 cells or the effector molecules produced as a consequence of activation of Th2 cells. That this approach might be feasible came to be accepted after evidence was obtained that Th2 responses are not critical for survival and protection:

(1) IL-4-deficient mice are better protected than wild-type animals against the majority of infections.

(2) Th2 cells are indeed more protective than Th1 cells only against nematodes.

(3) People homozygous for ε gene deletions appear perfectly normal.

Thus, modifying either Th2 cells or Th2-dependent effector molecules may be a reasonable form of immunotherapy in patients with severe atopic disorders.

The transcription factors that are critical for IL-4 production and/or Th2-cell development, such as the c-*maf* oncogene product and STAT6, may be useful targets for manipulating Th2 responses. Indeed, mice lacking the gene encoding STAT6 have deficient Th2 responses. IL-4 activity may be antagonized by soluble IL-4 receptors or by a human IL-4 mutant protein (which binds IL-4R but does not induce any signal). The latter protein also antagonizes the biological activity of IL-13, as well as the IL-4-driven differentiation of Th2 cells *in vitro*.

IL-5 activity can be antagonized by humanized antibodies to IL-5, which inhibit eosinophil infiltration and normalize airway hyperreactivity in monkeys challenged with *Ascaris suis*. Recently developed humanized anti-human IgE antibodies bind to IgE residues critical for receptor binding, not reacting with IgE bound to mast cells or basophils. Interestingly, in house-dust-sensitive mice IgE targeting not only blocks the IgE-mediated allergic inflammation, but also down-regulates Th2 responses and the subsequent infiltration of eosinophils into the airways.

8. Concluding remarks

Several very recent studies suggest that allergen-reactive Th2 cells play an essential role in the activation and/or recruitment of IgE antibody-producing B cells, mast cells and eosinophils, the cellular triad involved in allergic inflammation (Figure 1). The Th2 cells induce B cells to produce IgE in at least two ways:

- The Th2 cells produce soluble IL-4, which induces germline ε expression on the B cell.

- The Th2 cells physically interact with B cells, via binding of CD40L expressed on the activated T-helper cell and CD40 on the B cell; this CD40L/CD40 interaction is required for the expression of productive mRNA and for the synthesis of IgE.

Other soluble factors produced by both T cells and non-T-cells have also been shown to play negative or positive regulatory effects on human IgE synthesis. Th2 cells represent the polarized arm of the effector-specific response that plays some role in the protection against nematodes, and they act as cross-regulatory cells for chronic and/or excessive Th1-mediated responses. Th2 cells are generated from precursor naïve T-helper cells when they encounter the specific antigen in an IL-4-containing micro-environment. However, the source of IL-4

required at the initiation of response for the development of naïve T-helper cells into Th2 effectors is still unknown. The most likely possibility is that maturation into the Th2 pathway mainly depends upon the levels and the kinetics of IL-4 production by naïve T-helper cells themselves at priming. It is not known how these Th2 cells are selected in atopic patients. Both (i) the nature of the TCR signalling provided by the allergen peptide ligand and (ii) dysregulation of IL-4 production probably contribute to determining the Th2 profile of allergen-specific T-helper cells, but genetic dysregulation of IL-4 production is the major determinant.

Several gene products selectively expressed in Th2 cells or selectively controlling the expression of IL-4 have recently been described. This leads to the suggestion that up-regulation of genes controlling IL-4 expression and/or abnormalities in the regulatory mechanisms of Th2 development and/or function may be responsible for Th2 responses against common environmental allergens in atopic people (Figure 1). These findings provide exciting opportunities for the development of novel immunotherapeutic strategies for atopic diseases.

References

Aruffo, A., Farrington, M., Hollenbaugh, D., Li, X., Milantovich, A., Nonoyama, S. *et al.* (1993). The CD40 ligand, gp39, is defective in activated T cells from patients with X-linked hyper IgM syndrome. *Cell* **72**, 291–300.

Bradding, P., Feather, I.H., Wilson, S., Bardin, P.G., Heusser, C.H., Holgate, S.T., and Howarth, P.H. (1993). Immunolocalization of cytokines in the nasal mucosa of normal and perennial rhinitic subjects: the mast cells as a source of IL-4, IL-5, and IL-6 in human allergic mucosal inflammation. *J. Immunol.* **151**, 3853–60.

Cho, S.S., Bacon, C.M., Sudarshan, C., Rees, R.C., Finblom, D., Pine, R., and O'Shea, J.J. (1996). Activation of STAT4 by IL-12 and IFN-alpha. Evidence for the involvement of ligand-induced tyrosine and serine phosphorylation. *J. Immunol.* **157**, 4781–9.

Chu, R.S., Targoni, O.S., Krieg, A.M., Lehman, P.V., and Harding, C.V. (1997). CpG oligodeoxynucleotides act as adjuvants that switch on T helper 1 (Th1) immunity. *J. Exp. Med.* **186**, 1623–31.

Del Prete, G.F., Maggi, E., Parronchi, P., Chretien, I., Tiri, A., Macchia, D. *et al.* (1988). IL-4 is an essential factor for the IgE synthesis induced *in vitro* by human T cell clones and their supernatants. *J. Immunol.* **140**, 4193–8.

Del Prete, G.-F., De Carli, M., Almerigogna, F., Daniel, K.C., D'Elios, M.M., Zancuoghi, D. *et al.* (1995). Preferential expression of CD30 by human CD4⁺ T cells producing Th2-type cytokines. *FASEB J.* **9**, 81–6.

Donelly, J.J., Ulmer, J.B., Shiver, J.W., and Liu, M.A. (1997). DNA vaccines. *Annu. Rev. Immunol.* **15**, 617–48.

Egan, R.W., Athwahl, D., Chou, C.-C., Emtage, S., Jehn, C.-H.,

Kung, T.T. *et al.* (1995). Inhibition of pulmonary eosinophilia and hyperreactivity by antibodies to interleukin 5. *Int. Arch. Allergy Appl. Immunol.* **107**, 321–2.

Gauchat, J.-F., Henchoz, S., Mazzei, G., Aubry, J.-P., Brunner, T., Blasey, H. *et al.* (1993). Induction of human IgE synthesis in B cells by mast cells and basophils. *Nature* **365**, 340–3.

Gorham, J.D., Guler, M.L., Steen, R.G., Mackey, A.J., Daly, M.J., Frederick, K. *et al.* (1996). Genetic mapping of a locus controlling development of Th1/Th2 type responses. *Proc. Natl Acad. Sci. USA* **93**, 12467–72.

Ho, C.I., Hodge, M.R., Rooney, J.W., and Glimcher, L.H. (1996). The proto-oncogene c-*maf* is responsible for tissue-specific expression of interleukin-4. *Cell* **85**, 973–83.

Hoyne, G.F., Askonas, B.A., Hetzel, C., Thomas, W.R., and Lamb, J.R. (1996). Regulation of house dust mite responses by intranasally administered peptide: transient activation of CD4[+] T cells precedes the development of tolerance *in vivo*. *Int. Immunol.* **8**, 335–42.

Jutel, M., Pichler, W.J., Skrbic, D., Urwyler, A., Dahinden, C., and Muller, U.R. (1995). Bee venom immunotherapy results in decrease of IL-4 and IL-5 and increase of IFN-γ secretion in specific allergen-stimulated T cell cultures. *J. Immunol.* **154**, 4187–94.

Kopf, M., Le Gros, G., Bachmann, M., Lamers, M.C., Bluthmann, H., and Kohler, G. (1993). Disruption of the murine IL-4 gene blocks Th2 cytokine responses. *Nature* **362**, 245–8.

Manetti, R., Parronchi, P., Giudizi, M.G., Piccinni, M.P., Maggi, E., Trinchieri, G., and Romagnani, S. (1993). Natural killer cell stimulatory factor (interleukin-12) induces T helper type 1 (Th1)-specific immune responses and inhibits the development of IL-4-producing Th cells. *J. Exp. Med.* **177**, 1199–204.

Marsh, D.G., Neely, J.D., Breazeale, D.R., Ghosh, B., Freidhoff, L.R., Ehrlich-Kautzky, E. *et al.* (1994). Linkage analysis of IL-4 and other chromosome 5q31.1 markers and total serum immunoglobulin E concentrations. *Science* **264**, 1152–6.

Mosmann, T.R. and Coffman, R.L. (1989). TH1 and TH2 cells: different patterns of lymphokine secretion lead to different functional properties. *Annu. Rev. Immunol.* **7**, 145–73.

Parronchi, P., Macchia, D., Piccinni, M.-P., Biswas, P., Simonelli, C., Maggi, E. *et al.* (1991). Allergen- and bacterial antigen-specific T-cell clones established from atopic donors show a different profile of cytokine production. *Proc. Natl Acad. Sci. USA* **88**, 4538–42.

Robinson, D.S., Hamid, Q., Ying, S., Tsicopoulos, A., Barkans, J., Bentley, A.M. *et al.* (1992). Predominant Th2-like bronchoalveolar T-lymphocyte population in atopic asthma. *New Engl. J. Med.* **326**, 295–304.

Romagnani, S. (1994). Lymphokine production by human T cells in disease states. *Annu. Rev. Immunol.* **12**, 227–57.

Romagnani, S. (1994). Regulation of the development of type 2 T-helper cells in allergy. *Curr. Opin. Immunol.* **6**, 838–46.

Romagnani, S. (1997) *The Th1/Th2 paradigm in disease*. R.G. Landes, Austin, Texas.

Vercelli, D. and Geha, R.S. (1995). Regulation of immunoglobulin E synthesis. In Busse, W.W. and Holgate, S.T. (eds), *Asthma and Rhinitis*, pp. 437–49. Blackwell Scientific, Oxford.

Wenner, C., Guler, M.L., Macatonia, S.E., O'Garra, A., and Murphy, K.M. (1996). Roles of IFN-γ and IFN-α in IL-12-induced Th1 development. *J. Immunol.* **156**, 1442–7.

XXV Cytokines in organ transplantation

Margaret J. Dallman

1. Transplantation: an introduction

The treatment of choice for patients with end-stage disease of organs, including kidney, heart, liver and lung, is transplantation. Whilst graft survival under modern regimes of immunosuppression is very good in the short term, with one-year survival rates for most organs of up to 90%, the longer-term picture is not as good. Many patients have severe problems with infections or the development of cancers and lymphoproliferative disorders. These complications may necessitate withdrawal of immunosuppression with an almost inevitable consequent rejection of the organ. In addition, grafts that survive for longer periods of time may develop signs of chronic rejection which result in decrease of graft function and eventual graft loss. Indeed, the greatest loss of grafts is now due to chronic rejection.

1.1 The immune response to allo- and xeno-antigens

Our understanding of the immune response to organ transplants has grown with our ever-increasing knowledge of the immune system. The response to a transplant takes place in a series of relatively well defined stages (Figure 1). Initial harvest and transplantation of the graft results in the induction of cytokines, chemokines and upregulation of adhesion proteins within the transplant. This allows an early infiltrate of the graft by monocytes and macrophages which alone will not result in graft rejection, but which may influence the development of immunity to the graft. Shortly after transplantation bone-marrow derived cells of the graft, the so-called passenger leukocytes, migrate to the local lymphoid tissue, maturing en route, wherein they can activate antigen-specific T lymphocytes. In addition to this 'direct' route of host cell sensitisation, graft derived antigen may be processed and presented by recipient antigen presenting cells to recipient T cells, via 'indirect' antigen presentation. Cytokines are at this stage important not only in amplification of the immune response, but also in determining the nature of the emerging immunity.

Whilst we normally associate acute graft rejection with T-cell-mediated immunity, it has become clear that grafts may be rejected through a variety of means including through both humoral and non-T-lymphocyte cellular effector mechanisms. This is probably why it is so difficult to prevent graft rejection without disabling, at a central point, the immune system.

2. Current clinical immunosuppressive strategies

Of the almost bewildering array of new immunosuppressive drugs, cyclosporin A, usually in combination with azathioprine and steroids, remains the linchpin of immunosuppression in many clinical transplant programmes. Exceptions to this include immunosuppression for both liver and small bowel, in which the use of FK506 or tacrolimus may now be favoured. Importantly, both cyclosporin and FK506 interfere with T-cell-receptor associated signalling events and result in dramatically reduced cytokine production and leukocyte activation. Another newer, less widely used agent, rapamycin, also affects cytokine-mediated activation of the immune system, this time at a more advanced stage of immunity through the interruption of cytokine receptor signals. The signalling pathways blocked by these agents are not only used by leukocytes, however, and therefore there are a number of side-effects associated with their use. In addition, the

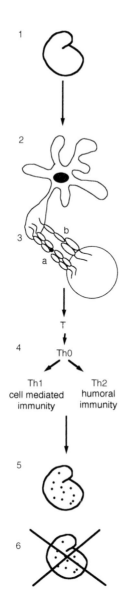

▮ Figure 1. The immune response to a transplant. (1) Following removal, perfusion and transplantation of an organ (e.g a kidney), pro-inflammatory cytokines are expressed and inflammatory cells such as macrophages are recruited into the graft. (2) Presentation of antigen to recipient T cells: passenger leukocytes migrate into the host lyphoid tissue and recipient leukocytes enter the transplant. (3) Activation signals for recipient T cells: T cells are stimulated by (a) TCR signal (MHC + peptide) and (b) co-stimulation (e.g. CD28). (4) Cytokine production leads to the generation of different types of immunity: cell-mediated and humoral. (5) The expression of MHC and adhesion proteins by the graft is up-regulated by cytokines like IFNγ and TNFα. Activated leukocytes are attracted into the graft by chemokines. (6) The graft is destroyed, in a process involving antibody, CTL, macrophages and cytokines.

non-specific nature of immunosuppression lays patients open to infections and cancers.

Since cytokines are so intimately involved in the generation of immunity and since our most potent immunosuppressive drugs affect cytokine pathways, there has been a great deal of interest in investigating the expression and function of cytokines in transplantation responses. Interest has focused on three main areas:

(1) A deeper understanding of the way in which grafts are rejected or tolerated has been sought.

(2) Monitoring the expression of cytokines has been used in an attempt to diagnose rejection episodes more precisely.

(3) Attempts have been made to manipulate the cytokine network with the aim of benefiting graft survival.

3. Cytokines in experimental models of transplantation

3.1 Expression patterns during rejection

3.1.1 Methods of analysis

A large amount of work has been performed on experimental transplant models in which graft rejection is unmodified by the use of immunosuppression. Many of these studies are based upon the analysis of cytokine transcripts within the grafts, most usually by reverse transcription–polymerase chain reaction (RT-PCR) with no attempt at isolating graft-infiltrating leukocytes or antigen-specific cells. There are several very good reasons for the emergence of such an approach:

(1) It has been very difficult to reliably identify or measure cytokine proteins within extracts of graft tissue (by, for instance, immunoassay or bioassay) or in tissue sections (by, for instance, immunohistology), although there has been some work in this area. The recent finding that it has been possible to obtain tissue staining with interleukin 4 (IL-4) antibodies in material derived from IL-4 knockout mice also questions the validity of the immunohistological approach.

(2) Where analysis of cytokine transcripts is performed, the short half-lives of such transcripts precludes the use of the lengthy cell

purification methods required to isolate cells from organ grafts.

(3) Cytokines produced by the parenchymal cells of the graft may well influence immunity and are thus of interest.

(4) In most situations it is not actually possible to identify antigen-specific cells and anyway the production of cytokines by bystander leukocytes may be very important.

(5) The level of cytokine transcripts as a percentage of the total mRNA isolated may be very low and difficult to measure by methods other than RT-PCR and the small amount of tissue often available precludes the use of methods such as Northern blotting or RNase protection analysis.

3.1.2 Expression in heart, kidney, skin and islets

Although results from different groups do vary, a consensus has emerged that in heart, kidney, skin or islet grafts it is usually possible to detect all pro-inflammatory cytokines tested and in unmodified rejection a predominance of cytokines like IL-2 and interferon gamma (IFNγ). However, in many people's hands it has been possible to detect not only these Th1-type cytokines, but also cytokines associated with a Th2 response. This suggests that graft rejection in unprimed, unmodified recipients can be driven by Th0 or a mixture of Th1 and Th2 cells. The peak of T-cell cytokine expression often occurs before the time at which deterioration in graft function can be observed by other means; it may be transient and indeed at the time of graft loss it may be very difficult to detect cytokines like IL-2 and IFNγ at all.

It has frequently been possible to detect pro-inflammatory cytokines like IL-1, IL-6 and tumour necrosis factor alpha (TNFα) within syngeneic grafts, in other words in grafts not undergoing rejection but which have been submitted only to the trauma of organ harvest, ischaemia and transplantation. In allografts, there is often a similar early peak of transcript shortly after transplantation, followed by a dip in expression before a final increase associated with graft loss.

3.1.3 Expression in other grafts

Not all allografts show the same patterns of cytokine expression: either the profile or the kinetics of the response may vary from that described above. Notably, in liver transplants, both Th1 and

Th2 cytokine transcripts are readily detectable throughout the evolution of the rejection response and, if anything, a predominance of IL-4 and IL-5 associates with rejection. There have been a limited number of studies of cytokine expression in xenografts, but it has been suggested that a Th2-type response is critical in rejection. In small bowel transplants an extremely early cytokine response, which peaks within 1–3 days after transplantation, can be observed in the mesenteric lymph nodes and Peyer's patches of the graft. Within the gut wall itself, the only cytokine whose expression clearly increases with rejection is IFNγ.

3.2 Manipulation of cytokines in experimental models

3.2.1 Studies with blocking antibody and antagonists

Antibodies raised against a variety of different cytokines have been administered to recipient animals following transplantation. The targeted cytokines are either those thought to be involved in development of an immune response (e.g. IL-2, IFNγ and IL-12) or those likely to mediate the tissue damage of rejection (e.g. TNFα, TNFβ).

Antibodies reactive with CD25, the IL-2 receptor p55 chain, have been used extensively in experimental transplantation and can induce long-term graft survival either alone or in combination with conventional immunosuppressive drugs like cyclosporin. Short-term use of such antibodies is attractive as they should target in a preferential fashion T lymphocytes that are activated at the time of treatment, thereby minimizing non-specific ablation of the immune system. These antibodies have been shown to be beneficial for a variety of transplants including heart, kidney, liver and neurone, and may mediate their effects either through the deletion of CD25-expressing cells or through simple blocking of IL-2 binding. Since T cells deprived of growth signals may become unresponsive to further antigen challenge and even acquire a regulatory function, blocking antibodies may be of greater interest in the context of transplantation than depleting antibodies.

Antibodies reactive with cytokines themselves have been fairly disappointing in their effects. Whilst antibodies against the TNFs may marginally prolong graft survival, those against IL-12 and IFNγ may even have a detrimental effect and accelerate graft rejection. A similar acceleration of cardiac transplant loss has observed in animals treated with IL-12 p40 homodimers to block IL-12 activity.

The naturally occurring soluble IL-4 receptor and an artificially manufactured soluble form of the IL-1 receptor have both been shown to bind their respective cytokines and also produce a modest prolongation of graft survival. Neither agent has a sufficient effect to merit further investigation in the transplant setting. The IL-1 receptor antagonist can ameliorate graft-versus-host disease following bone marrow transplantation, but has not been reported to have a beneficial effect in organ transplantation.

3.2.2 Overexpression or injection of cytokines

The augmentation of cytokines can have dramatic effects on transplant survival; indeed, injection of either IL-2 or IFNγ can prevent the induction of tolerance to heart or kidney in both rodent and large animal models. However, it has been much more difficult to show a beneficial effect of cytokine injection or overexpression. Efforts have concentrated on the manipulation of cytokines that may inherently be anti-inflammatory or immunoregulatory (such as TGFβ or IL-10) or which may shift the nature of immunity (e.g. IL-12 or IL-4). Early attempts at inducing effects through the injection of such cytokines were largely unsuccessful and so other strategies were developed in an attempt either to prolong the half-life of the cytokine or to allow continuous expression of the cytokine within the recipient or the graft itself. In the former approach, immunoglobulin Fc portions were engineered on to the cytokine, resulting in a much increased circulating half-life. Neither IL-10-Fc or IL-4-Fc engineered in this fashion, however, prolongs graft survival and indeed they may even result in accelerated rejection. Even injections of unmanipulated mammalian IL-10 can speed up the rejection kinetics.

There have now been several studies in which cells transfected with cDNAs encoding a variety of cytokines have been transplanted either as the test graft themselves or in conjunction with another graft. Transfection methods have included the use of viral vectors (retrovirus- or adenovirus-based) either *ex vivo* or *in vivo*. Mammalian IL-10 in such studies may have no effect or at best a moderately beneficial effect. Viral IL-10 appears to be of more use, probably as it lacks some of the pro-inflammatory effects of mammalian IL-10. TGFβ has been shown in one study to prolong graft survival in a mouse fetal heart transplant model in which grafts were transfected using plasmid DNA. IL-4-transfected L cells can prolong, but only slightly, the survival of mouse vascularized heart grafts and the survival of a myoblast cell line was modestly increased following transfection with IL-12 p40.

3.2.3 Genetic manipulation

The use of genetically manipulated animals in transplantation studies has become increasingly popular. As far as cytokines are concerned, lack of any single cytokine has very little or no effect on either graft survival or the induction of tolerance. Animals lacking IL-2, IL-12 (p40) or IFNγ are able to reject grafts, although this may occur later than usual. Conversely, in studies in which CTLA4-Ig has been used to block co-stimulation, IL-4 is not required for the induction of tolerance. It is very difficult to use this information to assess what role the targeted cytokines play in transplantation immunity in normal animals. Cytokines are so redundant in their activities that in the absence of any single cytokine another may take its place; indeed, lack of a cytokine from the beginning of development may be compensated for such that the leukocytes in these animals may be truly independent of the action of that cytokine. The use of animals in which the disruption is performed at a chosen time later during development, in the adult animal or at specific times during immune responses, would undoubtedly help to address these issues of cytokine redundancy and independence. Unfortunately such animals are not yet available, but the development of systems is an area of intense interest.

Mice overexpressing a protein from a transgene have not been used extensively in transplantation experiments. In one study, mice were manipulated to express IL-10 under the insulin promoter. Islet beta cells from such mice did not show enhanced graft survival when transplanted into normal animals, again indicating that IL-10 does not have a beneficial effect on allograft survival. Recipient mice expressing a soluble form of the IL-4 receptor under an inducible promoter (the metallothionein promoter) exhibited slightly prolonged survival of non-vascularized fetal heart grafts.

3.3 Th1 and Th2 cells — importance for rejection and tolerance

As discussed above, in many grafts rejection is associated with a predominance of Th1 cytokines and tolerance can be prevented by the injection of either IL-2 or IFNγ. There has consequently been a great deal of interest in the possibility that Th2 cells (or their cytokines) might not only be less damaging to a transplant than are Th1 cells, but also act as regulatory cells preventing the expansion or action of damaging Th1 cells. Initial evidence in support of this came from a number of studies in which prolonged graft survival was associated with a decrease in the expression of Th1 cytokines and increased or at least maintained production of Th2 cytokines. There are several studies in which true tolerance has been investigated and in several of these there is a complete down-regulation of cytokine expression although in some Th2 cytokines or even a mixture of Th1 and T2 cytokines may still be present. Since these data only show associative expression of cytokines with tolerance or prolonged graft survival, attention has turned to more direct ways of tackling this issue. It has not been possible to block graft rejection by either injecting or overexpressing cytokines, nor has the removal of any cytokine altered an ability to induce tolerance. These latter studies still do not completely address the questions surrounding the role of Th2 cells in the induction of tolerance since in many of the experiments it has not been possible to conclude that either a shift to or a block of the Th2 response has been induced. In one study in which the immune response clearly was shifted towards Th2 immunity using IL-12 p40 homodimers or in IL-12 knockout mice, rejection still occurred rapidly but unusually in association with eosinophilia. In other words the shift from a Th1 to a Th2 response allowed rejection to proceed, but through an alternative effector mechanism.

In many models of transplantation tolerance, it is possible to transfer tolerance into naive recipients using cells derived from the tolerant animal. Further, it has become clear that in the presence of tolerant cells, naive or even sensitized cells may also become tolerant through a process called infectious tolerance. These regulatory or suppressor cells are often of the CD4$^+$ phenotype but the mechanism of the transferable tolerance has been difficult to define. There have been several attempts to test whether isolated Th2 cells can perform such a regulatory role or whether the action of such cells can be prevented by blocking Th2 cytokines. Results from such studies are rather mixed, but it is generally now agreed that there is more to being a suppressor cell than having an ability to produce Th2 cytokines.

In summary, it would appear that whilst cell-mediated immunity driven by a Th1 response can almost certainly be damaging to a graft, Th2 cells and the immunity they induce may also be damaging but probably in a less aggressive fashion (with the possible exception of grafts like liver, in which Th2-driven immunity appears to dominate during rejection).

4. Cytokines in clinical transplantation

A variety of approaches have been used to monitor expression of cytokines following transplantation in the clinical setting, using material either derived from the graft itself or from the circulation. The common practice of taking needle-core biopsies during periods of graft dysfunction has made available tissue for histological, morphological and molecular analysis. However, the events leading up to tissue damage cannot be studied using such samples. Some information on events within grafts in stable function and before loss of function can be gained from protocol biopsies taken routinely during the early period after transplantation, but it is almost impossible to define the evolution of intragraft events using these biopsies. This presents a particular difficulty in the analysis of cytokine transcripts which, as described above, may be transient in their appearance and absent in grafts showing clinical signs of rejection. Nevertheless, many groups have reported the measurement of cytokine transcripts in biopsy material using *in situ* hybridization or, more commonly, RT-PCR. As with the experimental studies, there is no universal indicator of graft rejection, but, interestingly, cytokines like IL-4 and IL-5 appear frequently in liver graft rejection and less commonly in other types of transplant. While IL-2 and IFNγ are common precursors of rejection in animal models, an association with rejection has not always been found in the clinical setting. However, this may well be due to the low frequency of tissue sampling which is possible prior to signs of rejection in most studies. Fine-needle aspiration sampling is a relatively non-traumatic technique and studies using this approach from my own group have enabled an analysis of material obtained on a daily basis for up to 22 days after transplantation. When such material is available, close monitoring of the evolution of cytokine expression is possible and reiterates the transient appearance of cytokine transcripts and absence during periods of graft dysfunction observed for cytokines such as IL-2 and IFNγ in the experimental models. From two studies of this kind we have noted a close correlation of IL-2 and IFNγ transcripts with the events leading up to graft rejection in renal transplants. However, we have found that the presence of such transcripts is not always predictive of an ensuing episode of graft dysfunction since a percentage of transplants showing a significant infiltrate together with IL-2 and IFNγ transcripts do not progress to histologically and clinically defined rejection.

Because of the difficulties in obtaining samples for analysis from the graft itself, many workers have chosen to analyse, using enzyme-linked immunosorbent assay (ELISA) and related techniques, the presence of cytokines within the circulation or draining fluids of the transplant. The advantage of this approach over that described above is that protein is measured rather than mRNA. Since cytokines are largely regulated in their expression at the transcriptional rather than at the post-transcriptional level, this may not be such an advantage as it may initially seem. However, certain cytokines (for instance TNFα and TGFβ) are regulated at the translational or post-translational level and for these, measurement of protein or bioactive protein is essential. This work has shown a clear benefit of deriving material for study from the vicinity of the transplant since results using serum or plasma may show little relationship to events within the graft. For instance, in both experimental and clinical liver transplantation, while measurement of IL-6 in the serum has little value, measurement in the bile can predict subsequent rejection. This is perhaps not surprising, particularly for cytokines like IL-2, IFNγ, IL-4 and other T-cell derived cytokines, since these proteins have their most important effects when released locally, and may even require contact between the producer and target cell.

One finding from all of the clinical studies is that any relationship between cytokine expression and rejection can be confused in the face of an infectious episode and it remains unclear whether the measurement of any cytokine can accurately distinguish these two phenomena.

5. Immunotherapy in clinical transplantation

Given the results from experimental transplantation, it is perhaps not surprising that little effort has been devoted to the manipulation of cytokines in clinical transplantation as a strategy for immunosuppression. The one exception to this is the use of antibodies that bind CD25. Clinical trials first explored the use of mouse antibodies reactive with human CD25. Initial fairly promising results were marred by the development of antibodies reactive with mouse immunoglobulin, which would not only decrease the efficacy of the treatment but also preclude any future use of mouse antibodies. Chimeric or engineered antibodies are now available and early information indicates that they may be as potent in their effects as other agents like ATG and OKT3, but perhaps with fewer side-effects.

6. Conclusions

Analysis of cytokines during transplantation responses has undoubtedly increased our knowledge of the mechanisms involved in graft rejection. Because grafts can be rejected by several different types of immune response, it has not been possible to block completely the tissue-damaging response through altering the balance of Th1 and Th2 cytokines. For many types of graft, however, a Th2 response is apparently less damaging than a Th1 response, such that it may be possible, by blocking the Th1 response, to allow a window of opportunity in which regulatory cells may be induced. The only manipulation of the cytokine network that appears to have an immediate future in the clinic is the blocking or removal of CD25-positive cells. To date there have been no attempts clinically to treat graft rejection using cytokine-based gene therapy.

References

Brazelton, T.R. and Morris, R.E. (1996) Molecular mechanisms of action of new xenobiotic immunosuppressive drugs: tacrolimus (FK506), sirolimus (rapamycin), mycophenolate mofetil and leflunomide. *Curr. Opin. Immunol.* **8**, 710–20.

Dallman, M.J. (1993) Cytokines as mediators of organ graft rejection and tolerance. *Curr. Opin. Immunol.* **5**, 788–93.

Dallman, M.J. (1995) Cytokines and transplantation: Th1/Th2 regulation of the immune response to solid organ transplants in the adult. *Curr. Opin. Immunol.* **7**, 632–8.

Dallman, M.J. (1998) Immunobiology of graft rejection. In Thiru, S. and Waldmann, H. (eds), *Pathophysiology of Transplantation.* Blackwell Scientific, Oxford (in press).

Dallman, M.J. and Clark, G.J. (1991) Cytokines and their receptors in transplantation. *Curr. Opin. Immunol.* **3.**, 729–34.

Dallman, M.J. and Porter, A.C.G. (1991) Semi-quantitative PCR for the analysis of gene expression. In McPherson, M.J., Quirke, P. and Taylor, G.R. (eds), *PCR: A Practical Approach*, pp. 215–24. IRL Press, Oxford.

Dallman, M.J., Montgomery, R.A., Larsen, C.P., Wanders, A. and Wells, A.F. (1991) Cytokine gene expression: analysis using northern blotting, polymerase chain reaction and in situ hybridization. *Immunol. Rev.* **119**, 163–79.

Halloran, P.F., Broski, A.P., Batiuk, T.D. and Madrenas, J. (1993) The molecular immunology of acute rejection: an overview. *Transplant Immunol.* **1**, 3–27.

Hollander, G.A., Bierer, B.E. and Burakoff, S.J. (1996) Molecular mechanisms of immunosuppressive drugs: cyclosporin A, FK506, rapamycin. In Tilney, N.L., Strom, T.B. and Paul, L.C. (eds), *Transplantation Biology: Cellular and Molecular Aspects*, pp. 657–71. Lippincott–Raven, Philadelphia.

Kupiec-Weglinski, J.W., Diamantstein, T. and Tilney, N.L. (1988) Interleukin 2 receptor-targeted therapy-rationale and applications in organ transplantation. *Transplantation* **46**, 785–92.

LeMauff, B., Cantarovich, D., Jaques, Y. and Soulillou, J.-P. (1990) Monoclonal antibodies against interleukin-2 receptors in the immunosuppressive management of kidney graft recipients. *Transpl. Rev.* **4**, 79–92.

Morris, R.E. (1993) New small molecule immunosuppressants for transplantation: review of essential concepts. *J. Heart Lung Transpl.* **12**, S275–86.

Mottram, P.L., Purcell, L.J., Han, W.R., Maguire, J. and Stein-Oakley, A.N. (1997) Interleukin (IL) 4, the cytokine that isn't there: reactivity of IL-4 antibodies with cells in IL-4-/- mice. *Transplantation* **63**, 911–14.

Nickerson, P., Steurer, W., Steiger, J., Zheng, X., Steele, A.W. and Strom, T.B. (1994) Cytokines and the Th1/Th2 paradigm in transplantation. *Curr. Opin. Immunol.* **6**, 757–64.

Picotti, J.R., Chan, S.Y., VanBuskirk, A.M., Eichwald, E.J. and Bishop, D.K. (1997) Are Th2 helper T lymphocytes beneficial, deleterious, or irrelevant in promoting allograft survival? *Transplantation* **63**, 619–24.

Porter, A.C.G. and Dallman, M.J. (1997) Gene targeting: techniques and applications to transplantation. *Transplantation* **64**, 1227–35.

Rajewsky, K., Gu, H., Kuhn, R., Betz, U.A.K., Muller, W., Roes, J. et al. (1996) Conditional gene targeting. *J. Clin. Invest.* **98**, 600–603.

Strom, T.B., Roy-Chadhury, P., Manfro, R., Zheng, X.X., Nickerson, P.W., Wood, K. et al. (1996) The Th1/Th2 paradigm and the allograft response. *Curr. Opin. Immunol.* **8**, 688–93.

XXVI Cytokines and autoimmunity

A. Saoudi*, B. Cautain, I. Bernard, A. Badou, M. Savignac, P. Druet and L. Pelletier

1. Introduction

An understanding of how the immune system remains unresponsive to self antigens while retaining the capacity to respond to pathogenic organisms is central to any attempt selectively to control immunopathological manifestations. Self-tolerance has been attributed to two types of mechanism. The first type is dependent on physical elimination (by deletion) or functional elimination (anergy) of autoreactive T cells. The second type is dependent on the presence of T cells able actively to suppress potentially pathogenic autoreactive cells that have neither been clonally deleted nor rendered anergic. The discovery that CD4+ as well as CD8+ T cells are functionally heterogeneous, as a result of their different profile of cytokine production, has offered an explanation of why certain cells induce autoimmunity while others regulate autoreactive T cells (Figure 1). Type 1 T-helper cells (Th1 cells) are considered to be involved in the induction of many experimental autoimmune diseases. Evidence for this is based on adoptive transfer experiments demonstrating that CD4+ T cells producing Th1-type lymphokines can transfer disease in several models such as experimental autoimmune encephalomyelitis (EAE) and insulin-dependent diabetes mellitus (IDDM). The aims of this chapter are to review evidence for and against a direct role of the T-helper subset in the induction or prevention of autoimmunity and to give evidence that there are other 'regulatory cells', different from classical Th1 and Th2 cells, that control autoimmune manifestations.

2. Th1-mediated autoimmune diseases

In this section, we shall focus on the pathogenic role of Th1 cells, and the possible protective role of Th2 cells, in EAE and IDDM, but similar data have been obtained for other autoimmune diseases.

2.1 Experimental autoimmune encephalomyelitis

EAE is an inflammatory autoimmune disease of the central nervous system (CNS) and serves as an experimental model for multiple sclerosis. EAE is a T-cell-dependent, paralytic disease that can be induced in experimental animals by immunization with myelin-derived autoantigens emulsified in complete Freund's adjuvant (CFA) or following adoptive transfer of T-cell lines or clones specific for various myelin proteins. In Lewis rats the disease is transient and affected animals recover completely within 4–5 day of its onset. After recovery, attempts to induce further episodes of the disease are unsuccessful despite the fact that spleen cells from animals that have recovered can be shown to have the potential to cause the disease. It appears that in some way the pathological response to myelin-derived autoantigens is held in check in animals that have recovered from their single episode of paralysis.

Several arguments indicate that the induction phase of the disease is mediated by Th1 cells:

(1) T-cell lines and clones that transfer EAE in rodents produce interferon gamma (IFNγ), interleukin 2 (IL-2), tumour necrosis factor alpha (TNF-α) and TNF-β and these cytokines are detected in the CNS at the acme of the disease.

* Corresponding author.

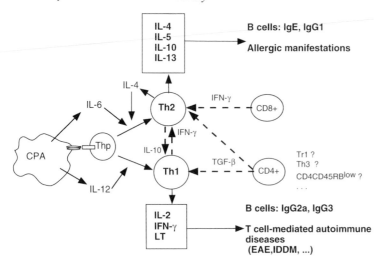

Figure 1. Role of cytokines in Th1/Th2 differentiation and in the induction or regulation of immunopathological manifestations. Solid arrows, stimulation; broken arrows, negative inhibition.

(2) Treatment with anti-lymphotoxin, anti-TNF-α or anti-IL-12 monoclonal antibodies (mAbs) has a beneficial effect on the active form of EAE.

(3) IL-12 enhances the pathology of EAE, and inhibition of endogenous IL-12 *in vivo* prevents EAE induced by transfer of stimulated immune lymph node cells.

While the induction phase of EAE is clearly due to Th1 cells, several experiments suggest that the regulatory phase is under the control of Th2 or Th3 (TGF-β-producing cells) cells. The remission phase is associated with IL-4, IL-10 and TGF-β expression in the CNS while mRNA for IFNγ is no longer detected. In keeping with this observation, it has been shown that resistant mice develop EAE when their IL-10 gene is knocked out. Furthermore, the prevention of EAE by anti-IL-12 mAb injections is also associated with an increase in IL-10 production, suggesting that, in the absence of IL-12, the IL-10-producing T cells are able to control autoimmunity. Based on the indication that Th2-like cells could have a protective effect, several groups have attempted to prevent EAE by skewing the Th1 pathogenic response to a Th2 protective response in several ways:

(1) Administration of IL-4, IL-10 or IL-13 has a beneficial effect.

(2) Treatment with retinoids improves the clinical course even when given after disease onset, probably due to the induction of Th2 cells.

(3) Targeting autoantigen to B cells favours a predominant Th2 response and results in protection; in agreement, B-cell-deficient mice are unable to regulate EAE, suggesting that these cells play a role in skewing the immune response towards Th2 during the regulatory phase of disease.

(4) Encephalitogenic T cells, transduced with a retroviral gene constructed to express IL-4, can delay the onset and reduce the severity of EAE when adoptively transferred to mice immunized with myelin basic protein (MBP).

(5) Th2 clones specific for myelin antigens, that secrete high levels of IL-4 and IL-10, inhibit the induction of EAE when transferred at the time of active immunization or at onset of clinical disease.

While all these results support the concept that, in EAE, Th1 cells are pathogenic and Th2/Th3 cells are regulatory, conflicting results have been reported. Injection of IFNγ may protect against EAE, while administration of anti-IFNγ antibody may exacerbate the disease. Similar unexpected results were observed in mice with a disrupted IFNγ gene; the lack of this cytokine converts an otherwise EAE-resistant mouse strain to become susceptible to disease. Although it may seem unexpected that a pro-inflammatory cytokine exerts apparently an anti-inflammatory effect, IFNγ is known to have anti-proliferative effect on T-cell responses *in vivo* and *in vitro*; removal of IFNγ may therefore tip the balance of pathogenic autoreactive cells in resistant mice and so predispose towards autoimmunity. IFNγ has also been reported not to be necessary for the development of EAE, since IFNγ-deficient mice are still susceptible to induction of this disease. It is possible that, in IFNγ-deficient mice, other cytokines and mediators may assume the functions of the inactivated genes. In this respect, it has been shown an increase in TNF-α levels in the CNS of IFNγ-deficient mice with EAE. Considering the large body of information suggesting that TNF-α and lymphotoxin α are important in the pathogenesis of EAE and multiple sclerosis, it is tempting to speculate that in IFNγ-deficient mice, TNF compensates for the defect of IFNγ.

Recently, it has been shown that Th2 cells can induce EAE-like disease in some circumstances. Th1 and Th2 cells were generated from MBP-specific T-cell receptor (TCR) transgenic mice and shown to be able to transfer EAE in immunocompromised mice whereas only Th1 cells were able to induce disease in normal recipients. However, the disease induced by Th2 cells is completely different from classical EAE and resembles allergic manifestations. These data show that Th2 cells may become pathogenic but their pathological effect is held under control by a distinct subset of regulatory $\alpha\beta$ T cells that exist in normal but not in immunodeficient mice. Several groups have suggested that CD8$^+$ T cells play a crucial role in regulating Th2 cells via IFNγ production but the role of this population has not been tested in this model.

2.2 Insulin-dependent diabetes mellitus

Autoimmune diabetes which appears spontaneously in non-obese diabetic (NOD) mice and in BB rats is a T-cell dependent autoimmune disease in which both CD4$^+$ and CD8$^+$ T cells are involved. Several studies have shown that this disease is mediated by Th1 cells:

(1) Th1-cell clones may passively transfer the disease or accelerate its appearance.

(2) IL-12 administration, which favours Th1 development, accelerates IDDM in genetically susceptible NOD mice but not in resistant BALB/c mice and this effect correlates with increased Th1 cytokine production by islet-infiltrating cells.

(3) Treatment with anti-IL2 receptor or anti-IFNγ mAbs, or IL-4 administration has a beneficial effect on IDDM.

(4) Double transgenic mice that express a TCR reactive to an islet neoantigen develop diabetes or do not depending upon the genetic background and the profile of cytokines produced; a Th1 pattern of cytokines is associated with the development of the disease while a Th2 pattern confers protection.

(5) Transgenic NOD mice that express IL-4 in their pancreatic β cells under the control of the human insulin promoter were completely protected from insulitis and diabetes.

Although all these data demonstrate that, in IDDM, Th1 cytokines are pathogenic and Th2 cytokines are protective, several studies suggest that this proposal is simplistic and show that the relation between the pro-inflammatory and Th2-associated cytokine is extremely complex. Th2 cells transgenic for a TCR derived from a clone able to transfer IDDM, when injected into neonatal NOD mice invaded the islets but did not provoke disease and neither did they provide substantial protection. Similar results were also obtained when Th1 and Th2 cells lines were cotransferred into neonatal mice. It has been shown in NOD mice that administration of recombinant IL-10, an inhibitory lymphokine of Th1 cells, ameliorates IDDM while in NOD mice expressing this cytokine under control of the insulin promoter, β cells were destroyed more rapidly than in control mice. These data do not support the concept that Th2 cells afford protection from IDDM. Rather, they are in agreement with a recent study showing that Th2 cells may induce diabetes in immunodeficient NOD/scid mice and that this disease is inhibited by anti-IL-10 but not anti-IL-4 antibodies. These findings resemble those described above in EAE and again suggest that in normal healthy individuals exists a distinct population of regulatory T cells able to control the pathological effect of Th2 cells and that this population is absent in immunocompromised animals.

3. Th2-mediated autoimmune diseases

3.1 Diseases associated with allogeneic reactions

Injection of parental lymphocytes (X) into F$_1$ (X x Y) hybrids normally induces the production of IFNγ and TNF by alloreactive T cells and leads to fatal, acute graft-versus-host disease (GVHD). However, in some genetic combinations, especially when X and Y differ only by MHC class II molecules or when CD8$^+$ cells are unable to mount a cytotoxic response, only alloreactive CD4$^+$ lymphocytes are activated, leading to chronic GVHD. These alloreactive CD4$^+$ cells produce Th2 cytokines and polyclonally activate F_1 B cells to produce several autoantibodies including antibodies against the nucleus, against double-stranded DNA and against glomerular basement membrane. The recipients display immune glomerulonephritis and hyperimmunoglobulinaemia affecting mainly IgG1 and IgE. Th2 cytokines play a major role in chronic GVHD, because administration of anti-IL-4 mAb has a beneficial effect, and treatment with IL-12 converts chronic GVHD into the acute form, probably due to an immune deviation towards Th1 responses. Similar immunopathological manifestations appear when spleen cells from an adult F_1 hybrid are injected neonatally in a recipient from one parental strain. A specific cytotoxic lymphocyte tolerance to alloantigens is induced but alloreactive Th2 CD4$^+$ cells from the recipient are primed by allogeneic donor B cells. These B cells are polyclonally activated to produce IgG1 and IgE as well as numerous autoantibodies. As in chronic GVHD, anti-IL-4 mAbs or recombinant IFNγ administration improve the disease.

3.2 Drug-induced autoimmunity

$HgCl_2$ and gold salts trigger Th2-mediated systemic autoimmunity in Brown Norway (BN) rats while $HgCl_2$ induces non-antigen-specific immunosuppression in Lewis rats and protects this strain against the development of autoimmune diseases such as EAE.

In BN and (Lewis x BN) F_1 hybrids, chronic injections of non-toxic doses of $HgCl_2$ or gold salts induce a T-cell-dependent B-cell polyclonal activation marked by an increase in serum IgE and IgG1 concentration. As in chronic GVHD, the animals produce various autoantibodies including antibodies against the glomerular basement membrane which are associated with the development of a glomerulopathy. Rats display linear IgG deposits along the glomerular basement membrane and develop proteinuria. Autoreactive anti-self major histocompatibility complex (MHC) class II T cells have been found in diseased BN rats, and T-cell lines have been derived from rats injected with gold salts. These lines produce IL-4 and IL-6 but not IL-2, IFNγ or transforming growth factor beta (TGF-$β$). They promote B-cell polyclonal activation *in vitro* and transfer the disease into CD8$^+$ cell-

depleted BN recipients. This suggests that CD8$^+$ cells are normally able to counteract pathogenic autoreactive T cells and that disease induced by gold salts or $HgCl_2$ is due both to the emergence of autoreactive Th2 cells and to a defect at the CD8$^+$ level.

Results of several studies indicate that IL-4 is involved in the early phase of the disease:

(1) In $HgCl_2$ injected BN rats, Th2 isotypes (IgE and IgG1) dominate.

(2) IL-4 mRNA is detected in lymph node and spleen cells early during the time course of the disease.

(3) Normal BN but not Lewis T cells express IL-4 mRNA when cultured with $HgCl_2$.

(4) $HgCl_2$ protects from a Th1-mediated autoimmune disease by skewing the immune response towards Th2.

(5) An anti-self MHC class II Th2 cell line may passively transfer the disease.

The immune manifestations induced by $HgCl_2$ or gold salts are spontaneously down-regulated from the third week of drug injections and the rats become resistant to rechallenge. Treatment with a mAb that recognizes the IL-2 receptor delays the down-regulation of these immune phenomena. Since such a treatment preferentially inhibits proliferation of Th1 cells, these results may implicate Th1 cells in the regulation phase of mercury disease. This is further supported by the finding that IL-12 mRNA is detected in the spleen of $HgCl_2$-injected BN rats during this phase. It is of note that this phase is also associated with a decrease in the frequency of autoreactive T cells and with the emergence of CD8$^+$ regulatory cells. These CD8$^+$ cells have also been shown to be important in the resistance to development of disease after $HgCl_2$ rechallenge but the cytokines involved are still unknown.

In Lewis rats, $HgCl_2$ induces non-antigen-specific immunosuppression and protects against several autoimmune diseases such as EAE. Autoreactive anti-self MHC class II CD4$^+$ T cells have been found in Lewis rats injected with $HgCl_2$. T-cell lines have been derived and have been shown to produce IL-2, IFNγ and TGF-$β$. When transferred into syngeneic recipients, these lines protect against the development of EAE in a CD8$^+$ cell-dependent manner. Transfer of these T-cell lines into F_1 (Lewis x BN) hybrids protects the recipient against the development of Th2-mediated autoimmunity induced by $HgCl_2$ in a CD8-independent but TGF-β-dependent manner. The role of regulatory CD4$^+$ T cells that differ from classical Th1 and Th2 cells is emphasized in this model. It is possible that this subset represents a minor but important subset, which is present in normal Lewis rats and which expands in HgCl$_2$-injected rats.

4. Systemic lupus erythematosus and myasthaenia gravis

4.1 Systemic lupus erythematosus

Some genetically predisposed strains of mice (e.g. NZB x NZW, BxSB, MRL *lpr/lpr*) develop spontaneously systemic lupus erythematosus (SLE) which is characterized by a T-cell-dependent production of various autoantibodies and various organ lesions including glomerulonephritis. There is now strong evidence that Th1 cells play a major role in the development of SLE in (NZB x NZW) or in other related SLE-prone mice. Nucleosome-specific Th1-cell lines have been derived from SLE-prone (NZW x SWR) mice. These cell lines provide B-cell help for autoantibody production *in vitro*.

In addition, the adoptive transfer of these Th1 lines induces severe lupus nephritis into young syngeneic recipients. Male (NZB x C57BL/6 *Yaa*) mice develop SLE; *Yaa* is a gene that is present on chromosome Y and that accelerates autoimmunity. Transgenic mice that express IL-4 at the B-cell level are protected, which suggests that downmodulation of Th1 cells has a protective effect. A deleterious role of Th1 cells in (NZB x NZW) mice is also supported by the fact that administration of anti-IFNγ antibodies or of soluble IFNγ receptor inhibits the onset of glomerulonephritis in that strain. However, it has recently been reported that Th2 cells are not innocuous in the development of SLE in (NZB x NZW) mice. Indeed, the transfer of IL-4- or IL-12-treated syngeneic splenocytes into (NZB x NZW) mice increases anti-DNA IgG production. Treatment with anti-IL-4 or anti-IL-12 mAbs decreases autoantibody production but only anti-IL-4 mAb administration prevents the development of lupus nephritis. It has also been reported that SLE is accelerated by IL-6 or IL-10 administration and inhibited by anti-IL-10 antibody therapy. In addition, administration of AS101, a drug that decreases IL-10 and enhances IFNγ and TNF-α levels, reduces the production of autoantibodies and prevents the development of nephritis. In fact, these data do not argue for a role of Th2 cells, since B cells and macrophages, but not T cells, are responsible for the excessive production of IL-10 in SLE.

A poor prognosis seems also to be associated with a Th1-dependent circuit in MRL *lpr/lpr* mice. IL-12 administration accelerates the nephritis. Autoreactive T-cell lines that produce both IL-4 and IFNγ have been derived from MRL *lpr/lpr* mice; they trigger a nephritis which is prevented by anti-IFNγ antibody treatment. Overall, a mouse line, derived from the MRL *lpr/lpr* and called MRL *lpr/lpr* l/l (l/l for long lived) produces predominantly IL-4 but not IFNγ and survives much longer than classical MRL *lpr/lpr* mice. A defective IFNγ production in MRL *lpr/lpr* l/l would be responsible for the low production of IgG3, an isotype known to be nephritogenic. A recent report has confirmed the role of IFNγ in the development of both lymphadenopathy and SLE disease by using IFNγ knockout MRL *lpr/lpr* mice but it has also shown that IL-4 knockout MRL *lpr/lpr* mice are also partially protected. This suggests that Th1 and, to a lesser degree, Th2 cells contribute to the disease in this strain.

4.2 Myasthaenia gravis

Myasthaenia gravis (MG) is one of the rare organ-specific autoimmune diseases for which the target autoantigen, the nicotinic acetylcholine receptor (AChR) of the neuromuscular junctions, has been well characterized. Experimental autoimmune myasthaenia gravis (EAMG) can be induced in susceptible mouse and rat strains by immunization with AChR from the electric organs of *Torpedo californica* emulsified in complete Freund's adjuvant. Animals develop a T-cell-dependent antibody response against *T. californica* AChR that cross-reacts with their own receptor resulting in a neuromuscular disease with clinical symptoms resembling human MG. Both Th1 cells and Th2 cells are implicated in this disease. A recent study in humans has indirectly suggested that Th2 or Th0 T cells could be involved, due to the presence in serum of MG patients of antibodies directed against the pro-inflammatory cytokines IL-12 and TNF-α. Several studies have suggested that IFNγ may play a pivotal role in EAMG in mice. First, it has been shown that expression of IFNγ transgene in the neuromuscular junction induces an MG-like syndrome, although the mechanism involved appears to be different from classical EAMG. Second, it has been shown that the failure of IFNγ-deficient mice to develop EAMG was associated with an impaired antibody response to self-AChR. However, studies based on kinetic analysis of cytokine mRNA synthesis by mononuclear cells during the course of EAMG in rats have suggested that both IFNγ and IL-4-producing cells might be involved in the genesis of this autoimmune syndrome. Our recent unpublished work is in agreement with this result and demonstrates that EAMG in rats may occur in the context of a Th1- or Th2-type immune response. Indeed, by using Lewis and BN rats that differ markedly in their susceptibility to develop either Th1- or Th2-mediated autoimmune manifestations, we showed that, although the incidence and severity of EAMG were comparable in the two strains, the immune response to *T. californica* AChR was polarized towards the Th1 or Th2 phenotype in Lewis and BN rats, respectively. In addition we showed that redirecting the response from a Th2- to a Th1-phenotype in BN rats by administration of IL-12 *in vivo* has little if any effect on the induction and severity of EAMG. In this respect, it has been shown that prevention of EAMG, using either oral or nasal tolerance or anti-CD8 mAb treatment, is associated with down-modulation of both Th1 and Th2 cytokines. It has been suggested that TGF-β might play a role in this phenomenon, indicating that this cytokine could inhibit the development of both Th1 and Th2 subsets in this model. Therefore, immunotherapeutic strategy aimed at favouring the development of TGF-β-producing cells might prove beneficial in the treatment of this antibody-mediated autoimmune disorder.

5. Immunopathology associated with perturbation of regulatory cytokines

Rat and mouse peripheral CD4$^+$ T cells can be subdivided, on the basis of expression of CD45 isoforms, into those that predominantly produce Th1 or Th2 cytokines. CD45, the leukocyte common antigen, is expressed in a number of different isoforms, generated by the differential use of three exons, A, B and C. The expression of different isoforms of CD45 by T cells is complex and varies depending on their state of maturation, activation and differentiation. However, the isolation of T-cell subsets on the basis of their expression of different CD45 isoforms has provided clear evidence for functional specialization among CD4$^+$ T cells. In the rat, it has been shown that CD45RC^{low} CD4$^+$

T cells secrete Th2 cytokines and provide B cells with help for secondary antibody responses while CD45RC^{high} CD4$^+$ T cells secrete Th1-like cytokines and mediate alloreactivity in GVHD. Similarly, subsets of mouse CD4$^+$ T cells have been studied and it appears that there are close similarities between the two species with exon B expression in the mouse paralleling that of exon C in the rat.

5.1 Wasting disease induced by transfer of congenic histocompatible T cells

Congenital athymic rats injected intravenously with CD45RC^{high} CD4$^+$ T cells, from histocompatible congenic euthymic donors, develop a severe wasting disease. This disease is characterized by mononuclear cell infiltration into different organs such as liver, stomach, pancreas and thyroid. In contrast, rats injected with the CD45RC^{low} subset of CD4$^+$ T cells remain healthy. The co-transfer of CD45RC^{low} with the CD45RC^{high} subset prevents the appearance of the wasting disease and the multi-organ mononuclear cellular infiltration. These data show that the normal T-cell repertoire contains pathogenic T cells (CD4$^+$ CD45RC^{high}) and regulatory T cells (CD4$^+$ CD45RC^{low}) and that this latter population has a dominant effect in regulating the former. Similar results were obtained when SCID mice were used as recipients of subsets of mouse CD4$^+$ T cells separated on the basis of their levels of CD45RB expression. SCID mice injected with CD45RB^{high} CD4$^+$ T cells developed a lethal wasting disease with severe mononuclear cell infiltration into the colon. In contrast, animals treated with CD45RB^{low} CD4$^+$ T cells or with unfractionated CD4$^+$ T cells did not develop any sign of disease. The co-transfer of both subpopulations prevented the wasting disease and colitis, suggesting that regulatory interactions occur between CD45RB^{high} CD4$^+$ and CD45RB^{low} CD4$^+$ T cells. It has been shown that CD4$^+$ cells from diseased mice display a Th1 pattern of cytokine production upon polyclonal activation *in vitro*. Anti-IFNγ mAb or systemic infusion of recombinant IL-10 given after T-cell transfer prevents colitis while treatment with recombinant IL-4 has no effect. Continual neutralization of TNF-α with anti-TNF-α mAb improves the disease but there is no beneficial effect when the antibody is given only until day 14 after transfer. These experiments clearly demonstrate that potentially autoaggressive Th1 cells are present in normal individuals.

5.2 Autoimmunity induced by rendering rats lymphopaenic

The BB rat, which has the RT1^{u}7HC haplotype (rat 7HC), is lymphopaenic for some genetically determined but unidentified reason. These rats spontaneously develop diabetes and have been shown to lack a subset of regulatory cells, RT6$^+$ T cells. The IDDM in these rats can be prevented by transfer of T cells from a non-diabetic congenic line. PVG(RT1^u) is a normal healthy strain of rat with the same MHC haplotype as the BB rat but the background genes of PVG. As a consequence of thymectomy and radiation-induced lymphopoenia, PVG rats develop IDDM. Diabetes in these lymphopenic rats is associated with selective destruction of the β cells of the pancreas involving both CD4$^+$ and CD8$^+$ T cells. The disease can be completely prevented by the intravenous injection of peripheral CD4$^+$ T cells from syngeneic normal donors. The phenotype of the disease-preventing T cells is CD4$^+$CD45RC^{low}. When a different rat strain, PVG-RT1^c, was rendered lymphopoenic by a regime of thymectomy and γ-irradiation similar to the one used to induce IDDM, this strain now developed thyroiditis. As in IDDM, transfer of CD4$^+$ CD45RC^{low} cells from a normal syngeneic donor prevents the development of this disease.

5.3 How do CD4$^+$ CD45RB^{low} or CD45RC^{low} T cells protect from autoimmunity?

In the experimental systems described above, it is evident that normal healthy laboratory animals (rats and mice), which do not develop any autoimmune disease, harbour T cells that are potentially capable of causing lethal inflammatory responses. Under normal circumstances these cells are held in check by a phenotypically distinguishable subpopulation of regulatory CD4$^+$ T cells. The cells that protect from radiation-induced IDDM have some of the characteristics of Th2-type T cells. Given that

∎ **Table 1.**   Regulatory cells able to downmodulate immunopathological manifestations

Cell	Specificity	Cytokines produced	Conditions of generation	Effect
Th3 (CD4+)	anti-MBP	IL-4, TGF-β*	oral tolerance to myelin	prevent EAE
CD4+CD45RB^{low}	?	IL-4, TGF-β* IL-10	present among CD4+CD45RB^{low} cells in normal animals,	prevent colitis induced by transfer of normal CD4+CD45RB^{high} cells into SCID mice
CD4+CD45RC^{low}	?	IL-4, TGF-β? IL-10	present among CD4+CD45RC^{low} cells in normal animals,	prevent wasting disease induced by transfer CD4+CD45RC^{high} cells into nude rat
Tr1 (CD4+)	anti-OVA	IL-10, IFN-γ TGF-$\beta\pm$	culture of OVA-specific T cells in the presence of IL-10	prevent colitis induced by transfer of normal CD4+CD45RB^{high} cells into SCID mice
CD4+	stimulated by normal B cells	IL-10	Normal animals treated by anti-IL-12 mAb	prevent EAE
CD4+	anti-self MHC class II	IFN-γ, TGF-β*	HgCl$_2$-injected LEW rats	prevent EAE and Th2-dependent HgCl$_2$-induced autoimmunity
CD4+	anti-self MHC class II	IFNγ, TGF-β*	prediabetic NOD mice	prevent IDDM
CD8+	anti-MBP	TGF-β	oral tolerance to myelin	prevent EAE

*anti-TGFβ mAb abolished the protective effect of these regulatory T cells; Ova, ovalbumin; MBP, myelin basic protein.

IL-4 has been shown to antagonize the development of Th1-type T cells that are involved in cell-mediated immunity, it was tempting to speculate that IL-4 could be involved in protection. However, the injection of a neutralizing mAb to rat IL-4 does not abrogate protection. In mice, it has been shown that CD4[+] CD45RB[low] T cells protect from Th1-mediated colitis in a TGF-β-dependent manner. IL-4 is not required for either the differentiation or function of protective cells, since CD4[+] CD45RB[low] cells from IL-4-deficient mice are fully effective. These data suggest that this immunoregulatory population is distinct from Th2 cells. Furthermore, it has been shown that the colitis induced by adoptive transfer of CD4[+] CD45RB[high] cells was completely abrogated in mice treated systemically with recombinant IL-10 but not with recombinant IL-4. Recently, regulatory CD4[+] T-cell clones (Tr1) were generated after chronic activation of both human and mouse CD4[+] T cells in the presence of IL-10. These cells have also been shown to inhibit the colitis induced in SCID mice. Altogether, these data suggest that TGF-β and IL-10, but not IL-4, play a crucial role in regulating the immunopathological manifestations induced by pathogenic cells.

6. Conclusion

The results described above demonstrate that IL-4 is not the key cytokine involved in immunoregulation, but that TGF-β and IL-10 play a crucial role. The importance of these cytokines is underscored by the finding that mice with TGF-β or IL-10 knocked out develop systemic cell-mediated inflammation due to mononuclear cell infiltration. This disease is reminiscent of the one observed in nude rats and in SCID mice given an injection of CD4[+] T cells lacking the protective subset. Furthermore, all the regulatory cells described so far in the literature have distinct cytokine profiles but produce TGF-β and/or IL-10 (Table 1). In addition, the regulation of the pathogenic responses by these cells apparently involves these cytokines. Taken together, these data suggest that deviation of the immune response will not be an adequate strategy to prevent autoimmune diseases and that it would be more beneficial to manipulate the immune system in order to favour the development of TGF-β- and IL-10-producing cells.

References

Berg, D.J., Davidson, N., Kuhn, R., Muller, W., Menon, S., Holland, G., Thompson-Snipes, L., Leach, M.W. and Rennick, D. (1996). Enterocolitis and colon cancer in interleukin-10-deficient mice are associated with aberrant cytokine production and CD4[+] Th1-like responses. *J. Clin. Invest.* **98**, 1010–20.

Bridoux, F., Badou, A., Saoudi, A., Bernard, I., Druet, E., Pasquier, R., Druet, P. and Pelletier, L. (1997). Transforming growth factor β (TGF-β)-dependent inhibition of T helper cell (Th2)-induced autoimmunity by self-major histocompatibility (MHC) class II-specific, regulatory T cell lines. *J. Exp. Med.* **185**, 1769–75.

Chen, Y., Kuchroo, V. K., Inobe, J.-i., Hafler, D.A. and Weiner, H. L. (1994). Regulatory T cell clones induced by oral tolerance: suppression of autoimmune encephalomyelitis. *Science* **265**, 1237–40.

Druet, P., Ramanathan, S. and Pelletier, L. (1996). Th1 and Th2 lymphocytes in autoimmunity. *Adv. Nephrol.* **25**, 217–41.

Fowell, D., McKnight, A.J., Powrie, F., Dyke, R., and Mason, D. (1991). Subsets of CD4[+] T cells and their role in the induction and prevention of autoimmunity. *Immunol. Rev.* **123**, 37–64.

Goldman, M., Druet, P. and Gleichmann, E. (1991). T$_{H}$2 cells in systemic autoimmunity: insights from allogeneic diseases and chemically-induced autoimmunity. *Immunol. Today* **12**, 223–7.

Groux, H., Bigler, M., de Vries, J. E. and Roncarolo, M.G. (1996). Interleukin-10 induces a long-term antigen-specific anergic state in human CD4[+] T cells. *J. Exp. Med.* **184**, 19–29.

Han, H. S., Jun, H.S., Utsugi, T., and Yoon, J.W. (1996). A new type of CD4+ suppressor T cell completely prevents spontaneous autoimmune diabetes and recurrent diabetes in syngeneic islet-transplanted NOD mice. *J. Autoimmun.* **9**, 331–9.

Liblau, R.S., Singer, S.M. and McDevitt, H.O. (1995). Th1 and Th2 CD4[+] T cells in the pathogenesis of organ-specific autoimmune diseases. *Immunol. Today* **16**, 34–8.

Lindstrom, J., Shelton, D. and Fujii, Y. (1988). Myasthenia gravis. *Adv. Immunol.* **42**, 233–84.

Mosmann, T.R. and Sad, S. (1996). The expanding universe of T cell subsets: TH1, TH2 and more. *Immunol. Today* **17**, 138–46.

Nakajima, A., Hirose, S., Yagita, H. and Okumura, K. (1997). Roles of IL-4 and IL-12 in the development of lupus in NZB/W F1 mice. *J. Immunol.* **158**, 1466–72.

Peng, S. L., Mosiehi, J. and Craft, J. (1997). Roles of interferon-g and interleukin-4 in murine lupus. *J. Clin. Invest.* **99**, 1936–46.

Powrie, F., Carlino, J., Leach, M.W., Mauze, S. and Coffman, R.L. (1996). A critical role for transforming growth factor-β but not interleukin 4 in the suppression of T helper type-1 mediated colitis by CD45RB[low] CD4[+] T cells. *J. Exp. Med.* **183**, 2669–74.

Santiago, M. L., Fossati, L., Jacquet, C., Muller, W., Izui, S. and Reininger, L. (1997). Interleukin-4 protects against a genetically linked lupus-like autoimmune syndrome. *J. Exp. Med.* **185**, 65–70.

Saoudi, A., Seddon, B., Heath, V., Fowell, D. and Mason, D. (1996). The physiological role of regulatory T cells in the prevention of autoimmunity: the function of the thymus in the generation of the regulatory T cell subset. *Immunol. Rev.* **149**, 196–216.

Segal, B. M., Dwyer, B.K. and Shevach, E.M. (1998). An interleukin (IL)-10/IL-12 immunoregulatory circuit controls susceptibility to autoimmune disease. *J. Exp. Med.* **187**, 537–46.

Shull, M.M., Ormsby, I., Kier, A.B., Pawlowski, S., Diebold, R.J., Yin, M. *et al.* (1992). Targeted disruption of the mouse transforming growth factor-β1 gene results in multifocal inflammatory disease. *Nature* **359**, 693–9.

XXVII Cytokine antagonists and autoimmunity

Marc Feldmann*, Ravinder Maini and Fionula M. Brennan

1. Background

While the aetiology of autoimmune diseases remains elusive, there is now considerable understanding of their pathogenesis. Genetic influences are clear, with the human leucocyte-associated antigen (HLA) complex being the best known predisposing factor. However, there are non-genetic risk factors, as concordance for disease in identical twins can be as low as 15% in the case of rheumatoid arthritis (RA). Of the genetic influences, there is increasing evidence that some may involve cytokines. For example, polymorphisms in the promoters of tumour necrosis factor alpha (TNFα) and interleukin 1 receptor antagonist (IL-1ra) have been associated with predisposition to, or severity of, various autoimmune diseases.

Relapses of autoimmune diseases such as RA and multiple sclerosis have been reported after cytokine therapy with IL-2 and interferon gamma (IFNγ), and the induction of autoimmune manifestations such as thyroiditis documented after IFNα therapy.

These phenomena and the up-regulated expression of HLA antigens, both class I and class II in sites of local autoimmune responses such as Graves' disease, Hashimoto's thyroiditis and RA led us to propose that cytokine-mediated up-regulation, in particular by IFNγ, of antigen-presenting function was at the core of the autoimmune disease process. This hypothesis was successfully tested by making transgenic mice which expressed IFNγ locally, in the islets of Langerhans of the pancreas under the control of the rat insulin promoter. These mice developed chronic autoimmune diabetes, complete with autoantigen-reactive

T cells. Subsequent work has revealed that local IFNγ at other sites such as the eye and neuromuscular junction, and IFNα, but not other cytokines can also induce autoimmunity also provoked autoimmunity. One interesting aspect was that cytokine up-regulation causes autoimmunity in mouse strains apart from those known to be genetically susceptible to that disease. This suggests that cytokine effects may be prominent risk factors in pathogenesis.

There is considerable heterogeneity in the clinical manifestation of autoimmune diseases. Some are localized, with prominent recruitment of cells of the immune and inflammatory systems to the site of disease, such as RA or insulin-dependent diabetes. In systemic autoimmune diseases, such as systemic lupus erythematosus, there is no abnormal site for the autoimmune response, which presumably takes place in the normal lymphoid tissues. These diseases are not as easy to study as the localized disease.

There is evidence that CD4$^+$ T cells are heterogeneous in their production of cytokines. Some, now termed the Th0 type, are able to synthesize a wide spectrum of cytokines including IL-2, IFNγ, IL-4, IL-5, lymphotoxin, etc. These cells are sometimes precursors of two types of cells with a more restricted cytokine profile: (i) cells that produce IL-2 and IFNγ but not IL-4 or IL-5, and are termed Th1, and (ii) cells that produce IL-4 and IL-5 but not IL-2 or IFNγ, and are termed Th2. It was initially reported that Th1 cells were chiefly responsible for delayed-type hypersensitivity (DTH) responses, and Th2 for antibody responses, but these are oversimplifications, as in the mouse Th1 cells promote IgG2 responses, and Th2 especially IgG1 and IgE. Another oversimplification is the concept that Th1 cells are pro-inflammatory and

* Corresponding author

Th2 anti-inflammatory. IL-10 production by T cells is found in both Th1 and Th2 subsets and is probably one of the major aspects of anti-inflammatory activity.

Because of the relative ease of sampling the disease site both early and late in the immune/inflammatory process, more is known about cytokine expression in RA than in other local autoimmune diseases. Hence this chapter will highlight findings obtained in RA and compare and contrast them with those of other diseases. Much more detailed reviews of the role of cytokines in autoimmunity have recently been published (see reference list).

2. Role of cytokines in rheumatoid arthritis

A wide spectrum of pro-inflammatory and anti-inflammatory cytokines have been found at the major site of the disease, in the synovial membrane. From mRNA analysis it is evident that most, if not all, of these cytokines are synthesized locally. Table 1 summarizes what is currently known about cytokine expression in RA joints. This listing of expressed cytokines does not provide clues concerning which, if any of these may be of major importance in the disease process, and hence could be targets for therapy. Other groups have used synovial fluid analyses, or *in situ* hybridization of synovial membrane to document that cytokine expression joints is up-regulated in RA joints.

During a normal immune or inflammatory response, cytokine expression is usually transient, lasting 1–3 days *in vivo* or *in vitro*. The consistently up-regulated expression of cytokines in the RA synovium suggests that, in contrast to normal tissue, cytokine expression may be prolonged in the diseased tissue. A model to test that hypothesis was devised by dissociating synovial membrane tissue into a cell suspension, which was cultured *in vitro*

in the absence of extrinsic stimulation. Confirming the above hypothesis, IL-1α mRNA and IL-1 bioactivity were detected for at least 6 days *in vitro*, whereas in normal control cells (PBMC), IL-1α mRNA induced with a mitogen cocktail only lasted 24 h.

The above result also provided us with an *in vitro* system to dissect the mechanism of up-regulated IL-1 production. It was not due to prolongation in mRNA half-life, and hence was due to up-regulation of mRNA production and consequently protein synthesis. Since IL-1 was at the time (in the late 1980s) the best known inducer of joint damage in RA, the signals driving IL-1 production were investigated. It was found using anti-TNFα antibody that, of the many signals in the synovium potentially capable of inducing IL-1, TNFα was dominant. This was our first clue that TNFα may be a useful therapeutic target in RA. These results also suggested that the plethora of cytokines expressed in RA synovium may not be independently regulated, but co-ordinated. Hence it was of interest to evaluate what other cytokines are regulated by TNFα in this human diseased tissue. Granulocyte–macrophage colony-stimulating factor (GM-CSF), another pro-inflammatory cytokine, was found to be TNFα-regulated and subsequently IL-10, IL-6 and IL-8 were also.

Cytokine antagonists are up-regulated in disease. There are three major types of cytokine 'antagonists':

* the shed extracellular domain of most cytokine receptors which are found in biological fluids at significant levels (3–50 ng/ml)
* IL-1 receptor antagonists and
* certain cytokines.

Not all cytokines are potentially pathogenic. Some are anti-inflammatory and anti-immune, such as TGFβ and IL-10. The expression of these anti-inflammatory cytokines in RA synovium has been ascertained. Both TGFβ and IL-10 are present at high and biologically active concentrations. A question of interest is whether these cytokines are actively contributing to the down-regulation of the disease activity. This was found to be the case for IL-10 as in RA synovial cultures anti-IL-10 antibody induced a two- to three-fold augmentation of IL-1 and TNFα levels, and additional exogenous IL-10 diminished IL-1 and TNFα production. In contrast, additional TGFβ did not have the same effect. There are a variety of other anti-cytokine activities in RA joints, such as the IL-1ra and

■ **Table 1.** Cytokine expression in rheumatoid joints

Abundant	— IL-6, TNFα, IL-1, IL-8, IL-11 GM-CSF, PDGF, IFNα, VEGF RANTES, G-CSF, M-CSF, Other chemokines IL-10, TGFβ, IL-13
Less Abundant	— IL-2, IL-3, LT, IFNγ, IL-12, IL-15
Sporadic	— IL-4

soluble TNF receptors. These are all up-regulated as judged by comparison of the matched synovial fluid and serum levels in RA patients, and their comparison with normal serum levels. It is likely that these endogenous cytokine antagonists are important in limiting the extent and degree of inflammation, and are important steps in the chronicity of the disease.

The concepts we have built up to help understand the mass of data concerning cytokine expression are summarized in Figures 1 and 2. The TNFα-dependent 'cytokine cascade' concept we proposed from analysis of rheumatoid synovial cultures *in vitro* has also been established *in vivo*, and is not a pathological process but part of normal immune physiology. Thus mice injected with Gram-negative bacteria sequentially make abundant TNFα, IL-1 and IL-6, assessable in serum. If injected with anti-TNFα antibody, the IL-1 and IL-6 serum peaks are diminished. In clinical trials of anti-TNFα antibody in RA, serum IL-6 levels fell precipitously after treatment, confirming that the TNFα-dependent cytokine cascade is operative *in vivo*, as noted *in vitro*.

3. Clinical trials of anti-cytokine therapy validate the critical role of cytokines in RA

Based on the anti-inflammatory effects of anti-TNFα antibodies on rheumatoid synovial tissue *in vitro*, the presence of up-regulated TNFα in joints, and supportive therapeutic data in animal models of arthritis, such as collagen-induced arthritis, we initiated clinical trials of anti-TNFα antibody in RA patients in 1992. The results were very interesting. Despite the late stage of the disease, in patients not responding to existing drugs, all 20 patients receiving a high concentration (20 mg/kg over 2 weeks) of the chimeric (human IgG1, mouse Fv) anti-TNFα antibody (cA2, produced by Centocor, Inc. now terned Remicade™) responded well in all parameters assessed, compared with their initial status. The duration of benefit ranged from 8 to 26 weeks. A key issue was whether, after patients had relapsed after anti-TNFα antibody therapy, an equivalent benefit could be reduced with further doses of anti-TNFα antibody. This was found to be the case, demonstrating that the pathological processes in RA remain TNFα-dependent.

The above trials were 'open', not placebo-controlled, and hence potentially susceptible to bias, so a necessary step towards formal proof of the effectiveness of anti-TNFα therapy, and hence of the importance of TNFα in the pathogenesis of RA, was a randomized, double-blind, placebo-controlled trial. With both low and high doses of cA2 (1 and 10 mg/kg), clear benefit was obtained in 44% and 79% of cases, compared with 8% with placebo. Clinical trials of other anti-TNFα biologicals (antibody or TNF receptor–IgG fusion protein) have also all (e.g. CDP571, Celltech), demonstrated their efficacy in RA (e.g. Enbrel™, Immunex). This

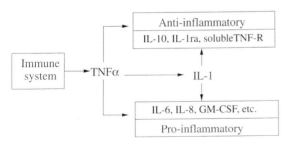

∎ **Figure 1.** Cytokine cascade in rheumatoid arthritis. Reprinted, with permission, from Feldmann *et al., Cell* **85**, 307–10, 1996.

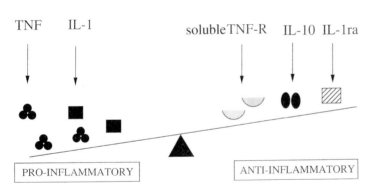

∎ **Figure 2.** Cytokine imbalance. Reprinted, with permission, from Feldmann *et al., Cell* **85**, 307–10, 1996.

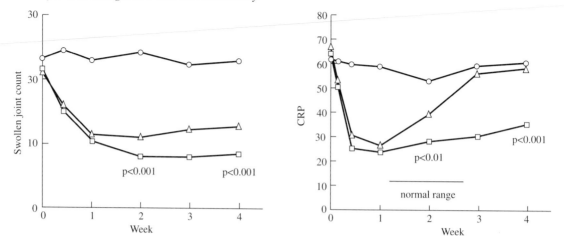

■ **Figure 3.** Randomized, placebo-controlled trial of cA2 (an anti-TNFα antibody) in rheumatoid arthritis. Left: swollen joint count; right: CRP. Key: {○}, placebo; {△}, 1 mg/kg cA2; {□}, 10 mg/kg cA2. Reproduced from Elliott, M.J. *et al.* (1994) *Lancet* **344**, 1105–10.

degree of consistency is unusual in clinical trials and attests to the importance of TNFα and of the cytokine network in RA.

Other anti-cytokine therapies have been tested in RA patients. Most extensively used has been the IL-1ra, but so far the results have been confusing. The anti-inflammatory effects have been insignificant, but in early RA patients a statistically significant reduction in the rate of joint destruction was noted.

Anti-IL-6 antibody has been used in a small number of patients with evidence of efficacy, and so has anti-IL-6 receptor antibody. These clinical results are all consistent with the cytokine cascade concept illustrated in Figure 1. Further *in vivo* evidence supporting the cytokine cascade comes from the rapid reduction of serum IL-6 in RA patients after anti-TNFα therapy.

4. Cytokines and tissue destruction

It is currently believed that matrix metalloproteinase enzymes are of major importance in joint destruction. The effect of cytokines on many of these enzymes is well known *in vitro*: pro-inflammatory cytokines such as IL-1 and TNFα up-regulate their synthesis, while anti-inflammatory cytokines such as IL-4, IL-10 and IL-11 augment the production of their inhibitor, TIMP-1. In animal

models of arthritis, such as collagen type II induced arthritis and TNFα transgenic arthritis, anti-TNFα therapy diminishes or prevents joint destruction. However, in human RA, there is a school of thought that proposes that inflammation and joint destruction are separate processes. The effects of anti-TNFα and IL-1ra in clinical trials do not yet resolve this issue, as conclusive evidence of joint protection is still lacking due to the short-term and limited nature of trials conducted so far. However, anti-TNFα therapy was shown to reduce the previously elevated serum levels of matrix metalloproteinase 1 and 3. This suggests that TNF blockade may reduce matrix metalloproteinase synthesis, but of course is not proof of what is happening at the eroding cartilage interface.

5. Cytokine therapy

There have been early attempts to use cytokines with anti-inflammatory properties in other autoimmune diseases. IL-10 has been used in Crohn's disease, with efficacy, and in RA it has just successfully completed phase I trials. TGFβ has been used in multiple sclerosis. IL-11 has successfully completed phase I trials in Crohn's disease and is in phase I trials in RA. IL-4 is in phase I trials for RA. Currently it is not possible to know if these cytokines will be therapeutically useful as direct anti-inflammatory agents, or for altering the Th1/Th2 ratio (e.g. IL-4).

6. Prospects for the future

As our understanding of the role of cytokines in autoimmune disease increases, it is to be expected that it will be possible to build on the current therapeutic success of anti-TNFα therapy which is now licensed for use clinically. Whether this will involve combination therapy with anti-immune drugs or antibodies, drugs with anti-cytokine properties, or gene therapy, or all of these options remains to be seen. It is clear that a wide spectrum of diseases is amenable to regulation, even in the late stages, by anti-cytokine therapy, and the challenge is to deliver this new knowledge to patients in a safe, effective, and — as is increasingly necessary — cost-effective manner.

Acknowledgements

The work of the authors was chiefly supported by the Arthritis Research Campaign, and is part of a longstanding collaboration with many current and former members of the Kennedy Institute of Rheumatology, including Drs R. Williams, M. Elliott and E. Paleolog. The clinical trials have been supported by Centocor, Inc.

References

Arend, W. P. (1993). Interleukin-1 receptor antagonist. *Adv. Immunol.* **54**, 167–227.

Atkins, M. B. *et al.* (1988). Hypothyroidism after treatment with interleukin-2 and lymphokine-activated killer cells. *New Engl. J. Med.* **318**, 1557–63.

Bottazzo, G. F. *et al.* (1983). Hypothesis: role of aberrant HLA-DR expression and antigen presentation in the induction of endocrine autoimmunity. *Lancet* **ii**, 1115–19.

Brennan, F. M. and Feldmann, M. F. (eds) (1996). *Cytokines in Autoimmunity*. R.G. Landes, Austin, TX.

Brennan, F. M. *et al.* (1989). Inhibitory effect of TNFα antibodies on synovial cell interleukin-1 production in rheumatoid arthritis. *Lancet* ii, 244–7.

Brennan, F. M. *et al.* (1997). Reduction of serum matrix metalloproteinase 1 and matrix-metalloproteinase 3 in RA patients following anti-TNFα (cA2) therapy. *Brit. J. Rheumatol.* **36**, 643–50.

Bresnihan, B. *et al.* (1995). Synovial pathology and articular erosion in rheumatoid arthritis. *Rheumatol. Eur.* **24(S)**, 158–160.

Buchan, G. *et al.* (1988). Interleukin-1 and tumour necrosis factor mRNA expression in rheumatoid arthritis: prolonged production of IL-1α. *Clin. Exp. Immunol.* **73**, 449–55.

Burman, P. *et al.* (1986). Thyroid autoimmunity in patients on long term therapy with leukocyte-derived interferon. *J. Clin. Endocrinol. Metab.* **63**, 1086–90.

Butler, D. M. *et al.* (1995). Modulation of proinflammatory cytokine release in rheumatoid synovial membrane cell cultures. Comparison of monoclonal anti-TNFα antibody with the IL-1 receptor antagonist. *Eur. Cytokine Netw.* **6**, 225–30.

Campion, G. V. *et al.* (1996). Dose-range and dose frequency study of recombinant human interleukin-1 receptor antagonist in patients with rheumatoid arthritis. *Arthr. Rheum.* **39**, 1092–101.

Cope, A. P. *et al.* (1992). Increased levels of soluble tumor necrosis factor receptors in the sera and synovial fluid of patients with rheumatic diseases. *Arthr. Rheum.* **35**, 1160–69.

Cutulo, M. *et al.* (1996). Loading/maintenance doses approach to neutralization of TNF by lenercept in patients with rheumatoid arthritis treated for 3 months: results of a double blind placebo controlled phase II trial. *Arthr. Rheum.* **39**, Suppl. 9, S243.

Elliott, M. J. *et al.* (1993). Treatment of rheumatoid arthritis with chimeric monoclonal antibodies to TNFα. *Arthr. Rheum.* **36**, 1681–90.

Elliott, M. J. *et al.* (1994). Randomised double blind comparison of a chimaeric monoclonal antibody to tumour necrosis factor α (cA2) versus placebo in rheumatoid arthritis. *Lancet* **344**, 1105–10.

Elliott, M. J. *et al.* (1994). Repeated therapy with monoclonal antibody to tumour necrosis factor α (cA2) in patients with rheumatoid arthritis. *Lancet* **344**, 1125–27.

Fava, R. *et al.* (1989). Active and latent forms of transforming growth factor β activity in synovial effusions. *J. Exp. Med.* **169**, 291–6.

Feldmann, M. *et al.* (1996). Role of cytokines in rheumatoid arthritis. *Annu. Rev. Immunol.* **14**, 397–440.

Fong, Y. *et al.* (1989). Antibodies to cachectin/tumor necrosis factor reduce interleukin 1β and interleukin 6 appearance during lethal bacteremia. *J. Exp. Med.* **170**, 1627–33.

Gregersen, P. K. *et al.* (1987). The shared epitope hypothesis. An approach to understanding the molecular genetics of susceptibility to rheumatoid arthritis. *Arthr. Rheum.* **30**, 1205–13.

Gu, D.-L. *et al.* (1995). Myastenia gravis-like syndrome induced by expression of interferon-gamma in the neuromuscular junction. *J. Exp.l Med.* **181**, 547–57.

Hanafusa, T. *et al.* (1983). Aberrant expression of HLA-DR antigen on thyrocytes in Graves' disease: relevance for autoimmunity. *Lancet* **ii**, 1111–5.

Haworth, C. *et al.* (1991). Expression of granulocyte–macrophage colony-stimulating factor in rheumatoid arthritis: regulation by tumor necrosis factor-α. *Eur. J. Immunol.* **21**, 2575–9.

Janossy, G. *et al.* (1981). Rheumatoid arthritis: a disease of T-lymphocyte/macrophage immunoregulation. *Lancet* **ii**, 839–42.

Katsikis, P. *et al.* (1994). Immunoregulatory role of interleukin 10 (IL-10) in rheumatoid arthritis. *J. Exp. Med.* **179**, 1517–27.

Klareskog, L. *et al.* (1982). Evidence in support of a self perpetuating HLA-DR dependent delayed type cell reaction in rheumatoid arthritis. *Proc. Natl Acad. Sci. USA.* **72**, 3632–6.

McDowell, T. L. *et al.* (1995). A genetic association between juvenile rheumatoid arthritis and a novel interleukin-1α polymorphism. *Arthr. Rheum.* **38**, 221–8.

Moreland, L. W. *et al.* (1997). Treatment of rheumatoid arthritis with a recombinant human tumor necrosis factor receptor (p75)–Fc fusion protein. *New Engl. J. Med.* **337**, 141–7.

Mosmann, T. R. and Coffman, R. L. (1989). Th1 and Th2 cells: different patterns of lymphokine secretion lead to different functional properties. *Annu. Rev. Immunol.* **7**, 145–73.

Oppenheim, J. J. and Saklatvala, J. (1993). Cytokines and their receptors. In Oppenheim, J. J. Rossio, J. L. and Gearing, A. J. H. (eds), *Clinical Applications of Cytokines: Role in Pathogenesis, Diagnosis and Therapy*, pp. 3–15. Oxford University Press.

Panitch, H. S. *et al.* (1987). Exacerbations of multiple sclerosis in patients treated with gamma interferon. *Lancet* **i**: 893–5.

Rankin, E. C. C. *et al.* (1995). The therapeutic effects of an engineered human anti-tumour necrosis factor alpha antibody (CD571) in rheumatoid arthritis. *Brit. J. Rheumatol.* **34**, 334–42.

Sarvetnick, N. *et al.* (1990). Loss of pancreatic islet tolerance induced by β-cell expression of interferon-γ. *Nature* **346**, 844–7.

Saxne, T. *et al.* (1988). Detection of tumor necrosis factor α but not tumor necrosis factor β in rheumatoid arthritis synovial fluid and serum. *Arthr. Rheum.* **31**, 1041–5.

Silman, A. J. *et al.* (1993). Twin concordance rates for rheumatoid arthritis: results from a nationwide study. *Brit. J. Rheumatol.* **32**, 903–7.

Stewart, T. A. *et al.* (1993). Induction of type-1 diabetes by interferon-γ in transgenic mice. *Science* **260**, 1942–7.

van Deventer, S. J. H. *et al.* (1997). Multiple doses of intravenous interleukin-10 in steroid-refractory Crohn's disease. *Gastroenterology* **113**, 383–9.

Watt, O. and Cobby, M. (1996). Recombinant human IL-1 receptor antagonist (rhIL-1ra) reduces the rate of joint erosion in rheumatoid arthritis (RA). *Arthr. Rheum.* **39**, S576.

Wendling, D. *et al.* (1993). Treatment of severe rheumatoid arthritis by anti-interleukin 6 monoclonal antibody. *J. Rheumatol.* **20**, 259–62.

Williams, R. O. *et al.* (1994). Synergy between anti-CD4 and anti-TNF in the amelioration of established collagen-induced arthritis. *Proc. Natl Acad. Sci. USA* **91**, 2762–6.

Wilson, A. G. *et al.* (1994). A genetic association between systemic lupus erythematosus and tumor necrosis factor alpha. *Eur. J. Immunol.* **24**, 191–5.

Wood, N. C. *et al.* (1992). *In situ* hybridization of interleukin-1 in CD14-positive cells in rheumatoid arthritis. *Clin. Immunol. Immunopathol.* **62**, 295–300.

Yssel, H. *et al.* (1992). IL-10 is produced by subsets of human CD4$^+$ T cell clones and peripheral blood T cells. *J. Immunol.* **149**, 2378–84.

XXVIII Cytokines in cancer

Lisa P. Seung, Donald A. Rowley and Hans Schreiber*

1. Introduction

Cancer research has contributed considerably to our knowledge of cytokines. For example, tumor necrosis factor (TNFα) and transforming growth factor beta (TGFβ) were originally discovered by cancer researchers. Further important information has been gained by the study *in vivo* and *in vitro* of cancer cells transfected to express certain cytokines. Some of this research has revealed previously unknown effects of cytokines and from this research numerous clinical protocols for cancer treatment have been developed that involve cytokines and cytokine gene-transfected tumor cells.

The distinction between molecules that are growth factors and those that are considered cytokines remains unclear. For our purposes, cytokines include growth factors. The effects of cytokines on anti-tumor immunity (mediated by T and B cells) and on immunotherapy are discussed in Chapter XXIX. In their recent thorough review, Qin and Blankenstein (see reference list) have tabulated the results of studies on cytokine gene transfection in cancer. Our purpose is to discuss new perspectives concerning the role of endogenous cytokines in cancer promotion or inhibition, the nature of key unresolved questions, and opportunities for future research.

Cytokines and chemokines are critical components of inflammation. There is convincing evidence that inflammation plays an important role in each stage of malignant disease from tumor promotion to progression and metastasis. In the stage of tumor promotion, cells carrying silent mutations (so called 'initiated cells') express these mutant genes and become cancerous. During tumor progression, cancer cells acquire additional mutations to become variants with increased malignant potential. Finally, in the stage of metastasis, further mutations allow cancer cells to spread and vascularize (Figure 1). Tumor cells can secrete cytokines which stimulate or inhibit their own growth (autocrine effects) and attract inflammatory cells. Once recruited, inflammatory cells can then produce cytokines that directly or indirectly stimulate or inhibit tumor growth (paracrine effects). The specific effects on tumor growth seem to vary depending on the cells and factors present in the local tumor environment. Increased knowledge of the very complex mechanisms involved should help us develop new ways for prevention and therapy of human cancer.

2. Cytokines and tumor promotion

Tumor promotion is the stage of tumor development that begins with an initiated cell. An initiated cell is a cell that has acquired mutations either due to germline transmission or due to carcinogen exposure. Tumors may not develop unless the initiated cell and subsequent daughter cells are exposed to tumor-promoting agents. Tumor promotion was first described in the skin as the second part of a two-stage process of carcinogenesis, the first stage being the initiation stage, sometimes consisting of a single topical application of carcinogen to the skin. Second-stage promotion was usually done by wounding or topical application of phorbol esters (phorbol 12 myristate-13-acetate or PMA). Several studies have shown that the effectiveness of promotion appears to correlate directly with the degree of the inflammatory reaction elicited in which granulocytes and macrophages are the dominant cells. Mouse strains that are resistant to the inflammatory effects of phorbol esters or wounding

*Corresponding author. Supported by grant R01-CA22677 from the U.S.P.H.S.

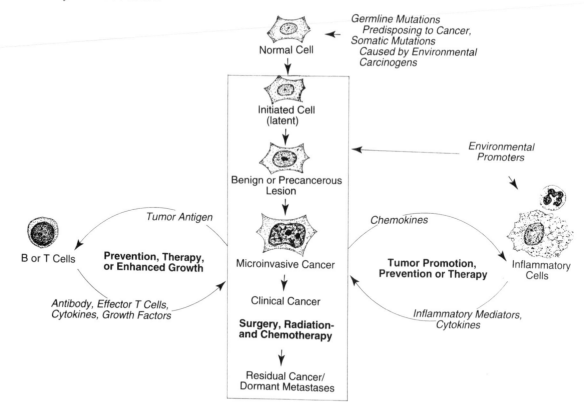

▮**Figure 1.** The stages of tumor progression from a normal cell to an initiated cell, a precancerous lesion, a micro-invasive cancer, clinical cancer and finally metastatic disease. At many stages, various factors enhance or inhibit tumor development. Modes of therapeutic intervention are indicated. T and B cells can be attracted and stimulated by tumor antigen or chemokines produced by the tumor cells. Similarly, inflammatory cells, such as neutrophils and macrophages, can be attracted by chemokines released by the tumor. Dependent upon the state of activation and molecules produced by these lymphocytes and inflammatory cells, tumor development can be enhanced (as in the result of treatment with chemical promoters) or inhibited (as may be achieved by specific T- or B-cell immunity). However, it is important to note that, dependent upon the conditions, the opposite results may also occur: certain responses of lymphocytes may also stimulate tumors to grow while certain responses of inflammatory cells may also inhibit the growth of cancer cells.

are more resistant to tumor promotion. The inflammatory reaction results in an environment of local growth factors and cytokines that appear to promote tumor development. Numerous pieces of evidence are consistent with this notion. For example, PMA-induced interleukin 1 (IL-1) and TNFα regulate the proliferation of keratinocytes, stimulate the migration and attraction of leukocytes, and contribute to their release of reactive oxygen intermediates. By contrast, inhibition of pro-inflammatory cytokine gene expression inhibits PMA-induced tumor promotion and skin tumor development in mice treated with chemical carcinogens. Thus, there appears to be an intricate relationship between the cytokines produced during

inflammatory reactions and the resulting effect on cellular growth. On one hand, inflammation may destroy malignant cells or arrest their growth; on the other hand, compelling evidence seems to be shifting the paradigm to one in which inflammatory cytokines favor tumor promotion.

A carcinogen that induces mutations as well as tumor promotion and contributes to subsequent progression is called a complete carcinogen. Ultraviolet (UV) light is considered to be such a complete carcinogen. The surprising new finding is that the generation of unrepaired photoproducts resulting from radiation-induced DNA damage causes not only the induction of the repair machinery and expression of p53, but also the generation of inflammatory and/or immunosuppressive cyto-kines. DNA damage caused by UV irradiation can trigger IL-10 production from keratinocytes, resulting in a state of immune suppression. This suppression can prevent the host from rejecting highly immunogenic tumor cells. Similarly induction of TNFα can be seen following treatment with certain chemical tumor promoters. Increasing DNA repair prevents these immunological alterations and prevention of TNFα expression can prevent tumor promotion. Thus, blocking the inflammatory response may be effective in reducing tumor induction by both UV light and chemical promoters.

TGFβ appears to be required for keeping the inflammatory response in check and seems to exist in a delicate balance with TNFα. TGFβ knockout mice are runts and die early of a 'hyperinflammatory state' characterized by lymphoproliferation and invasion of tumors by lymphocytes. This phenotype is simulated by transgenic mice overproducing TNFα. Furthermore, while tumor promoters increase TNFα production in keratinocytes, transgenic mice expressing the TGFβ-related molecule bone morphogenetic protein 4 (BMP-4) under the control of a cytokeratin promoter do not develop an inflammatory response to promoter application, nor do these mice develop any skin tumors.

Several other cytokines have been shown to have growth-promoting activities in tumors, including IL-2, IL-3, granulocyte–macrophage colony-stimulating factor (GM-CSF), macrophage colony-stimulating factor (M-CSF) and IL-9. These cytokines may be involved in both autocrine and paracrine loops, either by directly stimulating the growth of the initiated cells or by promoting an inflammatory response. For example, GM-CSF transcription is induced in keratinocytes, papilloma cells and leukocytes following application of tumor promoters. Since GM-CSF is a potent chemoattractant for neutrophils and macrophages, it may be critical in regulating early inflammation. Indeed, dermal inflammation can be reduced with anti-GM-CSF specific antibodies. Also, anti-IL-1 specific antibodies can inhibit neutrophil infiltration in the dermis when given prior to promoter application. Even though inflammation can lead to keratinocyte proliferation and hyperplasia, hyperplasia does not necessarily lead to neoplastic growth of initiated cells (i.e. cells carrying silent mutations). Inhibition of some pro-inflammatory cytokines with specific antibodies inhibits both keratinocyte hyperplasia and inflammation, while the inhibition of others only limits inflammation. The mechanisms whereby cytokines regulate tumor promotion need to be determined.

Tumor promoters such as phorbol esters and wounding lead to the release of cytokines as well as eicosanoids from epithelial cells and mesenchymal cells. Eicosanoids are precursors of prostaglandins. Prostaglandins are made by many cell types including the major inflammatory cell types, granulocytes and macrophages/monocytes. Overproduction of eicosanoids is observed in experimental skin carcinogenesis but can be inhibited with nonsteroidal anti-inflammatory drugs (NSAIDs) including aspirin. Interestingly, subsequent tumor development can be inhibited as well. Clinical trials have shown NSAIDs to be effective in reducing the size and number of colonic polyps in patients with familial adenomatous polyposis. The mechanism is not known but is likely to involve the inhibition of prostaglandin H-synthase 2, which is encoded by the gene *COX-2* (named from 'cyclooxygenase', an earlier name for prostaglandin synthase).

Prostaglandin-H synthase 1 (encoded by *COX-1*) is a housekeeping enzyme constitutively expressed in many but not all mammalian tissues. In contrast, the *COX-2*-encoded synthase is induced in macrophages/monocytes when stimulated with cytokines, particularly IL-1β and TNFα. Some of the production of these prostaglandins may result from the induction of the *COX-2*-encoded enzyme by reactive oxygen intermediates. The same enzyme may also lead to the oxidation of certain compounds, such as aflatoxin, to form mutagens. This may allow the initiated cells to acquire the additional mutations needed to become neoplastic. In addition to affecting tumor promotion, the generation of prostaglandins by this enzyme may cause increased growth of target cells which have surface receptors for prostaglandins. Prostaglandins may be produced either by the target cells themselves or by the interstitial cells. In a murine model in which colonic polyps develop, it was found by histochemistry that the primary activity of the *COX-2*-encoded enzyme was in interstitial cells. Mice in this model have a mutated *APC* gene as do individuals with familial adenomatous polyposis. Most importantly, inhibition of prostaglandin-H synthase 2, either by genetic knockout or by using pharmacological inhibitors of prostaglandin-H synthase 2, significantly reduced the formation of intestinal polyps in this cancer-prone mouse model. Taken together, these findings suggests a paracrine stimulatory pathway, possibly involving granulocytes and/or macrophages/monocytes in which the enzyme prostaglandin-H synthase 2 is known to be induced. Furthermore, this study suggests that prostaglandin-H synthase 2, which is characteristically induced in inflammatory cells by cytokines and tumor promoters, holds a central biochemical place in colon carcinogenesis and represents an important example of the importance of inflammatory cells in promoting cancer.

3. Cytokines and tumor progression

Tumor progression is usually described as the stage of cancer development that begins with an invasive cancer cell and ends with metastatic cancer. This period is characterized by the sequential development of variant subpopulations which show increased malignant behavior. The term 'tumor progression' was originally used for describing the transition of benign papilloma to invasive squamous carcinoma, and inflammatory cytokines may be important during this period since papillomas may regress, rather than progress to cancer, when the application of the chemical tumor-promoter is stopped too soon. In numerous studies, cytokines have been introduced into tumor cells by transfection to examine the effect on the malignant potential of tumors. Few studies, however, have addressed the role of endogenous cytokines on tumor progression.

Recent work has indicated that during tumor progression cancers may switch from being inhibited by such cytokines as TGFβ to a stage where they are stimulated by the same cytokines. In a mouse prostate cancer model, both loss of TGFβ growth inhibition and induction of collagenase activity are observed as the cells progress from primary to metastatic tumors. In addition, expression of TGFβ2 often correlates with the depth of invasion by melanoma cells. Targeted depletion of the TGFβ1 gene causes rapid progression of initiated keratinocytes to squamous cell carcinomas. Furthermore, human colon cancer cells may escape growth inhibition by TGFβ by mutational inactivation of their TGFβ receptors as a consequence of defective DNA repair mechanisms.

We have found that a murine tumor cell line transfected to secrete large amounts of latent TGFβ grows faster *in vivo* than the parental cells. *In vitro*, both this transfectant and another independent transfectant become sensitive to additional growth factor stimulation, including more TGFβ. Thus, a common mechanism may exist in which sensitivity

to stimulation by one cytokine may lead to an increased ability to be stimulated by other cytokines. A similar switch has been reported to occur during melanoma progression. Initially melanoma cells are inhibited by IL-6 but switch to a more malignant phenotype in which cells are stimulated by IL-6.

Some tumor cells not only become stimulated by factors that are usually inhibitory, but may also produce their own stimulatory factors. For example, some myeloma cell lines have acquired the ability to produce IL-6 which stimulates growth in an autocrine loop. Also, increased growth autonomy of melanomas has been associated with the production of TGFα, TGFβ, platelet-derived growth factor (PDGF), interleukins and basic fibroblast growth factor (bFGF). Some of these factors are likely to be responsible for the autocrine growth stimulation of tumor cells, while others may be responsible for the generation of a microenvironment that favors tumor survival and invasion. This could include the stimulation of surrounding fibroblasts, endothelial cells, inflammatory cells and/or keratinocytes to produce growth-promoting factors. Clearly, the combined effect of 'multi-cytokine resistance' and 'multi-growth factor independence' helps to provide tumor cells with the competence to grow in other tissue sites and thus primes them for metastatic spread.

As mentioned earlier, TGFβ can affect the growth of the tumor cells directly. In addition, TGFβ also affects various other cell types that are necessary for supporting or stimulating tumor growth, or that are needed for an effective immune response. Some of these cell types are, for example, responsible for the vascularization of the growing tumor. TGFβ may also suppress immune responses. For example, immunogenic tumors transfected to produce TGFβ are rejected less effectively than the parental cells. Finally, serum levels of latent TGFβ increase during continued tumor growth. However, it is not clear how and where this material is activated and biologically effective. Key questions concerning TGFβ, and possibly other cytokines released in an inactive (latent) form, are how and where they are activated. At present, the most likely mechanism would be activation of TGFβ by macrophages and/or neutrophils and other cell types that are in the environment into which latent TGFβ is released. A new and particularly important aspect of this activation is the selectivity for only particular cells to be regulated. Latent TGFβ molecules can be physically linked to other molecules which then allow cell-type-specific internalization and activation. The best studied example of such a mechanism is latent TGFβ coupled to IgG, which is one hundred to one thousand times more potent than TGFβ that has been activated by acidification *in vitro*. Other examples include TGFβ linked to human chorionic gonadotropin or to IL-4. Thus, pluripotent cytokines may preferentially affect specific cells through carrier proteins to which they are linked.

It is clear that the tumor stroma plays a critical role in tumor growth. Stromal components include the inflammatory cells, as well as new blood vessels, matrix components and the cells responsible for their production. Of the tumor-associated cells, macrophages are the most common (sometimes comprising up to half of the cell mass in breast carcinomas) and have been studied in the most detail. Murine and human tumor cells secrete chemotactic factors. The first that was identified molecularly is the monocyte chemotactic and activating factor (MCAF/MCP-1), which is specific for macrophages/monocytes. In addition, tumor cells have been shown to secrete other cytokines that are involved in attracting and regulating macrophages. These include the related molecules MCP-2 and MCP-3, as well as M-CSF, GM-CSF, IL-8, Groα, and vascular endothelial growth factor (VEGF; also called vascular permeability factor or VPF). Macrophages are extremely versatile cells, and can produce a variety of substances that stimulate tumor cell proliferation, including platelet-derived growth factor (PDGF), epidermal growth factor (EGF), and prostaglandin E_2 (PGE$_2$). Macrophages can activate latent TGFβ particularly when linked to IgG. Active TGFβ may initially be inhibitory, but later during tumor progression it may stimulate tumor growth. Other macrophage-derived substances initiate angiogenesis (discussed in the following section), modulate the host's immune responses and cause additional growth-enhancing mutations leading to tumor progression. For example, released reactive oxygen intermediates may be directly mutagenic, may activate carcinogens or induce prostaglandin-H synthase 2 which can activate carcinogens. On the other hand, macrophages when appropriately activated can be cytotoxic to tumor cells and/or can elicit tissue destructive reactions aimed at the tumor vasculature. Interestingly, normal cells are quite resistant to killing by activated macrophages. Whether macrophages have stimulatory and inhibitory effects appears to depend upon the intrinsic characteristics of the tumor cells and the state of macrophage activation, which depends upon other cells and factors in the tumor environment. Histological examination alone cannot reveal these important functional differences which seems to explain the contradictory results correlating inflammatory infiltrates in tumors with clinical prognosis.

Tumor progression is the result of Darwinian selection of heritable variants that grow more quickly and more aggressively in the host. A striking example of such tumor progression is the development of tumor variants that kill normal mice from parental UV-induced tumors that are regularly rejected by normal mice. Recent work using this parental/variant model has indicated that as the result of a selection process, parental tumor cells can change to heritable variants that induce and exploit the cellular and cytokine environment in which they grow. Thus, variant tumor cells appear to be selected not for their resistance to host effector function, but instead for their ability to use host cells to stimulate tumor growth through a paracrine loop (Figure 2). Inflammatory cells attracted by the variant tumors could directly stimulate the proliferation of malignant cells as shown by *in vitro* co-culture experiments. Neutrophils appear to play a significant role in the establishment of this paracrine stimulatory loop since elimination of neutrophils with neutrophil-specific antibodies reduces the rate of growth of the variant tumors in nude mice. Furthermore, inhibition of paracrine stimulation by anti-neutrophil antibodies leads to the rejection of a tumor challenge in immunocompetent mice by CD8$^+$ T cells, possibly because the rate of growth was reduced enough for T cells to be able to eliminate the tumor. It is also possible that elimination of neutrophils prevented effective tumor establishment since neutrophils play an important role in tumor vascularization. It is also not clear whether neutrophils are the actual stimulatory cells, or whether their role is to recruit macrophages which may be the actual stimulatory cells. In any case, the neutropenia that occurs as a side effect of radiation treatment or chemotherapy may have some indirect antitumor effects. Indeed, it has been shown that radiation may indirectly inhibit tumor growth which correlates with the elimination of peripheral leukocytes. Other important questions that remain include:

(1) What are the tumor-produced chemotactic factors that attract neutrophils and possibly other inflammatory cells to the growing tumor?

(2) What are the cytokines/growth factors produced by the inflammatory cells that stimulate malignant growth?

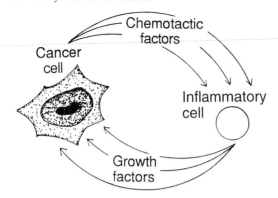

■ **Figure 2.** Diagram of a paracrine stimulatory loop. Inflammatory cells attracted to and stimulated by malignant cells produce factors that in turn stimulate tumor growth. Depletion of these cells by X-ray or neutrophil-specific antibodies arrests tumor growth, probably by interrupting this paracrine loop.

(3) How may we be able to interrupt the paracrine stimulatory loop?

With the development of highly specific inhibitors, these questions may be answered.

4. Cytokines and tumor angiogenesis

Recent extensive reviews have covered the complex interactions of cytokines with malignant and stromal cells driving the process of angiogenesis. We would like to focus on the key cytokines involved during the multiple rearrangements that occur in the tumor environment leading to the development of a vascularized tumor. In order for tumors to grow beyond a few cubic millimeters in size, new capillaries must be formed from the existing vascular network. These new blood vessels provide a route for providing essential oxygenation, nutrients and removal of metabolic products. In addition, new vessels provide a route for metastasizing tumor cells to enter the systemic circulation. Extensive new vascularization is a poor prognostic indicator in many types of human cancer. (Vascular counting is the second most effective indicator after lymph node metastasis in determining prognosis.) The process of vascularization involves the degradation of extracellular matrix by collagenases and proteases, and the proliferation and migration of endothelial cells, with differentia-

tion into functioning capillaries. Each stage is mediated by cytokines produced by various cell types within the tumor. Many of the cytokines have multiple direct and indirect effects on tumor progression by inducing other cytokines and altering the expression of cell-surface-bound and soluble cytokine receptors on both tumor and stromal cells. To complicate matters more, redundancy exits so that similar actions are regulated by different cytokines. In any case, the end effect is to regulate the proliferation of tumor cells by enhancing the development of a new blood supply and stimulating the deposition of matrix proteins to form an optimal stromal compartment in which the tumor can grow. Macrophages appear to play a significant role in angiogenesis also, as many of these cytokines and factors are produced by infiltrating macrophages.

Most tumor cells can persist for months to years without neovascularization. Usually these lesions remain undetectable until a subgroup of cells within the tumor switches to the angiogenic phenotype and the tumor becomes vascularized. It is important to note that the onset of neovascularization occurs independently of malignant transformation, although it may develop concurrently with other genetic alterations in the cells. Interestingly, tumor cells in the prevascular state may proliferate very rapidly. However, in the absence of a blood supply, the rate of proliferation is limited and eventually reaches a state of equilibrium with the rate of apoptotic cell death. Once the tumor becomes vascularized, the balance shifts so that the rate of cell death is reduced and the tumor begins to grow. Both positive and negative regulators of angiogenesis exist within the tumor environment. In order for tumors to switch to the angiogenic phenotype, this balance must be upset. Tumor cells may overexpress one or more angiogenic proteins, may recruit inflammatory cells to produce these factors, may mobilize an angiogenic factor from the extracellular matrix, or may use a combination of these mechanisms. However, simply shifting the balance of positive and negative regulators is not sufficient for tumor cells to become angiogenic. Powerful suppressors of angiogenesis, such as thrombospondin, angiostatin, and the recently described endostatin, exist in the circulation and extracellular matrix, and normally restrain endothelial cell proliferation. These inhibitors of vessel growth must be downregulated. It is not yet clear how the various cells involved are activated to release angiogenic cytokines, although considerable evidence suggests that hypoxia and possibly other factors associated with low oxygen saturation or inflammation such as low cellular pH and high lactate concentrations may be the trigger.

Of the angiogenic cytokines characterized so far, VEGF and bFGF are the two most commonly found in tumors. The action of VEGF and bFGF is synergic and both are directly stimulatory for endothelial cells. bFGF was the first endothelial cell growth factor to be identified. It is produced by a variety of cells including macrophages, fibroblasts, endothelial cells, smooth muscle cells and tumor cells. Since bFGF lacks a signal sequence, it is uncertain how it becomes available *in vivo* in the extracellular space. It may be released from dying cells during the inflammatory process. Heparinases from activated macrophages can then release bFGF from the extracellular matrix so that it can reach receptors on endothelial cells. Increased bFGF production has been correlated with poor prognosis in renal cell carcinoma and increased levels of bFGF have been found in the urine and sera of patients with various forms of cancer.

VEGF mRNA is expressed by many different human tissues, suggesting that this protein may have a role in maintaining vessel differentiation or endothelial cell repair following vascular injury. Tumor cells, macrophages, fibroblasts and smooth muscle cells have all been found to produce VEGF. However, as indicated by *in situ* hybridization with VEGF-specific probes, endothelial cells do not produce VEGF and are only stimulated in a paracrine fashion. Stimuli for VEGF expression include hypoxia, differentiation factors, growth factors and tumor promoters such as TPA. The involvement of VEGF in tumor angiogenesis was first indicated by observations that many tumor cell lines produce VEGF. In addition, VEGF is abundant in tumors that are highly vascularized. The levels of VEGF appear to increase with progression towards higher malignancy in several types of human tumor. For example, glioblastoma multiforme is highly malignant and usually very well vascularized while astrocytoma is a less aggressive form of brain tumor and usually poorly vascularized. Levels of VEGF have been reported to be low in astrocytoma and markedly increased in glioblastoma. The significance of VEGF as an angiogenesis factor has been supported by several other lines of evidence. Anti-VEGF monoclonal antibodies inhibit the growth of tumors in mice without affecting tumor cell proliferation *in vitro*. Also, growth of tumors in mice with dominant-negative mutants of the VEGF receptor, which have defective receptor function, is significantly reduced. Likewise, increas-ing VEGF expression increases the tumorgenicity and vascularization of various cell lines *in vivo* without altering cell growth *in vitro*. The mitogenic potential of VEGF seems to be specific for endothelial cells; however, it is a chemoattractant for both endothelial cells and macrophages. Since macrophages can produce VEGF, a positive feedback loop could exist between these two cell types. In addition, VEGF induces endothelial cells to release other factors involved in tissue remodeling, and fibrin deposition in the surrounding stroma is enhanced because of the increased vascular permeability. This provides a better environment for the growing tumor stroma and enables better entry of monocytes, fibroblasts and endothelial cells. It is easy to see why VEGF is such a pivotal factor in tumor angiogenesis.

The role of both TNFα and TGFβ in tumor angiogenesis remains unclear. Both appear to have stimulatory effects in models of angiogenesis *in vivo*, while inhibiting endothelial cell proliferation *in vitro*. The discrepancy may be due to an indirect effect of these cytokines that involves the attraction of other cells that produce angiogenic cytokines. Other studies have shown that TNFα inhibits angiogenesis *in vitro* and *in vivo* at high concentrations (microgram quantities) but is stimulatory at low concentrations (nanogram quantities). Because tissue levels of TNFα are unlikely to reach these high levels required for inhibition, TNFα is likely to stimulate angiogenesis *in vivo*. Both tumor-associated macrophages and epithelial cells can produce soluble TNFα receptors. This may protect tumors from the potential anti-angiogenic and tumoricidal effect of high TNFα levels. In addition, TNFα induces the release of extracellular plasminogen activator from endothelial cells, leading to proteolysis of the extracellular matrix and release of matrix-bound angiogenic growth factors.

In nude mice, tumors transfected with TGFβ show increased vascularization compared with untransfected tumors. TGFβ is able to induce expression of VEGF, bFGF and PDGF. However, at high concentrations (5–10 ng/ml), TGFβ no longer potentiates the angiogenic effect of these cytokines, but instead inhibits it. This illustrates that the action of a cytokine is context-specific, depending on the presence and concentration of other cytokines in the environment.

EGF and TGFα have been detected in several tumor types. Both stimulate endothelial cell proliferation *in vivo* and *in vitro*, binding to the same receptor. Their effects are similar, although TGFα is a more potent angiogenic factor. Like TNFα and

TGFβ, EGF and TGFα can induce plasminogen activator. Both stimulate secretion of VEGF and expression of EGF receptors. Receptor expression in solid tumors has been correlated with an aggressive phenotype and poor prognosis. TGFα has been localized to several types of solid tumors and is thought to be made by both tumor cells and macrophages.

Other cytokines, such as G-CSF, GM-CSF and IL-8, may exert their angiogenic effects primarily through chemoattraction and activation of inflammatory cells. IL-1 appears to have primarily inhibitory effects, possibly by decreasing the binding sites for bFGF on endothelial cells. Both stimulatory and inhibitory effects of IL-6 on angiogenesis have been reported. IFNγ can inhibit angiogenesis induced by bFGF *in vitro*, and IFNα has been successfully used in the treatment of hemangiomas in infants. Elucidating the role of numerous additional factors participating in the complex angiogenic process is beyond the scope of this discussion. Obviously, tumor angiogenesis is an extremely complicated process involving numerous cell types and their cytokines interacting in various combinations and concentrations; however, because of this complexity, multiple molecular targets may be available for therapeutic intervention in malignant growth.

5. Cytokines in tumor metastasis

The metastatic spread of tumor cells to secondary sites represents the most difficult clinical challenge to curative cancer treatment. Most patients die of metastatic disease rather than of local growth of the primary tumor. The acquisition of the metastatic phenotype is thought to be another step in the process of tumor progression. Initiation of metastasis requires the invasion of the surrounding normal stroma by tumor cells that have acquired increased motility. The next step involves extravasation into vascular or lymphatic channels where cells may grow or detach and be transported within the circulatory system. The majority of circulating cancer cells that enter the bloodstream die within a short time either because the cells do not come to be in a supportive environment and undergo apoptosis or because the cells are actively destroyed by the host's immune and non-immune defenses. Those cells that survive after lodging in the capillary bed of a compatible organ must extravasate

into the tissue to proliferate. Further growth of these small tumor lesions requires the establishment of a blood supply and continuous evasion of the immune system. In order for this metastatic process to be successful, tumor cells must synthesize a number of tumor cell molecules (such as adhesion molecules) that are necessary for each step of metastasis. In addition, tumor cells may down-regulate other molecules that make them more antigenic. Many of these properties required for invasion and metastasis may be lost after metastasis has occurred; this is either due to the transient acquisition of the metastatic phenotype by individual cells within the primary tumor and/or due to the fact that the further changes in gene expression occur at the secondary site. (These molecules and properties will not be discussed here. Rather we will focus on cytokines involved in attracting and regulating the growth of tumor cells at metastatic sites.)

During the initial stages of metastatic progression, many tumors have a tendency to spread to certain tissues preferentially. Substances promoting the growth of the tumor cells have been detected in conditioned media of these tissues. Thus, paracrine growth-stimulatory mechanisms are likely to determine the tissue/organ site to which tumor cells can metastasize successfully. At the secondary sites, tumor cells may be less dependent on their normal growth signals and more responsive to signals in the new environments. Thus, organ preference in experimental models is correlated with enhanced mitogenic responses to cytokines released from the target organs. For example, murine melanoma cell lines that colonize the liver and ovaries are stimulated by conditioned media from those target organs. In contrast, lung conditioned medium inhibits the growth of these cells. In addition to having growth-inhibitory molecules, conditioned medium from lung tissue has growth-stimulatory molecules for other tumor cell types. The primary stimulatory factor has been purified and found to be a transferrin. Bone is also an important site for metastatic growth, especially for prostatic carcinoma cells. Bone tissue has its own set of stimulatory cytokines, one of which has been found to be a transferrin similar to that found in lung tissue-conditioned medium.

There is not always such a tight correlation between preference for a particular organ and the ability of tumor cells to be stimulated by those tissues in culture. This emphasizes the fact that other factors, such as vascular and lymphatic drainage, cell adhesion, motility, and cell invasion, are involved in organ preference for metastatic colonization. Differences in vascular and lymphatic drainage of a tumor and other mechanical and anatomical considerations also account for differences in colonization patterns.

Oncogene-mediated conversion of benign tumor cells to highly malignant cells is often accompanied by increased levels of growth factor receptors. Both novel expression of receptors not previously present and an increased number and/or affinity of receptors already present would be important for tumors to be responsive to paracrine signals from secondary sites. This could also explain why certain tumors show preference for metastatic growth at particular sites and why tumor cells in culture frequently become more responsive to certain cytokines. EGF receptor is often overexpressed in tumors and is associated with enhanced metastasis and a poor prognosis. Likewise, the growth factor receptor p185 is also frequently overexpressed. Overexpression of each of these receptors leads to increased mitogenic signaling. Moreover, p185 appears to be involved in controlling tumor cell adhesion to capillary endothelial cells, invasion through a basement membrane, and cell motility in response to chemotactic factors. Transfection of transformed cells with p185 results in cells acquiring the metastatic phenotype. However, some caution must be used in interpreting these transfection data since the process itself may alter other already unstable genes.

Along with growth-stimulatory signals, tumor cells may also receive growth-inhibitory signals from the site to which they are disseminated and therefore be prevented from establishing a metastasis at such a site. As already mentioned, conditioned medium from certain tissue fragments can inhibit the growth of tumor cells that do not metastasize to those organs. Only a few of these inhibitors have been identified. One of the most potent inhibitors of metastatic growth identified so far is TGFβ. Interestingly, large amounts of TGFβ are released into kidney cell-conditioned medium and almost no tumors metastasize early to the kidney. An exception to this is melanoma. TGFβ stimulates metastatic melanoma cells to grow. However, the ability of a cancer cell to be stimulated by TGFβ is not sufficient to predict metastatic potential. For example, some clones from some renal and colorectal tumors are stimulated by TGFβ while others are inhibited, regardless of their metastatic potential, suggesting that no single growth factor is likely to be sufficient for a tumor to grow metastatically. Another cytokine connected to inhibition of metastasis is IL-6. IL-6 appears to be particularly critical in regulating the growth responses of early stage melanomas. Melanoma cells in the more advanced stage of vertical growth lose responsiveness to IL-6 as well as other growth inhibitors.

As cancers progress towards a more aggressive phenotype, there is a tendency for cells to become indifferent to paracrine signals. They are no longer additionally stimulated by exogenous cytokines, nor are they inhibited. Many tumors at advanced stages of metastatic growth produce a variety of growth factors which may be involved in a mostly autocrine regulation of tumor growth. This could explain the loss of preference for secondary growth sites and the widespread dissemination to multiple tissues and organs.

Very little is known about which cytokines are involved in the metastatic progression of tumors. The sources of these cytokines (both stimulatory and inhibitory) also remain unclear. Possible cell sources of such factors include fibroblasts, endothelial cells, macrophages, mast cells and parenchymal cells. Since tumor cells are able to release cytokines that stimulate surrounding cells (such as angiogenic factors for endothelial cells), a reciprocal relationship may exist in which different cell types feed each other. In addition, the extracellular matrix may be a source of growth factors which can be released by factors produced by macrophages, or by the consequences of wounding.

6. Cytokines in tumor dormancy and recurrence

After surgical removal of primary tumors, local recurrence can occur. Tumor recurrence rates are higher for larger tumors and those with more aggressive growth patterns, but recurrence can occur in all types of tumor. A new unified theory for tumor recurrence that emphasizes the importance of cytokine stimulation of persisting residual cancerous cells in the tumor environment is beginning to emerge from years of compiled data mainly with breast cancer patients.

For years it was thought that local relapse of breast cancer resulted from cancer cells left behind after surgery. This led to the use of radical surgery in an attempt to minimize local recurrence and optimize cures. However, recurrence rates in general did not decrease significantly and it was therefore suggested that recurrence results from systemic micrometastases already present at the time of diagnosis. Today many cancers are diagnosed very early. Therefore, it is probable that distant metastases may be the result from dissemination of cancer cells during surgery. Growth of residual cells left behind and/or the survival of tumor cells released during surgery could depend on the multitude of cytokines released as a result of injury and inflammation at the surgical site. Indeed, a high proportion of local recurrences occur at the surgical scar and, even when radical resection is performed, local recurrence of breast cancer at the chest wall can be seen without radiation therapy.

The fact that spillage of cancer cells occurs during surgery has been demonstrated by several lines of evidence. Manipulation of animal tumors increases the dissemination of cells into the bloodstream and the incidence of distant metastasis. Sensitive techniques, such as polymerase chain reaction (PCR) and monoclonal antibodies to tumor antigens, have been used to detect circulating cancer cells in the bloodstream after surgical removal of a tumor. Fortunately, most tumor cells are unable to proliferate at new sites or are destroyed by the immune system. However, the greater the number of cancer cells introduced either locally or systemically, the greater the chance for tumor escape.

The high proportion of tumors that recur at the surgical site may be attributable in part to cytokines present within the healing wound. Normally, these factors are secreted at basal levels to regulate normal cellular activity and maintain homeostasis, but growth factors are known to increase dramatically following injury, or in the presence of carcinogens or of malignant cells. Wounding has already been discussed as a powerful tumor-promoter following initiation. In this case, injury would start a cascade of events involved in wound healing. Increased cytokine levels would attract endothelial cells, fibroblasts and inflammatory cells. We have already discussed the importance of inflammation in establishing an environment rich in growth factors. Similarly, tumor cells and infiltrating stromal cells could make use of paracrine interactions. In support of this, the number of murine carcinoma cells required for establishment of intraperitoneal tumors can be reduced 100-fold by first inducing a macrophage-rich environment. Furthermore, when murine adenocarcinoma cells are injected intravenously, they will only implant in the spleen if it is first wounded; then implantation occurs at the wound site.

For reasons that are not understood, tumors can remain dormant for long periods and then suddenly proliferate rapidly. Dormancy has been described as a state of equilibrium from which tumors emerge if that state is disrupted. Angiogenesis may either initiate, or may be a consequence of, the termination of dormancy. Tumor metastases must not only receive stimulatory signals to survive and grow, but must also be able to establish a blood supply for growth to continue. As already discussed, positive and negative signals regulate this process. Removal of a primary tumor in cancer patients can spill tumor cells and create a rich growth factor environment. It can also result in depletion of a potent circulating inhibitors of angiogenesis, angiostatin and endostatin, that are produced by the host in response to the primary tumor. This suppression may be particularly important when stimulatory factors are present. Studies have shown that cells in distant micrometastases proliferate at a high rate but tumors remain small because the rate of cell death is even higher. Removal of angiostatin and/or endostatin allows tumors to vascularize, thus decreasing the rate of cell death and increasing tumor size.

7. Conclusions

Cytokines play a critical role during the many stages of tumor development. Several cytokines are produced by tumor cells, while others are produced by surrounding stromal cells. Wounding is common to both tumor promotion (leading to a cell to becoming malignant after being initiated by mutations) and recurrence of dormant tumor cells. Inflammation is likely to be the link. The numerous cytokines released by infiltrating cells and from the damaged extracellular matrix can stimulate tumor growth. In fact, stimulation by paracrine cytokines can result in rapid tumor growth that overcomes specific tumor-inhibitory immune responses. In addition, weak inefficient humoral or T-cell immunity may lead to a local cytokine environment that stimulates rather than suppresses malignant growth. Paracrine regulation of tumor growth by cytokines appears, however, to be less important at the end stages of tumor progression when extremely aggressive cells are found that are largely resistant to any exogenous signals and use autocrine stimulatory pathways.

It is often difficult to specify the effects of a particular cytokine since the effects vary with concentration, with the presence of other cell types and their factors, and with the receptiveness of the tumor cells. Different effects may involve alterations in cytokine receptors and/or signaling pathways as the genetically unstable tumor cells progress to become more malignant. Also, many cytokines act indirectly through recruiting and activating other cells. That is why results with isolated cytokines *in vitro* should be interpreted cautiously. An important concept concerning the influence of cytokines on cancer is that of balance. Throughout the process of tumor progression there is a balance between growth stimulation and growth inhibition. During tumor promotion and tumor progression, cytokines acting on tumor cells through a paracrine mechanism can potentiate or reduce the inflammatory response. Indeed, inhibition of this response often leads to reduced tumor formation. We have seen how positive and negative regulators control the switch to the angiogenic phenotype. Inflammatory cells can have either tumor-promoting or tumor-suppressing activity, depending on the activation state of the cells. Macrophages can produce many factors that both stimulate tumor growth and angiogenesis. They can also be directly cytotoxic, destroy vascular beds and activate specific antitumor immune responses. However, response to bacterial substances (such as lipopolysaccharide) or bacillus Calmette–Guérin (BCG) or other bacteria are usually needed to activate inflammatory cells to full tumoricidal or tumor-inhibitory activity *in vitro*. Replacing the signals given by the bacterial substances with cytokines has so far not been possible. However, blocking paracrine or autocrine stimulatory pathways by specific inhibitors may become a fruitful approach to cancer therapy and may allow specific tumor immunity to become effective. Clearly, however, a better understanding of how all the factors, cytokines and cells involved interact is needed to achieve better regulation and interruption of tumor progression.

References

Brattain, M.G., Howell, G., Sun, L.Z., and Willson, J.K. (1994). Growth factor balance and tumor progression. *Curr. Opin. Oncol.* **6**, 77–81.

Filmus, J. and Kerbel, R.S. (1993). Development of resistance mechanisms to the growth-inhibitory effects of transforming growth factor-beta during tumor progression. *Curr. Opin. Oncol.* **5**, 123–9.

Folkman, J. (1995). Tumor angiogenesis. In Mendelsohn, J., Howley, P.M., Israel, M.A., and Liotta, L.A. (eds) *The Molecular Basis of Cancer*, pp. 206–32. W.B. Saunders, Philadelphia.

Fujiki, H., Suganuma, M., Komori, A., Yatsunami, J., Okabe, S., Ohta, T., and Sueoka, E. (1994). A new tumor promotion pathway and its inhibitors. *Cancer Detection and Prevention* **18**, 1–7.

Kerbel, R.S. (1992). Expression of multi-cytokine resistance and multi-growth factor independence in advanced stage metastatic cancer. *Am. J. Pathol.* **141**, 519–24.

Leek, R.D., Harris, A.L., and Lewis, C.E. (1994). Cytokine networks in solid human tumors: regulation of angiogenesis. *J. Leukocyte Biol.* **56**, 423–35.

Mantovani, A. (1994). Tumor-associated macrophages in neoplastic progression: a paradigm for the *in vivo* function of chemokines. *Lab. Invest.* **1**, 5–16.

Nicolson, G.L. (1993). Cancer progression and growth: relationship of paracrine and autocrine growth mechanisms to organ preference of metastasis. *Exp. Cell Res.* **204**, 171–80.

O'Reilly, M.S., Boehm, T., Shing, Y., Fukai, N., Vasios, G., Lane, W.S., Flynn, E., Birkhead, J.R., Olsen, B.R., and Folkman, J. (1997). Endostatin: an endogenous inhibitor of angiogenesis and tumor growth. *Cell* **88**, 277–85.

Qin, Z. and Blankenstein, T. (1996). Influence of local cytokines on tumor metastasis: using cytokine gene-transfected tumor cells as experimental models. *Curr. Topics Microbiol. Immunol.* **213**, 55–64.

Reid, S.E., Scanlon, E.F., Kaufman, M.W., and Murthy, M.S. (1996). Role of cytokines and growth factors in promoting the local recurrence of breast cancer. *Brit. J. Surg.* **83**, 313–20.

Rowley, D.A., Bechen, E.T., and Stach, R.M. (1995). Autoantibodies produced spontaneously by young lpr mice carry transforming growth factor beta and suppress cytotoxic T lymphocyte responses. *J. Exp. Med.* **181**, 1875–80.

Seung, L.P., Rowley, D.A., Dubey, P., and Schreiber, H. (1995). Synergy between T-cell immunity and inhibition of paracrine stimulation causes tumor rejection. *Proc. Natl Acad. Sci. USA* **92**, 6254–8.

Seung, L.P., Weichselbaum, R.R., Toledano, A., Schreiber, K., and Schreiber, H. (1996). Radiation can indirectly inhibit tumor growth while depleting circulating leukocytes. *Radiat. Res.* **146**, 612–8.

Stach, R.M. and Rowley, D.A. (1993). A first or dominant immunization II. Induced immunoglobulin carries transforming growth factor beta and suppresses cytolytic T cell responses to unrelated alloantigens. *J. Exp. Med.* **178**, 841–52.

Uotila, P. (1996). The role of cyclic AMP and oxygen intermediates in the inhibition of cellular immunity in cancer. *Cancer Immunol. Immunother.* **43**, 1–9.

Yarosh, D.B. and Kripke, M.L. (1996). DNA repair and cytokines in antimutagenesis and anticarcinogenesis. *Mutat. Res.* **350**, 255–60.

XXIX Immunotherapy, gene therapy, cytokines and cancer

Giorgio Parmiani*, Mario P. Colombo and Cecilia Melani

1. Tumor immunology: historical background

Modern tumor immunology dates back to the time when tumor-specific antigens were first defined in animal models and the quest for their use in cancer immunotherapy began. Such antigens, also called tumor-rejecting or tumor-transplantation antigens due to their ability to elicit *in vivo* rejection of subsequent implants of the tumor, were found to be expressed in neoplasms induced by various means (chemically, virally or by ultraviolet (UV) or X-ray radiation) in various animal species; the induced tumors were of varying histological origin.

Initial findings in mice showed that, while a tumor generated cell-mediated immunity against itself, it often did not generate immunity against other tumors, even those induced by the same carcinogen in the same individual animal, leading to the suggestion that different tumors had different tumor-specific antigens. (An exception to this is virus-induced tumors, which have cross-reacting antigens because these antigens are encoded by the viral DNA either directly (in DNA oncoviruses) or indirectly (in RNA viruses or retroviruses).) At first the nature of these antigens was not known: it took several decades before the first mouse tumor antigen was molecularly defined, thanks to the pioneering work of Thierry Boon's group.

Until a few years ago, human tumor antigens had been little studied due to the lack of tools to recognize them and to investigate their molecular structure. In fact, many oncologists and immunologists doubted that such antigens existed at all. However, in the 1980s, peripheral blood T cells that recognized autologous tumor cells in the context of a specific class I HLA molecule were cloned from

cancer patients, and this gave new impetus to the search for well defined human tumor antigens. Using a cumbersome and, at that time, difficult approach, Thierry Boon's group was the first to demonstrate the existence of genes coding for human tumor (melanoma) antigens that could be recognized by autologous cytotoxic T lymphocytes (CTLs). This milestone observation proved beyond doubt that such antigens existed, and generated a flurry of reports by the same and other groups describing new T-cell-defined antigens in melanoma and, to a lesser extent, in other human tumors. Of particular interest was the demonstration that normal, lineage-related proteins of melanocytes (e.g. tyrosinase, gp100, Melan-A/MART-1) or normal proteins predominantly found in tumors of different histology (e.g. melanoma antigen (MAGE), BAGE or GAGE) can provide peptides recognized by autologous CTLs (see Table 1).

Several potential tumor antigens were also found to be generated by mutations of oncogenic proteins (Ras, p53) or from fusion proteins resulting from chromosomal translocation taking place during neoplastic transformation. In contrast to what had been found with mouse tumors, however, most of the antigens described in human neoplasms are cross-reactive, either among tumors of the same histology (e.g. tyrosinase, gp100, Melan-A/MART-1 of melanomas) or among tumors of different histology (e.g. MAGE, BAGE and GAGE). In the former case, these antigens are also expressed in the normal cells from which the neoplasm derived (i.e they are differentiation or lineage-related proteins). In the last two years, however, antigens (e.g. MUM-1, β-catenin) that are unique to tumor cells have been reported in human tumors, and a few cross-reacting antigens have been described in mouse neoplasms (e.g. P1A, gp70^{env}), suggesting that the types of tumor antigen in the two species

*Corresponding author.

∎ **Table 1.** Tumor Ag recognized by T-cells and mechanisms of their generation

Mechanisms	Murine tumor Ag		Human tumor Ag	
	Name	Histology	Name	Histology
Normally activated genes	TRP-2	Melanoma	Tyrosinase, gp 100, Melan-A/MART-1, TRP-1, -2	Melanoma
Activated, silent genes	P1A	Mastocytoma, plasmocytoma, fibrosarcoma	MAGE, BAGE GAGE, RAGE	Melanomas, Renal cancer, Other tumors
Activated retroviral gene	gp70 env	Thymoma, melanoma, carcinoma, fibrosarcoma	—	—
Activated oncogenes	pRL1, ras	Leukemia Sarcoma	RAS, P53, Her-2/Neu	Breast, colon, lung, ovary, pancreatic cancers
Translocation and gene fusion	—	—	bcr/abl, pml/RARα	Chronic myelogenous leukemia, acute promyelocytic leukemia
Transcription from cryptic promoter	—	—	N-acetylglucosamyl transferase V	Melanoma
Post-translational modification	—	—	Tyrosinase	Melanoma
Translation of alternative ORF	—	—	TRP-1	Melanoma
Incomplete splicing	—	—	MUM-1, gp100	Melanoma
Mutation of expressed genes	P91A, P35B, P198	Mastocytoma Mastocytoma	CDK4 MUM-1	Melanoma Melanoma
	Connexin 37	Lung carcinoma	β-catenin	Melanoma
	L9 ribosomal protein	Skin tumors	CASP-8	Head and Neck Cancer
Transposition of genes into a new genomic location	LEC-A	Leukemia	—	—

are similar and that the previously reported discrepancies were a reflection of the fact that fewer studies had been done in the mouse system than in the human system. The two groups of antigens and their different mechanisms of origin are compared in Table 1. It is clear that common mechanisms of antigen generation exist in mouse and human tumors, although more mechanisms have been reported for the latter because it has been studied more extensively.

A new strategy, serological identification of antigens by recombinant expression cloning (SEREX), has been developed that allows the serological identification of human tumor antigens. A systematic search and molecular definition of tumor antigens by this technique has been carried out by the group of Pfreundschuh using the antibody repertoire of cancer patients. This approach resulted in the definition of several distinct antigens expressed by tumor cells, some of which are identical to those previously identified by biochemical or genetic approaches using T-cells.

Before the molecular nature of tumor antigens was understood, most approaches to immunotherapy were based on the hope that human tumors might, like mouse tumors, express the 'tumor-rejection' type of antigen and that such antigens could then be used to immunize the host against the tumor. Other trials investigated the possibility that non-specific stimulation of the immune system might also boost anti-tumor immunity, by analogy with the many experimental studies carried out in the mouse system. There was no clear idea of the antigens involved and, more importantly, no reliable assay that could measure a specific immune response supposed to be induced by these immunological manoeuvres, so it is not surprising that these early attempts to use vaccination in cancer therapy proved disappointing.

A first step forward came from the discovery of the lymphokine activated killer (LAK) cell phenomenon, whereby peripheral blood lymphocytes (PBLs) of normal individuals or cancer patients cultured in the presence of interleukin 2 (IL-2) become able to kill *in vitro* a large array of neoplastic or virus-infected target cells while sparing normal cells (with the possible exception of lymphoblasts). This observation prompted the use of such LAK cells along with their growth factor, IL-2, in cancer patients after an extensive pre-clinical investigation in rodents on the feasibility, toxicity and mechanism of action of this therapeutic approach. Although the clinical response obtained was limited, this study showed that manipulation of the immune system could result in partial and, in a few cases, complete long-term regression of metastatic tumor lesions growing in distant, visceral organs of the patient (see Table 2). These clinical responses were probably due both to non-specific cytotoxic mechanisms mediated by the LAK cells and tumor-cytotoxic cytokines released by them or by the endogenous lymphoid cells and by activation of specific T-cell responses by IL-2 and/or other cytokines released at the tumor site and throughout the body. The limited clinical response rate, the high toxicity and the complexity of this approach made it unacceptable as therapy, but systemic IL-2 was later approved for treatment of metastatic renal cancer after a metanalysis showed that this improved their overall survival.

Another approach that has been tried used tumor-infiltrating lymphocytes (TILs), i.e. T cells obtained from the tumor infiltrate and expanded *in vitro* in the presence of autologous tumor cells (i.e. the antigens) and IL-2. It was shown that these cells included a small proportion of anti-tumor-specific T-cells and, when given systemically in large amounts (10^{10}–10^{11}), that they caused regression of metastatic melanoma in 34% of cases. Clinical response in melanoma patients adoptively transferred with TILs was shown to depend on the level of *in vitro* antitumor activity of the infused lymphocytes. However, these clinical responses did not translate into a better overall survival and the complexity and cost of the whole procedure prevented its application on a large scale.

The discovery of genes that encode human tumor antigens recognized by T cells has opened an entirely new approach to the specific immunotherapy of cancer. When the first peptides became available for clinical use, trials began in which melanoma patients were vaccinated with different peptides (e.g. tyrosinase, Melan-A/MART-1, gp100 or MAGE-1 and -3) alone or mixed with adjuvants. Though these studies are in an early phase, clinical responses have been reported in a group of metastatic melanoma patients vaccinated with the MAGE-3.A1 peptide.

▌ **Table 2.** Clinical response in metastatic patients treated with intravenous IL-2 at high doses either in bolus or by continuous infusion

Tumor	N. of patients	N. Responses and %		
		CR*	PR*	Total
Bolus				
Melanoma	342	11 (3%)	40 (12%)	51 (15%)
Renal cancer	288	15 (5%)	34 (12%)	49 (17%)
Continuous infusion				
Renal cancer	204	5 (2%)	29 (14%)	34 (16%)
Melanoma	144	0	16 (11%)	16 (11%)

Source: Marincola, F.M. and Rosenberg, S.A. (1995). *Biological therapy of cancer.* V. De Vita, S. Hellman, and S. Rosenberg, eds., Lippincott, pp. 250–62, and Young, J.C. and Rosenberg, S.A., pp. 262–9.
* CR, complete response; PR, partial response.

Important developments have also occurred in the last few years in the use of monoclonal antibodies (mAbs) in cancer therapy. These reagents have been widely used in diagnosis of cancer, either on tissue sections (immunohistochemistry) or to evaluate tumor markers that may help both in diagnosis and follow-up of cancer patients. Less relevant appears to be their use in tumor radio-imaging, although consistent progress has been made in this area of investigation. mAbs have also been repeatedly tested as carriers of different tumor-cytotoxic compounds such as toxins, radionuclides or anticancer drugs, with variable success. B-cell lymphomas seem to respond best to immunotoxins, while solid tumors are intractable. The major limitations of this approach are:

- the immunogenicity of mouse mAbs;

- the low penetration of the antibody molecule into tumor tissues (particularly in solid neoplasms); and

- the toxicity due to the expression of the target molecule by normal tissues.

With advances in genetic engineering techniques many of these problems may well be overcome in the near future.

An interesting new area of therapeutic investigation with mAbs has been opened by the results of a phase III clinical study that showed a statistically significant increase in overall survival in colon cancer patients given a mAb directed against a colon cell-surface glycoprotein in an adjuvant sitting. This therapeutic effect may result from specific elimination of tumor cells in the bone marrow or circulating in the biological fluids of the patients. Several similar studies in other types of cancer are ongoing to evaluate the potential advantages of such a biological, non-toxic therapy in comparison with other therapeutic options.

2. Role of cytokines in antitumor immunity

2.1 Non-specific immunity

Non-specific approaches to cancer immunotherapy probably date back to the nineteenth century, when sporadic, spontaneous remission of tumors was observed in patients suffering severe bacterial infections. Such an observation prompted William B. Coley to begin, in 1891, treating patients with soft tissue sarcoma using a mixture of Gram-positive and Gram-negative bacteria (Coley's toxins). That

such an approach might be fruitful was enforced by Shear's discovery that endotoxin was the active component of Gram-negative bacteria, inducing tumor hemorrhagic necrosis.

The finding that bacillus Calmette–Guérin (BCG) may increase the resistance to tumor transplants in mice, led to the widespread and indiscriminate clinical application of BCG. This compound is still used today, but only for treating superficial bladder cancers. In mesenchymal tumors, the antitumor effects of BCG and its derivatives are largely dependent on the so-called Koch phenomenon, in which there is indiscriminate necrosis of tissues containing mycobacteria, whereas other tumors may be affected by non-specific amplification of the T-cell-mediated immune response, although this latter mechanism has never been convincingly demonstrated. The discovery of cytokines explained most of the phenomena induced by microbial products, so cytokines were then used in the hope that they would still have the positive effects of the bacterial products while lacking the negative ones.

More recently, the discovery of Th1 and Th2 distinct pathways of T-cell differentiation helped to understand protective and non-protective BCG-induced cell-mediated immune reaction in tuberculosis, a phenomenon that may occur also in the immune response against cancer. In the Th2 type of immune response, inflamed tissue is sensitive to tumor necrosis factor alpha (TNF-α), and extensive necrosis can occur, but in the presence of a Th1-deflected immune response, TNF-α cannot cause this necrosis. Thus cytokines that deflect the immune response to Th1 (e.g. IL-2 and IL-12) or to Th2 (e.g. IL-4 and IL-10) may influence the type of immune response to cancer cells. A strong Th1 response and a strong Th2 response may both be able to induce tumor destruction and immune memory with the same efficacy, though by different mechanisms (see below).

2.2 Specific immunity

The activity of microbial product mainly consists of local effects which may be reproduced and improved by local injection of recombinant cytokines. Experiments in non-tumoral systems had shown that IL-2 offsets defective antigen recognition and overcomes tolerance, thus suggesting the use of cytokines not only to stimulate tumor destruction but also to overcome possible tolerance or immunological ignorance towards tumor-associated antigen (TAA) and activate their effective and specific immune recognition. Identification and cloning of gene(s) coding for the long-elusive TAA recognized by MHC-restricted T cells, especially in human tumors, allowed a better definition of the role of cytokines themselves in an anti-tumor immune response against specific antigens presented either on tumor cells or as peptide. As expected, cytokines involved in an anti-tumor immune response are essentially identical to those that mediate recognition and response to conventional antigens. It should be noted, however, that tumor cells may constitutively release several cytokines that either suppress or stimulate the T-cell response, thereby interfering with its regular development. For example, transforming growth factor beta (TGF-β), which is known to inhibit early steps of lymphocyte activation, is produced by several types of human tumors, including melanomas and ovarian carcinomas. Antigen-presenting cell (APC) activity can be up-regulated by granulocyte–macrophage colony-stimulating factor (GM-CSF) and down-regulated by IL-10, two cytokines that can be released by at least a fraction of tumors.

2.3 Cytokine at the tumor site

2.3.1 Cytokine gene transfer

Even highly tumorigenic tumor cells, like those induced by chemical carcinogens, grow unrestricted in the majority of syngeneic normal, unimmunized mice whereas previous immunization with irradiated tumor cells or surgical resection of a tumor nodule results in the rejection of a subsequent challenge of an otherwise lethal dose of the same cells. To explain such a paradox, it was hypothesized that neoplastic cells lack factors that can activate T-helper lymphocytes necessary to trigger an effective T-cell-mediated response against antigenic tumor cells in an unprimed host. Jan Bubenik and Guido Forni and their co-workers therefore injected exogenous cytokines, initially IL-2, close to the tumor site or to the draining lymph nodes to generate specific T-cell reactions against the tumor. This approach has been called lymphokine-activated tumor inhibition (LATI). Tumor growth was inhibited in these studies, and was thought to be mediated by inflammatory cells recruited at the tumor site by the cytokine. As well inhibiting tumor growth, this treatment often conferred systemic immunity upon mice, making them resistant to a subsequent challenge of the tumor cells injected without any accompanying cytokine. The disadvantage of this approach is that the cytokine needs to be administered continuously in order to prevent

tumor growth. However, *in situ* application of cytokines can avoid the toxicity associated with systemic administration.

Another, more elegant way of releasing cytokines at the tumor site was then devised by using molecular techniques to transfer genes coding for the given cytokine into tumor cells themselves. This procedure allowed constant release of the cytokine at the tumor site and resulted in tumor growth reduction or inhibition accompanied by the same immunological consequences observed with the LATI approach. Such gene transfer can be achieved using viral or non-viral vectors. The choice of vector depends on the cells that need to be transfected and on whether they will be used *in vitro* or *in vivo* (Table 3). In the majority of *in vitro* procedures, when *ex vivo* normal or neoplastic recipients cells were used, retroviral vectors have been successfully employed. Retroviruses infect only replicating cells, but use of a selectable gene marker in the vector allows efficient *in vitro* selection and/or cloning so that one can obtain a homogeneous cell line in which theoretically all the cells express the gene. Adenoviral vectors have also been used, and have the advantage that they also infect non-replicating cells with high efficiency, resulting in elevated but transient expression of the transgene. Replication-defective herpes simplex virus (HSV) can also be used, in particular to prepare autologous vaccines, because they rapidly transfer genes (e.g. that encoding IL-2) to fresh tumor cells.

In vivo gene transfer has been accomplished with viral and non-viral (e.g. liposomes or plasmid DNA) vectors. The *in vivo* approach is usually limited by the low efficiency of tumor tissue targeting, except when neoplastic lesions can be directly injected with the vector.

2.3.2 Cytokine–antibody fusion proteins

Taking advantage of the targeting ability of tumor-specific mAbs, fusion proteins have been constructed that allow cytokines to be targeted specifically to the tumor site. A number of strategies have been used to construct 'immunoconjugates', or expression vectors that produce fusion proteins between cytokine and either the single-chain Fv of the F(ab′) portion of the antibody, or the entire immunoglobulin molecule. A variety of systems have been used for the expression of the fusion proteins, including bacteria, yeasts, baculoviruses and myeloma cells, all resulting in biologically active molecules, although the efficiency of

∎ **Table 3.** Different types of vectors and their features

Vector	Cell range	Duration of expression	Level of expression	Advantages	Disadvantages
Retrovirus	Replicating cells	Stable	Moderate	Genome integration, no recombination with host viruses	Limited insert size, low titer
Adenovirus	Replicating and non-replicating cells	Transient	High	Efficient infection, high titer, high expression, no host DNA integration	Limited insert size, immunogenicity, generation of replication competent viruses
Adeno-associated virus	Replicating and non-replicating cells	Transient/stable	Moderate	Possible genomic insertions at specific sites	Lack of permanent producer lines, limited insert size, host immunity
Pox virus	Replicating and non-replicating cells	Transient	High	Large insert size, high titer, no DNA integration	Immunity in vaccinated individuals, immunogenicity
Herpes virus	Replicating and non-replicating cells	Transient	Moderate	Large insert size, high titer	Permanent producer lines not available, host immunity, possible host recombination

their production and complexity of purification varies widely among the systems.

Owing to its ability to activate the immune system, IL-2 has been the cytokine most frequently fused to anti-tumor mAbs. Extensive analysis of the therapeutic use of such fusion proteins has been conducted by Reisfeld and co-workers who constructed fusion proteins between IL-2 and an atibody against GD2 ganglioside or against EGF receptor, and expressed them in hybridoma cells. Adoptive immunotherapy with human LAK cells and repeated administration of either mAb fused to IL-2 in SCID mice bearing liver or lung human melanoma metastases that express such antigens, resulted in the given fusion protein accumulating in the neoplastic lesions, and leading to tumor eradication. In this xenogeneic system, the anti-tumor effect was likely to be mediated by the activation of non-specific LAK cells due to the presence of IL-2 at the tumor site. However, the same authors showed that, in an entirely syngeneic mouse tumor system, CD8+ T cells were also involved in the rejection of distant metastases that occurred even when treatment was up to 35 days after tumor cell injection. Similar results were obtained in a syngeneic model of colon carcinoma expressing the human Ep-CAM antigen where the

tumor IL-2 targeting was mediated by its fusion to the humanized antibody huKS1/4. Interestingly, in both experimental systems, treatment with the IL-2–mAb fusion protein did not necessarily induce the recognition of the targeted tumor antigens recognized by the antibodies, but could activate anti-tumor cytotoxic effector cells primed against other TAA, resulting in the eradication of heterogeneous metastasis.

Other groups have tried to exploit the vasoactivity of IL-2 and produced a Lym1–IL-2 fusion protein that targeted specifically human B cell lymphomas. This recombinant protein increased the vascular permeability of tumors *in vivo*, thus allowing better uptake of therapeutic molecules, such as radiolabeled anti-tumor antibody, by the tumors.

TNF-α was fused to the single-chain Fv of a mAb in order to detoxify the cytokine and re-target its toxicity towards tumor cells expressing the antigens recognized. This fusion protein was less cytotoxic to antigen-negative, TNF-sensitive cells, while antigen-expressing cells resistant to TNF-α were not affected by the treatment. In this and similar studies, TNF was used more as toxin than as a modulator of the inflammatory or immune response. Fusion proteins between anti-tumor antibodies and bacterial or plant toxins have displayed similar results, with specific targeting but non-specific killing that failed to generate an anti-tumor immune response.

The immunostimulatory properties of cytokines have been also used quite effectively to improve vaccination against murine B-cell lymphomas. Idiotype-cytokine fusion proteins, given systemically or produced *in vivo* upon vaccination with their coding plasmid DNA, act as potent immunogens that can elicit a significant anti-tumor immune response *in vivo*. Such an effect, however, does not depend on *in situ* accumulation of immune cells or on the level of anti-idiotypic response but is the result of an adjuvant effect of the cytokine.

In general, the approach of fusing cytokines to molecules that help their targeting to tumors or utilize their activity as immune adjuvant was successful and provided interesting tools for a less toxic and more efficient immunotherapy. Clinical trials are now being planned to exploit this technology in different human neoplasms.

2.3.3 Mechanisms of tumor rejection, systemic immunity and vaccination

Local accumulation/release of cytokines can thus be obtained by various approaches, including:

- direct injection at the tumor site;
- insertion of the cytokine gene into normal somatic cells to be injected at the tumor site or into tumor cells themselves before implantation; or
- targeting with mAb.

Cytokine accumulation at the tumor site can reduce tumor size and sometimes eradicate the tumor. This tumor shrinkage is often followed by establishment of systemic tumor-specific immune memory depending on the antigenic strength of the tumor and on the presence of appropriate co-stimulatory factors in its microenvironment. However, tumors may be reduced by non-specific immune reactions in such a short time that antigen release and processing by APCs and, therefore, efficient T-cell priming, are prevented. This process may mimic the dichotomy described for BCG: indiscriminate necrosis *versus* protective immunity.

Are the effector cells recruited at the tumor site by cytokines and responsible of tumor shrinkage responsible for the described effects? The role of polymorphonucleated cells (PMN) in this process has been addressed. PMN are the first cells to reach the sites of pathogen entry and activate inflamma-tion. Their sensitivity and prompt response are not limited to pathogen killing (through the generation of reactive oxygen intermediates and the release of lytic enzyme): they can also produce different cytokines which eventually influence the type of immune response that may follow PMN local activation. In the rat, PMN depletion abrogates CD8[+] CTL induction as well as priming and elicitation of a delayed-type hypersensitivity (DTH) reaction to sheep red blood cells. PMN were the first cells found to infiltrate murine tumors transduced with cytokines like G-CSF and IL-2 (neutrophils) as well as IL-4 (eosinophils). PMN depletion can also abrogate tumor rejection. This is documented in Table 4, which summarizes studies carried out with the spontaneous mammary adenocarcinoma TSA, which has low immunogenicity. It is clear that depletion of PMN allows tumor growth in most cases, suggesting that, with the majority of cytokines released at tumor site, a common pathway exists in which PMN are crucial in determining the inflammation-like reactions that ultimately result in tumor growth inhibition. Although a role as APCs has been suggested, PMN are certainly more apt to influence cellular and possibly humoral immune responses through the release of cytokines deflecting the immune response to a Th1 or Th2 type. However, PMN do not express interferon gamma (IFNγ) or IL-4, while they do release IL-12 and IL-10, so their activity on Th cells remains unclear.

What, then, is the role of PMN at the tumor site? Experiments performed in mouse tumor models may offer some insights. For example, *in vitro* studies revealed that, depending on the target cell type, PMN may disrupt a monolayer architecture of tumor cells with or without cell killing. Insensitivity to PMN-mediated killing may explain why, even in the presence of large numbers of PMN, some tumors grow progressively.

In the case of murine colon carcinoma C-26 transduced with the human G-CSF gene (C-26/G-CSF), and injected as a cell suspension, tumor growth was completely inhibited because of the PMN recruited and activated at the tumor site. C-26/G-CSF cells, however, form large tumors when injected into sublethally irradiated mice; these tumors regress when leukocyte functions are reconstituted. This system provided an opportunity to study the mechanisms responsible for inhibition of tumor establishment in syngeneic animals and of regression of an established tumor in sublethally irradiated mice injected with these cells.

■ **Table 4.** Role of PMN in the *in vivo* inhibition of cytokine gene-transduced TSA/cell growth

Tumor cells	Tumor take (%) after immunosuppression with		
	None	Anti-ASGM1	Anti-PMN
TSA-control	100	100	100
TSA-IL-2	0	5	60
TSA-IL-4	20	25	100
TSA-IL-7	25	40	40
TSA-IL-10	20	100	80
TSA-IL-12	5	85	95
TSA-IFN-α	35	20	55
TSA-IFN-γ	75	75	100
TSA-TNF-α	20	35	75
TSA-cytosin deaminase + exogenous 5-FC†	0	0	67

Modified from Musiani *et al.* 1997.
*ASGM1 is a marker of NK and of a fraction of T-cells.
†5-FC, 5-fluorocytosine that is metabolized into 5 fluorouracil by the enzyme cytosin deaminase.

Inhibition of tumor establishment and regression of an established tumor occurred through different mechanisms: the former was dominated by PMN expressing IL-1α, IL-1β and TNF-α, whereas the latter was characterized by PMN, macrophages and T-cells, including CD8$^+$ cells which are required for IFNγ-mediated tumor regression. Thus, transfer of a single cytokine gene, e.g. G-CSF, into tumor cells is sufficient to trigger the cascade of cell interactions and cytokine production necessary to destroy a tumor nodule. The endothelial cells within the C-26/G-CSF regressing tumor expressed ICAM-1, E-selectin and VCAM-1 in addition to CD31 and laminin. Expression of vascular adhesion molecules indicates that the cytokines are released by the infiltrating leukocytes as biologically active proteins depending on the type of leukocyte. PMN are attracted to the tumor because of the chemotactic activity of G-CSF. IL-1α, IL-1β and TNF-α released by PMN contribute to induce expression of endothelial VCAM-1 and E-selectin and thus, indirectly, to attract T cells. At that point, the vascular endothelium appears to be injured, with loss of erythrocytes, possibly targeted for leukocyte attack, and tumor regression follows. At the site of inflammation, endothelium is highly susceptible to PMN-mediated injury which com-promises the function of the vasculature and of the underlying tissues. Within the regressing C-26/G-CSF tumor, both tumor and endothelial cells, which express ICAM-1 at similar levels, are targeted by activated PMN. Thus, tumor destruction is probably the result of the action of direct cytolysis and hypoxia due to the loss of tumor vasculature. PMN appear then to play a major role in the tumor regression associated with hemorrhagic necrosis.

Newly discovered cytokines are now being evaluated for their possible application in the immunotherapy of cancer. IL-13, a Th2-secreted factor which shares properties with IL-4 has some specific immunological activities that account for its use in tumor immunotherapy. Release of IL-13 at the site of the non-immunogenic Lewis lung tumor cell line 3LL either reduced or inhibited tumor growth depending on the dose, though it did not afford protection against subsequent challenge with 3LL cells. In contrast, injection of immunogenic P815 tumor cells engineered to release IL-13 resulted in initial tumor growth followed by regression and establishment of immunological memory against parental P815 cells. Tumor rejection appeared to be mediated by a non-specific immune response, with granulocytes and monocytes largely infiltrating the regressing nodule. Since a direct stimulatory effect of IL-13 on T lymphocytes is unlikely, since T cells lack IL-13 receptor, the IL-13-mediated anti-tumor immune response could be mediated by a cascade of cytokines secreted by other cells, such as monocytes. In fact, IL-13 is chemotactic for monocytes and macrophages, up-regulating their MHC class II expression *in vitro*. Although IL-13 cannot stimulate macrophage and NK anti-tumor cytotoxicity *in vitro*, it can improve tumor antigen presentation, resulting in better sensitization of anti-tumor immune T lymphocytes.

IL-18, which is produced by mature macrophages, augments human and mouse NK activity *in vitro* and induces the production of $IFN\gamma$ by T lymphocytes. In experimental models, IL-18 pre-treatment induces tumor rejection involving NK cells and the generation of $CD4^+$ cytotoxic effectors followed by an immunological memory against the rejected tumor. The mechanism underlying this anti-tumor immune reaction analyzed *in vitro* involves early activation of NK with increased production of GM-CSF and IL-10, followed by a reduction in NK activity and increase in IL-2 production accompanying the enhancement of cytolytic cell activity.

Many of the functions of IL-15 overlap with those of IL-2, since receptors for these two cytokines share the β and γ subunits. Although, *in vitro*, IL-15-activated cytolytic effector cells seem to be less powerful than IL-2-activated cells, IL-15 is less toxic *in vivo*. IL-15 has been shown to induce LAK cells, from PBMC of melanoma patients, that lysed autologous melanoma cells though to a lesser extent that IL-2 LAKs. Differential expression of the IL-15 receptor complex would allow IL-15 to display its activity where IL-2 receptor α expression is low or IL-2 secretion is absent due to lack of T-cell activation.

(a) Does tumor shrinkage correlate with induction of tumor immunity?

Vaccination experiments have been performed in which live, replicating or non-replicating tumor cells (irradiated or mitomycin-C treated), engineered to produce cytokines, have been used to immunize mice against a subsequent challenge with the parental tumor. The results consistently indicate that the strongest protection is obtained following rejection of replicating tumor cells engineered to produce cytokines. This suggests that strong immunogenicity derives from the ability of live cells to give rise to an initial tumor that regresses when

the secreted cytokine reaches the threshold necessary to trigger a local reaction. This threshold probably varies among different cytokines, but its exact value is hard to determine because the number of tumor cells in a given tumor mass is usually not known. To dissect the role of tumor shrinkage after initial growth from that of cytokines, Forni and co-workers have transfected the TSA mammary carcinoma with the gene coding for cytosine deaminase (CD), an enzyme converting the non-toxic drug 5-fluorocytosine (5-FC) into the cytotoxic 5-fluorouracil. In mice injected with TSA/CD cells and treated with 5-FC, tumor becomes established but is then rejected through a mechanism similar to that elicited by cytokine-transduced tumors (see Table 4). Tumor regression renders mice able to reject a subsequent challenge of parental TSA cells. This suggests that in at least some tumors regression is by itself sufficient to trigger a systemic immune response; however, such a response was undetectable in mice which rejected C26/G-CSF (see above). Therefore, tumor regression may be an immunogenic event, such as surgical resection of some murine fibrosarcomas, whereas cytokines not only induce tumor regression but also regulate the mechanism that eventually results in a memory immune response.

(b) What are the mechanisms of such regulation and how does immunization occur?

Engineered tumor cells produce the transduced cytokine but are also the source of antigen. In fact, tumor cell debris are likely to be internalized by APCs and the related antigen presented to T-cells. It has been suggested that host bone-marrow derived APCs are responsible for induction of CTL-mediated protective immunity against tumors following immunization with irradiated tumor cells, so transfer of exogenous TAA to host are and its processing for presentation to MHC class I restricted T cells by host APCs may be possible *in vivo*. What remains unclear is whether T-cell priming can occur at the tumor site. Immuno-histology and *in situ* hybridization show that tumor shrinkage provoked by certain cytokines is characterized by the presence of secondary cytokine production, cell–cell interaction among different leukocyte types, endothelium and fibroblasts, which together create an environment resembling that of an activated lymph node. For the purpose of immunization, however, the site of priming may not be important if the response is equally efficient.

(c) Does induction of systemic immunity mean that the elicited reactivity will stop or hamper the growth of an established tumor?

Immune mechanisms that efficiently protect against a subsequent challenge of monodispersed tumor cells often do not cure mice with established tumors. An induction period is required in immunization, and during this time murine tumors, which grow rapidly, may become too large to be eradicated. Furthermore, tumor cell organization, stromal vessels and extracellular matrix can limit the efficacy of many anti-tumor immune reactions. In fact, analysis of different experimental systems suggests at best a weak correlation between CTL activity *in vitro* and their ability to eradicate tumors. Possible explanations for this lack of correlation may include (i) impaired migration of CTLs to the tumor site, because of V-CAM-mediated down-modulation of tumor blood vessels, or (ii) inability of CTL to penetrate the stroma deeply (which becomes more evident as the tumor progresses). These observations should be taken into account in designing clinical protocols with gene-modified cellular vaccines aimed at generating anti-tumor CTLs in cancer patients.

2.3.4 Tumor-associated blood vessels: a secondary target of anti-tumor immune response

A striking correlation appears to exist between host-induced immune response and tumor destruction by cytokines, and their ability to induce both tumor-associated blood vessel injury and inhibition of angiogenesis. As described above, blood vessel injury is associated with local inflammation and is dominated by PMN and by cytokines like TNF-α and IL-1. Angiostasis, i.e. inhibition of angiogenesis, mainly results from the activity of secondary chemokines activated by the IFNγ-like protein IP-10 (interferon-inducible protein 10) and MIG (monokine induced by IFN-γ). IL-12-dependent tumor rejection is largely due to the ability of IL-12 to stimulate IFNγ and, therefore, IP-10 and MIG. Rejected tumors are characterized by a large necrotic area, similar to that caused by the local presence of IL-10.

Thus there are a number of possible ways in which the same end, tumor rejection, can be achieved; activation of different immunological pathways may have the same effect. The above studies also stress the importance of tumor blood vessels in tumor progression and regression, hence making them an important target of immunotherapy.

3. Immunotherapy of human tumors with cytokines

3.1 Cytokine therapy: models and clinical trials

Some cytokines, notably IFNα, IL-2 and, more recently, IL-12, have shown remarkable antitumor activity in animal models. However, their activity in clinical trials has been disappointing (see below). Why were these models such poor predictors of the effectiveness of these cytokines in the clinical situation? There are two main reasons for the discrepancy between the results from animal models and clinical trials, of which only few investigators appear to be fully aware. First, most of the animal tumors studied were intrinsically and homogeneously immunogenic, long-term transplanted neoplasms, so it is likely that destruction of even a small fraction of their cells could have caused effective systemic immunization. This leads on to the second major reason, namely that most of the animal experiments make use of healthy, non-tumor-bearing individuals whose immune system is probably functionally different from that of patients who only come to the attention of doctors after the tumor has had months or, more frequently, years to adapt and to bypass all the body's potential defence mechanisms. This slow, dynamic process of reciprocal interactions between the growing tumor and the body's defence systems has practically never been taken into consideration in animal studies, and is only now becoming appreciated. Thus the failure to replicate in clinical trials what has been observed in animals, is not because 'mice are different from man' but because the tumors and the immune system of these mice are different from those of cancer patients. Awareness of this may lead to the design of better animal models which will be more accurate predictors of the potential antitumor activity of cytokines or other biologicals to be applied in cancer therapy.

An example of this discrepancy between animal and human studies is the application of LATI in the latter system. In the largest study of LATI carried out by Cortesina and co-workers in humans, IL-2 was given around the major cervical lymph node of patients with head/neck cancer in an attempt to activate tumor-cytotoxic lymphocytes which, by travelling to the tumor lesion, were supposed to destroy neoplastic cells. Such as multicenter randomized clinical trial, failed to show a significant therapeutic benefit of local treatment with IL-2, even though an early phase I–II clinical study had shown a 60% response rate. Similar findings have been reported from the studies of cytokine gene transfer (see below).

We will now briefly review the results of using cytokines given systemically or locally in the therapy of human tumors.

3.1.1 Interferon α/β

Interferon was the first cytokine to be used in the therapy of tumors and its use spread rapidly to treat many types of cancer, yet the mechanism of its antitumor activity is still far from being understood. *In vitro*, it shows antiproliferative activity against some but not all tumors, but *in vivo* it has pleiotropic effects on many organs and functions, making it difficult to dissect which mechanisms are operating *in vivo* in the presence of histologically different tumors like melanoma or myeloid leukemia.

The mechanisms by which interferon influences the functions of normal and neoplastic cells have been the subject of many investigations during the last few years, leading to the discovery of molecules involved in the signal transduction of this and other cytokines (see Chapter VIII). Of the various biological properties of the interferons, those which seem most relevant for their antitumor activity are:

- their direct antiproliferative activity, which requires expression of appropriate receptors on tumor cells;

- their activation of macrophages and other immune cells (NK cells and T cells);

- their differentiation-inducing and anti-angiogenetic properties on tumor and certain normal cells; and

- their modulatory activity on the metabolism of anticancer drugs.

None of these mechanisms has been specifically identified as an explanation of the therapeutic results obtained by the use of interferon in humans, whereas animal studies have mainly suggested that the immunomodulatory activity was instrumental in their antitumor activity.

A vast number of clinical trials have been conducted on cancer patients during the last 10 years or so in many different cancer centers, and it is now possible to draw clear conclusions on the therapeutic activity of interferons, particularly IFNα 2a/b, in different neoplastic diseases. A summary of these clinical results is presented in Table 5.

■ **Table 5.** Summary of antitumor activity of interferons

Tumor	Response rate	CR%
Hairy cell leukemia	80–90*	10
Chronic myelogenous leukemia	80–90	20
Non-Hodgkin's low grade lymphoma	40–50*	
Mycosis fungoide and Sézary syndrome	90	
Multiple myeloma	28 (11–50)*	
Kaposi's sarcoma	40	30
Melanoma	16	2
Renal cancer	14	2
Bladder carcinoma *in situ*	43	43

* Range, see text.

The major therapeutic impact of IFNα is in the treatment of hairy cell leukemia, although chemotherapy can now efficiently complement the effect of the cytokine. Given subcutaneously as a single agent at relatively low doses (2–3 megaunits three times per week), IFNα significantly improves the disease state, although only 40% of patients show complete, lasting responses; the combination of IFNα and chemotherapy gives a much higher response rate. In chronic myelogenous leukemia (CML), 40–60% of patients show complete response after 6–18 months of interferon treatment. Combination with chemotherapy or bone marrow transplantation is important to eradicate the disease, but the cytokine has been reported to have a specific cytotoxic activity against CML blasts, maybe because they express the chimeric protein Bcr/Abl, which may be a specific target of IFNα.

IFNα has been used also for treating non-Hodgkin lymphomas and cutaneous lymphomas (mycosis fungoides, Sézary syndrome), as shown in Table 5. A French group, coordinated by Solal-Celigny, has reported, at the 1996 Meeting of the American Society of Clinical Oncology, on a multicenter randomized study showing that adding subcutaneously administered IFNα to poly-chemotherapy significantly improved the overall survival of patients with advanced follicular lymphoma, and with only mild toxicity. IFNα has also shown considerable activity in preventing recurrences and thereby increasing the overall survival in patients with myeloma in remission after conventional chemotherapy. The anti-angiogenetic activity of IFNα is probably the reason for its activity on Kaposi sarcoma (see Table 5) and in other blood vessel-derived neoplasms like pulmonary hemangiomas in children.

Interferons have frequently been used against solid tumors, even though, in the many phase I–II studies carried out, the clinical response rate rarely exceeds 15–20%. However, a recent prospective randomized, adjuvant study on melanoma patients has shown that high doses of IFNα 2a, given subcutaneously for 6 months to 2 years, results in a significant increase in the overall survival of the treated patients. Interferon has now been approved in the USA and other countries for the treatment of such patients. Local application of interferon to bladder carcinomas has been shown to be highly effective and less toxic than local administration of BCG (see Table 5).

3.1.2 Interleukin 2

IL-2 (or T-cell growth factor) is a 15 kDa glycoprotein that is produced by T-helper cells as a result of antigenic activation. During an immune response the release of IL-2 allows differentiation of CTLs and activates NK cells, B cells and macrophages which express IL-2 receptor (see Chapter I for the details of the IL-2 receptor). As a pleiotropic factor, IL-2 influences the function of several organs and systems. Some carcinoma lines produce IL-2 constitutively and, because they also express IL-2 receptors, their growth can be inhibited by adding exogenous IL-2.

Exposure of lymphocytes to IL-2 *in vitro* for a few hours, or up to several days, has been shown to result in cytotoxic activation of some of the lymphocytes, which acquire the capacity to destroy, *in vitro*, tumor cells of a variety of histological origin without any HLA restriction. Rosenberg and co-workers called these activated lymphocytes 'lymphokine-activated killers', or LAK. Many groups have shown that IL-2 alone, or together with LAK, can, when given intravenously, result in the delay or even the rejection of tumors growing in various organs of mice and rats. The best responses were generally found when highly immunogenic tumors were used in animals with a limited tumor burden. The mechanism of this antitumor effect is not completely understood, but it is probably mediated by the direct killing of neoplastic cells due to activation of endogenous LAK cells by IL-2 administered to animals, or (in the case of injection of LAKs prepared *ex vivo*) to the combined effect of exogenous and endogenous LAK. The cytotoxic

activity of several cytokines (including TNF-α, IL-6, IFNγ and GM-CSF), large quantities of which are released by the activated lymphocytes, also appears to be involved in the therapeutic activity. However, these cytokines also have significant systemic toxicity, causing capillary leaking syndrome which results in the extravasation of both lymphocytes and fluid in most organs, and can lead impair their function. The toxic and therapeutic effects are both dose-related. The administration of IL-2 also results in the appearance of soluble cytokine receptors (e.g. soluble IL-2R and soluble TNFR) in the circulation, and this may affect the availability of cytokines in the tissues and, possibly, at the tumor site.

These *in vitro* and animal studies provided the rationale for clinical protocols in which patients with advanced tumors, particularly melanomas and kidney cancers, were given either IL-2 (initially, at least, in high doses, in an attempt to induce a high response rate) or LAKs and IL-2. Several hundred patients were treated worldwide and the results can be summarized as follows (Table 2). When IL-2 was used as a single agent at a relatively high dosage and given as bolus, complete or partial clinical response was achieved in approximately 17% of patients with metastatic kidney cancer, and 15% of those with metastatic melanoma; complete response was obtained in 5% and 3% of renal cancer patients and melanoma patients, respectively; most of the complete responses were long-lasting, a few of them now for more than 8 years. In randomized studies to compare the efficacy of IL-2 alone with that of IL-2 plus LAKs, no significant differences were found in response rate or in overall survival; even a combination of cytotoxic drugs with IL-2 therapy did not give a better response rate than the single agents.

Clinical responses in other types of tumors, e.g. colon cancer, non-Hodgkin lymphoma or NSCLC, were much less frequent and it can be concluded that IL-2 has no significant activity against these neoplasms. However, it has been observed that in advanced acute myelogenous leukemia, administration of IL-2 induced complete and long-lasting remission in eight of 14 patients with a low proportion of residual bone marrow blasts. Despite early promising results, a combination of IL-2 and IFNα failed to show an additive or synergic effect in melanoma and renal cancer patients, and was associated with considerable toxicity.

In renal cancer patients, survival appears to depend on (i) the performance status, (ii) the duration from diagnosis to onset of therapy, and (iii) the number and location of metastatic sites, those in bone and liver being the most resistant to therapy. Reliable biological markers that can predict clinical response have not been identified. Local administration of IL-2 (e.g. through the hepatic artery to liver cancers, through an Ommaya catheter to glioblastomas, or intraperitoneally to ovarian cancers), does not give a better response rate than other therapies, and is accompanied by local and systemic toxicity. More recently, however, natural IL-2 was given as an aerosol five times per day to 14 patients with pulmonary metastases of renal cancer, and one complete response and one partial responses was detected, suggesting that this approach warrants further evaluation.

In other studies, IL-2 has been given subcutaneously or intramuscularly in an attempt to achieve slower release, a more durable plasma concentration and lower toxicity. Low doses administered by such routes were better tolerated by the patients, achieved a response rate of 10–15% and, more importantly, at least in the renal cancer patients, a significant increase in overall survival. In a prospective randomized trial, Rosenberg and his associates compared high (720 000 IU/kg every 8 h) and low doses (72 000 IU/kg every 8 h) of IL-2 in renal cancer patients, and found a that the former was much more toxic but that the two groups showed statistically similar response rate (20% and 15% respectively). Therefore, despite the lack of randomized phase III studies to compare IL-2 with other types of treatment, IL-2 was been approved for the treatment of metastatic kidney cancer in several European countries and USA in the early 1990s.

The adverse effects of IL-2/LAK administration can be severe, and depend on the dose and route of administration, but are reversible upon cessation of therapy. Capillary leaking syndrome occurs in almost all patients receiving IL-2, with its severity depending on the dosage. Toxicity may include hypotension (requiring vasopressor drugs in most patients), weight gain (up to 10% increase), renal failure, pulmonary congestion, neurological symptoms, cardiac toxicity, hepatic dysfunction, and haematological and dermatological alterations. Awareness of such symptoms and of the underlying mechanism now enables the clinician to handle these patients without major problems.

Despite the limited success of these early trials, they have shown that in patients with advanced cancer, considered incurable by standard therapies, a significant clinical response can be obtained, albeit in a minority of cases, using immunotherapy.

This represents a milestone in the development of a fourth type of cancer therapy, in addition to surgery, chemotherapy and radiotherapy.

The main reason for the limited success of the IL-2/LAK therapy of tumors appears to be the low tumor-infiltrating capacity of endogenous and/or exogenous LAK cells. This is a consequence of the mechanical and biological properties of LAK cells: they are larger and less deformable than normal lymphocytes, and so are less able to enter small capillary vessels, and they lack appropriate homing adhesion molecules, preventing extravasation of sufficient LAKs necessary to cope with the millions or even billions of tumor cells that constitute a visible tumor mass. Other pathophysiological features of the tumor, such as hypoxia, necrosis, low availability of nutrients, or high hydrostatic pressure within the tumor, may contribute to the inefficiency of LAK and of other transferred lymphocytes (see below).

3.1.3 Other cytokines

Several other cytokines have been tested for antitumor activity in animal models and in the clinic, including IL-1, 4, -6 and -12, and TNF-α. None of them has yet got beyond phase I, because of either high toxicity (in the case of IL-1 and TNF-α, for example) or negligible antitumor activity (IL-4 and IL-6). IL-12 appears to be more promising. It has powerful antitumor activity in animal models, as a result of two main mechanisms: activation of the immune system and anti-angiogenesis. In our pilot study, patients with metastatic melanoma received 0.5 μg/kg of IL-12 for two identical 28-day cycles, with injections given on days 1, 8 and 15 of each cycle. Side effects consisted of a 'flu'-like syndrome and a transient increase in transaminases and triglycerides. Transient reduction in circulating $CD8^+$ and $CD16^+$ lymphocytes and neutrophils was observed after the first administration of the cytokine. High levels of IFNγ and IL-10 were detected in all patients within 24–48 h. Tumor regression of subcutaneous nodules occurred in two patients, and regression of adenopathies and liver metastasis each in one patient, but the level defined as partial response (>50% of all lesions) was not achieved. IL-12 is now being studied in phase II trials to ascertain its potential antitumor activity in humans. It is also used as immunological adjuvant in vaccination protocols.

3.1.4 Combination of cytokines

Although many experimental studies *in vitro* and in animal models have shown that combinations of different cytokines (e.g. IL-2 and IFNα or IFNγ) can show additive or even synergic effects in anti-tumor activity, no such effect has been found in several clinical studies. As one of us suggested a few years ago, the lack of additive effects in clinical trials with cytokines is due to the high amounts of such biologicals that are given to patients. Each single cytokine can generate a cascade of other cytokines released by the body cells, and this cascade cannot be further modulated by adding another exogenous agent.

3.2 Cytokine gene-transduced neoplastic or normal cells in the immunotherapy of cancer

Based on findings from animal models, a series of clinical protocols have been designed to vaccinate cancer patients with autologous or allogeneic tumor cells into which genes have been transferred encoding cytokines (e.g. IL-2, -4, -6, -7 or -12, IFNγ or GM-CSF) or lymphocyte co-stimulatory molecules like B7-1, or both. Some authors have used cytokine gene-transduced fibroblasts mixed with fresh autologous tumor cells to avoid the need to generate cell lines from each tumor. Autologous and allogeneic cells each have advantages and disadvantages, so there is debate over which to use. In our own studies we decided to use allogeneic melanoma cells, since they can be selected with certain phenotypes considered to make them more immunogenic, such as high expression of classes I and II HLA, or of adhesion molecules or, particularly with human melanomas, of well defined antigens recognized by T cells (e.g. Melan-A/MART-1, tyrosinase, gp100, MAGE, BAGE, GAGE). However, this selection will be difficult at the level of individual patients and will significantly reduce the number of elegible patients because several types of tumors often have less than the normal amount of one or more of these crucial molecules, or sometimes even lack them completely. In addition, safety tests have to be conducted, which not only are costly but also require weeks or even months to perform, during which time some patients inevitably die as a result of tumor progression. The availability of genes encoding new tumor antigens or other important molecules (e.g. B7-1) now allows their transfection into tumor lines to make them even more immunogenic, but this procedure will be difficult to carry out successfully for every individual patient.

Such tumor cell vaccines have been prepared based on the premise that neoplastic cells can themselves present tumor antigens to T lymphocytes. That tumor cells may function as APCs once they express the appropriate molecules (B7-1, class II HLA, adhesion molecules) has been shown in different systems. However, recent data obtained in the mouse suggest that antigen presentation usually occurs indirectly by the host bone-marrow derived APCs that process antigens released by tumor cells. If so, transfer of genes other than those encoding TAA into tumor cell vaccines may be pointless. However, cytokine release upon transduction at the tumor cell level may increase the inflammation-like reactions mediated by NK cells and granulocytes, thus augmenting the release of tumor antigen that can then be presented by professional APCs. Even the confirmed increased immunogenicity of tumor cells expressing B7-1 is likely to be mediated mainly by activation of NK cells and not of T cells, which may be absent from the tumor site at least in the early phase of vaccination. It should be noted, however, that tumor-bearing patients may already be primed to tumor antigen and usually receive multiple immunizations that may result in a rapid infiltration of vaccine site by circulating tumor-specific T lymphocytes which can be then efficiently re-stimulated by the antigenic, B7-1-expressing neoplastic cells.

Based on these considerations and rationale, several clinical protocols have been proposed and initiated during the last few years to evaluate the hypothesis that cytokine gene-transduced tumor cells may be more effective vaccines than those used in previous clinical studies, which were based on unmodified human tumor cells and adjuvants. Although these protocols all used tumor cells transduced with cytokine genes as vaccines, they varied in a number of respects, including the histotype of the tumor, the type of cytokine gene transduced, the amount of cytokine released and the *in vitro* assay adopted to document the immune response. So far, the results of these trials have been rather disappointing, with only few cases of partial response or mixed response and, more importantly, with no more than 30% of patients showing an increased tumor-specific T-cell response after vaccination.

A few clinical studies have made use of fibroblasts transduced with IL-4, IL-2 or IL-12 genes mixed with fresh tumor cells to vaccinate patients with renal, colon or prostate cancers, but early available results are no better than those obtained by gene-transduced tumor cells.

How can the therapeutic activity of such vaccines be improved? It will be important to establish the amount of cytokine that, when released locally, can confer immunogenicity since mouse studies have shown that an excess of IL-2 may be detrimental to vaccination and a too low amount of it will not enable the cellular vaccine to stimulate T lymphocytes. Thus clinical protocols need to be designed to compare patients vaccinated with the same tumor line releasing different doses of cytokine. The number of cells injected, the route of immunization and the time elapsing between injections may all be important in stimulating the immune system, and these variables need to be tested in humans. The most important issue, in our opinion, is to fully understand whether, and how, subcutaneously or intradermally injected, irradiated tumor cells stimulate the host's immune system.

3.3 Cytokines as adjuvants in expanding an immune response induced by tumor antigen

The availability of molecularly defined antigens from human tumors, or of mutated proteins operationally considered as potential tumor antigens (e.g. p53, Ras), has prompted several new clinical trials in which these antigens have been used as peptide epitopes to stimulate the immune system of patients whose tumor expresses the antigen. These peptides have been given with cytokines like IL-2, GM-CSF or, more recently, IL-12 to increase or expand the immune response activated by the vaccine. The immunodominant epitope of Melan-A/MART-1, gp100 tyrosinase, mixed with incomplete Freund's adjuvant (IFA) or adenovirus vector carrying these genes, have been administered into metastatic melanoma patients with or without IL-2 or GM-CSF. A frequent increase in the specific immune response, at least in the patients injected with Melan-A/MART1 or gp100 in IFA, was obtained by Rosenberg's group, but there was no significant clinical response. The addition of IL-2, or GM-CSF, however, appears to increase the likelihood of generating partial or even complete regression of tumor lesions through a mechanism (s) that is under investigation.

4. Concluding remarks

Cytokines are important molecules that regulate many physiological processes. They have been used in a variety of approaches in cancer therapy. Some of them have shown a powerful anti-tumor activity in animal models, but this is seldom reflected in humans due to the biological differences between healthy mice and cancer patients. Despite these limitations, IL-2 is now considered the treatment of choice for inoperable metastatic renal cancer, while IFNα has been shown to be effective, in an adjuvant setting, in the treatment of melanoma patients. Cytokines are also being used as immunological adjuvants in several ongoing studies of peptide- or protein-based cancer immunotherapy. Moreover, cytokine gene-transduced normal or neoplastic cells have been and are being used as vaccines in an attempt to improve the immunogenicity of molecularly defined antigens recognized by T cells and borne by such tumor cells. Thus, cytokines seem to be directly or indirectly involved in the control of tumor growth and progression. Better knowledge of these mechanisms can only improve the use of cytokines in cancer therapy. Further laboratory and clinical studies are needed on cytokines, including the newly available ones, in an attempt to increase the number and types of cancer patients that can benefit of treatment with cytokines.

References

Arienti, F., Sulè-Suso, J., Belli, F., Mascheroni, L., Rivoltini, L., Melani, C., Maio, M., Cascinelli, N., Colombo, M.P., and Parmiani, G. (1996). Limited antitumor T cell response in melanoma patients vaccinated with interleukin-2 gene-transduced allogeneic melanoma cells. *Human Gene Ther.* **7**, 1955–63.

Boon, T. and Van der Bruggen, P. (1996). Human tumor antigens recogized by T lymphocytes. *J. Exp. Med.* **183**, 725–30.

Cormier, J.N., Salgaller, M.L., Prevette, T., Barracchini, K.C., Rivoltini, L., Restifo, N.P., Rosenberg, S.A., and Marincola, F.M. (1997). Enhancement of cellular immunity in melanoma patients immunized with a peptide from MART-1/Melan A. *Cancer J.* **3**, 37–44.

De Stefani, A., Valente, G., Forni, G., Lerda, W., Ragona, R., and Cortesina, G. (1996). Treatment of oral cavity and oropharinx squamous cell carcinoma with perilymphatic interleukin-2: clinical and pathological correlations. *J. Immunother.* **19**, 125–33.

Frankel, A.E., Ftzgerald, D., Siegall, C., and Press, O.W. (1996). Advances in immunotoxin biology and therapy: a summary of the fourth international symposium on immunotoxins. *Cancer Res.* **56**, 926–32.

Hu, P., Hornick, J.L., Glasky, M.S., Yun, A., Milkie, M.N., Khawli, L.A., Anderson, P.M., and Epstein, A.L. (1996). A chimeric Lyn/interleukin 2 fusion protein for increasing tumor vasculature permeability and enhancing antibody uptake. *Cancer Res.* **56**, 4998–5004.

Huang, H.Y., Golumbeck, P., Ahmadzadeh, M., Jaffee, E., Pardoll, D., and Levitsky, H. (1994). Role of bone marrow-derived cells in presenting MHC class I-restricted tumor antigens. *Science* **264**, 961–5.

Jaffee, E.M. and Pardoll, D.M. (1996). Murine tumor antigens: is it worth the search? *Curr. Opin. Immunol.* **8**, 622–7.

Kirkwood, J.M., Strawderman, M.H., Ernstoff, M.S., Smith, T.J., Borden, E.C., and Blum, R.H. (1996). Interferon α2b adjuvant therapy of high-risk resected cutaneous melanoma: The Eastern Cooperative Oncology Group Trial EST 1684. *J. Clin. Oncol.* **14**, 7–17.

Marchand, M., Weynants, P., Rankin, E., Arienti, F., Belli, F., Parmiani, G. *et al.* (1995). Tumor regression responses in melanoma patients treated with a peptide encoded by the gene MAGE-3. *Int. J. Cancer* **63**, 883–5.

Meloni, G., Foa', R., Vignetti, M., Guarini, A., Fenu, S., Tosti, S., Tos, A.G., and Mandelli, F. (1994). Interleukin-2 may induce prolonged remission in advanced acute myelogenous leukemia. *Blood* **84**, 2158–63.

Micallef, M.J., Tanimoto, T., Kohono, K., Ikeda, M., and Kurimoto, M. (1997). Interleukin 18 induces the sequential activation of natural killer cells and cytotoxic T lymphocytes to protect syngeneic mice from transplantation with Meth A sarcoma. *Cancer Res.* **57**, 4557–63.

Musiani, P., Modesti, A., Giovarelli, M., Cavallo, F., Colombo, M.P., Lollini, P.L., and Forni, G. (1997). Cytokines, tumour-cell death and immunogenicity: a question of choice. *Immunol. Today* **18**, 32–6.

Ostrand-Rosenberg, S. (1994). Tumor immunotherapy: the tumor cell as antigen-presenting cell. *Curr. Opin. Immunol.* **5**, 722–7.

Parmiani, G. (1990). An explanation of the variable clinical response to interleukin 2 and LAK cells. *Immunol. Today* **2**, 113–5.

Parmiani, G., Colombo, M.P., Melani, C., and Arienti, F. (1997). Cytokine gene transduction in the immunotherapy of cancer. *Adv. Pharmacol.* **40**, 259–307.

Reisfeld, R.A., Becker, J.C., and Gillies, S.D. (1997). Immunocytokines: a new approach to immunotherapy of melanoma. *Melanoma Res.* **S52**, S99–106.

Riethmuller, G., Schneider-Gadicke, E., Schlimok, G., Schmiegel, W., Raab, R., Hoffken, K., *et al.* (1994). Randomised trial of monoclonal antibody for adjuvant therapy of resected Dukes' C colorectal carcinoma. *Lancet* **343**, 1177–83.

Rosenberg, S.A. (1996). Development of cancer immunotherapies based on identification of the gene encoding cancer regression antigens. *J. Natl Cancer Inst.* **88**, 1635–44.

Sahin, U., Tureci, O., and Pfreunschuh, M. (1997). Serological identification of human tumor antigens. *Curr. Opin. Immunol.* **9**, 109–16.

Abbreviations

AcP	accessory protein
AChR	acetylcholine receptor
ACTH	adrenocortinotrophic hormone
ADCC	antibody dependent cellular cytotoxicity
AF-1	accessory factor 1
AIDS	acquired immune deficiency syndrome
ALS	amyotrophic lateral sclerosis
ALV	avian leukosis-sarcoma virus
AML	acute myeloid leukaemia
AP-1	activator protein 1
APC	antigen-presenting cell
APRF	acute-phase response factor (another name for STAT3)
β-APP	β-amyloid precursor protein
ARDS	acute respiratory distress syndrome
ARE	activin-responsive element
ARNT	aryl hydrocarbon nuclear receptor translocator
ATL	adult T-cell leukaemia
AVP	arginine vasopressin
BAL	bronchoalveolar lavage
BCG	bacillus Calmette–Guérin
BCGF	B-cell growth factor
BCR	B-cell receptor
BCGF 1	B-cell growth factor 1 (an old name for IL-4)
bFGF	basic fibroblast growth factor
BFU	burst-forming unit (suffixed -E indicates erythroblast)
BMP	bone morphogenetic protein
BN	brown Norway
bp	base pair
BSF-1	B-cell stimulating factor 1 (early name for IL-4)
BSF-2	B-cell stimulating factor 2 (early name for IL-6)
[Ca^{2+}]$_i$	intracellular calcium concentration
CAD	caspase-activated deoxyribonuclease
CAR	cytopathic ALSV receptor
CBP	CREB-binding protein
CCR	CC (Cys–Cys) chemokine receptor
CD	cluster of differentiations
CD28RE	CD28-responsive element
C/EBP	CCAAT/enhancer-binding protein
CFU	colony-forming unit (may have suffixes indicating cell type, e.g. CFU-E for erythroblasts)
CHO cells	Chinese hamster ovary cells
CIS	cytokine-inducible SH2
CLA	cutaneous lymphocyte-associated antigen
CLE1	conserved lymphokine element 1
cM	centimorgan
CMLF	cytotoxic lymphocyte maturation factor (early name for II.-12)
CNDF	cholinergic neuronal differentiation factor (early name for LIF)

CNS	central nervous system
CNTF	ciliary neurotrophic factor
ConA	concanavalin A
CR1	complement receptor type 1
CRE	cyclic AMP responsive element
CREB	CRE binding protein
CRF	corticotrophin-releasing factor
CRH	corticotropin-releasing hormone *or* cytokine receptor homologous [domain]
CRP	C-reactive protein
CsA	cyclosporin A
CSF-1	colony-stimulating factor 1 (another name for M-CSF)
CT-1	cardiotrophin 1
CTAPIII	connective tissue activating protein III
CTL	cytotoxic T lymphocyte
CTLA	cytotoxic T lymphocyte-associated antigen
CXCR	CXC (Cys–Xaa–Cys) chemokine receptor
DC-CK1	dendritic cell-derived CC chemokine 1
DD	death domain
DED	death effector domain
DEMA	dehydroepiadroster
DFF	DNA fragmentation factor
DMSO	dimethylsulfoxide
DP	Dermatophagoides pteronyssinus
DR	death receptor
DSE	double symmetry element
DTH	delayed-type hypersensitivity
EAE	experimental autoimmune encephalomyelitis
EAMG	experimental autoimmune myasthaenia gravis
EBNA	Epstein–Barr nuclear antigen
EBS	Ets binding site
EBV	Epstein–Barr virus
EDN	eosinophil-derived neurotoxin
EGF	epidermal growth factor
ELC	EBI1-ligand chemokine
ELISA	enzyme-linked immunosorbent assay
ENA-78	epithelial cell-derived neutrophil attractant 78
Epo	erythropoietin
ERK	extracellular signal-regulated kinase
ES cell	embryonal stem cell
EST	expressed sequence tag
Fas-L	Fas ligand
FAST-1	forkhead activin signal transducer 1
5-FC	5-fluorocytosine
fMLP	formyl-methionyl-leucyl-phenylalanine
GABP	GA-binding protein
GGAA	GGAA-binding protein
GAP	GTPase-activating protein
GAS	gamma-activated sequences
GC	germinal centres *or* glucocorticoids
GCP	granulocyte chemotactic protein (alternative name for IL-8)
GCP-2	granulocyte chemotactic protein 2
G-CSF	granulocyte colony-stimulating factor
GH	growth hormone

GHR	growth hormone receptor
GITR	glucocorticoid-induced TNFR family-related gene
GM-CSF	granulocyte–macrophage colony-stimulating factor
gp	glycoprotein
GPI	glycosyl phosphatidylinositol
GRE	glucocorticoid responsive element
GS domain	glycine–serine domain
GVHD	graft-versus-host disease
HCP	haematopoietic cell phosphatase
5-HETE	5-hydroxy-eicosatetraenoic acid
HGF	hepatocyte growth factor
HIF-1	hypoxia-inducible factor-1
HILDA	human interleukin for DA cells (early name for LIF)
HIM	hyper-IgM
HIV	human immunodeficiency virus
HLA	human leucocyte-associated antigen
HMG	high mobility group protein
HNF-4	hypoxia nuclear factor-4
HPAA	hypothalamus–pituitary–adrenal axis
HPGF	hybridoma plasmocytoma growth factor (early name for IL-6)
HSA	heat stable antigen
HSF	hepatocyte-stimulating factor
HSV	herpes simplex virus
HTLV	human T-cell lymphotropic virus
HVEM	herpes virus entry mediator
HVS	herpes virus Saimiri
IBD	inflammatory bowel disease
ic	intracellular receptor agonist
ICAD	inhibitor of CAD
ICAM-1	intercellular adhesion molecule 1
ICE	IL-1-converting enzyme
IDDM	insulin-dependent diabetes mellitus
IFN	interferon
IFNβ2	interferon beta 2 (early name for IL-6)
IgE (etc.)	immunoglobulin E (etc.)
IGF	insulin-like growth factor
IGIF	interferon gamma inducing factor
IKK	IκB kinase
IL-1 (etc.)	interleukin 1 (etc.)
IL-2R (etc.)	interleukin-2 receptor
IL-1ra	interleukin-1 receptor antagonist
iNOS	inducible nitric oxide synthase
IP-10	interferon-γ-inducible protein 10
IRAK	IL-1-activated serine/threonine kinase
IRF-1	interferon regulatory [*or* response] factor 1
IRS	insulin receptor substrate
ISRE	interferon-stimulated response element
ITAM	immunoreceptor tyrosine-based activation motif
Jak	Janus kinase
JH domain	Jak homology domain
JNK	c-Jun N-terminal kinases
KI	kinase insert
KL	c-Kit ligand (another name for SCF)

KS	Kaposi's sarcoma
-L (suffix)	ligand
LACK	*Leishmania* homologue receptor for activated C kinase
LAF	lymphocyte-activating factor
Lag	lymphocyte activation gene
LAK	lymphokine-activated killer [cells]
LAP	latency-associated peptide
LARC	liver and activation-regulated chemokine
LATI	lymphokine-activated tumour inhibition
LCF	lymphocyte chemoattractant factor (alternative name for IL-16)
LCMV	lymphocytic choriomeningitis virus
LDGF	leukocyte-derived growth factor
LEM	leukocytic endogenous mediator (early name for IL-1)
LIF	leukaemia inhibitory factor
LP-1	lymphopoietin-1
LPS	lipopolysaccharide
LT	lymphotoxin
LTB$_4$	leukotriene B$_4$
LTBP	latent TGF-β-binding protein
mAb	monoclonal antibody
MACH/FLICE	MORT1-associated CED-3 homologue/FADD-like ICE
MAF	macrophage-activating factor
MAP kinase	mitogen-activated protein kinase
MCA	methylcholanthrene
MCP	monocyte chemotactic protein
M-CSF	macrophage colony-stimulating factor
MDC	macrophage-derived chemokine
MDNCF	monocyte-derived neutrophil chemotactic factor (alternative name for IL-8)
MEK	MAP kinase–ERK kinase
MG	myasthaenia gravis
MGF	mammary gland factor (also called STAT5)
MGSA	melanoma growth-stimulatory activity
MHC	major histocompatibility complex
Mig	monokine induced by IFNγ
MIP	macrophage inflammatory protein
MMP	matrix metalloproteinase
MONAP	monocyte-derived neutrophil activating peptide (alternative name for IL-8)
MORT1/FADD	mediator of receptor-induced toxicity/Fas-associated death domain protein
MPIF	myeloid progenitor inhibitory factor
MPLV	myeloproliferative leukaemia virus
MSH	melanocyte-stimulating hormone
NAP-2	neutrophil activating protein 2
NF-AT	nuclear factor of activated T cells
NF-IL2	nuclear factor IL-2
NF-IL6	IL-6 transcription factor
NF-κB	nuclear factor kappa enhancer binding protein
NGF	nerve growth factor
NIK	NF-κB-inducing kinase
NIP	nuclear inhibitory protein
NK cell	natural killer cell

NKSF	natural killer stimulatory factor (early name for IL-12)
NLS	nuclear localization signal
NMMA	N-monomethyl-L-arginine
NMR	nuclear magnetic resonance
nod	non-obese diabetic
NOS	nitric oxide synthase
NSAID	non-steroidal anti-inflammatory drugs
Oct	octamer
OSM	oncostatin M
OVLT	organum vasculosum laminae terminalis
PAF	platelet-activating factor
PAI-1	plasminogen activator inhibitor 1
PALS	periarteriolar lymphoid sheath of the spleen
PARC	pulmonary and activation-regulated chemokine
PARP	poly(ADP–ribose) polymerase
PAS protein	proline–alanine–serine protein
PBL	peripheral blood lymphocyte
PBMC	peripheral blood mononuclear cell
PBP	platelet basic protein
PBSF	pre-B-cell stimulatory factor
PC-PLC	phosphatidylcholine-specific phospholipase C
PCR	polymerase chain reaction
PDGF	platelet-derived growth factor
PDK-1	phosphoinositide-dependent kinase 1
PF-4	platelet factor 4
PGE$_2$	prostaglandin E$_2$
PHA	phytohaemagglutinin
PH domain	pleckstrin homology domain
PI domain	phosphotyrosine interaction domain
PI3K	phosphoinositide 3-OH kinase
PI-PLC	phosphatidylinositol phospholipase C
PKC	protein kinase C
PKR	protein kinase R
PLA$_2$	phospholipase A$_2$
PLC	phospholipase C
PLP	proteolipid protein
PMA	phorbol 12-myristate 13-acetate
PMN	polymorphonuclear cell(s)
PPD	purified protein derivative
PRR	promoter regulatory region
PTB domain	phosphotyrosine-binding domain
PTK	protein tyrosine kinase
PVN	paraventricular nucleus
PWM	pokeweed mitogen
-R (suffix)	receptor
RA	rheumatoid arthritis
RAIDD	RIP-associated ICH-1/CED-3-homologous protein
RANTES	regulated on activation, normally T-cell expressed and secreted
RGS	regulator of G-protein signalling
RIP	receptor interacting protein
RSD	Rel similarity domain
RTK	receptor tyrosine kinase
RT-PCR	reverse transcription–polymerase chain reaction

S	soluble receptor agonist
SAPK	stress-activated protein kinase
sCD30	soluble CD30
SCF	stem cell factor
SCID	severe combined immunodeficiency
SCR	short consensus repeat
SDF-1	stromal cell-derived factor 1
SEREX	serological expression vector
SH2 domain	Src homology 2 domain
SIRS	systemic inflammatory response syndrome
SIV	simian immunodeficiency virus
SLC	secondary lymphoid-tissue chemokine
SLE	systemic lupus erythematosus
Smad	Sma and Mad-related [proteins]
SRE	serum responsive element
SRF	serum response factor
STAM	signal-transducing adaptor molecule
STAT	signal transducer and activator of transcription
TAA	tumour-associated antigen
TACE	TNF-alpha-converting enzyme
TACI	transmembrane activator and CAML interactor
TAK-1	TGF-β-activated kinase 1
TARC	thymus and activation-regulated chemokine
TCR	T-cell receptor
TD	thymus-dependent *or* T-cell-dependent
TECK	thymus-expressed chemokine
TGF	transforming growth factor
Th cell	T-helper cell
TI	thymus-independent *or* T-cell-independent
TIL	tumour-infiltrating lymphocyte
TIMP	tissue inhibitor of metalloproteinases
TNF	tumour necrosis factor
TPA	tissue plasminogen activator
Tpo	thrombopoietin
TRADD	TNF receptor-associated death domain protein
TRAF	TNF receptor-associated factor
TRAIL	TNF-related apoptosis-inducing ligand
TRAMP	TNF receptor-related apoptosis-mediating protein
TRE	TPA responsive element
TRF	T-cell-replacing factor (an early name for IL-5)
TSLP	thymic stroma derived lymphopoietin
UPS	upstream promoter site
UTR	untranslated region
VCAM	vascular cell adhesion molecule
VEGF	vascular endothelial cell growth factor
VPF	vascular permeability factor (another name for VEGF)
X-SCID	X-linked severe combined immunodeficiency syndrom

Index